THE PHYSICAL MEASUREMENT OF BONE

Series in Medical Physics and Biomedical Engineering

Series Editors:
C G Orton, Karmanos Cancer Institute and Wayne State University, Detroit, USA
J A E Spaan, University of Amsterdam, The Netherlands
J G Webster, University of Wisconsin-Madison, USA

Other books in the series

Therapeutic Applications of Monte Carlo Calculations in Nuclear Medicine
H Zaidi and G Sgouros (eds)

Minimally Invasive Medical Technology
J G Webster (ed)

Intensity-Modulated Radiation Therapy
S Webb

Physics for Diagnostic Radiology
P Dendy and B Heaton

Achieving Quality in Brachytherapy
B R Thomadsen

Medical Physics and Biomedical Engineering
B H Brown, R H Smallwood, D C Barber, P V Lawford and D R Hose

Monte Carlo Calculations in Nuclear Medicine: Applications in Diagnostic Imaging
M Ljungberg, S-E Strand and M A King (eds)

Introductory Medical Statistics 3rd Edition
R F Mould

Ultrasound in Medicine
F A Duck, A C Barber and H C Starritt (eds)

Design of Pulse Oximeters
J G Webster (ed)

The Physics of Medical Imaging
S Webb

Series in Medical Physics and Biomedical Engineering

THE PHYSICAL
MEASUREMENT OF BONE

Edited by

C M Langton

Centre for Metabolic Bone Disease
Hull and East Yorkshire Hospitals NHS Trust
and University of Hull
Hull
UK

C F Njeh

Department of Radiology
University of California, San Francisco
San Francisco
USA

Io**P**

Institute of Physics Publishing
Bristol and Philadelphia

© IOP Publishing Ltd 2004

British Library Cataloguing-in-Publication Data

A catalogue record for this book is available from the British Library.

ISBN 0 7503 0838 9

Library of Congress Cataloging-in-Publication Data are available

Series Editors:
C G Orton, Karmanos Cancer Institute and Wayne State University, Detroit, USA
J A E Spaan, University of Amsterdam, The Netherlands
J G Webster, University of Wisconsin-Madison, USA

Commissioning Editor: John Navas
Production Editor: Simon Laurenson
Production Control: Sarah Plenty
Cover Design: Victoria Le Billon
Marketing: Nicola Newey and Verity Cooke

Published by Institute of Physics Publishing, wholly owned by The Institute of Physics, London

Institute of Physics Publishing, Dirac House, Temple Back, Bristol BS1 6BE, UK

US Office: Institute of Physics Publishing, The Public Ledger Building, Suite 929, 150 South Independence Mall West, Philadelphia, PA 19106, USA

Typeset by Academic + Technical, Bristol
Printed in the UK by MPG Books Ltd, Bodmin, Cornwall

The Series in Medical Physics and Biomedical Engineering is the official book series of the International Federation for Medical and Biological Engineering (IFMBE) and the International Organization for Medical Physics (IOMP)

IFMBE

The International Federation for Medical and Biological Engineering (IFMBE) was established in 1959 to provide medical and biological engineering with a vehicle for international collaboration in research and practice of the profession. The Federation has a long history of encouraging and promoting international cooperation and collaboration in the use of science and engineering for improving health and quality of life.

The IFMBE is an organization with membership of national and transnational societies and an International Academy. At present there are 48 national members and two transnational members representing a total membership in excess of 30 000 worldwide. An observer category is provided to give personal status to groups or organizations considering formal affiliation. The International Academy includes individuals who have been recognized by the IFMBE for their outstanding contributions to biomedical engineering.

Objectives

The objectives of the International Federation for Medical and Biological Engineering are scientific, technological, literary and educational. Within the field of medical, clinical and biological engineering its aims are to encourage research and the application of knowledge, and to disseminate information and promote collaboration.

In pursuit of these aims the Federation engages in the following activities: sponsorship of national and international meetings, publication of official journals, cooperation with other societies and organizations, appointment of commissions on special problems, awarding of prizes and distinctions, establishment of professional standards and ethics within the field, as well as other activities which in the opinion of the General Assembly or the Administrative Council would further the cause of medical, clinical or biological engineering. It promotes the formation of regional, national, international or specialized societies, groups or boards, the coordination of bibliographic or informational services and the improvement of standards in terminology, equipment, methods and safety practices, and the delivery of health care.

The Federation works to promote improved communication and understanding in the world community of engineering, medicine and biology.

Activities

The IFMBE publishes the journal *Medical and Biological Engineering and Computing*, which includes a special section on *Cellular Engineering*. The *IFMBE News,* published electronically, keeps the members informed of the developments in the Federation. In cooperation with its regional conferences, IFMBE issues a series of the IFMBE Proceedings. IFMBE's official book series, *Medical Physics and Biomedical Engineering* is published by the Institute of Physics Publishing in cooperation with IOMP and represents another service to the Biomedical Engineering Community. The books in this series describe applications of science and engineering in medicine and biology and are intended for both graduate students and researchers. They cover many topics in the field of medical and biological engineering, as well as medical physics, radiology, radiotherapy and clinical research.

The Federation has two divisions: *Clinical Engineering* and *Technology Assessment in Health Care*. Additional special interest groups are the regional working groups: *Africa/ICHTM*, *Asian-Pacific*, *Coral*, *Developing Countries*, and the scientific working groups: *Cellular Engineering*, *Neuro-engineering*, and *Physiome*.

Every three years the IFMBE holds a World Congress on Medical Physics and Biomedical Engineering in cooperation with the IOMP and the IUPESM. In addition, annual, milestone and regional conferences are organized in different regions of the world, e.g. in the Asia-Pacific, Nordic-Baltic, Mediterranean, African and South American regions.

The Administrative Council of the IFMBE meets once a year and is the steering body for the IFMBE. The council is subject to the rulings of the General Assembly, which meets every three years at the occasion of the World Congress.

Information on the activities of the IFMBE are found on its website at http://www.ifmbe.org.

IOMP

The IOMP was founded in 1963. The membership includes 64 national societies, two international organizations and 12 000 individuals. Membership of IOMP consists of individual members of the Adhering National Organizations. Two other forms of membership are available, namely Affiliated Regional Organization and Corporate members. The IOMP is administered by a Council, which consists of delegates from each of the Adhering National Organizations; regular meetings of council are held every three years at the International Conference on Medical Physics (ICMP). The Officers of the Council are the President, the Vice-President and the Secretary-General. IOMP committees include: developing countries, education and training; nominating; and publications.

Objectives

- To organize international cooperation in medical physics in all its aspects, especially in developing countries.
- To encourage and advise on the formation of national organizations of medical physics in those countries which lack such organizations.

Activities

Official publications of the IOMP are *Physiological Measurement, Physics in Medicine and Biology* and the *Series in Medical Physics and Biomedical Engineering*, all published by the Institute of Physics Publishing. The IOMP publishes a bulletin *Medical Physics World* twice a year.

Two council meetings and one General Assembly are held every three years at the ICMP. The most recent ICMPs were held in Kyoto, Japan (1991), Rio de Janeiro, Brazil (1994), Nice, France (1997) and Chicago, USA (2000). These conferences are normally held in collaboration with the IFMBE to form the World Congress on Medical Physics and Biomedical Engineering. The IOMP also sponsors occasional international conferences, workshops and courses.

For further information contact: Hans Svensson, PhD, DSc, Professor, Radiation Physics Department, University Hospital, 90185 Umeå, Sweden. Tel: (46) 90 785 3891. Fax: (46) 90 785 1588. Email: Hans.Svensson@radfys. umu.se. WWW: http://www.iomp.org.

Contents

List of contributors

Dr Jean Elizabeth Aaron
School of Biomedical Sciences, Worsley Building, The University of Leeds, Leeds LS2 9JT, UK

Prof Alun Beddoe
Medical Physics Department, Queen Elizabeth Hospital, Egbaston, Birmingham B15 2TH, UK

Dr James L Cunningham
Department of Mechanical Engineering, University of Bath, Bath BA2 7AY, UK

Dr Harry Genant
Osteoporosis and Arthritis Research Group, University of California San Francisco, 505 Parnassus Avenue, M392, San Francisco, CA 94143, USA

Dr Christopher Gordon
Hamilton Health Sciences Corporation, Department of Nuclear Medicine, Henderson Site (Box 2000), 711 Concession Street, Hamilton, Ontario L8V 1C3, Canada

Dr Didier Hans
Head of Research and Development, Nuclear Medicine Division, Geneva University Hospital, 1211 Geneva 14, Switzerland

Dr Yebin Jiang
Osteoporosis and Arthritis Research Group, Department of Radiology, University of California San Francisco, 513 Parnassus Avenue, HSW-207A, San Francisco, CA 94143-0628, USA

Mr Alan P Kelly
School of Biomedical Sciences and Safety Advisory Services, Worsley Building, The University of Leeds, Leeds LS2 9JT, UK

Dr Thomas F Lang
Associate Professor, Radiology, University of California San Francisco, 533 Parnassus Avenue, U368E, San Francisco, CA 94143-1250, USA

Dr Christian M Langton
Centre for Metabolic Bone Disease, Hull Royal Infirmary, Anlaby Road, Hull HU3 2RW, UK

Dr Sharmila Majumdar
Magnetic Resonance Science Center, Box 1290, AC 109, 1 Irving Street, University of California San Francisco, San Francisco, CA 94143, USA

Dr Patrick H Nicholson
29 Gensing Road, St Leonards on Sea, TN38 0HE, UK

Dr Christopher F Njeh
The John Hopkins University, School of Medicine, Division of Radiation Oncology, The Harry and Jeanette Weinberg Building, 401 North Broadway, Suite 1440, Baltimore, MD 21231-1240, USA

Dr Laurent Pothuaud
Magnetic Resonance Science Center, Box 1290, AC 109, 1 Irving Street, University of California San Francisco, San Francisco, CA 94143, USA

Dr Jae-Young Rho (deceased)
University of Memphis, Department of Biomedical Engineering, ET330, Memphis, TN 38152, USA

Dr Clifford Rosen
Director, The Maine Center for Osteoporosis Research and Education, St Joseph Hospital, 268 Center Street, Bangor, ME 04401, USA

Prof John A Shepherd
Associate Technical Director, Osteoporosis & Arthritis Research Group, Department of Radiology, 350 Parnassus Ave., Suite 205, University of California San Francisco, San Francisco, CA 94143-1349, USA

Dr Patricia Shore
Hard Tissue Biology Laboratory, School of Biomedical Sciences, Worsley Building, The University of Leeds, Leeds LS2 9JT, UK

Dr Ian Stronach
Medical Physics Department, Queen Elizabeth Hospital, Egbaston, Birmingham B15 2TH, UK

Dr Jon A Thorpe
Centre for Metabolic Bone Disease, Hull Royal Infirmary, Anlaby Road, Hull HU8 8PY, UK

Dr Bert van Rietbergen
Eindhoven University of Technology, Department of Biomedical Engineering, PO Box 513, 5600 MB Eindhoven, The Netherlands

Dr Jenny Zhao
Osteoporosis and Arthritis Research Group, Department of Radiology, University of California San Francisco, 513 Parnassus Avenue, HSW-207A, San Francisco, CA 94143-0628, USA

Preface

The British scientist Lord Kelvin (William Thomson 1824–1907) stated in his lecture to the Institution of Civil Engineers on 3 May 1883 that 'I often say that when you can measure what you are speaking about, and express it in numbers, you know something about it; but when you cannot express it in numbers your knowledge is a meagre and unsatisfactory kind; it may be the beginning of knowledge but you have scarcely, in your thoughts, advanced to the stage of science, whatever the matter may be.' This famous remark emphasizes the importance that physical measurement has in science, medicine and patient management.

The skeleton is the fundamental framework of the human body. In addition to contributing to its shape and form, bones perform several important functions including support, protection, movement and chemical storage. Understanding the physical integrity of bone is essential to understanding its function. Knowledge of the physical properties of bone help us to predict the load the skeleton can bear, which may be further used to predict and monitor the effects of ageing and disease on bone. The physical measurement of bone is also of value in a number of other avenues including the clinical management of osteoporosis, hip replacement, rheumatoid arthritis and osteomalacia.

In this book, various physical measurement techniques for bone are discussed. Emphasis is placed on the fundamental principles of measurement, sample preparation and sources of error. This is the first book that has addressed in one volume the various physical measurement techniques for bone. We hope it will be a useful text for a wide range of specialists including physicists, biologists, biomechanical engineers and clinicians.

Christian M Langton, Christopher F Njeh
September 2003

has become our key to understanding how bones fail and how treatments can prevent bone loss and structural damage.

1.2. BONE MORPHOLOGY AND ORGANIZATION

It should be noted that the three essential elements of the skeleton (bone cells, bone matrix and bone mineralization) must be considered within the framework of how bone is organized. The skeleton is composed of two parts: an axial skeleton which includes the vertebrae, pelvis and other flat bones such as the skull and sternum; and an appendicular skeleton, which includes all the long bones. The vertebrae are principally composed of cancellous or trabecular bone, i.e. finely striated bone with a large surface area made up of branching lattices oriented along lines of stress and interspersed with the bone marrow. Trabecular bone has a very high rate of turnover and an extremely large surface area. In contrast, cortical bone, which is the principal component of long bones, is four times more dense than trabecular bone, but is metabolically less active and has less surface area. In contrast to trabecular bone, cortical bone is subject to bending and torsional forces as well as compressive loading.

The long bones of the appendicular skeleton are divided into three parts: the epiphysis, metaphysis and diaphysis. The epiphysis is the portion of long bone found at either end and develops from ossification centres distinct from the rest of the bone. It is separated by a layer of growth cartilage. The metaphysis is the region where remodelling occurs during growth and development. The diaphysis comprises the length of the long bone.

There are essentially two types of bone tissue, woven and lamellar. Woven bone is immature or primitive bone found only in the embryo, the newborn, in fracture calluses or rarely, in the metaphysis. It is also found in patients with Paget's disease (see below). It is composed of coarse fibres, is poorly mineralized and has no uniform orientation. Lamellar bone replaces woven bone by 1 year of age, and by age 4 almost all bone tissue will be lamellar. This type of tissue exhibits anisotropic properties, i.e. the mechanical behaviour of bone differs according to the orientation of the bone fibres.

Macroscopically, bone is organized in a complex but efficient manner. Haversian bone is the principal structure. Vascular channels are arranged circumferentially around lamellae of bone to form an osteon, an irregular braching cylinder composed of a neurovascular canal surrounded by layers of cells within the bony matrix. Osteons are connected to each other by Volkmann's canals which are oriented perpendicularly to the osteon. The vascular canals resemble capillaries which allow or limit transport of ionic material to bone. Unlike trabecular bone, which is intimately associated with the surrounding bone marrow, cortical bone has two surfaces: one on the inner side which faces the marrow and is called the endosteal surface, and one on the outer side facing soft tissue and muscle, known as the periosteal surface. The endosteal cells interact with a number of factors (see below) which are involved in bone formation and resorption. The periosteum has an outer fibrous layer, and an inner layer composed of chrondrocyte and osteoblast precursor cells. Appositional bone growth occurs in this layer.

At an ultrastructural level, bone is composed of individual structural units or bone metabolic units (BMUs). These units control remodelling, which is a surface process associated with the periosteal, endosteal, Haversian canal and trabecular envelopes. Up to 10% of the adult skeleton is remodelled per year, principally at trabecular sites, although cortical bone can show active remodelling especially during the first two years of life. In cortical bone, the osteon represents the BMU, and is a cylinder of approximately 250 µm diameter, running parallel to the long axis of the bone. In trabecular bone, the BMU is represented by thin crescents about 60 µm deep and 600 µm long. Remodelling takes place in a highly regulated sequence that begins with activation of bone forming cells, followed by recruitment of bone resorbing cells, the actual resorption of bone by osteoclasts, and the reversal and formation phase in which new bone is laid down. The net result is a balanced remodelling cycle with no change in bone mass.

1.3.   BONE TISSUE I: THE ROLE OF BONE CELLS

There are three major types of bone cell: the osteoblast, the osteoclast and the osteocyte. Each has a unique role in modelling/remodelling, mineralization and matrix formation. Unlike other tissues, however, their origins and physiological destiny have been difficult to elucidate, in part because of the nature of bone, and in part because the model systems necessary for *in vitro* analysis are more complex.

1.3.1.   *The osteoclast*

Osteoclasts are multinucleated (up to five nuclei) giant cells that arise from bone marrow mononuclear precursor cells of the hemopoietic lineage [1, 2]. Mature cells live less than 35 days, and their principal, if not their only, function is the resorption of bone. From birth in uncommitted progenitor populations within the marrow, to programmed cell death, the osteoclast is destined for one goal [1]. These mature giant cells are equipped with several proteolytic enzymes necessary for digestion of protein matrix and have a complex proton pump system which is required for the complete dissolution of apatite mineral matrix [2, 3].

Actively resorbing osteoclasts are highly polarized and multinucleated. When the actin-containing membrane seals this cell to bone, a 'ruffled border' is formed, permitting the process of bone dissolution to begin [2–5]. Resorbing multinucleated osteoclasts appear in groups at the bone surface in trabecular bone (i.e. cancellous bone) excavating a lacuna or cavity. In cortical bone, these cells dig Haversian canals. In both types of bone, the release of calcium into the extracellular space is accompanied by proteolysed collagen in the form of cross-links and collagen fragments.

The osteoclasts are well suited for these activities. Several structures acidify the lacunae (or resorption pit) to a pH of 4 or less [6]. In the ruffled border the proton pump, a vacuolar ATPase, has been identified [6–8]. Carbonic anhydrase in the cytoplasm enhances the conversion of bicarbonate into carbon dioxide and protons. Besides the secretion of hydrogen ions, there is a plethora of enzymes involved in matrix degradation, including cysteine proteinases, cathepsin K and neutral collagenases, that are released by the osteoclast [8–10].

Since osteoclasts do not reside directly on the bone surface (in contrast to osteoblasts), an elaborate system of directional signals is required during the course of their maturation, to guide them to their final destination. Osteoblasts and stromal cells residing in the bone marrow elaborate a series of cytokines which direct osteoclast differentiation and homing to the bone surface. In fact, differentiation in the bone marrow of both cell types, at approximately the same time, suggests that coupling occurs earlier than was originally thought, possibly during the first phases of differentiation [1]. This phenomenon was first observed when, in the presence of osteoblast 'feeders' and 1,25-dihydroxyvitamin D, mononuclear monocytes from the granculocyte macrophage colony in marrow differentiated *in vitro* into multinucleated giant cells [11]. Osteoblast-originated signals have been reported to include tumour necrosis factor-alpha, transforming growth factor beta, prostaglandin E2, and the interleukins -1, -6 and -11 (figure 1.2) [1, 11–13]. More recently, it has become apparent that macrophage colony stimulating factor-1, derived from granulocyte–macrophage precursors (mCSF-1) and another factor, RANK ligand (RANKL or alternatively called TRANCE or osteoprotogerin ligand [OPGL]), are the two cytokines elaborated by osteoblasts or their precursors that are absolutely essential for osteoclast maturation [14–18]. mCSF is critical in the early phase of osteoclast precursor recruitment, and RANKL is necessary for completion of osteoclastic differentiation. In fact, gene knockouts of mCSF, RANKL and c-src (another protein essential for osteoclast function) lead to the syndrome of osteopetrosis, a condition in mice and man associated with the absence of osteoclasts but greatly enhanced cortical bone mineral, and a nearly absent marrow cavity [16, 19].

The interaction of multinucleated osteoclasts with the bone matrix is an intricate process requiring landing, attachment and then elaboration of proteolytic enzymes and protons to dissolve the skeletal matrix. However, the matrix is not a passive element in this process. During development,

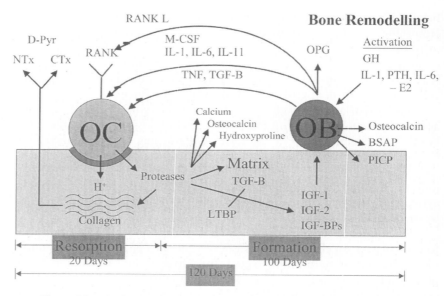

Figure 1.2. *The bone remodelling process occurs in adult bone as a result of osteoblast activation by several signals including the interleukins, growth hormone, oestrogen withdrawal, PTH and others. The OB (osteoblasts) secrete soluble and membrane bound factors in response to several of these stimuli (e.g. PTH stimulates membrane bound OPGL expression and release of IL-6), which in turn enhance the differentiation of OC (osteoclasts). The entire remodelling sequence takes about 120 days with resorption taking up a very small proportion of that time (2 weeks). The release of proteolytic enzymes and protons not only dissolve the matrix but also provide a mechanism to release growth factors from their binding proteins (LTBP-1 for TGF-b and IGFBPs for IGFs).*

the matrix can direct cell migration and differentiation, and during mechanical loading can deform and transmit this information to cells. The primary adhesion receptors through which cells attach to the matrix and thereby receive these signals are called the integrins [20]. These compounds are heterodimeric transmembrane receptors, with intracellular domains that interact with the cytoskeleton. There are at least nine distinct integrins, and several of these are reported to be found in osteoclasts and osteoblasts. The major osteoclast integrins are $\alpha v\beta3$ and $\alpha v\beta1$, the vitronectin receptors which bind ostepontin and bone sialoprotein, and $\alpha2\beta1$, the collagen receptor [21]. Antibodies to $\alpha v\beta3$ have illustrated the importance of integrins in osteoclast function, since it can be shown that these proteins completely block osteoclast action, either by inhibiting osteoclast attachment or blocking the process of bone resorption.

1.3.2. The osteoblast

Osteoblasts are plump cuboidal cells which line the osteoid or non-mineralized bony matrix (figure 1.3) [22]. Although bone is embryonically derived from sclerotomes, branchial arches and the neural crest, osteoblasts originate from mesenchymal precursor cells which can then give rise to other cell types such as fibroblasts, adipocytes, myoblasts and tendon cells. The differentiated function of osteoblasts is to secrete matrix components such as collagen type I. A large nucleus with plump endoplasmic reticulum, these cells represent the bone-forming component of the bone remodelling unit (figure 1.2). In addition to type I collagen, osteoblasts also produce non-collagenous proteins such as osteopontin, osteocalcin, osteonectin, bone sialoprotein, and alkaline phosphatase [23, 24].

In general, there are three specific 'programmes' which are followed by the osteoblast over its life span: proliferation, differentiation, and mineralization. As noted in figure 1.1, the interaction of all three components of bone is best exemplified by the life cycle of the osteoblast. After originating from stem cells, osteoblasts are committed towards a specific bone phenotype by both transcription factors and growth factors, some of which are tissue specific, such as bone morphogenic protein-2 and Cbfa1, and others which are non-specific such as *c-fos*, and *egr* [24–29]. The expression of bone-specific markers in osteoblasts is time-specific. After a period of proliferation, which is accelerated by several growth factors including the insulin-like

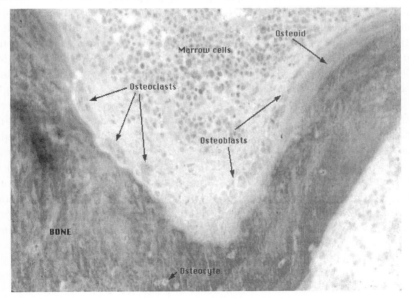

Figure 1.3. *The fine structure of trabecular bone as it relates microscopically to marrow cells, osteoblasts and osteoclasts.*

growth factors (IGFs), transforming growth factor-beta, fibroblast growth factor and the bone morphogenic proteins (BMPs), differentiative markers are expressed [30–35]. These include alkaline phosphatase and bone sialoprotein as early indices. Later on, as osteoblasts line the bone surface (figure 1.2) osteopontin and osteocalcin are released. *In vitro*, mature osteoblasts can form nodules and begin the process of mineralization. Osteoblasts are also thought to regulate the local concentrations of calcium and phosphate in such a way as to promote the formation of an apatite matrix. The production of heteropolymeric matrix fibrils from collagen synthesized by osteoblasts is likely to be a key step prior to mineralization, although the precise mechanism is not well delineated (see section 1.5).

Bone marrow stromal cells destined to become osteoblasts were once considered to be structural components of the marrow, but relatively minor players in the bone remodelling drama. However, it is now clear that stromal cells are necessary for the coupling of formation to resorption, i.e. the elaboration of cytokines (as shown in figure 1.2), by supporting osteoclastogenesis [1, 26]. Indeed, these less differentiated osteoblastic progeny elaborate OPGL and mCSF that are essential for the recruitment and differentiation of marrow precursors of the osteoclast lineage [17, 18, 36]. Finally, and somewhat surprisingly, it should be noted that terminally differentiated osteoblasts on the bone surface lose the ability to recruit osteoclasts.

There are several fates for the osteoblast. First, these cells can undergo programmed cell death or apoptosis [37–39]. Pharmacologic agents, including glucocorticoids which are used for the treatment of various inflammatory conditions including rheumatoid arthritis, hasten this process, although it is suppressed by agents such as the bisphosphonates [39–41] which are used for the treatment of osteoporosis. Second, the osteoblasts could become lining cells whose function in bone remains unclear. Quiescent bone cells that line or cover mineralized osteoid are situated in sites where bone remodelling is not taking place. However, there is some evidence that these cells can secrete collagenases, which are necessary to initiate bone resorption by allowing osteoclast attachment. Thus, although these osteoblasts are metabolically less active, and terminally differentiated, it is conceivable that lining cells may signal or 'home' osteoclasts. Third, the osteoblast may be encased in the very bone it has synthesized, resulting in an osteocyte (see below). Overall, the balance of cell life, starting from early stromal cell generation and proliferation, through the process of differentiation and osteoclast support, to apoptosis or osteocyte generation, determines the overall metabolic fate of the bone remodelling unit and the net amount of mineralized matrix.

1.3.3. *The osteocytes*

Osteocytes and to some degree lining cells are in a unique morphologic position in bone to sense mechanical strain [1, 42]. Osteocytes are buried deep within cortical bone and are part of the BMU or osteon. The cells are characterized structurally by striking stellate morphology, closely resembling the dendritic network of the nervous system. Interestingly, osteocytes are by far the most abundant cell in bone. Through an extensive network of canaliculi, osteocytes buried within bone can communicate with both surface bone cells and marrow stromal cells that, in turn, have cellular projections into endothelial cells. The location of the osteocyte, and this elaborate network of canaliculi, have provided clues as to the function of these entombed osteoblast-like cells. Although still theoretical, interstitial fluid changes flowing through canaliculi as a result of mechanical strain and differences in circulating levels of steroid hormones could lead to osteocyte signalling and relays from there to active osteoblasts and marrow stromal cells. A unified hypothesis linking the osteocyte to mechanical strain and in turn bone remodelling appears to be much more plausible and, if true, marks a major paradigm shift in our understanding of skeletal physiology.

1.4. BONE TISSUE II: THE BONY MATRIX

Bone substance, the multiphasic composite material that is living tissue, in reality is composed of cells, matrix, mineral and growth factors. The inter-relationship of these components provides structural stability and a ready source of calcium for overall homoeostasis of the organism. Bone cells, and in particular osteoblasts, produce matrix proteins that are integral for skeletal integrity. In addition, these cells also secrete several growth factors that are stored within the skeletal matrix, and are liberated during the resorptive phase of modelling and remodelling (see figure 1.2). The three major factors that are latently attached to binding proteins in bone are IGF-I, IGF-II and transforming growth factor-beta. Each of these proteins plays a unique role in:

1.   the recruitment of young osteoblasts to the bone surface,
2.   the differentiation of those osteoblasts,
3.   the promotion of collagen biosynthesis, and
4.   the maintenance of the remodelling cycle [30–33].

As such, it is this continuous modelling and remodelling of bone that provides the skeletal milieu with the necessary growth factors to stimulate further cell recruitment and differentiation.

In addition to these growth factors, the bone matrix also contains skeletal specific proteins such as osteocalcin, and other connective tissue molecules like osteonectin and osteopontin, derived from differentiated osteoblasts [43–46]. But, by far, the major structural component of the skeletal matrix is collagen type I, a large protein composed of three separate peptide chains organized in parallel and synthesized by mature osteoblasts (see section 1.5). In fact, 90% of unmineralized osteoid is collagen type I [47, 48]. Individual subunits of the collagen helix are connected terminally to each other by cross-linking amino acids, added as a post-translational modification of the entire collagen molecule. During resorption, cross-links are catalysed initially, and are subsequently liberated from the matrix proper [43]. Some of these enter the circulation as telopeptide fragments, and are eventually filtered by the renal tubules. Both qualitative and quantitative defects in collagen synthesis, modification or mineralization can lead to chronic disorders characterized by enhanced skeletal fragility, such as osteoporosis.

Type I procollagen is composed of three separate peptide chains encoded by two different genes. Each mature collagen molecule comprises two alpha(I) 1 chains and one alpha(I) 2 chain encoded by the COL1A1 and COLIA2 genes [22, 47–49]. Following translation, individual collagen chains undergo extensive modification within the Golgi with hydroxylation and glycosylation of proline and lysine residues. A triple helix is formed before secretion from bone cells and, in the extracellular milieu, collagen fibrils are self assembled. As the collagen matures, the fibrils undergo further modification with cross-linking through specialized covalent bonds (pyridinium cross links) that enhance the stability, elasticity and strength of the mature collagen.

After maturation of collagen, individual fibrils are organized, according to their site of synthesis, into 'woven' or lamellar bone. In the former, collagen fibrils are poorly organized and offer less structural competency than lamellar bone in which the fibrils are tightly organized and lie in parallel. As noted in section 1.2, woven bone is present in the embryo and is rapidly synthesized during repair of bony defects, such as fracture healing, or in some early postnatal bone structures such as the calvariae. Woven bone is also found in certain bone tumours, in patients with osteogenesis imperfecta and in patients with Paget's disease. The strength of woven bone is modest compared with cancellous bone, which is more mature and results from the remodelling of woven bone or from pre-existing bone tissue. In compact, or cortical bone, collagen fibres are organized in an axial rotation to form osteons. Cancellous bone is equally well organized with respect to the orientation of the collagen fibrils, albeit from a different perspective (table 1.1). The three-dimensional network of trabeculae on the endosteum of bone provides a huge surface area for bone turnover and for meeting the mechanical needs of such a system. Plates and sheets of collagen that compose cancellous bone also provide protection against mechanical stresses that are more multi-directional, such as twisting and bending. Indeed, directional generation of collagen fibrils in a cancellous bone in part relate to the mechanical load applied to that bone.

The best example of the diverse nature of the skeletal matrix as a framework, and its relationship to mechanical loading, is the femur (figure 1.4). On inspection, either radiologically or microscopically, the femur is

Table 1.1. *Types of bone in the adult skeleton.*

Bone characteristics	Remodelling characteristics	Predominant sites in the skeleton
Cortical bone	Cutting cones, via osteoclasts	Mid-shaft of long bones, proximal radius
Mixed cortical/trabecular		Proximal femur, proximal humerus, pelvis
Trabecular bone	Surface remodelling—specific site	Lumbar vertebrae, calcaneus, ribs

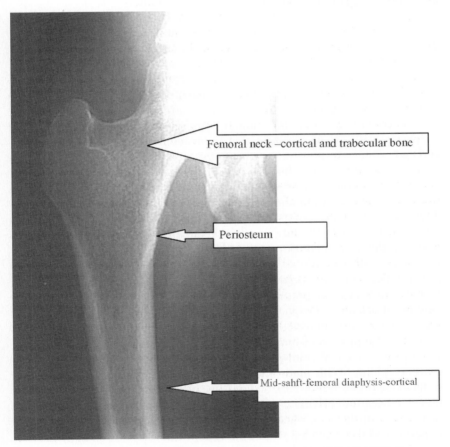

Figure 1.4. *Normal human femur with reduced bone mineral density. This is a radiograph from a postmenopausal woman with previous osteoporotic fractures of the vertebrae.*

composed of both cancellous and compact bone depending on the particular region. For example, there are spatially organized elements of cancellous tissue which fan out in the femoral neck and correspond to the mechanical loads imposed in that particular direction (i.e. twisting or bending) on that region of the hip. In contrast, just down the femoral shaft at the diaphysis, lamellar or compact bone predominates, principally to resist directional forces acting directly on the bone. The femoral diaphysis is a hollow cylinder rather than a solid mass, in part because distribution of bone mass to the outer circumference in that region requires less material and therefore is lighter but more able to resist force. Thus it very obvious that bone, at all levels of organization from shape to size, is a structural material efficiently designed to perform complex biological functions as well as meeting the mechanical needs of the organism.

1.5. BONE COMPOSITION: MINERALIZATION OF BONE MATRIX

From a structural perspective, the mechanical properties of bone as a tissue relate to the mineralization of a soft organic matrix into a hard rigid material. The process of this mineralization is complex and has defied a complete delineation [50, 51]. It is certain that the osteoblast and some of its differentiated products are essential elements required for the growth of the apatite crystal. It is also evident, like the other major components of bone such as tissue, that calcium and phosphate ions are needed not only for hardening tissue, but also for maintaining physiological needs and for facilitating the interaction of organic components of the matrix with bone cells. Hence, as noted in figure 1.1, the interface of all three components is necessary to maintain the functional properties of bone.

Spatial distribution of calcium and phosphate crystals (as calcium apatite), not unlike the organic matrix organization, remains a central part of the mineralization process. These crystals are composed of a specific amount of calcium, phosphate and carbonate (but do not contain hydroxyl groups, hence are not strictly considered hydroxyapatite) in a ratio that is critical to the process of mineralization [52–54]. During the early phases of this process, a Ca–P solid phase is produced which is amorphous rather than crystalline. With maturation, crystalline development occurs but age of the bone (in respect of remodelling), age of the organism and other factors make delineation of the crystal component difficult and confusing. The problem is further confounded by the intimate relationship of very small mineral crystals embedded within the collagen fibrils.

The time course and maturation of the crystalline phase is still debated, although it is clear that this process is coupled in the remodelling cycle to matrix deposition [55]. Most of the initial crystals deposited in the newly developed collagen fibrils are located in hole zones or channels between connecting fibrils. Further calcification occurs both by primary hetero-geneous nucleation and by secondary and tertiary nucleation from crystals already formed and propagated in the collagen pores [52, 55, 56]. In addition, collagen fibrils expand in the regions of the hole zones, permitting additional deposition of crystals. Eventually, all the available space within the fibril becomes a continuous hard substance. The time course of this crystallization is, as noted, tied intimately to matrix apposition and is most rapid in the first 10–15 days of the remodelling cycle (1–1.5 μm/day) [51]. As with matrix deposition, mineralization slows over time as the osteoblasts become flatter and more extended in shape. Eventually by 90 days (figure 1.2) the osteo-blasts become lining cells, and the osteoid seam disappears because all the matrix has become mineralized.

Matrix vesicles formed from bone cells were once thought to be the controlling determinants of crystallization during modelling and remodelling [57]. These extracellular vesicles, which are pinched away from osteoblasts, contain enzymes such as alkaline phosphatase which are essential for proper mineralization, and other proteins such as bone sialoprotein, which appears very early in the course of osteoblast mineralization *in vitro*. Indeed, deficiencies in alkaline phosphatase expression result in syndromes of osteomalacia, or 'soft bone', characterized by large amounts of osteoid that is not mineralized. However, it is now apparent, notwithstanding the phenotype of deficient alkaline phosphatase, that matrix vesicles are a func-tion of mineralization only during the earliest phases of bone development, when so-called woven bone is produced. During later stages of postnatal development, the bone tissue containing matrix vesicles is resorbed and new tissue is formed which does not contain these vesicles, but which is appropriately mineralized. Hence, only during a specific developmental time period, or in the case of fracture during the production of woven bone, are the matrix vesicles important in mineralization [58, 59].

Deficiencies in mineralization of collagen fibrils result in the syndromes of osteomalacia. Histologically, these are characterized by accumulation of osteoid that is not mineralized. Vitamin D deficiency has been considered the most prominent osteomalacia syndrome, although it is clear that several other disorders unrelated to vitamin D metabolism can lead to this histo-logical appearance. Notwithstanding, the clinical phenotype of 'soft' bones with poor mineralization and deformed extremities is shared by all these disorders. Clinically, these patients also present with low serum calcium, reduced serum phosphorus in some instances, abnormal vitamin D metabo-lites in the majority of cases (approximately 75% of all osteomalacia is related to vitamin D deficiency or resistance), increased alkaline phospha-tase, bone pain, and severely reduced bone mineral density. Treatment must be individualized to the specific syndrome but it can lead to reversal of many of the symptoms and signs of this heterogeneous disorder.

1.6. METABOLIC DISORDERS OF BONE

1.6.1. Introduction

Although it is beyond the scope of this chapter to discuss in detail all metabolic bone diseases, it should be quite apparent that skeletal disorders can affect any or all of the three major components of bone substance, i.e. bone cells, matrix or mineral. As noted above, mineralization defects define the osteomalacia syndrome. Numerous diseases target bone cells and there are primary and secondary collagen disorders which can affect the quality and quantity of bone matrix. Mutations in specific growth factors or cytokines can have a profound effect on the skeleton of mammals. For example, a deletion of the gene encoding mCSF results in severe and potentially lethal osteopetrosis due to the absence of osteoclasts [16]. Mice with an induced Cbfa1 knock-out are completely lacking in new bone [29]. A deletion in exon 5 of the IGF-I gene leads to low bone density and failure to grow. Despite this ever-growing list, it is clear that the majority of individuals who suffer from the number one cause of metabolic bone disease, i.e. osteoporosis, do not have single mutations but rather have impairment in peak bone mass as a result of polygenic influences and environmental interactions.

The osteoporosis syndrome represents a heterogeneous disorder related to low bone density and skeletal fragility. Bone strength is determined by the size of bone, the rate of bone turnover, and the microarchitecture. As noted in figure 1.5, loss of bone mass is associated with drop-out of trabecular networks, increased trabecular spacing, reduced connectivity and enhanced skeletal fragility. A common pathophysiological component of this syndrome is the disordered bone remodelling cycle [1, 47–49]. Osteoporosis can thus be considered a disease in which there is uncoupling of resorption

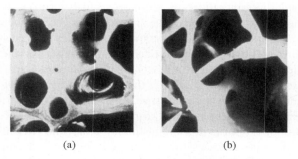

(a)   (b)

Figure 1.5. *Microstructure of (a) normal bone and (b) osteoporotic bone in an iliac crest biopsy sample obtained from a healthy older woman and an osteoporotic postmenopausal woman. The figure highlights differences in bone mass and architecture.*

from formation, such that more bone is resorbed than formed. This can be a result of multiple alterations in cell–cell interactions. For example, during oestrogen deficiency states in older women, there is a marked increase in elaboration of several cytokines including the interleukins, RANKL and TGF-b [1, 34]. Bone resorption is markedly enhanced and bone turnover is stimulated. However, since formation and mineralization of new bone takes considerably more time than does resorption (figure 1.2), over a prolonged period there is a net deficit in the remodelling cycle. Another scenario, less often seen but certainly appreciated, is related to impaired bone formation [39]. Immobilization, weightlessness, some age-related syndromes and glucocorticoid-induced osteoporosis are all characterized by a pronounced reduction in osteoblast-mediated bone formation. In most of these situations the lag in bone formation is exacerbated by a pronounced increase in bone resorption. The end result is a significant loss of bone over a relatively short time frame. Moreover, to add insult to injury, the uncoupling of resorption and formation results in a biomechanically unstable structure, enhancing the risk of subsequent fracture. In all cases, acquired changes in the osteoblast and/or the osteoclast result in a defined skeletal phenotype. It would be fair to conclude that, for most bone

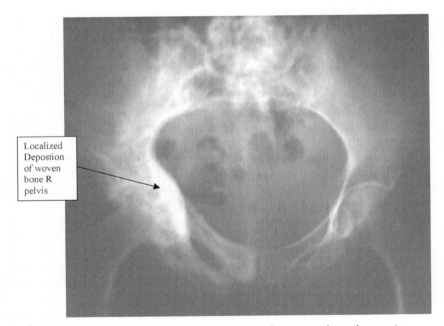

Localized
Depostion
of woven
bone R
pelvis

Figure 1.6. *Paget's disease of bone. The bone undergoes accelerated resorption and this leads to the deposition of primary or woven bone which is structurally compromised.*

disorders, an alteration in bone cell function results in subsequent alterations in the other compartments.

Other disorders of bone are associated with abnormalities in the BMU. For example, Paget's disease is an uncommon, sometimes heritable, focal disorder characterized by accelerated bone resorption. The remodelling unit is disordered, with a massive increase in bone resorption accompanied by a significant increase in bone formation. Although the key pathophysiological component of this disease relates to osteoclastic recruitment, bone formation is accelerated so dramatically that the result is the production of primary or woven bone rather than lamellar bone. These changes can be detected on plane radiographs and are frequently found in the spine, pelvis and skull (figure 1.6). Since woven bone is abnormal in respect to orientation and strength, it is also more fragile. This results in an increased likelihood of spontaneous fracture at sites of Paget's involvement. In addition, localized pain, nerve entrapment and malignant degeneration can also occur. Treatment with agents such as the bisphosphonates, which slow bone turnover, are considered the first line of therapy for this disorder.

1.7. OSTEOPOROSIS

1.7.1. Introduction

To date osteoporosis is the main area where the assessment of bone status has a profound clinical utility. The term 'osteoporosis' was first used in the mid-nineteenth century in France and Germany as a histological description of aged human bone, emphasizing its apparent porosity [60]. Over the years, many definitions of osteoporosis have been suggested according to its nature and causes, as well as its specific skeletal abnormalities. In recent years, however, more consistent definitions have been developed, with definitions covering the spectrum of manifestations, from the reduced amount of bone present to some of the consequences of bone loss [61]. A panel from the National Institute of Health Consensus Conference defined osteoporosis as 'a disease characterized by low bone mass and microarchitectural deterioration of bone tissue, leading to enhanced bone fragility and a consequent increase in fracture risk' [62]. Osteoporosis is generally categorized as primary or secondary, depending on the absence or presence of associated medical diseases, surgical procedures or medications known to be associated with accelerated bone loss.

Osteoporosis is sometimes termed the 'silent epidemic' because early osteoporosis is asymptomatic, and significant bone loss may become evident only after a hip or vertebral fracture has occurred. Fractures, especially of the spine, hip and wrist, are the clinical manifestation of osteoporosis. Initially,

Figure 1.7. *Grading of osteoporotic fractures of the spine. There are more than 1.5 million vertebral fractures in the US alone each year. This figure represents the grading and severity of crush fractures of the vertebrae. Although some may be asymptomatic, and only detected on lateral chest X-rays, the majority of woman report significant back pain and height loss. The presence of a previous osteoporotic spine fracture on X-ray represents a major risk factor in an individual for subsequent fractures.*

spine fractures tend to be asymptomatic but they are associated with significant morbidity as the severity and number of fractures increase (figure 1.7). The most serious fractures are those of the hip, which contribute substantially to morbidity, mortality and health care cost. Within a year of a hip fracture the mortality rate is as high as 20% with reduced functional capacity in 50% of patients [63].

The pathophysiology of osteoporosis is multifactorial and complex, the risk of fracture depending on a variety of factors including the propensity to fall, visual acuity and response to falling and bone strength [64, 65]. However, studies have shown that bone mass is the most important determinant of bone strength. Reduced bone mass is therefore a useful predictor of increased fracture risk [66]. Many prospective studies have shown that a decrease in bone density at the spine or hip of one standard deviation increases the risk by a factor of two to three [67, 68]. Methods of measuring bone mineral density are therefore pertinent to the detection of osteopenia, identification of those individuals at risk of atraumatic fracture, and assessment of the efficacy of either prevention or treatment of osteoporosis.

1.7.2. Pathophysiology of osteoporosis

Despite mounting evidence indicating that genetic factors have a strong role in determining peak bone mass and influencing the rates of change of bone mass at particular sites during ageing, lifestyle and environment appear to interact with these powerful genetic regulators of bone metabolism to determine net bone mass [69, 70]. For example, low calcium and protein intake, deficient intake of other nutrients, endocrine dysfunction, chronic illness, inactive lifestyle and immobilization may all result in a suboptimal peak bone mass and thereby increase the risk of developing osteoporosis in later life [71, 72]. The most dramatic determinant of and contributor to osteo-porosis is rate of bone loss.

1.7.2.1. Bone loss

Bone changes occurring during normal ageing are a universal phenomenon of biology regardless of sex, race, lifestyle, economic development, geo-graphic location and historical epoch [73, 74]. The changes are both quantitative and qualitative in nature. Bone mass rises through puberty to reach maximal peak adult bone mass during the second to third decade [75, 76]. Thereafter, a gradual loss begins, which is faster in the trabecular bones and is further accelerated in women during menopause [77, 78].

Involutional osteoporosis, that is, gradual, progressive bone loss, can be categorized into two types: postmenopausal osteoporosis and senile osteo-porosis [77]. Postmenopausal osteoporosis mostly occurs between the ages of 50 and 65 years. Here trabecular bone accelerates resorption related to oestrogen deficiency and is often manifest in a fracture in the spine or wrist. Senile osteoporosis usually occurs in both men and women at the age of 75 or older, with a disproportionate loss of both trabecular and corti-cal bone, seen in fractures in the hip, proximal end of the humerus, tibia and pelvis [79]. If the sex difference in the fracture rate in elderly people is mainly the result of postmenopausal bone loss, reducing fracture incidence implies different strategies for men and women. On the other hand, the environ-mental risk factors for bone loss probably do not differ greatly between men and women [80].

It is well established that the age-related decrease in bone mineral content (BMC) or bone mineral density (BMD) structurally weakens bone and thus predisposes bone to fractures. However, old bone is more highly mineralized than young bone, and thus bone substance density changes much less with age than bone matrix volume [73, 81]. Since BMD accounts for about 75–85% of the variance in bone strength in normal individuals [82, 83], the remaining 15–25% could be accounted for by other factors such as bone size, shape, collagen, the amount and orientation of the organic matrix [81, 84].

It is well known that the mechanism of bone loss varies in different types of bone tissue. A disruption of the trabecular network in vertebrae with age is mainly caused by perforation of horizontal supporting struts. These struc-tural changes cannot be reversed [85]. Cortical bone, on the other hand, has a different pattern of loss. Cortical bone is lost mostly from the endosteal surfaces so that the marrow cavity of bone is expanded, leading to a net reduction of as much as 30–50% in the thickness of cortical bone in women [82, 86]. Such changes are much greater in the lower extremities than in the upper and may vary according to age, sex and skeletal site [82]. Moreover, there is evidence that cortical thickness may be preferentially maintained in bone segments subjected to high bending and torsional stresses, while metaphyseal bone becomes progressively weaker [87].

Previous studies [88–91] have shown that blacks have a higher trabecular bone density than whites, which is due to a greater trabecular thickness and lower bone turnover. Blacks also have greater cortical thickness than whites. Black men have 7% higher subperiosteal and 30% higher medullary areas than white men; the corresponding figures for women are 14% and 49% higher [92]. Despite the differences between blacks and whites in bone density, there are few differences in their trabecular and cortical bone loss. Before the menopause, the rate of bone loss is not greater in white women than black women [90]. However, around the menopause, white women lose bone more rapidly at the spine and radius than black women, but the rate of loss does not differ in the late postmenopausal periods [93]. The ethnic difference in cortical bone loss is not clear. A study by Han *et al* [94] showed no difference in cortical bone loss rate. The magnitude of the differences between blacks and whites was the same throughout the age range 20–74, indicating differences between blacks and whites are due to peak adult bone values, but not the rates of bone loss with age.

Trabecular bone loss with age, as measured with quantitative computed tomography (QCT), is similar in men and women. However, there is a clear difference in cortical bone loss between men and women [95]. This gender difference in cortical bone loss is due to greater endocortical loss and smaller periosteal apposition in women [95–97].

1.7.3.   Etiologic factors in osteoporosis

Regulation of the bone-remodelling process is complex. It involves a number of cellular functions directed towards the coordinated resorption of old bone and formation of new bone. There are numerous systemic hormones, such as parathyroid hormone, 1,25-dihydroxy-vitamin D (calcitriol), calcitonin, oestrogens and androgens which serve in part to regulate the process. In adults, approximately 25% of the trabecular bone is resorbed and replaced every year, in contrast to only 3% of cortical bone [23]. This difference suggests that the rate of remodelling is controlled primarily by both local and systemic factors. However, in most cases, the exact molecular mechanisms are unknown, and even the identities of primary target cells are often unclear.

1.7.3.1.   Oestrogen deficiency

Oestrogen deficiency is a major etiological component which is caused by either menopause or ovariectomy, resulting in accelerated bone loss and osteoporosis [77, 98]. In these conditions the bone turnover increases, but bone resorption far exceeds bone formation [99]. Previous studies have shown oestrogen receptors to be involved in osteoblastic lineage [100, 101] and in avian osteoclasts [102]. This suggests that oestrogen may have direct effects on both of these bone cell types, although none have yet been found [103]. On the other hand, evidence from recent studies showed that oestrogen acts indirectly on bone cells via medium factors like interleukins (IL-1 and IL-6), which are present in the bone microenvironment and play a role in the stimulation of bone resorption [14, 104, 105]. Oestrogen deficiency is also found to directly and indirectly decrease the efficiency of intestinal and renal calcium absorption and re-absorption, respectively [106]. Oestrogen deprivation may be an important pathophysiological component of bone loss over the entire postmenopausal life of a woman. Studies have shown that oestrogen replacement can maintain bone density after menopause [107], but only current exposure to oestrogen protects against hip fractures [108].

1.7.3.2. Growth hormone

Certain transforming growth factors seem to play a role in the stimulation of bone formation. Although the direct effect of growth hormone (GH) on bone formation is limited [109], GH secretion is critical to both longitudinal growth and acquisition of peak bone mass during adolescence. Studies showed that GH is important for maintenance of adult bone mass. A low bone mass and a small bone size may be a result of GH deficiency in humans [110]. Age-related bone loss in relation to decline in GH secretion in the elderly is not, however, clear [111]. Nevertheless, there is a growing increased use of GH therapy in patients with GH deficiency and the treatment has had positive results [112, 113]. GH therapy seems to have a primary effect on compact bone [114, 115]. It increases calcium absorption in the gastrointestinal tract which is mediated by an increase in 1,25-dihydroxy-vitamin D_3 production [109]. However, GH therapy in elderly men and women has not been shown to have significant effects on bone mass other than activation of remodelling sequences [116, 117]. Decreases of GH and insulin-like growth factor (IGF) with ageing may be responsible for the increase of body fat that occurs with ageing [118].

1.7.3.3. Parathyroid hormone

Parathyroid hormone (PTH) is a peptide that stimulates bone through its ability to activate the PTH/PTHrp (parathyroid-hormone-related protein) receptor located on the surface of osteoblasts [119]. PTH plays a direct and important role in bone remodelling. Bone histomorphometric studies have demonstrated that the administration of PTH *in vivo* stimulates bone formation [120, 121]. However, the effects on bone formation are complex. It can stimulate and inhibit bone collagen and matrix synthesis [122]. PTH stimulates differentiation of committed progenitors to fuse, forming mature multinucleated osteoclasts. However, PTH does not directly stimulate bone resorption because the osteoclast does not respond to PTH [109]. It is probably mediated through cells in the osteoblast lineage such as the lining cells [123].

PTH modulates the activity of specific cells in bone and kidneys to regulate the levels of calcium phosphate in the blood. PTH stimulates bone to release calcium and phosphate into the blood circulation. It stimulates reabsorption of calcium and inhibits reabsorption of phosphate from glomerular filtrate. PTH also stimulates the renal synthesis of 1,25-$(OH)_2$D [124].

It has been found that serum PTH levels increase with age as a result of impaired calcium absorption and age-related decline in renal function [125]. However, whether this causes age-related bone loss is not clear. A French study found a weak correlation between PTH and BMD of the femoral neck after age adjustment in older institutionalized women [126]. Other studies have showed that changes in PTH do not predict changes in bone density [127].

1.7.3.4. Risk factors

Osteoporosis is a complex multicausal disease and not all its causes are known. However, certain factors have been identified to be linked to the development of osteoporosis or contribute to an individual's likelihood of developing the disease. These are referred to as risk factors. While many people with osteoporosis have several risk factors, there are others who develop osteoporosis with no identifiable risk factors (idiopathic osteoporosis).

The susceptibility to fracture depends on a variety of factors including the propensity to fall, visual acuity, response to falling and bone mass [64, 65]. However, studies have shown that bone mass is the most important determinant of bone strength and accounts for up to 80% of its variance [128]. Reduced bone mass is therefore a useful predictor of increased fracture risk [66]. Many prospective studies have shown that a decrease of one standard deviation in bone density at the spine or hip increases the risk by a factor of two to three [68]. Methods of measuring BMD are pertinent to the detection of osteopenia, identification of those individuals at risk of atraumatic fracture, and assessment of the efficacy of either prevention or treatment of osteoporosis. Additionally, the presence of clinical risk factors such as lifestyle, diet and family history of osteoporosis are relatively insensitive in predicting the presence of osteopenia [129].

1.7.4. Epidemiology

1.7.4.1. Prevalence of osteoporosis

The clinical endpoint of osteoporosis is fracture, predominantly of the wrist, spine and hip. The number of people considered to have osteoporosis depends on the way the condition is defined in practice [130]. Using WHO's definition of osteoporosis, from the data of the Third National Health and Nutrition Examination Survey, it was shown that the prevalence of osteoporosis of the hip was 21% among white women compared with 16% among Hispanic women and only 10% among African–American women [131]. The updated estimates in women in the four femur regions using the WHO diagnostic criteria showed that 4–6 million women after age 50 have osteoporosis and 13–17 million women have osteopenia [131]. Age-adjusted prevalence of both osteopenia and osteoporosis were higher in non-Hispanic white women than non-Hispanic black women; prevalence in Mexican–American women was similar but slightly lower than in non-Hispanic white women. Epidemiologically, elderly white women have the highest rate of osteoporotic fractures compared with other races and men [132, 133].

The prevalence of osteoporosis with age increases rapidly after age 50 years. Within 5–10 years of menopause, Colles' fractures of the wrist and forearm first occur. About a decade later, vertebral fractures begin to be observed, and then hip fractures occur 10 years after that. The National Osteoporosis Foundation of the United States in 1997 reported the osteoporosis attribution probabilities in 72 categories comprised four specific fracture types (hip, spine, forearm, and all other sites combined), stratified by three age groups (45–64, 65–84 and 85+ years), three racial groups (white, black and all others), and both genders. The osteoporosis attribution probabilities for any fracture increase with age in either race or gender group. The highest rate of fractures were hip and spine. It was estimated that at least 90% of all hip and spine fractures among elderly white women were attributed to osteoporosis.

Hip fracture is the most serious complication of osteoporosis and is associated with considerable morbidity and mortality. In the United States, annually, there are 250 000 hip fractures and of these 12–20% die of related complications [130, 134]. The excess deaths occur primarily in the first six months following a hip fracture [135].

As previous data have shown, there is a marked racial variation in prevalence of fractures. The lower fracture rate in blacks may a result of high peak bone mass, lower rates of bone loss and a lower risk of falls. The lower fracture rates in Asians may be partly due to shorter stature and altered hip geometry [136].

There is also marked variation in fracture occurrence between geographic locations and countries. In the United States, the age-adjusted fracture rate in white women increases with socio-economic deprivation, decreased January sunlight, decrease in water hardness and access to fluoridated water and decrease in the percentage of land in agricultural use [60]. In Europe, Scandinavian countries have the highest fracture occurrence. However, regional variation did not correspond with obesity, smoking, alcohol consumption or Scandinavian heritage.

In general, data concerning the prevalence of hip, wrist, and other non-vertebral fractures are more reliable than vertebral fracture data. This is due to the fact that many vertebral fractures are not clinically evident and there is a lack of clear radiographic definition [135]. Early definitions of vertebral fracture relied on subjective assessment of wedge, crush and biconcave deformities, a technique with poor reproducibility. More recently, definitions based on vertebral morphometry with fixed cut-off values, which have increased specificity [137], have been proposed. The age-adjusted prevalence of radiological vertebral fractures has been estimated at between 8 and 25% in women aged over 50 years [138, 139].

1.7.4.2.   *Incidence of new fractures resulting from osteoporosis*

Incidence refers to the number of new fractures occurring in a population within a specified time. Estimations of future fracture rates are based on projections of the size, age and sex distribution of the world population, and of the age-adjusted incidence rates for fracture. Worldwide, the number of hip fractures is increasing in both women and men [139, 140]. This increase is mostly due to the increasingly older population. For example, in the United States, the number of persons aged 65 years and over is expected to rise from 32 million to 69 million between 1990 and 2050, while those aged 85 years and over will increase from 3 million to 15 million. Approximately 275 000 new osteoporotic hip fractures will occur each year in the United States [141] and 5% of 80-year-old women will experience a new vertebral fracture each year among Japanese–American women in Hawaii [142]. The female:male ratio of the incidence is 2:1. The other races have much lower incidence than whites.

Worldwide, the population over 65 years old is estimated to increase almost fivefold from 323 million to 1555 million by 2050 [130]. This could lead to an increase in the number of hip fractures from 1.66 million in 1990 to 6.26 million in 2050. The greatest increase of fractures is likely to be seen in underdeveloped countries, where the life expectancy is increasing exponentially.

1.7.4.3.   *Economic costs*

The health care expenditures attributable to osteoporotic fractures in 1995 were estimated at $13.8 billion in the United States. Approximately 75.1% of the cost was for treatment of Caucasian females, 28.4% Caucasian males, 5.3% for non-Caucasian females and 1.3% for non-Caucasian males over the age of 45. In England and Wales, the cost of osteoporotic fractures in 1990 was estimated at £742 million per year. In France an estimated 56 000 hip fractures alone cost about FF 3.5 billion per year. The vast majority of these costs are accounted for by direct costs, including in-patient and out-patient hospital care and nursing home care [60]. They include an estimated 3.4 million hospital bed days per year, resulting from 253 000 hospitalizations. Indirect costs such as loss of earnings are not a major component of the total costs of hip fracture because most patients are retired.

1.8. SUMMARY

The anatomy and physiology of bone as a tissue relates to the three major compartments: bone cells, bone mineral and bone matrix. The health of these three components depends to a great degree on each other and, as shown, disorders in one can result in profound changes in the others. More work is needed to understand the basic mechanism of bone mineralization, to define the genetic regulation of bone cells and their functions, and the increasingly important interaction between matrix and cell. Once these goals are attained, it is likely that treatment of metabolic bone diseases will also become more successful.

REFERENCES

[1] Manolagas S C 2000 Birth and death of bone cells basic regulatory mechanisms and implications for the pathogenesis and treatment of osteoporosis *Endocr. Rev.* **21**(2) 115–37
[2] Suda T, Nakamura I, Jimi E and Takahashi N 1997 Regulation of osteoclast function *J. Bone Miner. Res.* **12**(6) 869–79
[3] Suda T, Takahashi N, Udagawa N, Jimi E, Gillespie M T and Martin T J 1999 Modulation of osteoclast differentiation and function by the new members of the tumor necrosis factor receptor and ligand families *Endocr. Rev.* **20**(3) 345–57
[4] Reddy S V and Roodman G D 1998 Control of osteoclast differentiation *Crit. Rev. Eukaryot. Gene Expr.* **8**(1) 1–17
[5] Salo J, Metsikko K, Palokangas H, Lehenkari P and Vaananen H K 1996 Bone-resorbing osteoclasts reveal a dynamic division of basal plasma membrane into two different domains *J. Cell. Sci.* **109** (Pt 2) 301–7
[6] Blair H C, Teitelbaum S L, Ghiselli R and Gluck S 1989 Osteoclastic bone resorption by a polarized vacuolar proton pump *Science* **245**(4920) 855–7
[7] Sly W S and Hu P Y 1995 Human carbonic anhydrases and carbonic anhydrase deficiencies *Ann. Rev. Biochem.* **64** 375–401
[8] Bekker P J and Gay C V 1990 Biochemical characterization of an electrogenic vacuolar proton pump in purified chicken osteoclast plasma membrane vesicles *J. Bone Miner. Res.* **5**(6) 569–79
[9] Baron R, Neff L, Louvard D and Courtoy P J 1985 Cell-mediated extracellular acidification and bone resorption: evidence for a low pH in resorbing lacunae and localization of a 100-kD lysosomal membrane protein at the osteoclast ruffled border *J. Cell Biol.* **101**(6) 2210–22

[10] Tezuka K, Nemoto K, Tezuka Y, Sato T, Ikeda Y, Kobori M *et al* 1994 Identification of matrix metalloproteinase 9 in rabbit osteoclasts *J. Biol. Chem.* **269**(21) 15006–9

[11] Manolagas S C and Jilka R L 1995 Bone marrow, cytokines, and bone remodeling. Emerging insights into the pathophysiology of osteoporosis *N. Engl. J. Med.* **332**(5) 305–11

[12] Kelly KA, Tanaka S, Baron R and Gimble JM 1998 Murine bone marrow stromally derived BMS2 adipocytes support differentiation and function of osteoclast-like cells in vitro *Endocrinology* **139**(4) 2092–101

[13] Teitelbaum S I, Tondravi M N and Ross F P 1996 Osteoclast biology. In *Osteoporosis* eds F Marcus, D Feldman and J Kelsey (San Diego: Academic Press) pp 61–94

[14] Horowitz M C 1993 Cytokines and estrogen in bone anti-osteoporotic effects *Science* **260**(5108) 626–7

[15] Takahashi N, Udagawa N, Akatsu T, Tanaka H, Shionome M and Suda T 1991 Role of colony-stimulating factors in osteoclast development *J. Bone Miner. Res.* **6**(9) 977–85

[16] Felix R, Cecchini M G and Fleisch H 1990 Macrophage colony stimulating factor restores in vivo bone resorption in the op/op osteopetrotic mouse *Endocrinology* **127**(5) 2592–4

[17] Lacey DL, Timms E, Tan H L, Kelley M J, Dunstan C R, Burgess T *et al* 1998 Osteoprotegerin ligand is a cytokine that regulates osteoclast differentiation and activation *Cell* **93**(2) 165–76

[18] Yasuda H, Shima N, Nakagawa N, Yamaguchi K, Kinosaki M, Mochizuki S *et al* 1998 Osteoclast differentiation factor is a ligand for osteoprotegerin/osteoclastogenesis-inhibitory factor and is identical to TRANCE/RANKL *Proc. Natl. Acad. Sci. USA* **95**(7) 3597–602

[19] Soriano P, Montgomery C, Geske R and Bradley A 1991 Targeted disruption of the c-src proto-oncogene leads to osteopetrosis in mice *Cell* **64**(4) 693–702

[20] Hynes R O 1992 Integrins: versatility, modulation, and signaling in cell adhesion *Cell* **69**(1) 11–25

[21] Nesbitt S, Nesbit A, Helfrich M and Horton M 1993 Biochemical characterization of human osteoclast integrins. Osteoclasts express alpha v beta 3, alpha 2 v beta 1, and alpha v beta 1 integrins *J. Biol. Chem.* **268**(22) 16737–45

[22] Raisz L G and Rodan G A 1998 Embryology and cellular biology of bone. In *Metabolic Bone Diseases and Related Disorders* eds L V Avioli and S M Krane (San Diego: Academic Press) pp 1–22

[23] Parfitt A M 1994 Osteonal and hemi-osteonal remodeling: the spatial and temporal framework for signal traffic in adult human bone *J. Cell Biochem.* **55**(3) 273–86

[24] Stein G S, Lian J B and Stein J L 1996 Mechanisms regulating osteoblast proliferation and differentiation. In *Principles of Bone Biology* eds L G Raisz and J P Bilezikian (San Diego: Academic Press) pp 69–86

[25] Bennett J H, Joyner C J, Triffitt J T and Owen M E 1991 Adipocytic cells cultured from marrow have osteogenic potential *J. Cell. Sci.* **99**(1) 131–9

[26] Friedenstein A J, Chailakhyan R K, Latsinik N V, Panasyuk A F and Keiliss-Borok I V 1974 Stromal cells responsible for transferring the microenvironment of the hemopoietic tissues. Cloning in vitro and retransplantation in vivo *Transplantation* **17**(4) 331–40

[27] Ducy P, Zhang R, Geoffroy V, Ridall AL, and Karsenty G 1997 Osf2/Cbfa1: a transcriptional activator of osteoblast differentiation *Cell* **89**(5) 747–54

[28] Gao YH, Shinki T, Yuasa T, Kataoka-Enomoto H, Komori T, Suda T *et al* 1998 Potential role of Cbfa1, an essential transcriptional factor for osteoblast differentiation, in osteoclastogenesis: regulation of mRNA expression of osteoclast differentiation factor (ODF) *Biochem. Biophys. Res. Commun.* **252**(3) 697–702

[29] Komori T, Yagi H, Nomura S, Yamaguchi A, Sasaki K, Deguchi K *et al* 1997 Targeted disruption of Cbfa1 results in a complete lack of bone formation owing to maturational arrest of osteoblasts *Cell* **89**(5) 755–64

[30] Bonewald L F and Dallas S L 1994 Role of active and latent transforming growth factor beta in bone formation *J. Cell Biochem.* **55**(3) 350–7

[31] Mohan S and Baylink D J 1994 The role of IGF-1 in the coupling of bone formation to bone resorption. In *Modern Concepts of IGFs* ed EM S (New York: Elsevier) pp 169–84

[32] Noda M and Camilliere J J 1989 In vivo stimulation of bone formation by transforming growth factor-beta *Endocrinology* **124**(6) 2991–4

[33] Bonewald L F and Mundy G R 1990 Role of transforming growth factor-beta in bone remodeling *Clin. Orthop.* **250** 261–76

[34] Rosen C J and Donahue L R 1998 Insulin-like growth factors and bone: the osteoporosis connection revisited *Proc. Soc. Exp. Biol. Med.* **219**(1) 1–7

[35] Bolander M E 1992 Regulation of fracture repair by growth factors *Proc. Soc. Exp. Biol. Med* **200**(2) 165–70

[36] Hall T J and Chambers T J 1996 Molecular aspects of osteoclast function *Inflamm. Res.* **45**(1) 1–9

[37] Hughes D E and Boyce B F 1997 Apoptosis in bone physiology and disease *Mol. Pathol.* **50**(3) 132–7

[38] Tomkinson A, Reeve J, Shaw R W and Noble B S 1997 The death of osteocytes via apoptosis accompanies estrogen withdrawal in human bone *J. Clin. Endocrinol. Metab.* **82**(9) 3128–35

[39] Weinstein R S, Jilka R L, Parfitt A M and Manolagas S C 1998 Inhibition of osteoblastogenesis and promotion of apoptosis of osteoblasts and osteocytes by glucocorticoids. Potential mechanisms of their deleterious effects on bone *J. Clin. Invest.* **102**(2) 274–82

[40] Hock J M 1999 Stemming bone loss by suppressing apoptosis *J. Clin. Invest.* **104**(4) 371–3

[41] Jilka R L, Weinstein R S, Bellido T, Roberson P, Parfitt A M and Manolagas S C 1999 Increased bone formation by prevention of osteoblast apoptosis with parathyroid hormone *J. Clin. Invest.* **104**(4) 439–46

[42] Aarden E M, Burger E H and Nijweide P J 1994 Function of osteocytes in bone *J. Cell Biochem.* **5b5**(3) 287–99

[43] Delmas P D 1995 Biochemical markers of bone turnover *Acta Orthop. Scand. Suppl.* **266** 176–82

[44] Harris H 1990 The human alkaline phosphatases: what we know and what we don't know *Clin. Chim. Acta* **186**(2) 133–50

[45] Kim R H, Shapiro H S, Li J J, Wrana JL and Sodek J 1994 Characterization of the human bone sialoprotein (BSP) gene and its promoter sequence *Matrix Biol.* **14**(1) 31–40

[46] Boskey A L 1996 Matrix proteins and mineralization: an overview *Connect. Tissue Res.* **35**(1) 357–63

[47] Dawson P A and Marini J C 1999 Osteogenesis Imperfecta. In *The Genetics of Osteoporosis and Metabolic Bone Diseases* ed M Econs (Towana, NJ: Humana Press) pp 75–93

[48] Prockop D J and Kivirikko K I 1984 Heritable diseases of collagen *N. Eng. J. Med.* **311**(6) 376–86

[49] Willing M C, Pruchno C J and Byers P H 1993 Molecular heterogeneity in osteogenesis imperfecta type I *Am. J. Med. Genet.* **45**(2) 223–7

[50] Glimcher M J 1959 Molecular biology of mineralized tissues with particular reference to bone *Rev. Mod. Phys.* **31** 359–93

[51] Glimcher M J 1998 The nature of the mineral phase in bone: Biology and clinical implications. In *Metabolic Bone Diseases and Related Disorders* eds L V Avioli and S M Krane (San Diego, CA: Academic Press) pp 23–50

[52] Sauer G R and Wuthier R E 1988 Fourier transform infrared characterization of mineral phases formed during induction of mineralization by collagenase-released matrix vesicles in vitro *J. Biol. Chem.* **263**(27) 13718–24

[53] Fratzl P, Fratzl-Zelman N and Klaushofer K 1993 Collagen packing and mineralization. An x-ray scattering investigation of turkey leg tendon *Biophys. J.* **64**(1) 260–6

[54] Rey C, Miquel J L, Facchini L, Legrand A P and Glimcher M J 1995 Hydroxyl groups in bone mineral *Bone* **16**(5) 583–6

[55] Traub W, Arad T and Weiner S 1992 Origin of mineral crystal growth in collagen fibrils *Matrix* **12**(4) 251–5

[56] Glimcher M J 1968 A basic architectural principle in the organization of mineralized tissues *Clin. Orthop. Rel. Res.* **61** 16–36

[57] Genge B R, Sauer G R, Wu L N, McLean F M and Wuthier R E 1988 Correlation between loss of alkaline phosphatase activity and accumulation of calcium during matrix vesicle-mediated mineralization *J. Biol. Chem.* **263**(34) 18513–9

[58] Rey C, Beshah K, Griffin R and Glimcher M J 1991 Structural studies of the mineral phase of calcifying cartilage *J. Bone Miner. Res.* **6**(5) 515–25

[59] Kim H, Rey C and Glimcher M J 1996 X-ray diffraction, electron microscopy, and Fourier transform infrared spectroscopy of apatite crystals isolated from chicken and bovine calcified cartilage *Calcif. Tissue Intl.* **59**(1) 58–63

[60] Arden N and Cooper C 1998 Present and future of osteoporosis epidemiology. In *Osteoporosis Diagnosis and Management* ed P J Meunier (London: Martin Dunitz) pp 1–16

[61] Kanis J A, Melton L J I, Christiansen C, Johnston C C and Khaltaev N 1994 The diagnosis of osteoporosis *J. Bone Miner. Res.* **9**(8) 1137–41

[62] Anonymous 1993 Consensus development conference: diagnosis, prophylaxis and treatment of osteoporosis *Am. J. Med.* **94** 646–50

[63] Cooper C, Atkinson E J, Jacobsen S J, O'Fallon W M and Melton L J D 1993 Population-based study of survival after osteoporotic fractures *Am. J. Epidemiol.* **137**(9) 1001–5

[64] Prudham D and Evans J G 1981 Factors associated with falls in the elderly: a community study *Age Ageing* **10**(3) 141–6

[65] Kelsey J L and Hoffman S 1987 Risk factors for hip fracture [editorial] *N. Eng. J. Med.* **316**(7) 404–6

[66] Ross P D, Davis J W, Vogel J M and Wasnich R D 1990 A critical review of bone mass and the risk of fractures in osteoporosis *Calcif. Tissue Intl.* **46**(3) 149–61

[67] Marshall D, Johnell O and Wedel H 1996 Meta-analysis of how well measures of bone mineral density predict occurrence of osteoporotic fractures [see comments] *Br. Med. J.* (Clinical Research Edn) **312**(7041) 1254–9

[68] Cummings S R, Black D M, Nevitt M C, Browner W, Cauley J, Ensrud K *et al* 1993 Bone density at various sites for prediction of hip fractures. The Study of Osteoporotic Fractures Research Group *Lancet* **341**(8837) 72–5

[69] Kelly P J and Eisman J A 1993 Osteoporosis genetic effects on bone turnover and bone density [editorial] *Ann. Med.* **25**(2) 99–101

[70] Krall EA and Dawson-Hughes B 1993 Heritable and life-style determinants of bone mineral density *J. Bone Miner. Res.* **8**(1) 1–9

[71] Cormier C 1991 Physiopathology and etiology of osteoporosis *Curr. Opin. Rheumatol.* **3**(3) 457–62

[72] Edelson G W and Kleerekoper M 1995 Bone mass, bone loss, and fractures *Phys. Med. Rehab. Clinics N. Am.* **6**(3) 455–64

[73] Parfitt AM 1988 Bone remodelling relationship to the amount and structure of bone, and the pathogenesis and prevention of fractures. In *Osteoporosis: Etiology, Diagnosis, and Management* eds B L Riggs and L J I Melton (New York: Raven Press) pp 45–93

[74] Kiebzak GM 1991 Age-related bone changes *Exp. Gerontol.* **26**(2) 171–87

[75] Snow-Harter C and Marcus R 1991 Exercise, bone mineral density, and osteoporosis *Exercise and Sport Sci. Rev.* **19** 351–88

[76] Barquero LR, Baures M R, Segura J P, Quinquer J S, Majem L S, Ruiz P G *et al* 1992 Bone mineral density in two different socio-economic population groups *Bone and Mineral* **18** 159–68

[77] Riggs B L and Melton LJ I 1986 Involutional osteoporosis *New Engl. J. Med.* **314**(26) 1676–84

[78] Wasnich R D, Ross P D, Vogel J M and Davis J W 1989 How and where is BMC measured? In *Osteoporosis: Critique and Practicum* eds R D Wasnich, P D Ross, J M Vogel, J M Davis (Honolulu: Banyan Press) pp 114–23

[79] Genant H K 1993 Radiology of osteoporosis. In *Primer on the Metabolic Bone Diseases and Disorders of Mineral Metabolism* 2nd edn ed M J Favus (New York: Raven Press) pp 229–40

[80] Slemenda C W, Christian J C, Reed T, Reister T K, Williams C J and Johnston C C 1992 Long-term bone loss in men: effects of genetic and environmental factors *Ann. Internal Med.* **117** 286–91

[81] Hayes W C and Gerhart W C 1985 Biomechanics of bone: applications for assessment of bone strength. In *Bone and Mineral Research 3* ed W A Peck (Amsterdam: Elsevier) pp 259–94

[82] Melton L J I, Chao E Y S and Lane J 1988 Biomechanical aspects of fractures. In *Osteoporosis: Etiology, Diagnosis, and Management* eds B L Riggs and L J I Melton (New York: Raven Press) pp 111–31

[83] McCalden R W, McGeough J A, Barker M B and Court-Brown C M 1993 Age-related changes in the tensile properties of cortical bone The relative importance of changes in porosity, mineralization, and microstructure *J. Bone Joint Surg.* [American volume] **75**(8) 1193–205

[84] Suominen H 1993 Bone mineral density and long term exercise. An overview of cross-sectional athlete studies *Sports Med.* **16**(5) 316–30

[85] Mosekilde L 1993 Vertebral structure and strength in vivo and in vitro *Calcif. Tissue Intl.* **53** Suppl 1 S121–5 [discussion S125–6]

[86] Einhorn T A 1992 Bone strength: the bottom line *Calcif. Tissue Intl.* **51** 333–9
[87] Ruff C B and Hayes W C 1982 Subperiosteal expansion and cortical remodeling of the human femur and tibia with aging *Science* **217** 945–8
[88] Meier D E, Luckey M M, Wallenstein S, Lapinski R H and Catherwood B 1992 Racial differences in pre- and post-menopausal bone homeostasis: association with bone density *J. Bone Miner. Res.* **7**(10) 1181–9
[89] Kleerekoper M, Nelson D A, Flynn M J, Pawluszka A S, Jacobsen G and Peterson E L 1994 Comparison of radiographic absorptiometry with dual-energy x-ray absorptiometry and quantitative computed tomography in normal older white and black women *J. Bone Miner. Res.* **9**(11) 1745–9
[90] Luckey M M, Wallenstein S, Lapinski R and Meier D E 1996 A prospective study of bone loss in African–American and white women—a clinical research center study *J. Clin. Endocrinol. Metab.* **81**(8) 2948–56
[91] Perry H M R, Horowitz M, Morley J E, Fleming S, Jensen J, Caccione P *et al* 1996 Aging and bone metabolism in African American and Caucasian women *J. Clin. Endocrinol. Metab.* **81**(3) 1108–17
[92] Garn S M, Nagy J M and Sandusky S T 1972 Differential sexual dimorphism in bone diameters of subjects of European and African ancestry *Am. J. Phys. Anthropol.* **37**(1) 127–9
[93] Seeman E 1997 From density to structure: growing up and growing old on the surfaces of bone *J. Bone Miner. Res.* **12**(4) 509–21
[94] Han Z H, Palnitkar S, Rao D S, Nelson D and Parfitt A M 1996 Effect of ethnicity and age or menopause on the structure and geometry of iliac bone *J. Bone Miner. Res.* **11**(12) 1967–75
[95] Kalender W A, Felsenberg D, Louis O, Lopez P, Klotz E, Osteaux M *et al* 1989 Reference values for trabecular and cortical vertebral bone density in single and dual-energy quantitative computed tomography *Eur. J. Radiol.* **9**(2) 75–80
[96] Schnitzler C M, Pettifor J M, Mesquita J M, Bird M D, Schnaid E and Smyth A E 1990 Histomorphometry of iliac crest bone in 346 normal black and white South African adults *Bone Miner.* **10**(3) 183–99
[97] Aaron J E, Makins N B and Sagreiya K 1987 The microanatomy of trabecular bone loss in normal aging men and women *Clin. Orthop.* **215** 260–71
[98] Riggs B L and Melton L J 1983 Evidence for two distinct syndromes of involutional osteoporosis *Am. J. Med.* **75** 899–901
[99] Eriksen E F, Hodgson S F, Eastell R, Cedel S L, O'Fallon W M and Riggs B L 1990 Cancellous bone remodeling in type I (postmenopausal) osteoporosis: quantitative assessment of rates of formation, resorption, and bone loss at tissue and cellular levels [see comments] *J. Bone Miner. Res.* **5**(4) 311–9
[100] Komm B S, Terpening C M, Benz D J, Graeme K A, Gallegos A, Korc M *et al* 1988 Estrogen binding, receptor mRNA, and biologic response in osteoblast-like osteosarcoma cells *Science* **241**(4861) 81–4
[101] Eriksen E F, Colvard D S, Berg N J, Graham M L, Mann K G, Spelsberg T C *et al* 1988 Evidence of estrogen receptors in normal human osteoblast-like cells *Science* **241** 84–6
[102] Oursler M J, Osdoby P, Pyfferoen J, Riggs B L and Spelsberg T C 1991 Avian osteoclasts as estrogen target cells *Proc. Natl. Acad. Sci. USA* **88**(15) 6613–7
[103] Ciocca D R and Roig L M 1995 Estrogen receptors in human nontarget tissues: biological and clinical implications *Endocr. Rev.* **16**(1) 35–62

[104] Jilka R L, Hangoc G, Girasole G, Passeri G, Williams D C, Abrams J S *et al* 1992 Increased osteoclast development after estrogen loss: mediation by interleukin-6 *Science* **257**(5066) 88–91

[105] Pacifici R 1996 Estrogen, cytokines, and pathogenesis of postmenopausal osteoporosis *J. Bone Miner. Res.* **11**(8) 1043–51

[106] Edelson G W and Kleerekoper M 1996 Pathophysiology of osteoporosis. In *Osteoporosis Diagnosis and Treatment* ed D J S (New York: Marcel Dekker) pp 1–18

[107] Lindsay R and Tohme J F 1990 Estrogen treatment of patients with established postmenopausal osteoporosis *Obstet. Gynecol.* **76**(2) 290–5

[108] Felson D T, Zhang Y, Hannan M T, Kiel D P, Wilson P W and Anderson J J 1993 The effect of postmenopausal estrogen therapy on bone density in elderly women [see comments] *N. Eng. J. Med.* **329**(16) 1141–6

[109] Canalis E 1996 Regulation of bone remodeling. In *Primer on the Metabolic Bone Diseases And Disorders of Mineral Metabolism* 3rd edn ed M J Favus (New York: Raven Press) pp 29–34

[110] Donahue L R and Beamer W G 1993 Growth hormone deficiency in 'little' mice results in aberrant body composition, reduced insulin-like growth factor-I and insulin-like growth factor-binding protein-3 (IGFBP-3), but does not affect IGFBP-2, -1 or -4 *J. Endocrinol.* **136**(1) 91–104

[111] Rudman D, Feller A G, Nagraj H S, Gergans G A, Lalitha P Y, Goldberg A F *et al* 1990 Effects of human growth hormone in men over 60 years old [see comments] *N. Eng. J. Med.* **323**(1) 1–6

[112] Rosen T, Johannsson G and Bengtsson B A 1994 Consequences of growth hormone deficiency in adults, and effects of growth hormone replacement therapy *Acta Paediatr. Suppl.* **399** 21–4 [discussion 25]

[113] Finkenstedt G, Gasser R W, Hofle G, Watfah C and Fridrich L 1997 Effects of growth hormone (GH) replacement on bone metabolism and mineral density in adult onset of GH deficiency: results of a double-blind placebo-controlled study with open follow-up *Eur. J. Endocrinol.* **136**(3) 282–9

[114] Bravenboer N, Holzmann P, de Boer H, Roos J C, van der Veen E A and Lips P 1997 The effect of growth hormone (GH) on histomorphometric indices of bone structure and bone turnover in GH-deficient men *J. Clin. Endocrinol. Metab.* **82**(6) 1818–22

[115] Sass D A, Jerome C P, Bowman A R, Bennett-Cain A, Ginn T A, LeRoith D *et al* 1997 Short-term effects of growth hormone and insulin-like growth factor I on cancellous bone in rhesus macaque monkeys *J. Clin. Endocrinol. Metab.* **82**(4) 1202–9

[116] Holloway L, Kohlmeier L, Kent K and Marcus R 1997 Skeletal effects of cyclic recombinant human growth hormone and salmon calcitonin in osteopenic postmenopausal women *J. Clin. Endocrinol. Metab.* **82**(4) 1111–7

[117] Gonnelli S, Cepollaro C, Montomoli M, Gennari L, Montagnani A, Palmieri R *et al* 1997 Treatment of post-menopausal osteoporosis with recombinant human growth hormone and salmon calcitonin: a placebo controlled study *Clin. Endocrinol. (Oxf.)* **46**(1) 55–61

[118] Toogood A A, Adams J E, O'Neill P A and Shalet S M 1996 Body composition in growth hormone deficient adults over the age of 60 years *Clin. Endocrinol. (Oxf.)* **45**(4) 399–405

[119] Quarles L D and Siddhanti S R 1996 Guanine nucleotide binding-protein coupled signaling pathway regulation of osteoblast-mediated bone formation [editorial] *J. Bone Miner. Res.* **11**(10) 1375–83

[120] Mosekilde L, Sogaard C H, Danielsen C C and Torring O 1991 The anabolic effects of human parathyroid hormone (hPTH) on rat vertebral body mass are also reflected in the quality of bone, assessed by biomechanical testing: a comparison study between hPTH-(1–34) and hPTH-(1–84) *Endocrinology* **66**(4) 1432–9

[121] Liu C C and Kalu D N 1990 Human parathyroid hormone-(1–34) prevents bone loss and augments bone formation in sexually mature ovariectomized rats *J. Bone Miner. Res.* **5**(9) 973–82

[122] Canalis E, Hock J M and Raisz L G 1994 Parathyroid hormone: Anabolic and catabolic effects on bone and interactions with growth factors. In *The Parathyroids* eds J P Bilezikian, R Marcus and M A Levine (New York: Raven Press) pp 65–82

[123] McSheehy P M and Chambers T J 1986 Osteoblastic cells mediate osteoclastic responsiveness to parathyroid hormone *Endocrinology* **118**(2) 824–8

[124] Kronenbery H M 1996 Parathyroid hormone: mechanism of action. In *Primer on the Metabolic Bone Diseases and Disorders of Mineral Metabolism* 3rd edn ed M J Favus (New York: Raven Press) pp 68–70

[125] Nussbaum S R, Zahradnik R J, Lavigne J R, Brennan G L, Nozawa-Ung K, Kim L Y *et al* 1987 Highly sensitive two-site immunoradiometric assay of parathyrin, and its clinical utility in evaluating patients with hypercalcemia *Clin. Chem.* **33**(8) 1364–7

[126] Meunier P J, Chapuy M C, Arlot M E, Delmas P D and Duboeuf F 1994 Can we stop bone loss and prevent hip fractures in the elderly? *Osteoporos. Intl.* **4** Suppl 1 71–6

[127] Dawson-Hughes B, Harris S S, Krall E A, Dallal G E, Falconer G and Green C L 1995 Rates of bone loss in postmenopausal women randomly assigned to one of two dosages of vitamin D *Am. J. Clin. Nutr.* **61**(5) 1140–5

[128] Hodgskinson R, Njeh CF, Currey J D and Langton C M 1997 The ability of ultrasound velocity to predict the stiffness of cancellous bone in vitro *Bone* **21**(2) 183–90

[129] Cooper C, Shah S, Hand D J, Adams J, Compston J, Davie M *et al* 1991 Screening for vertebral osteoporosis using individual risk factors. The Multicentre Vertebral Fracture Study Group *Osteoporos. Intl.* **2**(1) 48–53

[130] Melton L J I 1996 Epidemiology of osteoporosis and fractures. In *Osteoporosis: Diagnosis and Treatment* ed D J Sartoris (New York: Marcel Dekker) pp 57–78

[131] Looker A C, Orwoll E S, Johnston C C Jr, Lindsay R L, Wahner H W, Dunn W L *et al* 1997 Prevalence of low femoral bone density in older US adults from NHANES III *J. Bone Min. Res.* **12**(11) 1761–8

[132] Jacobsen S J, Goldberg J, Miles T P, Brody J A, Stiers W and Rimm A A 1992 Race and sex differences in mortality following fracture of the hip *Am. J. Public Health* **82**(8) 1147–50

[133] Baron J A, Barrett J, Malenka D, Fisher E, Kniffin W, Bubolz T *et al* 1994 Racial differences in fracture risk *Epidemiology* **5**(1) 42–7

[134] Cummings S R, Black D M, Nevitt M C, Browner W S, Cauley J A, Genant H K *et al* 1990 Appendicular bone density and age predict hip fracture in women The Study of Osteoporotic Fractures Research Group [see comments] *J. Am. Med. Assoc.* **263**(5) 665–8

[135] Cooper C, O'Neill T and Silman A 1993 The epidemiology of vertebral fractures European Vertebral Osteoporosis Study Group *Bone* **14** Suppl 1 S89–97

[136] Nakamura T, Turner C H, Yoshikawa T, Slemenda C W, Peacock M, Burr D B *et al* 1994 Do variations in hip geometry explain differences in hip fracture risk between Japanese and white Americans? *J. Bone Miner. Res.* **9**(7) 1071–6

[137] McCloskey E V, Spector T D, Eyres K S, Fern E D, O'Rourke N, Vasikaran S *et al* 1993 The assessment of vertebral deformity: a method for use in population studies and clinical trials [see comments] *Osteoporos. Intl.* **3**(3) 138–47

[138] Spector T D, McCloskey E V, Doyle D V and Kanis J A Prevalence of vertebral fracture in women and the relationship with bone density and symptoms: the Chingford Study *J. Bone Miner. Res.* 1993 **8**(7) 817–22

[139] Melton L J I, Lane A W, Cooper C, Eastell R, O'Fallon W M and Riggs B L 1993 Prevalence and incidence of vertebral deformities *Osteoporos. Intl.* **3**(3) 113–19

[140] Riggs B L and Melton L J R 1995 The worldwide problem of osteoporosis: insights afforded by epidemiology *Bone* **17**(5 Suppl) 505S–511S

[141] Jacobsen S J, Goldberg J, Miles T P, Brody J A, Stiers W and Rimm A A 1990 Regional variation in the incidence of hip fracture. US white women aged 65 years and older *J. Am. Med. Assoc.* **264**(4) 500–2

[142] Wasnich R D 1996 Epidemiology of osteoporosis. In *Primer on the Metabolic Bone Diseases and Disorders of Mineral Metabolism* 3rd edn ed M J Favus (New York: Raven Press) pp 249–54

Chapter 2

Biological safety considerations

Alan P Kelly

2.1. INTRODUCTION

Health and safety legislation varies considerably from one country to another. Within the EU, legislation is based on framework directives that determine common minimum standards throughout the community. Unlike trading standards, individual member states are free to introduce regulations that set higher standards; hence there is a certain amount of variation between the member states.

In most countries modern health and safety legislation is 'risk assessment' based whereby every work process involving a significant degree of risk must be assessed in advance of any practical work. The old style of prescriptive legislation (where a series of dos and don'ts are derived retrospectively from accident data) is now considered too restrictive, impractical and impossible to enforce. However, elements of prescriptive legislation remain where it cannot be avoided (such as reporting accident statistics). Reactive (prescriptive) legislation may still be the norm in some countries.

For the above reasons it would be impossible to offer health and safety advice on a specific topic such as bone analysis that would ensure compliance with the legislation regardless of the country where the work is to be carried out.

The following advice is, by necessity, based on the UK/EC risk assessment requirements and while it may not meet specific legislative requirements in certain countries it will undoubtedly offer a template to ensure that risk of harm to workers and third parties will be minimized.

2.2. DUTIES AND RESPONSIBILITIES

Regardless of statutory requirements we have a moral duty to avoid placing people at risk from our work activities.

The following (common law of England and Wales 'duty of care') was introduced by King Athelstan in the 10th century and has come down to us with little modification since 1272.

'We all have a duty of care to our neighbours to ensure that our actions do not cause harm or nuisance.'

There is a duty to provide a competent workforce, adequate materials, safe systems of work and effective supervision.

The standard of care includes foreseeing the existence of risk, assessing the magnitude of risk and devising reasonable precautions.

The much more recent Health and Safety at Work etc. Act (1974) [1] requires almost the same things. The principal difference is that under common law the injured party must prove that a duty of care was owed and that negligence had taken place. If proven, compensation will be awarded by the court. A breach of statutory legislation is a criminal offence for which society demands retribution in the form of fines or imprisonment of the offender. The two are not mutually exclusive, i.e. a single incident may give rise to both a criminal prosecution and a civil claim for negligence.

The aims of the Health and Safety at Work etc. Act (1974) are to secure health, safety and welfare of persons at work, to protect the public from work activities, to control the keeping of explosives, highly flammable or otherwise dangerous substances and to control the emission of dangerous, noxious or offensive substances.

The act places specific legal duties on employers and employees. The employer has a duty to provide

- a safe place of work
- safe plant and equipment
- safe systems of work
- instruction, information and training
- a written statement of safety policy
- periodic audits of systems of work.

Employees (workers, who may also be students) have a duty

- to take every reasonably practicable precaution to avoid putting themselves, their colleagues or members of the public at risk from their work activities
- to abide by the institution's policy and report any shortfalls in working practices or equipment.

Further advice on general safety legislation is available in *Essentials of Health and Safety at Work* [2]. For specific advice and policy guidance relating to biomedical laboratories *Health and Safety Policy and Guidance* [3] provides a good template.

2.3. ENVIRONMENTAL PROTECTION

In addition to health and safety legislation, in most countries consideration must be given to environmental protection legislation. This is particularly important with regard to toxic or offensive chemical waste and the disposal of clinical waste.

In the UK this is covered by the *Environmental Protection Act* [4] which places legislative controls on the correct disposal of waste products. In relation to laboratory work there are three main considerations.

1. Disposal of solid waste: this must be properly segregated, packaged and labelled for disposal by a reputable contractor.
2. Disposal of liquid waste via the public drain is often permitted for material of low toxicity where it does not persist in the environment; however, local authorities differ in the application of specific local regulations. Laboratory leaders and principal scientists are strongly advised to consult their local water authority before assuming drain disposal is permitted. Where drain disposal is permitted the material must be diluted with at least 100 volumes of water. Where appropriate, prior detoxification or neutralization may be required (e.g. neutralization of acids with calcium carbonate or soda ash).
3. Liquid that may not be disposed of down the public drain must be disposed of via a reputable contractor with a dangerous goods adviser who is qualified to attend to the required notifications before transport.

Disposal of gases is not normally an issue with regard to bone analysis, but it has had repercussions for clinical laboratories since on-site incineration is now virtually prohibited. Incinerators are subject to strict emission control standards which are difficult to meet with small scale units. Disposable laboratory plastic ware produces a number of toxic gases during incineration and must be processed through an extremely high temperature plant.

Whilst this may not be considered a health and safety issue *per se*, health and safety implications are often dominant in the discussion when ethical approval is considered for projects involving human subjects. For example, to carry out measurements on human volunteers, or patients, that involved exposure to X-rays would have major health and safety implications.

Ethical approval and informed consent is a prerequisite for any work involving human subjects, patients or human material derived from patients. This is covered under international law by the Declaration of Helsinki [5] and has been introduced into statute law in many countries (e.g. The Human Rights Act 1998 [6] in the UK).

Basically, you may not conduct any research using human subject volunteers or hospital patients without having first submitted a full protocol and risk assessment to a properly constituted ethical committee and received its approval. Additionally, prior 'informed' consent is required from the patient if material, e.g. a biopsy taken for diagnostic purposes, is also to be used for research.

Further advice is available in *Research Involving Patients* [7] and the Declaration of Helsinki [5].

2.4. RISK ASSESSMENT

Risk assessment is the current basis of modern health and safety legislation. The requirement to carry out risk assessment has been the cause of much anxiety and complaint in recent years; however, the concept should not be too difficult for almost anyone to grasp. We all assess risk each time we cross the road and possibly hundreds of times during a car journey of moderate length. The principles of risk assessment are the same for almost any task imaginable. The first step is to assess the 'hazard'.

Hazard is the potential to cause harm

e.g. A loose carpet presents a trip hazard but the severity of a resulting injury may depend upon location: the potential for harm is much greater if it is at the top of a stairway rather than in the corner of a room.

The hazards involved with the preparation of a decalcified thin section of a vertebral body are identical with respect to chemical and mechanical hazards regardless of the source of the sample. However, the difference in severity of biological hazard between an ox bone taken from the human food chain and a human bone infected with tuberculosis is obvious.

Hazard assessment is relatively easy, the potential for harm from most chemicals and biological agents is known, the only exceptions being novel drugs and some genetically modified organisms, but even in these cases it is usually not too difficult to make an educated guess based on comparable data

The next step is to assess the probability of harm from the hazard being realized.

Risk = harm × probability of harm being done

Assessing probability of harm being done seems to be the most problematic part of the process within academic circles. It requires the assessor to make an educated guess, sometimes in the absence of sufficient data.

e.g. A loose carpet on a level floor presents a similar hazard provided the person who trips over it does not fall downstairs or hit his head on a hard object during the fall. The probability of harm being realized depends upon a number of factors. If it is immediately in front of a doorway it is much more likely to cause harm than in the corner of a room. Probability of harm is increased proportionally as the number of users of the room increases. Familiarity of the occupants with their surroundings is an important factor; regular users know to step over it, strangers don't.

The use of the loose carpet example is not an attempt to oversimplify the risk assessment process. In virtually all work environments slips, trips and falls account for the majority of reported accidents. Imagine the potential for harm from faulty flooring in a histopathology laboratory where work with sharp knives and potentially dangerous machinery such as powered microtomes is carried out.

2.5. QUANTIFYING RISK

The simplest method of quantifying risk is to identify the most likely accident that would occur in a given work process. The second stage is to identify the maximum credible accident likely to occur during the same process.

e.g. In preparing histological sections on a microtome the most likely accident is a minor cut or a finger being trapped in the mechanism. The maximum credible accident may well be the amputation of a digit.

There are a number of sophisticated matrix systems published for quantifying risk; the most precise ones tend to be rather cumbersome and difficult to understand by the assessors. The simplest systems tend to allocate severity of hazard to high, medium and low; likewise with probability.

A simple risk assessment matrix.

Hazard severity	Priority rating		
	High	Medium	Low
Low	6	8	9
Moderate	3	5	7
Severe	1	2	4
		Probability	

Adapted with the permission of the originator from a system devised for the University of Leeds by Dr B Singleton.

There are a number of variations on this theme, many having five categories for both probability and severity of hazard and even up to ten. Most managers find it much easier to group things into three categories. However, it is subjective and the precision of judgment will vary from one person to another. Variation in assessment standard is not really an issue because the usefulness of the tool is to allocate a priority rating for reduction of risk.

Regardless of your perception of what constitutes a high severity or probability, the priority rating for that combination is 1. Procedures with a low hazard severity and a low probability are likely only to result in a trivial accident, very infrequently. For any given laboratory, managers can allocate health and safety resources according to the risk priority within their area.

2.6.   ACCEPTABLE RISK

For most circumstances one wishes to reduce risk to the lowest 'reasonably practicable' level. Elimination of risk is virtually impossible. Where hazards are extreme and the consequences of harm are widespread (e.g. a laboratory escape of an organism such as Ebola virus) risk must be reduced to a level considered to be 'effectively zero'. This is not quite the same as total elimination of risk.

The level of acceptable risk will vary according to the importance of the task being carried out. In scientific study it is difficult to imagine tasks involving a calculable number of deaths being acceptable; but this was certainly the case in the aftermath of the Chernoble disaster. The death rate among vulcanologists is somewhat above that which would be deemed reasonable by biomedical scientists. With regard to biomedical research there could be occasions where risk from infection by a fatal, incurable disease is acceptable if the research is pivotal in finding a cure or vaccine for the population at large, particularly where a pandemic is feared.

Most laboratory researchers would consider a risk of one individual requiring hospital treatment or being absent from work for four days every three to four years as being the maximum tolerable risk.

2.7.   RISK REDUCTION

Having assessed the magnitude and potential severity of the hazard and made some attempt to quantify the probability of harm and hence 'risk', the next logical step is to employ risk reduction measures. The hierarchy of risk reduction provides a formula which enables the reduction, or control, of risk.

2.8.   HIERARCHY OF RISK REDUCTION

- *Avoidance*. Can you use a safer method? Can you substitute a safer substance?
- *Reduction*. Use minimum concentrations, minimum quantities, and minimize exposure time.

- *Containment and control*. Avoid direct handling: use fume cupboards, isolators, microbiological safety cabinets etc.
- *Training*. Train all personnel in correct handling techniques and methods of disposal.
- *Personal protective equipment* (used as a last resort). Gloves, goggles, safety glasses etc.
- *Monitoring*. Where exposure to a substance with assigned exposure limits cannot be eliminated (may involve occupational health or routine air contaminant measurements). Occupational health surveillance where exposure cannot be satisfactorily controlled (e.g. exposure to animal allergens).

2.9.   SPECIFIC RISKS ASSOCIATED WITH THE PROCESSING OF BONE

Even within the limited topic of processing samples of bone it would be impossible to cover every possible hazard and quantify risk. Different methods of analysis introduce an almost unlimited scope of activities. The following is an attempt to identify the major areas for scrutiny but it cannot be considered by any means comprehensive.

2.9.1.   Hazard identification

Three broad categories of hazard come to mind:

- physical injury or electric shock arising from the use of machinery
- chemical hazards arising from processing
- biological hazards from the material itself.

Remember that the first step in risk reduction is to reduce the hazard at source.

2.10.   MECHANICAL HAZARDS

Machinery used to prepare bone sections or to measure compression or tensile strength share similar hazards to those encountered in many manufacturing industries. There is a risk from

- cuts from sharp blades
- entrapment and entanglement in moving machinery
- chips and splinters breaking away from the specimen.

Many years ago (before health and safety legislation applied to universities and hospitals) I worked as a recently qualified technician in a histopathology laboratory. I managed to cut my thumb to the bone on an old, communally used, sledge microtome. Several observations can be made from this accident: I was working alone at the weekend catching up on a backlog of work because a colleague was on long-term sick leave. I was suffering from a heavy cold at the time. The Perspex blade guard had long since broken off; each technician in the lab, including myself, was perfectly capable of making and fitting a replacement guard, but none of us had bothered to do so.

Communally used equipment tends to have no sense of ownership associated with it: the need for regular maintenance checks is most important. There was no policy on out-of-hours working on potentially dangerous procedures. I should not have attempted to carry out this work whilst running a fever.

When purchasing new equipment one should pay due attention to the first step of the risk reduction hierarchy 'removal of the hazard at source'. Wherever possible the equipment should be 'intrinsically safe by design'. This means that any moving parts or blades are adequately guarded to prevent injury and guards are interlocked in such a way that the machinery stops before access can be gained to the moving parts.

It is difficult to believe that a microtome could be intrinsically safe by design, but modern instruments are considerably safer than the older ones.

Whilst working as a diagnostic pathology technician in a hospital laboratory I was the sole user of an old Spencer microtome. I would still argue that this machine was one of the best microtomes ever made. However, it had one serious design fault that could have been easily avoided: it had three chuck orientation screws, one of which was located underneath the chuck, out of sight. The only way of adjusting this was to rest both hands on the chuck and adjust the screw with one finger from either side. If one had neglected to lock the mechanism the carriage would drop and the operator was likely to amputate both thumbs simultaneously. Many of the technicians had experienced near misses with this machine and chose to use a microtome of inferior quality but one that allowed chuck adjustment from above and the sides.

Where the instrument cannot be made safe by design you require a written 'safe system of work' which, if followed correctly, will minimize the possibility of an accident. Much of the machinery in a laboratory will require operators to undergo comprehensive training before being allowed to operate the equipment without supervision.

Training records should be kept and machinery that is more obviously dangerous such as large power-driven microtomes should be subject to an 'authorized user system' whereby the names of individuals considered to be sufficiently trained and experienced are recorded in an authorized user log, access being forbidden to anyone else unless they are under direct supervision during training.

It should be noted that prior to the 1974 Health and Safety at Work Act in the UK, the old Factories Act did not apply to universities and hospitals. The Factories Act listed a number of machines associated with woodworking, bookbinding and catering to which access was forbidden by young persons (under 18 years of age). Some of this has been carried forward into the modern legislation; although it does not apply directly in the laboratory context, it should be borne in mind that any form of microtome or powered saw for cutting bone would have been included in this category had it applied at the time: this could be used as an example in court should an accident occur.

- It would be singularly ill-advised to allow a work experience school leaver, under 18 years of age, to use any form of microtome unless under the most controlled, and directly supervised, circumstances.

Maintenance of equipment is an essential part of machine safety: many accidents occur as a direct result of poor maintenance; running repairs or adjustments must never be made whilst the machine is running.

Records of maintenance, whether carried out in house or by an external contractor, should be kept for each item, hence a maintenance history is available for each instrument in the case of an accident caused by a mechanical or electrical failure.

2.10.1.   Sawing bone

Wherever possible, sawing of bone should be done by hand. The use of power driven mechanical saws introduces a number of additional risks.

Bones are of an irregular shape and are extremely difficult to hold on a saw table. If a circular saw or band saw must be used it is advisable to embed the bone in a rectangular plastic block where feasible, or prepare wooden carriers that allow the bone to travel through the blade without twisting away from the operator's hand if the blade snags.

Chips of bone are likely to fly from the blade into the operator's face: impact resistant eye protection is a must.

Aerosols generated during sawing of unsecured bone on a power saw vastly increase any biological hazard that may be present.

Mechanical saws present a serious challenge with regard to cleaning and disinfecting after processing un-fixed biological samples. Moreover, the fixatives used for fixing biological materials may have a serious corrosive effect on the machinery.

A summary of reduction of mechanical hazards is as follows.

- Wherever possible avoid the use of powered tools.
- Wherever possible use a modern machine that is safe by design.
- Devise a written safe system of work incorporating the correct placement of guards and use of mechanism locks whilst adjusting chucks and blades etc.

- Ensure that adequate eye protection is provided where bone has to be sawn, chiselled or drilled.
- Ensure that all personnel are adequately trained on potentially dangerous machinery before they are allowed to use it unsupervised.
- Do not allow any out-of-hours working unless another competent person is present and the alarm can be raised easily in the case of an accident.
- Ensure that the machinery is adequately maintained and serviced and that maintenance records are kept.

Advice on preparation of un-fixed human tissue is available in Safe Working and the Prevention of Infection in the Mortuary and Post-mortem Room [8].

Advice on mechanical hazards is available in *Safe Use of Work Equipment: Provision and Use of Work Equipment Regulations* [9].

2.10.2. Electrical hazards

Biological material is usually wet; moreover it contains a number of electrolytes. Salt water and electricity are a dangerous combination. Electrical hazards can be reduced by

- purchasing double insulated equipment wherever possible
- the use of residual current circuit breakers in high-hazard areas
- the use of low-voltage equipment wherever possible
- routine maintenance, inspection and testing of the equipment.

All electrical equipment must be subjected to a thorough visual inspection. In a wet science laboratory the minimum reasonable frequency would be annual. In the case of a microscope in a dedicated dry microscopy area this could be reduced to every two years.

Integrity of the plug, mains cable and chassis entrance of the mains cable must be examined. The switches should be examined for evidence of scorch marks and tried for smooth spark-free operation. Casings should be examined for cracks and damage that could allow access to exposed metal conductors or allow water ingress.

Insulation resistance (with the mains switch on) and earth continuity should be measured with a portable appliance tester (PAT). Additionally other tests may be performed with the PAT such as earth leakage and power consumption. It is usually left to each institution to define the range and depth of tests required.

The following publications provide useful information on the testing and inspection of electrical equipment: *Memorandum of Guidance on the Electricity at Work Regulations* [10] and *Code of Practice for In-Service Testing of Electrical Equipment* [11].

2.10.3. Chemical hazards

Chemical suppliers are obliged by law to supply hazard information on their products at the time of purchase. Most of them provide a summary of the hazard assessment in their catalogues and information must be printed on the containers.

The user must assimilate the hazard information for all the chemicals used in a given procedure and come up with a comprehensive risk assessment for the method.

Chemicals can be grouped into three principal hazard areas: toxicity, corrosiveness, and reactive/flammability hazards.

Within the UK there is a mandatory requirement to carry out a COSHH (Control of Substances Hazardous to Health Regulations 1999) [12] risk assessment for all processes where potentially hazardous chemicals are used. Strictly speaking the COSHH regulations are only concerned with toxicity, however, it is common practice to incorporate flammability and reactive hazards if only to avoid carrying out another risk assessment to cover these aspects independently.

Within the EU chemicals are ascribed to a particular hazard category depending upon the nature and severity of the hazard. In most North American States they disapprove of this system on the basis that each combination of chemicals presents a unique combined risk and should, therefore, be assessed individually.

Advice on the UK regulations is available in COSHH Regulations (1999) Approved Codes of Practice [12].

2.11. HAZARD IDENTIFICATION

2.11.1. Toxicity hazard

Within the EU, chemicals are grouped according to toxicity into four sub-divisions:

- Very toxic (usually <5 mg/kg oral rat LD_{50}).
- Toxic (approximately 5–100 mg/kg oral rat LD_{50}).
- Harmful (>100 mg/kg oral rat LD_{50}).
- No hazard label, not considered harmful.

The actual toxicity level is not legally defined in each case. One must also consider the route of entry and likelihood of entry into the human body. It is self evident that each of these categories is rather broad. There is a considerable difference in toxicity between sodium cyanide, approximately on the threshold of the very toxic band, and botulinus toxin which is one of the most toxic substances known. It is for this reason that the Americans do not believe in banding hazards.

As a rule of thumb, a lethal dose of a substance marked very toxic may not be visible to the naked eye. A lethal dose of a substance marked toxic is visible, therefore a fatal accident is much less likely, but extreme care is still required in handling it. A lethal dose of a harmful substance would be virtually impossible to take accidentally.

With regard to very toxic substances and to a lesser extent to toxic substances the principal concerns are **acute effects**, i.e. of accidental poisoning. Within most laboratories the use of such substances is not usually extensive.

The vast majority of chemicals in common use are 'harmful' where the principal concerns arise from **chronic effects**, which may be

- cumulative
- reproductive
- carcinogenic
- teratogenic.

The third factor to consider when assessing toxicity hazards is to determine the route of entry into the body. Routes of entry can be by

- the digestive system
- the respiratory system
- the eyes and/or skin absorption.

Manufacturers of chemicals must supply information on acute toxicity, chronic effects and routes of entry where they are known.

Once you are aware of the nature and extent of the hazard it is relatively simple to devise risk reduction measures to ensure there is little probability of the harm being realized.

2.11.2.   Corrosive hazards

Corrosive substances are divided into two categories, corrosive and irritant. The logic is similar to that used for toxic hazards inasmuch as the principal concern with 'corrosive' substances are acute effects such as chemical burns, either externally or internally. The principal concern with substances labelled irritant are usually chronic.

Irritant substances may have cumulative effects, i.e. cause sensitization and allergies as a result of low-level long-term exposure. Others may cause rapid sensitization. Again it is important to determine whether inhalation of dust or skin absorption is the route of entry.

2.11.3.   Exposure limits

Within the UK and the other EU countries substances that are harmful, toxic or irritant by the respiratory route are ascribed exposure limits. These are expressed in quantities in the room atmosphere.

Within the UK we have a small number of substances ascribed with a maximum exposure limits (MEL) and a relatively large number of substances ascribed with an occupational exposure standard (OES), the latter usually being less serious.

Within the UK the use, and exposure, to a substance ascribed an MEL must be minimized as far as is reasonably practicable and to exceed the limit is a serious offence (no realistic defence, rather like exceeding a statutory speed limit).

Exposure to substances ascribed an OES must be minimized as far as is reasonably practicable but to exceed the limit is not an offence in itself if you can prove that you are taking active steps to control the situation.

MELs and OESs are given as long-term and acute limits. Long-term limits provide a figure in ppm or in mg/m^3 of breathable air for an 8 hour time-weighted average (TWA): this assumes a safe limit for someone working in an atmosphere of the substance for 8 hours per day, 5 days per week. Acute exposure limits are usually given as a 15 min maximum level. Effectively, in the average tissue laboratory the simplest solution is to handle all these substances, in the concentrated form, in the fume cupboard.

In biological laboratories the most problematic chemical in this respect is formaldehyde. Within the UK, formaldehyde is ascribed an MEL of 2 ppm or $2.5 \, mg/m^3$ 8 hour TWA and also an acute 15 min limit of 2 ppm or $2.5 \, mg/m^3$. This usually means monitoring the atmosphere regularly with a formaldehyde meter, unless fixatives can always be handled in the fume cupboard.

Most solvents used for tissue preparation also have exposure limits. Where usage is extensive (as in diagnostic laboratories) purpose-made staining benches with built-in fume extraction are available.

UK exposure limits are published in *Occupational Exposure Limits* [13].

2.11.4. *Reactive hazards*

When assessing a method it is important to look for combinations of chemicals that may cause adverse reactions. Particular attention must be given to the quantities used and dissipating heat from exothermic reactions.

The most pertinent example with regard to bone preparation is where the bone is to be embedded in plastic. **Methyl methacrylate** and **benzoyl peroxide**, if mixed in incorrect quantities, result in an explosion. Great care must be taken to ensure that bulk stocks of either are stored separately and only small 'in use quantities' are kept in the laboratory.

Fortunately, most tissue processing methods require washing in tap water or solvent between the different chemical treatments. Moreover, the quantities and concentrations used are such that there is little danger from adverse reactions during processing. The greatest risk arises from the concentrated chemicals during preparation and storage.

- Oxidizing agents must be kept away from acids and flammables.
- Acids must be kept away from solvents.
- Chemicals with particular incompatibilities such as methyl methacrylate and benzoyl peroxide must be segregated in storage.

There is an often quoted story among histopathologists about the spontaneously combusting laboratory coat: Bouin's fixative contains picric acid which is both flammable and explosive; silver nitrate is commonly used for staining nervous tissue. The tale relates to a laboratory technician who managed to spill both onto his laboratory coat at the same time. Nothing happened until the coat dried out in his locker overnight: the next morning, smoke was seeping from the locker and a full scale fire was only just avoided. Whether true or not, this is more than feasible—the experiment works on filter paper.

Apart from obvious reactive hazards one must also consider reactions with other chemicals that may be present from spills on benches and, most particularly, those remaining in drains.

If chlorine-releasing disinfectants are used, great care must be taken to ensure that they do not come into contact with acids or formaldehyde. I once witnessed chlorine gas emerging from a drain during a microbiology class. More importantly the reaction may produce an extremely potent carcinogen as a by-product.

Any hazardous chemical compounds produced during a process must be assessed, as well as the raw materials, along with any general purpose chemicals used for cleaning or other routine processes that take place in the laboratory.

2.11.5. *Flammability hazards*

Strictly speaking, under the UK legislation flammability hazards should be considered separately from chemical hazards. However, most flammable substances used in tissue processing also have toxicity hazards and many have exposure limits. It is common practice to include an assessment of flammability with the chemical hazard assessment.

In the UK the 'Highly Flammable Liquids and Liquefied Petroleum Gases Regulations (1972) [14] defines highly flammable liquids as those with a flash point below 32 °C. The regulations state that no more than 50 litres may be stored within a workplace and must be contained in a fire retardant cupboard. The definition of workplace is problematic: in a university building with many biological laboratories it would be impossible to limit the amount to 50 litres for a whole building. However, if every small laboratory held a working stock of 50 litres the hazard would be considerable. A common sense approach must be used to minimize laboratory stock as far is possible without unduly impeding the work. Bulk stocks must be kept in an external, specially constructed, solvent store. If necessary working stocks should be replenished from the bulk stock on a daily basis.

The most useful system for classifying flammable solvents with respect to tissue processing employs three hazard categories based on their flammability.

- *Flammable*: Flash point between 21 and 55 °C. These materials will ignite, but not readily at room temperature unless a wick is present or during a heating process: e.g. vegetable oil and paraffin.
- *Highly flammable*: Flash point between 0 and 21 °C. These materials will ignite readily at room temperature, e.g. ethanol, methanol.
- *Extremely flammable*: Flash point below 0 °C and boiling point below 35 °C. These materials will ignite readily in a refrigerator or even a freezer, e.g. acetone, anaesthetic ether.

The reason that these divisions are sensible in this context is that new workers only have to remember that extra precautions may be required for flammable material when it is to be heated or it is likely to contact absorbent combustible material.

The highly flammable category material is likely to ignite at room temperature and therefore must be kept away from naked flames and other sources of ignition.

The extremely flammable category material may form an explosive mixture in an enclosed space at refrigerator or freezer temperatures.

Note: *Domestic refrigerators and freezers must never be used in laboratories* unless they have been modified so that all switches, thermostats or other electrical contact breakers are outside the cabinet. Serious explosions have been reported in the past where anaesthetic ether was stored in a domestic refrigerator: the contact breaker for the internal light is more than sufficient to cause ignition. Acetone, another commonly used solvent, could explode in a freezer at -25 °C.

It should also be noted that in the UK confusion could arise over the definition of highly flammable liquid, since the regulations define them as substances with a flash point below 32 °C, approximately half way up the band graded as flammable under the above classification. It would be advisable to ensure that all staff are aware that the three-band system provides useful categorization for risk assessment relating to processes, whereas the legal definition relates to permissible quantities and storage conditions.

2.12. EXTINGUISHERS

With respect to solvents, it is vitally important that all laboratory workers are familiar with the nature of the solvents and the appropriate fire extinguisher to use should they ignite.

Water is the extinguishing medium of choice for miscible solvents because it takes heat from the fire and quickly dilutes the solvent to a level where it is most unlikely to re-ignite.

To use water on a fire involving a non-miscible solvent would be catastrophic: the water jet would simply spread the fire. Foam extinguishers should be used to smother the fire.

Provided that only small quantities of highly flammable liquids are in use, it is unlikely that an accidental ignition could result in a fire that could not be dealt with simply by covering with a fire blanket.

Within a solvent storage area, dry powder extinguishers are advised: these are probably the most universally effective type of extinguisher; unfortunately they cause a great deal of mess and are unsuitable for indoor use on small fires.

2.13.   RISK REDUCTION AND CONTROL: CHEMICALS

The first steps in risk reduction involve reducing the hazard at source, as follows.

- *Avoidance*: Where very toxic or corrosive substances are involved, can you use an alternative method which is much safer? Can you substitute a safer substances without compromising the method?
- *Reduction*: Purchase made-up reagents, i.e. avoid using concentrates. Purchase minimum quantities. Keep larger stocks in designated stores, not in the laboratory. Minimize exposure time for individual workers.

Once these possibilities have been explored remaining risks must be controlled.

- *Containment and control*: Avoid direct handling. Use fume cupboards for substances with ascribed exposure limits. Keep lids on containers (especially Coplin jars and containers of fixative).
- *Training*: Train all personnel in correct handling techniques and methods of disposal.
- *Personal protective equipment* (used as a last resort): Gloves, goggles, safety glasses etc.
- *Monitoring*: Where physical control measures are inadequate (e.g. when relying on room ventilation to carry away solvent or fixative vapour) regular monitoring of the atmosphere is required to ensure exposure limits are not exceeded.

2.13.1.   Fume cupboards

Where fume cupboards are relied upon to reduce exposure it is a requirement that they are tested annually by a 'competent person'. This is usually an independent engineer employed by the institution's insurers.

Additionally the users should record monthly face velocity measurements.

2.13.2. *Biological hazards*

The principles of reducing the hazard at source and the employment of control measures and monitoring of risks that remain can be applied equally well to biological hazards. The only problem arises in the identification of hazards that may be present. However, there is sufficient internationally recognized guidance to simplify the procedure.

The WHO publish lists of micro-organisms that have been allocated to hazard 'categories'. In the UK comprehensive details on how to classify and control biological hazards can be found in *Categorisation of Biological Agents According to Hazard and Categories of Containment* [15, 16]. Local codes of practice vary a little from one country to another. There are differences in the allocation of categories to some organisms based on risks related to prevalence in the community, the environmental conditions and local immunity, e.g. a tropical parasite is unlikely to spread to the local community in northern Europe if the insect vector does not occur there and the environment is simply too cold for the disease to persist. Conversely an organism may be considered to be more dangerous to some ethnic populations where there has been no historical exposure to the disease and therefore little immunity in the population.

In addition to human disease one must also be aware of the possibility of animal pathogens being present. Samples may harbour organisms that pose a threat to local agriculture or wildlife. Most countries have strict regulations regarding the import of animal material including material designated for the human food chain: you must consult your local Ministry for Agriculture (in the UK, DEFRA) before importing any animal bone. The recent devastation caused to the British agricultural industry, reputedly due to the import of dried meat from China for the restaurant trade, highlights the importance of these controls. Hopefully, a similar situation would not arise from the import of bone for laboratory analysis because the final disposal must be by incineration: unfortunately it is not unknown for mistakes to occur and such specimens have been found on landfill sites. In the UK the Environment Agency has threatened a number of universities with prosecution for failing to segregate and dispose of biological waste correctly

Exposure to allergens from laboratory animals and latex gloves is becoming an increasingly important issue: evidently latex allergy is about to become the most frequently reported occupation related disease. Rodent allergy is also very common.

Old animal bones, from excavation sites, may contain viable anthrax spores: no-one really knows how long anthrax remains viable in the soil but it is certainly viable for up to 60 years. Such bones are also likely to contain *Clostridium perfingens* (gas gangrene) and *Clostridium tetani* (tetanus).

2.14.   HAZARD CATEGORIES OF BIOLOGICAL AGENTS

Category 1   Agents are most unlikely to be pathogenic to humans (species not listed).

Category 2   May cause disease to a laboratory worker. They are unlikely to spread to the community but treatment and/or prophylaxis is available.

Category 3   Agents are likely to cause serious disease. They may spread to the community but treatment or prophylaxis is usually available.

Category 4   Are likely to cause severe disease, are likely to spread to the community and, normally, there is no available treatment or prophylaxis.

2.15.   HAZARD IDENTIFICATION AND HAZARD REDUCTION AT SOURCE

When planning a project you should consider the following points.

2.15.1.   For human bone

Substitution:

- Could you use animal bone (derived from the human food chain) without compromising the study?
- Could you use fixed material, i.e. a biopsy dropped into fixative in the operating theatre area before removal to the laboratory?

If you must use fresh human material, consider the following.

- Are the samples from a random population within your own community (e.g. for the UK within the UK or EU)? If so they must be contained at Category 2 unless risk can be reduced further.
- Have the samples been screened? If you have assurance from the hospital which provided the material that the individuals were screened prior to surgery/autopsy and are free from blood borne viruses and tuberculosis, you could probably consider category 1 containment measures adequate.
- Could you obtain samples from screened patients instead?

- If the samples are known to be diseased you must, as a minimum standard, contain to the category indicated by the disease, however, you must not rule out the possibility of other equally dangerous pathogens also being present.

2.15.2.   For animal bone

- Can you use bone from animals passed fit for the human food chain?
- If not, are they from domestic herds (accredited free of regulated animal pathogens) or laboratory-bred animals?
- Animal bone derived from the human food chain in the EU or from laboratory-bred stocks may be regarded as suitable for category 1 containment.
- Animal bone from wild stock must be assessed for the risk of carrying pathogens cited in the Specified Animal Pathogens Order [17].
- Animal bone known to be infected or from condemned meat must be contained at a level commensurate with the disease: advice and permission must be sought from the Ministry of Agriculture (in some cases, in the UK, a DEFRA licence may be required which will state the precise control measures as a condition of the licence).

2.16.   PRION DISEASES

During recent years there has been a great deal of concern regarding the possibility of encountering prion diseases. Categorization of prions as biological agents has in itself been problematic: non-variant prions are considered to be simply proteins and, although they may have to be considered in the chemical section of a risk assessment, are not considered to be biological agents.

Variant prions associated with disease such as Creutzfeld Jacobs disease, Kuru, New variant CJD (mad cow disease) and scrapie are considered to be biological agents and, regardless of their current hazard categorization, should be handled at containment level 3.

Unless the research project is specifically concerned with variant prions it is sensible to ensure that the probability of encountering them in random samples is reduced to effectively zero. The extremely low incidence of these diseases plus their relatively long incubation periods exacerbates the risk assessment process to some extent; however, a number of precautions can be taken.

With regard to animal bones one should look carefully at the source material, paying particular attention to history. If the material is obtained from herds that have remained free from prion disease there should be very little, if any, risk.

Animal bones, wherever possible, should be free from central nervous tissue (there is little risk associated with bone itself). This may be difficult to achieve where intact vertebrae are required, in which case additional care may be required in the containment of aerosols during processing. However, the precautions specified for containment level 2 would be satisfactory with regard to reducing risk from samples except those obtained from herds with a history of infection.

It must be remembered that fixation does not render variant prions harmless: in fact formaldehyde is reported to increase their stability.

Further information can be obtained from *Transmissible Spongiform Encephalopathy Agents: Safe Working and the Prevention of Infection* [18].

2.17. BIOLOGICAL CONTROL MEASURES

Having established the appropriate level of containment the required control measures are to a large extent prescriptive.

Level 1   Largely based on good microbiological practice in a laboratory fitted to a high standard.

Level 2   Relies on good microbiological practice in a laboratory fitted to a high standard but with additional managerial controls over authorized entry, training requirements etc. Where organisms that can be transmitted via the airborne route are involved, specific aerosol reduction/containment measures must be in place plus containment within a microbiological safety cabinet.

Level 3   By definition an escape could lead to the infection spreading to the community: control measures are based on semi-isolation including negative pressure inward air flow, extract via a HEPA filter, cabinet containment with the whole room sealable for fumigation. Strict protocols for the training and authorization of staff are required, entry and exit procedures, detailed protocols for dealing with spills and routine disinfection.

Level 4   Requires complete isolation. Since very few institutions have such facilities, work at this level of risk is limited to a small number of institutions, normally direct government funded.

For projects involving a higher level of biological risk it is important to be familiar with the specific control measures before embarking on the project. Many control measures involve physical barriers which may be expensive to install.

Precise details of the requirements are too extensive for inclusion in this chapter. Researchers who are likely to be affected by these requirements should, in the UK, refer to *The Management, Design and Operation of Microbiological Containment Laboratories* [19]. Outside the UK similar guidance or local regulations should be in place and available as a government publication, since the UK regulations are based on international legislation.

2.17.1. Allergens: control of exposure

Control of exposure to animal allergens and latex allergy are rapidly becoming high profile with regard to concern from the enforcement agencies. Reported cases of latex allergy are rising exponentially, but it is not clear whether this reflects an increased susceptibility to allergic conditions in the community at large, increased awareness of the problem or the increase in usage of latex gloves.

Workers who are exposed to animal allergens for more than 6 hours in any week per year should receive occupational health surveillance. All workers should wear gloves and face masks whilst handling animals, particularly rodents. Workers who are considered to be susceptible by the occupational health staff must be provided with additional protective equipment such as filtered air flow hoods.

Latex allergies are exacerbated by the use of powdered gloves because the allergens are adsorbed onto the powder which is then released into the atmosphere when the gloves are removed. This results in sensitization by the respiratory route as well as local skin irritation.

The current advice is that powdered gloves should be avoided altogether. Low allergen gloves such as thin nitrile gloves must be provided for workers who are considered particularly susceptible or are already sensitized. Unfortunately these gloves do impede manual dexterity to some extent, but there is no alternative solution for workers who are sensitized other than to cease working in a biological laboratory.

2.17.2. Microbiological safety cabinets

Microbiological safety cabinets are used for two different purposes: to protect the samples from contamination from the room air or to protect personnel from infection from the samples. It is imperative that personnel are aware of the reason behind using the cabinet.

If used for the protection of personnel the cabinet must be tested upon installation and annually thereafter. The tests are normally carried out by an independent contractor and include a smoke test to ensure the effectiveness of the HEPA filter (filter challenge), face velocity (and downward air current in Class 2 cabinets) and an operator protection factor (KI discus) test. In some countries determination of operator protection factor by KI discus test is not permitted. A microbiological challenge must be carried out using an aerosol of *Seratia marcescens* or *Bacillus subtilis*: this introduces a further complication due to the necessity of fumigation both prior to the tests and afterwards.

The HEPA filter must remove 99.997% of the smoke particles for category 2 containment and 99.9997% for category 3.

The minimum permitted operator protection factor is 10^5, i.e one organism per 100 000 may escape containment. The majority of modern cabinets are considerably better than this, but some of the older ones are abysmal, so beware!

Face velocities must be measured during the annual engineer checks but, in addition, should be measured and recorded every month by the user. This entails measuring the inward face velocity with a hand held anemometer at the four corners and centre of the aperture (0.5–0.7 m/s) and the downdraught (0.25–0.5 m/s) for Class 2 cabinets.

Further information can be found in British Standard EN 12469: 2000, Biotechnology: performance criteria for microbiological safety cabinets [20].

2.17.3. *Disinfectants*

In all laboratories where un-fixed biological samples are handled there should be a written protocol, 'standard operating procedure', for routine disinfection of work surfaces, floors, sinks and equipment.

It is essential that the disinfectant selected is effective against the organisms present, or suspected to be present.

Disinfectants are toxic substances (otherwise they would not work); it is necessary to ensure that the chemical hazard identification and risk reduction process is carried out as well as assessing the effectiveness of the disinfectant.

For most purposes chlorine-releasing disinfectants are generally effective against most organisms—hence a well known domestic brand that claims to kill 99% of all known germs. In the laboratory, liquid hypochlorite should never be used: tablet forms such as Precept and Haz-Tabs offer a much safer, and more accurate, alternative.

Where HBV and HIV are the principal concerns, Virkon is usually the disinfectant of choice. It is deemed to be safer to humans than chlorine releasers as well as being a more effective viricide. Additionally the slight pink coloration means that the area covered is visible: hence one is less likely to miss areas when disinfecting large areas such as benches.

Where *M. tuberculosis* is a concern, the effectiveness of chlorine releasers and Virkon has been questioned, although the latter claims to be effective at a high concentration. Currently phenolic disinfectants are advised, but how much longer this will be permitted is debatable. Phenol persists in the environment and new environmental legislation is moving towards elimination of phenolic products. One assumes that at some time in the future all tuberculosis infected/suspected material will require sterilization by autoclave.

Disinfection of equipment is particularly problematic: chlorine based products are unsuitable because they corrode metal. Virkon is commonly used but has, evidently, been known to cause damage to centrifuge rotor bearings (although the precise circumstances are not known). Glutaraldehyde solutions such as CIDEX are often recommended but the very low maximum exposure limit of glutaraldehyde means that complete containment within an isolator or fume cupboard is required. For most routine disinfection of metallic equipment 70% alcohol is probably the most satisfactory solution, but one should be mindful of the intoxicating effects and flammability when using on large areas. Also, its effectiveness is limited when compared with other solutions.

The protocol for disinfectant use must include both the optimum concentration and the minimum contact time to be employed. It is amazing how many people will wipe a bench with disinfectant and then wipe it off immediately—at least 10 minutes' contact time will be required for most routine surface disinfection.

Another, frequently overlooked, factor in disinfection is the penetration of the disinfectant. Proteins prevent adequate penetration of many disinfectants. If the surfaces are dirty an effective kill of organisms may not be achieved. Surfaces should be clean prior to disinfection; if this is not possible, concentrations and contact times must be adjusted accordingly.

2.17.4.   *Disinfection of cryostats*

Cryostats, particularly old ones, present great difficulties with regard to disinfection. The choice of disinfectants is limited to those suitable for metallic surfaces and the chamber must be thawed prior to disinfection because the effectiveness of the disinfectants will be much reduced at a cold temperature.

The necessity to disinfect can be reduced by the use of disposable knives. Models that incorporate a removable catch tray under the cutting mechanism allow removal of the parts likely to become contaminated.

Some of the newer machines can be sealed for fumigation with formaldehyde; however, the machine must be at room temperature.

By reducing the disinfection protocol to the minimum necessary, and on removable parts, it would only be necessary to disinfect the whole machine at infrequent intervals and in the case of a spill within the chamber. Where access is required to a cryostat on a daily basis (e.g. in diagnostic laboratories) it is recommended that two machines are employed, each on an alternating defrost/disinfection cycle.

2.17.5.   *Fumigation*

Fumigation may be the only option when large items need to be disinfected. In the case of microbiological safety cabinets used for containment at level 2 or above it is a requirement prior to maintenance work being carried out. In the case of a category 3 laboratory, it may be necessary to fumigate the whole room following a spillage.

Fumigation relies on the area/item to be disinfected being completely sealed whilst formaldehyde is boiled in a moist atmosphere. Commercially available apparatus may be acquired for carrying out this process and it is simply a matter of ensuring the correct amount of formaldehyde is used for the particular application.

The principal problem from fumigation is the removal of the formaldehyde gas at the end of the process. Where cabinets are evacuated through an integral duct the apparatus is de-gassed simply by switching the extractor fan on. Unfortunately many cabinets recirculate the filtered air to the room; in this case extra provision must be made for de-gassing (formaldehyde must not exceed 2 ppm in the room air in the UK and in some countries must not exceed 0.5 ppm). Some cabinets have hose kits that allow the cabinet to be de-gassed through an open window or fume cupboard. Some cabinet manufacturers produce fumigation kits that employ a double boiler, whereby the formaldehyde is neutralized with ammonia and then discharged to room air through a carbon filter and an ammonia filter. Under no circumstances should carbon filters be relied on to remove formaldehyde: they are only partially effective and soon become saturated.

Whenever fumigant is to be exhausted to room air, regardless of filtering systems, the air should be monitored with a formaldehyde meter throughout the process. If the MEL is exceeded the affected area should be evacuated.

When fumigation procedures are required it is essential to provide a step-by-step standard operating procedure including sealing off the surrounding area and restricting access. Two competent persons must be involved in carrying out the procedure. Where category 3 organisms are involved all personnel should be able to demonstrate familiarity with the emergency fumigation procedure—failure to do so could result in prosecution.

A template for a safe system of work for fumigation is provided by Bennet [21].

2.17.6.   *Disinfection of mechanical testing equipment and machine tools*

One aspect of bone analysis that always gives rise to difficulty is where samples of un-fixed bone must be tested/or machined on equipment designed for metallurgy (such as tensile testing machines). Band saws have already been mentioned with regard to physical dangers but, unless the material is very low hazard for which 70% ethanol is a suitable disinfectant, disinfection of these machines is almost impossible. Where there is a risk of infection a stainless steel band saw (as supplied to the catering industry) should be used. Cleaning and disinfection is then much less problematic: blades should be regarded as disposable and not left on the machine (they should be sent for incineration after use). Catering machines are much easier to clean and the stainless steel surface would permit disinfection by ethanol or quaternary ammonium compounds, and could even be wiped down with Virkon without causing significant harm. A cast metal saw cannot be cleaned adequately, let alone disinfected (without sealing in a chamber for fumigation), and should not be used for un-fixed material.

With some imagination and availability of engineering skills it is possible to design and make detachable sample containment units to facilitate stress and tensile testing. In one of our local laboratories vertebrae are tested for impact resistance on a metallurgical testing machine. Needless to say the machine is a fixture in a large mechanical engineering workshop. The chief bio-engineering technician produced a number of stainless steel test chambers capable of completely containing the sample during the test impact without danger of aerosol contamination. At the end of the test run the chambers are removed to the bio-containment area for autoclaving, i.e. contamination of the test machine is not an issue.

For the range of mechanical engineering machines likely to be used for bone analysis in one context or another it is impossible to provide more than general advice, but the prime consideration should be to contain the biological hazard without having to disinfect the machine. If it is not possible to eliminate the possibility of contamination completely, e.g. in the case of a primary container bursting under stress, it will be necessary to compile a protocol for emergency cleaning and disinfection.

2.17.7. *Autoclaves*

Sterilization by autoclave has long been regarded as the only certain way of killing dangerous micro-organisms, but this tenet should only be accepted with care. Effective autoclaving relies on a high temperature achieved by steam. If all the air is not excluded from the chamber a safe kill temperature may not be achieved. The temperature in the centre of the load may be entirely different from the temperature indicated on the dial. Plastic containers used in laboratories often melt and seal an insulating air pocket over the material.

Tests have shown that in the worst cases a chamber temperature of 121 °C for 20 min resulted in a maximum load temperature of 65 °C for less than 5 min, which would hardly have been sufficient to kill vegetative growth let alone resistant organisms or spores.

Modern 'state of the art' autoclaves have a load probe that records the load temperature for each cycle, vacuum displacement to ensure prior evacuation of air from the chamber and a rapid cooling cycle allowing a faster cycle time. Some have pulsed steam pressure which increases steam penetration and they all have temperature/pressure interlocks on the door to prevent the operator from opening the door before it is safe to do so.

Many of us have to make do with an autoclave that does not reach these standards. In such cases, safe systems of work must be established by carrying out thermocouple tests on typical loads: such tests must be repeated annually. This is normally done by a contracted autoclave engineer. Having established cycle protocols to ensure adequate sterilization and sufficient cooling of the load, protocols should be written up as standard operating procedures and kept with the machine.

All autoclaves must be tested annually by a 'competent person' to satisfy the legislation governing the use of pressure vessels. Even domestic pressure cookers, if used in the laboratory, must be registered and tested annually. The latter are often found in biological laboratories; however, they should only be used for small-scale sterilization of media—never for disposal of micro-organisms (the chamber temperature, let alone the load temperature, cannot be guaranteed).

Personnel should be trained to use autoclaves by a member of staff recognized as competent (usually an experienced technician) before being authorized to use it, and such training and authorization should be recorded.

Where autoclaves are used to sterilize biologically hazardous waste, prior to disposal, a record of the time and temperature regime as well as the contents of the bag, the name of the operator and date of disposal should be logged.

Further information on disinfectants and sterilization can be found in *Safe Working and Prevention of Infection in Clinical Laboratories* [8].

2.17.8.   *Disposal of biological waste*

Disposal of biological waste is one of the most obvious routes for infection to spread beyond the laboratory. Additionally there are ethical issues with regard to the disposal of human remains, even very small amounts of biopsy or autopsy material.

All waste must be segregated at source into colour-coded bags.

Black bags are used for general waste destined for landfill sites. Under no circumstances should bio-hazardous waste be disposed of via this route; for that reason it is often the best policy to eliminate black bags from laboratories generating biological waste. Biological material, particularly bones, found on landfill sites generate a great deal of public concern and can easily result in prosecution by the Environment Agency.

Category 3 organisms or infectious material containing them must be autoclaved within the laboratory suite before disposal. Providing the material is neither human nor animal tissue it is then permissible to dispose of it via an ordinary industrial waste collection. However, it is advisable to consult your waste removal contractor before doing so. Some areas will not accept it and others insist on a dedicated skip for autoclaved waste.

The simplest solution is to send all biological waste, and any soft materials such as paper waste, gels etc. directly for incineration as clinical waste. Yellow bags must be used for this purpose, all bags must be labelled with an identification tag and the contents recorded.

Clinical waste must be held in a secure area to await collection (usually a locked clinical waste store). The route to the waste store should be planned and recorded in a standard operating procedure. It must not be left unattended at any time during the transfer and the bags must be placed in a durable secondary container capable of containing the whole contents should a bag burst in transit. Only a recognized contractor who is licensed to dispose of clinical waste must be used for its final removal.

Animal tissue discarded from laboratories, even if it was originally deemed fit for the human food chain, is regarded as 'offensive waste' and again it is an offence to dispose of this as general industrial waste.

Special consideration must be given to human tissue. Biopsy material is usually disposed of by incineration via the clinical waste stream. Autopsy material may have to be returned to the family for disposal, if it so wishes. Donated human cadaveric material will have to be returned to the mortuary staff for inclusion with the remains when sent for cremation or burial.

Contaminated broken glass must be disinfected or autoclaved (whichever is the safer under the circumstances) and then disposed of via the standard industrial waste route.

Contaminated sharps must be placed in an incineration container (cinbin) for disposal via the clinical waste route: it is important not to fill these containers to more than three quarters full.

Further details are available in *Safe Disposal of Clinical Waste* [22].

2.17.9. *Removal of equipment*

In areas handling specimens at containment level 2 or above, all equipment must be rendered safe before it leaves the laboratory.

Re-usable plastic or glassware must be disinfected or autoclaved before being sent to a general washing up area.

Equipment being sent for repair or disposal must be disinfected by a validated method before removal from the laboratory: a permit to work certificate must be signed by the individual who carried out the disinfection and given to the maintenance staff/contractors at the time of removal.

Refrigerators, freezers and cryostats must be sent for de-gassing before final disposal.

2.18. USE OF PERSONAL PROTECTIVE EQUIPMENT

A white coat is the standard minimum item of protective equipment in most laboratories. It must be borne in mind that a laboratory coat is simply an overall to prevent clothing becoming contaminated. It is not a 'badge of office' worn by scientists. The coat should always be buttoned up and must never be worn outside the laboratory; wearing laboratory coats in common areas such as seminar rooms, staff rooms and canteens should be prohibited.

In laboratories handling biological agents at Category 2 or above Howie style coats should be worn. There must be suitable arrangements for autoclaving them before they are sent for laundering.

Additional personal protective equipment should be used only as a last resort.

Latex gloves are commonly used for extended periods in biological laboratories because they provide good protection against biological agents, they protect sensitive samples such as cell cultures from commensal organisms shed by workers and they preserve manual dexterity. Problems arise from allergies due to prolonged use of latex gloves and these are covered in section 2.16.1. It should be remembered that latex does not provide protection against a number of toxic chemicals and solvents. Nitrile or vinyl gloves should be provided, as appropriate, where chemicals are in use for which latex is unsuitable.

Respiratory protection is another area where care must be taken in the selection of equipment. Standard dust masks are of little use against toxic dusts and do nothing to remove solvent vapour. Where these substances are used outside the fume cupboard the correct type of mask to provide adequate protection must be on hand. For large spills of toxic substances or solvents more durable respiratory equipment should be provided, e.g. double cartridge face masks or self-contained breathing apparatus. Manufacturers produce a range of cartridges (usually colour coded) to suit a number of situations; it is important that the appropriate cartridges are available and all personnel are trained to select and fit the correct cartridge. Where self-contained breathing apparatus is deemed necessary it is essential that personnel are fully trained in its use.

Eye protection may also be problematic in areas dealing with bone since safety glasses and goggles tend to offer either impact resistance or chemical resistance. They are seldom suitable for both purposes. However, it is often simpler to adopt a policy of using impact resistant glasses on the basis that they will provide protection against chemical splashes even if the chemicals destroy the plastic. Regular replacement of a pair of safety glasses is simpler than trying to restore eyesight.

Where protective equipment of any kind is required the method of use should be clearly stated in the control measures section of the risk assessment or in standard operating procedures for the laboratory. The equipment must be maintained, with maintenance records kept, and staff should receive specific training in use where correct usage is not self evident.

2.19. GENERAL MANAGERIAL CONSIDERATIONS

Principal scientists who are responsible for funding and running research laboratories have considerable legal responsibilities under the health and safety legislation to ensure that subordinate workers or third parties are not put at risk.

Overcrowding in laboratories increases the probability of an accident considerably, particularly where harmful chemicals and mechanical equipment is used. There are no strict regulations (in the UK) for the amount of space to allocate per worker, this is variable depending upon the nature of the work. The only guidance comes from ACDP where a minimum space per worker is suggested as $22\,m^3$; assuming a maximum ceiling height of $2\,m$ this implies $11\,m^2$ of floor space per worker. This is only a 'rule of thumb'; particular attention should be given to the layout of the laboratory, space between benches and space around equipment (particularly microtomes, saws, and microbiological safety cabinets) to allow free access, egress and movement around the laboratory without endangering workers who are using potentially dangerous equipment.

Training of personnel is an important issue. If prospective workers do not fulfil the minimum entry requirements and adequate training, or direct supervision, cannot be provided they should not be allowed to work in the laboratory.

Personnel should not be allowed to work with organisms, or specimens, at containment level 2 until they have had sufficient experience at containment level 1. Work in a containment level 3 area is prohibited unless the individual has been specifically trained and authorized to work at this level. This usually involves working at containment level 2 for a reasonable length of time (around one year) and then completing their training at level 3 under the level 3 unit manager (usually a very experienced senior technician or MLSO). Training records, at this level, are required by the law.

2.19.1. Restricted access and permits to work

Where there is significant danger, access should be restricted to authorized personnel only. At biohazard containment levels 2 and 3 this is mandatory: level 3 areas must be kept locked at all times; the names of authorized users should be posted at the entrance and they should be the only personnel with access to the lock codes.

Entry to higher hazard areas by maintenance staff is only allowed under a 'permit to work' system. The permit is issued to the maintenance staff by the unit manager when the area has been suitably decontaminated and is safe to enter. The permit must be signed by the person who carried out the decontamination when this is not the unit manager, and work must not recommence until the permit is signed off (to indicate that maintenance work is complete).

2.19.2. Occupational health screening

Personnel who are likely to be exposed to biological agents, toxic chemicals or animal allergens should be screened by an occupational health professional before they begin working in the laboratory. In the case of biological agents, the control measures at the four levels of containment are based on the individual being healthy and having a normal level of immunity. Particular susceptibility to an organism, or allergen, may be indicated by family or personal history, medication or pre-existing disease. Particular care must be taken with regard to pregnant women and nursing mothers, for whom a special risk assessment to cover the pregnancy and lactation must be prepared.

This is a difficult area to manage as it crosses the boundary of medical confidentiality and ethics; however, a line manger is entitled to know whether or not an individual worker is suitable to work in their area and the nature of any extra precautions that may be required. They are not allowed to know the underlying reasons unless the individual wishes to volunteer the information.

2.19.3.   *Prophylactic treatment*

Technically, as a laboratory manager, you have no right to insist that a member of your staff attends occupational health surveillance or receives any form of prophylactic treatment. However, a manager is not obliged to offer employment or a research student place to any individual applicant.

Principal scientists should make it clear to staff from the outset that prophylaxis and screening are a requirement at the time of appointment. If they later refuse to comply, their position can be terminated without contravention of employment legislation etc.

Prophylactic treatment should always be offered to personnel where it is known to be effective, or even partially effective, over a long term and without significant adverse side effects.

The obvious ones with regard to work with human and animal tissue are hepatitis B and tetanus. Tuberculosis should be considered where the nature of the samples indicates risk.

With regard to human specimens, the overwhelming consideration is the risk from HBV: from a single needle stick involving known positive material there is a 1:6 probability of infection. Although there is no vaccine against HIV the probability of infection under similar circumstances is less than 1:1000.

2.20.   CONTENTS OF A RISK ASSESSMENT

The name of the institution/building

The names of the personnel involved

The area where the process is carried out

A brief description of the process (title of procedure and its purpose).

A list of the hazardous substances and hazard classification categories including the following.

- A description of acute hazards and the route by which they are hazardous.
- A description of chronic hazards, e.g. carcinogenic, teratogenic etc., and the routes by which they are hazardous.
- A description of any adverse reactions that may result from combining material and harmful waste products that may be generated.
- Biological agents, animal allergens and flammability hazards should be included here unless they are subject to a separate risk assessment.

The sources of hazard information

With respect to standard reagents, this need not necessarily be from the actual supplier provided the specifications of the substance are the same.

A statement as to whether safer alternatives and/or pre-diluted material could be obtained from the suppliers

A statement as to why safer alternatives are not used if this is the case

A valid argument may be that the safer method has not been validated or accepted, or that prepared stains may have a short shelf life.

Containment and control measures employed under normal working conditions

Standard operating procedures including specification of the use of fume cupboards and protective equipment such as gloves and eye protection.

Containment and control required in an emergency

The concept of the maximum credible accident is useful here: clearing of chemical spills or the containment of a biological agent must be specified in a step-by-step standard procedure.

It must contain specific information on any additional personal protective equipment required, neutralizing or detoxification of spilled materials and the correct method of clearing up, labelling and disposal.

First aid measures

Standard laboratory first aid measures should be committed to memory by all personnel. Where there are specific additional requirements, such as hydrofluoric acid burns or exposure to cyanide, clear instructions must be posted in the laboratory and all personnel must be made familiar with them.

Accident and incident reporting procedures should be clearly posted by the first aid facilities along with a list of current trained first aiders.

Fire fighting equipment to be used

Most laboratories have only water, carbon dioxide or fire blankets available as a means of extinguishing fires; in some cases these may be supplemented by a chemical foam extinguisher. It is essential that all personnel are familiar with the range of fire extinguishers available to them, that they are aware of correct usage and are aware of when and where it is appropriate to use a particular type.

It is useful, therefore, to include the recommended fire extinguisher type for each of the flammable materials used. This may be done generically, i.e. 'alcohols—use water' 'petroleum ether—use foam or blanket'.

Any special storage requirements for

- reagents involved
- products of the method
- storage of waste.

This must include information on segregation, the types of containers, limits on amounts and correct labelling.

Waste disposal

Declaration

Finally the assessment must be signed and dated by the originator and the person nominated to authorize it: this may be the individual laboratory leader or it may be a nominated health and safety officer within the institution.

2.20.1.   *Conveying the information to personnel*

The contents of a risk assessment must be conveyed to all personnel who may be affected by the work process. Moreover, it must be conveyed in language that they will understand. This is particularly relevant to non-scientific personnel such as cleaners and maintenance fitters. It is not appropriate to provide cleaners with long erudite descriptions of hazards and risks arising from processes that they could not be expected to understand. They must be given clear instructions as to the nature of the hazards they are likely to encounter in the laboratory during their work processes, along with clear instructions as to how to avoid exposure. Colour coded waste disposal systems, restricted entry systems posted on doors, clear instructions not to clean a particular area unless a senior technician is present, are all examples of simple methods by which this may be achieved.

2.20.2. Who should compile a risk assessment?

The simple answer is the person who is responsible for initiating a work process. In manufacturing industries it is normally a management responsibility because production workers tend not to be involved in decisions as to what is manufactured. However, they should be fully involved in modification of working methods in order to reduce risk. So, even in industry, managers should compile risk assessments in consultation with employees.

In a research laboratory, particularly in a university, the person who initiates a work process may well be an undergraduate student (usually final year working towards a BSc dissertation).

My own observations, having conducted a number of safety audits and inspections in university laboratories, are that the approach in laboratories tends to be one of two extremes. It is quite common to find an exemplary risk assessment file, fully indexed, written by an experienced laboratory technician: sadly the students and postdoctoral researchers usually have a scant knowledge of its contents. The other extreme is a set of rather inadequate assessments made by the students themselves. The latter is preferable because, in this case, the students are familiar with the hazards and are aware of the risk reduction process.

The ideal solution is to make the students responsible for compiling their own risk assessments but with help and advice available from an experienced technician. Unfortunately this is perhaps more time consuming, but it is by far the most effective approach: you should be mindful of the fact that your student is a manager or independent researcher of the future.

At Leeds we introduced a COSSH risk assessment assignment into one of our final year practical modules: over a period of three years the mean standard of assessments written by students somewhat overtook those written by the staff. It must be admitted, however, that the student assessments compiled for their assignment tended to be of a somewhat higher standard than their usual laboratory work assessments. The fact remains that they were able to perform the task competently when required and would have no difficulty in meeting a future employer's criteria.

2.21. TRANSPORT, PACKAGING AND LABELLING OF BIOLOGICAL SAMPLES

Transport of biological samples both nationally and internationally is a legal minefield. Since the events of 11 September 2001 it has also become a political minefield. There is an international agreement (UM 602) which stipulates minimum packaging and labelling of biological samples in a number of categories that may be applicable. Section 6.2, which deals with infectious substances and clinical samples, is probably the most relevant. Section 9 includes genetically modified material that is not considered harmful to humans or animals (UN 3245), and solids or liquids that may cause environmental damage (UN 3077 and UN 3082 respectively).

All goods must be triple packed in containers of a high specification that ensure against leakage in the most extreme conditions. Moreover, the transportation has to be supervised by a qualified dangerous goods adviser. Most of this is beyond the average laboratory. The only sensible solution is to employ the services of a specialist contractor.

Where animal material is to be imported there may be a requirement for an import licence from a regulating authority (DEFRA, formerly MAFF, in the UK). This largely applies to material from domestic animals or meat products as they may harbour disease transmissible to domestic animals. Inquiries and permission should be sought when any material of this nature is to be imported.

Within the UK, low risk biological samples may be sent via the postal services provided they are adequately packaged and labelled (again in accordance with UN 602 specifications). With regard to specimens in hazard categories 2 or 3, samples may only be sent with the prior approval of the dangerous goods officer of the postal service. Postal services throughout Europe are finding the compliance with these regulations to be uneconomical with respect to the volume of trade generated. In the UK the postal services no longer accept clinical samples packed with dry ice (perhaps the largest amount) and it seems likely that they may cease this side of their operations unless current attempts to simplify the regulations are successful.

It should also be noted that transport of material in hazard categories 2 and 3 (and 4) require prior notification (in the UK) to the Health and Safety Executive. However, with respect to category 2 samples this may be in the form of a blanket cover agreement, in the case of category 3 it is by a separate notification for each shipment.

Recent problems have also arisen from litigation. Where samples have caused contamination in transit, or upon receipt, the transport companies have been prosecuted as well as the sender and, in some cases, the recipient (for not having suitable systems of work for the receipt of dangerous goods). As a result many specialist transport companies insist that their staff supervise the packaging and labelling of any material they carry. Many of them are now offering training courses, either in-house or distance learning, in packaging of biological samples to UN 602 standard. Laboratories that routinely send material by carrier may find it more economical to have someone trained, and qualified, in packaging the material to avoid the necessity of a company adviser being present at the time of packaging.

2.22. IONIZING AND NON-IONIZING RADIATION

Adequate cover of regulations and restrictions on the use of radio-chemicals, radiation sources, lasers and ultraviolet sources would require a complete chapter in itself. However, with regard to ionizing radiation sources, every worker must be fully trained, registered and monitored.

Where work of this nature is envisaged it is essential for the laboratory manager or principal scientist to contact the local Radiation Protection Officer to ensure that both the laboratory and the workers fulfil the requirements.

2.22.1. *Ultraviolet light sources and lasers*

Ultraviolet light sources such as mercury vapour lights, if viewed directly through microscope optics, would cause considerable damage to the retina. Most modern instruments are safe by design, i.e. it is not physically possible to view the light without the necessary filter combinations in place. Older instruments may not be designed to the same standard and in this case it is important to use the instrument in accordance with the manufacturer's instructions and ensure that every user is familiar with the required filter combinations for each application.

In-house repairs to UV light sources or built-in laser equipment must never be attempted by staff not specifically qualified to do so. Screening around light sources must never be removed while the light is on, i.e. the source must never be looked at directly.

Lasers used in confocal microscope systems are relatively recent innovations and are, on the whole, well protected by design. It is important to remember that the classification of the laser is in the context of normal use, i.e. that the design of the equipment is such that it is impossible to view the laser beam whilst looking down the microscope. In some models it is possible to detach the laser source from the microscope without disarming the laser power supply, i.e. a worker could view the beam directly with the laser detached from the instrument. In this case, clear instructions must be posted adjacent to the instrument to ensure that this is not done.

2.22.2. *Genetic modification*

Any work involving genetically modified organisms (GMOs) or genetically modified micro-organisms (GMMOs) are subject to stringent regulations regarding their 'contained use'.

It should be borne in mind that tissue cultures are considered to be micro-organisms within the legal definition. Also 'contained use' applies to any activity involving the organism including examination, storage or even transport and disposal, i.e. the legislation does not just cover the genetic manipulation process itself.

Genetically modified material may only be used in registered premises and all activities must be approved by a local genetic modification safety committee. Genetic modification work is categorized into four 'classes' that roughly parallel the ACDP categories. All activities involving Class 2 and above GM work must be notified to the Health and Safety Executive and acknowledged by them before commencement of the work. Activities in Class 3 require specific approval from HSE for each activity.

It is imperative to seek advice from the local biological Safety Officer and local GM committee before planning any work involving genetic modification. Moreover, inspection of laboratory containment and notification takes time: the activities should be planned well in advance. An overview of the legislative controls is provided in A Guide to the Genetically Modified Organisms (Contained Use) Regulations 2000 [23] and detailed advice on risk assessment is given in Compendium of Guidance from the Health and Safety Commission's Advisory Committee on Genetic Modification [24].

REFERENCES

[1] HMSO 1975 *Health and Safety at Work etc. Act 1974* (London: HMSO) ISBN 0 10 543774 3
[2] Health and Safety Executive 1999 *Essentials of Health and Safety at Work* (London: HSE Books) ISBN 0 71 76071 6
[3] Medical Research Council (MRC) *Health and Safety Policy and Guidance*. From MRC, 20 Park Crescent, London W1B 1AL, UK
[4] Environmental Protection Act 1990
[5] Human Experimentation (Declaration of Helsinki) 1964 Code of Ethics of the World Medical Association *Br. Med. J.* 177–80
[6] Human Rights Act 1998
[7] Royal College of Physicians 1990 Research Involving Patients *J. Royal College of Physicians* **24**(1)
[8] Health Services Advisory Committee 1991 *Safe Working and Prevention of Infections in Clinical Laboratories* (London: HSE Books) ISBN 0 11 885446 71
[9] Health and Safety Executive 1998 *Safe Use of Work Equipment: Provision and Use of Work Equipment Regulations* 2nd edn (London: HSE Books) ISBN 0 776 1626 6
[10] Health and Safety Executive 1989 *Memorandum of Guidance on the Electricity at Work Regulations* (London: HSE Books) ISBN 0 11 883963
[11] The Institution of Electrical Engineers 1994 *Code of Practice for In-service testing of Electrical Equipment* (London: IEE) ISBN 0 85296 884 2
[12] Health and Safety Executive 1999 *Control of Substances Hazardous to Health Regulations: Approved codes of practice* (London: HSE Books) ISBN 0 7176 1670 3
[13] Health and Safety Executive 2001 *Occupational Exposure Limits 2002 (EH40 2002)* (London: HSE Books) ISBN 0 717 6203 2
[14] In: St John-Holt 1997 *Principles of Health and Safety at Work* (Institute of Occupational Safety and Health)

[15] Advisory Committee on Dangerous Pathogens 1995 *The Categorisation of Biological Agents According to Hazard and Categories of Containment* 4th edn (London: HSE Books) ISBN 0 471 92274 9

[16] ACDP 1998 Supplement to *The Categorisation of Biological Agents According to Hazard and Categories of Containment* 4th edn (London: HSE Books)

[17] *Specified Animal Pathogens Order (1998)* Available from DEFRA, Government Buildings (Toby Jug), Hook Rise, South Surbiton, Surrey KT6 7DX, UK

[18] Advisory Committee on Dangerous Pathogens 1998 *Transmissible Spongiform Encephalopathy Agents: Safe Working and the Prevention of Infection* (London: HMSO) ISBN 0 11 322166 5

[19] Advisory Committee on Dangerous Pathogens 2001 *The Management, Design and Operation of Microbiological Containment Laboratories* (London: HSE Books) ISBN 0 717 62034 4

[20] British Standards Institute 2000 British Standard EN12469: 2000, Performance criteria for microbiological safety cabinets.

[21] Bennet A 1977 Gaseous disinfection methods. In *Proceedings of the European Biosafety Association (EBSA) 1st Annual Conference*, November 1997

[22] Health Services Advisory Committee 1999 *Safe Disposal of Clinical Waste* (London: HSE Books) ISBN 0 7176 2492 7

[23] Health and Safety Executive 2000 *A Guide to the Genetically Modified Organisms (Contained Use) Regulations* (London: HSE Books) ISBN 0 7176 1758 0

[24] Health and Safety Executive 2000 *Compendium of Guidance from the Health and Safety Commission's Advisory Committee on Genetic Modification* ISBN 07176 1763 7 or http://www.hse.gov/hthdir/noframes/acgmcomp/acgmcomp.htm

Chapter 3

Radiation safety considerations

Christopher F Njeh

3.1. INTRODUCTION

Considerable progress has been made in the development of non-invasive methods for the assessment of the skeleton. Current techniques include radiographic absorptiometry (RA), single X-ray absorptiometry (SXA), dual X-ray absorptiometry (DXA), quantitative computed tomography (QCT) and quantitative ultrasound (QUS). Some of these techniques involve ionizing radiation. Hence, it is worth introducing the safety aspects involved with their use.

Ionizing radiation is any radiation capable of releasing an electron from its orbital shell. Ionizing radiation encountered in bone measurements are X-rays and γ-rays. It is worth mentioning that people have been exposed to naturally occurring ionizing radiation since the beginning of time. Today, it is estimated that 82% of the exposure of the US population to radiation comes from natural background sources. Natural background radiation comes from three sources: cosmic rays, terrestrial radiation that comes from radioactive materials naturally occurring in the earth and internal deposits of radionuclides in our bodies. On the other hand man-made sources include medical X-rays, nuclear medicine procedures, consumer products (TV, tobacco) and nuclear reactors.

It has been shown that the effect of ionizing radiation is stochastic. This means that any exposure to radiation carries a risk. Recent ICRP publications [1, 2] recommend the application of the ALARA (as low as reasonably achievable) principle when ionizing radiation is used for measurement. The last section of this chapter will discuss aspects of radiation protection.

3.1.1. Units of radiation measurement

3.1.1.1. Exposure

When X-rays or γ-rays interact in a volume of air, excitation and ionization of the air molecules occur. Consequently, the air can conduct electrical current. If the electrical conductivity of this air is measured, a value for the quantity of radiation causing the ionization is obtained. Exposure therefore can be defined as the amount of ionization produced by photons in air per unit mass of air. The traditional unit for exposure is the roentgen (R). One roentgen of exposure creates 2.58×10^{-4} coulomb of charge per kilogram of air. The SI unit of exposure is C/kg.

3.1.1.2. Absorbed dose

This is the amount of energy absorbed from the incident beam by a medium as a result of ionizing radiation passing through that medium. The traditional unit for absorbed dose is the rad, defined as 100 erg of energy absorbed per gram of absorbing material. The comparable SI unit is the gray (Gy) which is defined as 1 joule of energy absorbed per kilogram of absorbing material (1 Gy = 100 rad). One gray is a large dose of radiation and, in diagnostic radiology, absorbed dose is usually expressed in milligray (mGy).

3.1.1.3. Entrance surface dose (ESD)

An important quantity in diagnostic radiology is the entrance surface dose. This is the absorbed dose to the skin at the point where the X-ray beam enters the patient's body. Reported ESD for some of the DXA systems are presented in table 3.1.

3.1.1.4. Dose equivalent (DE)

This takes into account the fact that different types of radiation produce different amounts of biological damage. Alpha particles, for example, are high linear energy transfer (LET) radiation and therefore have a greater biological effect than X-rays. Thus a 0.2 Gy absorbed dose of alpha particles would be more damaging to a given mass of tissue than a 0.2 Gy absorbed dose of X-rays. To account for these differences in biological response, each type of radiation is assigned a quality factor (QF). The QF is a measure of the relative ability of the ionizing radiation to do biological damage. For the various different types of ionizing radiations (α-, β-, γ-rays and neutrons), it is always measured relative to X-rays. Thus, for X-rays, QF = 1.0 by definition, and the DE is always numerically equal to the absorbed dose. The traditional unit, the rem, is defined as the product of the absorbed dose in rads times the QF. The SI unit is the Sievert (Sv), and it is defined as the absorbed dose in grays times the QF. As with the

Table 3.1. *Manufacturer reported entrance skin dose (ESD) for some of the more common densitometers.*

Technique	Manufacturer	Model	Site	ESD (μGy)	Comments
DXA	Hologic	QDR 1000/1500	AP spine/femur Total body	<43 <13	70 and 140 dual output, 43 and 110 keV
		QDR 2000	AP spine/femur Lateral spine Total body	192 237 8	Fan beam mode
DXA	Lunar	DPX-L/DPX-IQ	AP spine/femur Lateral spine Total body	11* 77 0.2	76 kVp with cerium filter, 38 and 70 keV
		Expert XL	Spine/femur-5 mA Fast Total body 1.5 mA Fast Forearm/hand	530 50 120	Fan beam
DXA		PIXI	Calcaneus/forearm	<200	
SPA	Lunar	SP2	Forearm	150	7.4 GBq ^{125}I
DXA	Norland	XR-36	AP spine Femur Lateral spine Total body	0.9–44.4† 1.8–89† 123 0.4	100 kVp with samarium filter 80 and 46.8 keV
pDXA	Norland	pDEXA	Radius and ulna	<25	60 kVp, tin filter, 28 and 48 keV
SXA	Norland	SXA300	Calcaneus	13	36 kV, tin filter, 27 keV
DXA	Sopha Medical	L-XRA	AP spine Lateral spine	20 40	80 kVp with Neodymium filter, 43 and 70 keV
pQCT	Stratec	XCT 960	Forearm	30	45 kVp, 0.3 mA, 38.5 keV
pDXA	Osteometer	DX-200	Forearm		55 kVp, 0.3 mA, tin filter
DXA	DMS	Challenger	Spine/femur	<3.3	Samarium filter, 35 and 75 keV

* ESD for DPX-IQ is similar to DPX-L, and is also dependent on tube current, resolution and scan speed used. The ESD is linearly proportional to these parameters if everything else remains constant.
† The Norland XR-36 has a dynamic filtration which compensates for different tissue thickness, so the ESD is dependent on the body thickness of the patient.

gray, 1 Sv is a relatively large dose. Hence, in diagnostic radiology, DE is usually expressed in millisievert (mSv).

3.1.1.5. *Effective dose (ED)*

ESD does not permit a direct comparison of the risk associated with patient doses associated with absorptiometric and other radiological examinations. The radiation hazard to the patient is often expressed in terms of the effective dose (ED). So, the purpose of the effective dose is to relate exposure to risk. ED is defined as the sum of the absorbed doses to each irradiated organ weighted for the radiation type and the radio-sensitivity of that organ [1]. The ED is equivalent to the uniform whole body dose that will put the patient at equivalent risk from the carcinogenic and genetic effects of radiation [1].

The ED is obtained by multiplying the dose equivalent for each organ by a weighting factor proportional to the sensitivity of that organ to the stochastic effects of radiation, and summing over all the organs exposed. ED is defined as

$$ED(Sv) = QF \sum_{Tt} w_T \, D_T$$

where QF is the radiation quality factor, w_T is a tissue specific weighting factor for each organ or tissue irradiated, and D_T is the absorbed dose by the specific type of tissue T. The absorbed dose for a particular tissue, D_T, is in units of Gray (Gy) energy absorbed by the tissue per unit mass of tissue and $1\,Gy = 1\,J/kg$. The radiation QF is in units of Sv/Gy and the tissue-weighting factor is dimensionless. The above scheme for calculating effective dose was established by the International Commission on Radiological Protection (ICRP) [1]. The list of tissue weighting factors derived by the ICRP is summarized in table 3.2. These weighting factors replace an earlier set published in 1977 [2] and are based on updated information about cancer induction by radiation. Calculations using the old 1977 weighting factors are distinguished by using the term effective dose equivalent instead of effective dose.

For low doses, as encountered in DXA, the principal risks to patients are the stochastic process of carcinogenesis and genetic effects. For a given procedure, summing over all tissues irradiated and using the weighting factors gives the effective dose. Due to the low radiation doses encountered in BMD measurements, there have been limited studies on doses incurred in this examination. Table 3.3 is a summary of typical effective dose values for different regions of interests.

3.1.2. *Radiation detector*

Many instruments and devices can be used to detect radiation. However, the types of devices have their specific applications and limitations. For

Table 3.2. *Tissue weighting factors from ICRP-60 [1].*

Tissue type	Weighting factor
Ovaries	0.20
Bone marrow (red)	0.12
Colon	0.12
Lung	0.12
Stomach	0.12
Bladder	0.05
Breast	0.05
Liver	0.05
Oesophagus	0.05
Thyroid	0.05
Skin	0.01
Bone surfaces	0.01
Remainder*	0.05

* The following 10 tissues are included in the remainder with a weighting factor of 0.005 each: adrenal, brain, upper large intestine, small intestine, kidney, muscle, pancreas, spleen, thymus and uterus.

diagnostic applications, radiation detectors include gas-filled detectors, thermoluminescent dosimeters (TLD) and film.

Gas filled detectors have a chamber filled with a gas that is ionized in part or whole when radiation is present. Either the total quantity of electrical charge is measured or the rate at which charge is produced is measured.

A TLD uses phosphors that emit light when heated. Whenever the material is irradiated, electrons are released from the bound states in the valence band and become free to migrate in the conduction bands and

Table 3.3. *Scanning times, entrance surface doses and effective doses for post-menopausal women for pencil beam mode scans performed on the QDR 1000 (Hologic). Values in parentheses include the ovaries, therefore representing pre-menopausal women. Data from Lewis et al [4].*

Scan type	Scan time (min)	Entrance surface dose (μGy)	Effective dose (μSv)
Total body	17	18	3.6 (4.6)
AP spine (L1–L4)	8	60	0.5 (0.5)
Lateral spine (L2–L4)	20	238	0.6 (0.6)
Proximal femur	6	60	0.1 (1.4)
Distal forearm	6	113	0.07 (0.07)

subsequently into the forbidden band. If the crystal is heated to 100–200 °C, the electrons will receive enough thermal energy to move back into the valence band with release characteristic photons. Examples of TLD include LiF, whose atomic number is close to that of tissue, and hence is useful for patient absorbed dose measurement.

After development, X-ray film exposed to radiation turns black. The amount of blackness is called the optical density, and the optical density is related to the amount of radiation received by the film. The actual relationship is not linear and depends on the type of film, the type and energy of the radiation and the details of processing the film.

3.2. RADIATION DOSE TO THE PATIENT

3.2.1. Introduction

The usefulness of X-ray-based techniques in the assessment of bone integrity are presented in chapters 8–11. Studies of the radiation dose to patients from absorptiometry scans have confirmed that patient exposure is small compared with many other sources of exposure including most radiological investigations involving ionizing radiation [3–6]. Due to the low radiation doses encountered in bone mineral density (BMD) measurements, there have been limited studies on doses incurred in this examination. In cases where doses have been measured, only the entrance surface dose (ESD) has been reported (table 3.1).

For low doses, as encountered in DXA, the principal risks to patients are the stochastic process of carcinogenesis and genetic effects. The following sections will discuss doses incurred in BMD measurements from different techniques.

3.2.2. Patient doses from dual X-ray absorptiometry

The entrance radiation dose from a DXA examination is determined by machine-dependent parameters such as source-detector geometry, focal spot size, source collimation, scan speed, tube current and source spectra. The operator can only change the mA and the scan speed (resolution). However, the mA and scan speed use can affect the precision of BMD measurement. Effective dose from DXA measurements have been reported in the literature [3, 7, 8]. Lewis *et al* [4] presented the first extensive effective dose estimates for Hologic QDR 1000 and QDR 2000 DXA scanners. Similar studies have been reported by Njeh *et al* [5] and Bezakova *et al* [9] for Lunar machines. Due to differences in the production of X-rays by these systems, their respective EDs are expected to be different. Lewis *et al* reported the highest ED for the total body of 3.6 μSv (4.6 μSv including the ovaries)

Table 3.4. *Entrance surface dose (ESD) and effective dose (ED) for Lunar DPX-L DXA scanner.*

Reference	Mode	Entrance surface dose (µGy)	Effective dose (µSv) PA spine	Femur
[5]	Medium, 0.75 mA	10.25	0.21	0.08 (0.15)
[9]	Medium, 0.75 mA	11.5	0.19	0.023 (0.14)

(table 3.3). ESDs were found to be proportionate to the speed of the scanning arm of QDR 1000. Njeh *et al* [5] and Bezakova *et al* [9] observed lower values of effective dose for a Lunar DPX-L machine (table 3.4). Clinically patients on one visit may receive scans of a number of different regions (generally two but not more than three) such as the posterior–anterior (PA) spine, lateral spine, left and right proximal femur, hand, forearm, calcaneus and total body. The total effective dose per visit will therefore be a summation of the regions scanned. For example, a post-menopausal patient undergoing scans of the spine and hip performed on a QDR 1000 would receive an effective dose of 0.6 µSv (0.5 spine and 0.1 femur). However, a pre-menopausal woman scanned on the same scanner would receive an effective dose of 1.9 µSv because of inclusion of the ovaries. On the other hand, a post-menopausal woman and pre-menopausal woman scanned on a Lunar DPX-L will receive an effective dose of 0.29 and 0.36 µSv, respectively.

In lateral DXA scans, milliamp-second (mAs) values and therefore surface dose values are increased typically by a factor of 2–4. However, due to the fact that organ doses are significantly lower relative to the ESD for the lateral beam and that the length of the spine scanned is shortened in most cases, effective dose values for lateral scans are comparable with PA projection scans. Lewis *et al* [4] reported EDs of 0.5 and 0.6 µSv for AP and lateral spine respectively.

DXA measurement, if accompanied by a lateral radiograph, will significantly change the magnitude of the dose involved. For a conventional lateral X-ray film examination, Kalender [3] estimated an ED of 700 µSv for 90 kVp and radiation field height of 25 cm. For larger fields, values of 2000 µSv can be expected [10]. Thus an effective dose of up to 2000 µSv will result from a DXA scan plus X-ray film. On the other hand, Huda and Morin [6] estimated ED for 80 kVp AP and lateral X-ray examination to be in the range of 40–120 and 60–190 µSv, respectively. They assumed an ESD range of 2–4 mSv and an X-ray beam area between 400 and 600 cm^2 at the patient entrance.

The dose associated with DXA BMD measurement can be considered low or even insignificant in comparison with natural background radiation levels. It is well below the background value of about 7 µSv per day.

Table 3.5. *Scan times and doses (where values in parentheses include the ovaries) for a Hologic QDR 2000 fan beam mode reported by Lewis et al [4].*

Scan type	Scan time (min)	Entrance skin dose (μGy)	Effective dose (μSv)
PA spine (L1–L4)	0.1	57	0.4 (0.4)
	1	138	0.9 (0.9)
	2	271	1.8 (1.8)
	3	432	2.9 (2.9)
Lateral spine (L2–L4)	3	684	1.2 (1.2)
	6	1390	2.5 (2.5)
Proximal femur	1	138	0.3 (3.0)
	2	271	0.6 (5.9)
Total body	6	11	2.7 (3.6)

3.2.3. Patient doses from fan beam DXA

Fan beam DXA scanners were introduced to reduce scanning time and increase image resolution. A consequence has been an increase in radiation dose due to the increase in photon flux required. In the Lunar Expert there is an increase in photon flux by using a higher tube potential (134 instead of 76 kVp) and a higher tube current (5 instead of 0.75 mA). Hologic QDR 4500 operates at 140 and 100 kVp compared with 140 and 70 kVp for QDR 1000. Entrance surface doses have been reported for the Lunar Expert, Expert XL, Hologic QDR 2000 and QDR 4500. Stewart *et al* [11] reported an ESD of 0.7, 1.1 and 0.12 mSv for spine/femur, lateral spine and peripheral sites respectively for the Lunar Expert XL. Effective doses for these fan beams have been reported in the literature [4, 5, 12] (tables 3.5 and 3.6). The fan beam has led to about a 100-fold increase in dose for the Lunar and tenfold increase for the Hologic scanner. One of the reasons for the difference in effective doses between the Hologic QDR 4500 and the Expert XL is that for the Expert XL the spine measurements are

Table 3.6. *Doses from fourth generation (fan beam) DXA scanner.*

DXA system	Mode	Entrance surface dose (μGy)	Effective dose (μSv)		Reference
			PA spine	Femur	
QDR 4500	Array (1 min)	–	6.7	–	[40]
QDR 4500	–	295	8	–	[41]
Expert	5 mA fast	700	73 (max. scan size)	–	[11]
Expert XL	Medium (40 s)	895	31	4.7 (32.0)	[5]

Table 3.7. *Effective dose for vertebral morphometry scans performed on DXA bone densitometer and on conventional radiographs (Note: scan times are given in parentheses. For QDR 4500, study requires the acquisition of both a PA centreline and a lateral morphometry scan). Data from [14] and [42] for QDR 2000.*

System	Scan mode	Effective dose (μSv)	
		PA centreline scan	MXA scan
QDR 2000	–	–	34
QDR 4500	Dual-energy fast	4 (30 sec)	20 (5 min)
QDR 4500	Dual-energy medium	10 (90 sec)	41 (10 min)
QDR 4500	Dual-energy hi-res*	20 (3 min)	20 (10 min)
QDR 4500	Single-energy	–	2 (10 sec)
Expert XL	5 mA fast	–	38 (40 sec)
Lateral radiograph	–	–	800

* hi-res = high resolution.

performed in the anterior–posterior (AP) direction. Hence the distance to the sensitive organs such as the stomach is reduced and results in approximate doubling of ED.

Despite the increases associated with advances in technology, patient dose from DXA remains very low when compared with natural background radiation and from other common radiological procedures. For example, the mean ED per radiograph for AP lumbar spine X-ray examination has been quoted as 690 μSv [13] which is about tenfold higher than fan beam ED. On the other hand the fan beam DXA EDs are comparable with 17 μSv reported for chest PA X-ray examination [13].

3.2.4. Doses from vertebral morphometry using DXA

Lateral radiographs have a better spatial resolution than DXA vertebral morphometry or MXA. However, one advantage of MXA is its significantly lower radiation dose to the patient. Estimates of the effective dose to a patient from MXA have been reported by Lewis *et al* [4]. Blake *et al* [14] observed that although the scan length for a MXA study was approximately three times greater than a bone density scan of the lumbar spine, patient dose is approximately ten times higher (assuming the same scan mA, collimation and scan speed) because of the inclusion of the lungs and breast in the scan field. Table 3.7 shows that the highest dose of 51 μSv is incurred from the medium mode using QDR 4500. This is significantly lower than an approximate dose of 600–800 μSv from lateral radiographs (combined thoracic and lumbar) [13].

3.2.5. Paediatric doses from DXA

Different diseases such as juvenile rheumatoid arthritis [15] may directly or indirectly decrease the bone mineral content in children. Therefore, an assessment of bone integrity may assist in the choice of the right therapeutic strategy and best follow-up for patients with diseases affecting bone metabolism. Also, bone mass accrued during childhood has been proposed as a determinant of an individual's susceptibility to osteoporotic fractures in adulthood [16]. DXA is currently the most widely used technique for the assessment of bone integrity. As previously stated, DXA can also be used to study other aspects of body composition such as percentage of fat and lean body mass [17]. Njeh *et al* [18] assessed the dose incurred from DXA scanning of a 5-year-old child and a 10-year-old child using a Lunar DPX-L scanner. They measured the entrance surface dose and percentage depth doses for the total body and PA spine scan modes using lithium borate thermoluminescent dosemeters located at the surface and distributed throughout various organ locations in anthropomorphic child phantoms. The EDs were calculated from the percentage depth doses, amount of each organ irradiated and tissue weighting factors [1]. The results are represented in table 3.8. The EDs are more than two orders of magnitude lower than reported ESDs and EDs for paediatric chest X-rays. Fewer studies have reported DXA paediatric dose. Koo *et al* [19] reported an ESD of 3 μGy for paediatric total body using the Hologic 1000/W. Faulkner *et al* [20] reported 10 μGy ESD and effective dose equivalent of 2.7 μSv for total body mode using the Hologic QDR 2000 (fan beam mode). This is similar to adult total body ED reported by Lewis *et al* [4] for the same type of scanner. Therefore, one could extrapolate that similar EDs could be derived for the paediatric mode as were reported for the adult scanning mode, due to proportionate reduction in the scan field for the paediatric mode.

Few studies have reported the effective doses for common paediatric radiological examinations. However, ESDs for common paediatric radiological examinations have been reported [21–24]. ESDs are in the range of 50–400 μGy for PA chest X-ray and 420–1990 μGy for an abdominal X-ray examination. Njeh *et al* [18] used these ESD values and applied

Table 3.8. *Effective dose for a paediatric scan mode using a lunar DPX-L [18].*

Scan	Mode	Patient size	Age	Scan time (min)	ESD (μGy)	ED (μSv)
PA spine	–	6–16 cm	5	5	6.0	0.28
PA spine	–	–	10	–	6.0	0.20
Total body	Medium	15–25 kg	5	9	0.12	0.03
Total body	Large	25–35 kg	10	12	0.1	0.02

Table 3.9. *Reported mean ESDs per radiograph and calculated EDs using Monte Carlo simulation for two common radiological examination for comparison with table 3.8 [18].*

Reference	Age (years)	PA chest ESD (µGy)	PA chest ED (µSv)	Abdomen ESD (µGy)	Abdomen ED (µSv)
[21]	1–5	50	5.2	470	89.8
[22]	1–4	70	7.3	420	73.2
[23]	1–5	200	21.6	750	137
[24]	1–5	260	37.8	1200	242
[21]	6–10	80	7.9	770	132
[22]	5–9	60	6.1	670	106
[24]	6–10	310	43.2	1280	245
[22]	10–15	80	5.7	1060	125
[24]	10–14	400	38.8	–	–

NRPB-SR279 [25] Monte Carlo data to estimate the EDs. The results are presented in table 3.9, which in comparison with table 3.8 shows that doses encountered in DXA are more than two orders of magnitude less than common radiological examinations.

Exposure of children to ionizing radiation gives more grounds for concern than that for adults, because for children there is a longer period available for the delayed effects of radiation to manifest. This means that the lifetime risk of radiation-induced cancer per unit dose to the child is likely to be higher than for the adult. The probability of aggregated radiation detriment can be related to ED and for children this is age and sex related. For a 10-year-old female, the total lifetime risk of radiation-induced cancer is estimated as $16 \times 10^{-2}\,\mathrm{Sv}^{-1}$ compared with $5.9 \times 10^{-2}\,\mathrm{Sv}^{-1}$ for a 30-year-old adult [26]. A paediatric subject referred for a bone densitometry examination may receive a PA spine and total body scan. The total effective dose will then be 0.31 µSv, giving a lifetime risk of 5×10^{-8} for fatal cancer. This is negligible compared with the possible benefits from the scan. In addition to carcinogenesis, genetic effects are also of concern in the total body examination because of the irradiation of the gonads. The estimated dose to the ovaries was 0.002 µSv, which is 9% of the total dose. Even though these doses are quite low, good radiological practice should be implemented.

3.2.6.   *Patient doses from QCT*

As mentioned earlier, spinal QCT for BMD measurement can use standard CT machines or those with specialized software. As discussed by Jessen *et al*

CT has problems with specification of radiation doses imparted. The main problem is the non-uniformity of dose distribution within the patient. This is because in most CT examination a number of CT slices may be contiguous or overlapping. The radiation dose to patients undergoing QCT depends upon the following factors: (1) scan parameters (kVp, mAs, slice thickness, number and spacing of slices), (2) patient size, (3) X-ray source/detector geometry, and (4) detector collimation and calibration. Hence the dose delivered is highly scanner specific. The X-ray beam of the QCT systems has a slice thickness in the range of 8–10 mm; the radiation received by the patient is therefore limited to a narrow band. BMD assessment using CT involves in most cases three slices.

The radiation dose from QCT has been reported over the years, but there has been little consistency in the parameter reported—entrance surface dose, CT dose index (CTDI), mean depth dose, etc. Banks and Stevenson [27] reported the approximate mean surface dose (multislice, 125 kVp, 230 mAs) as 16 mGy. However, Kalender [3] reported a surface dose of 2.5 mGy for a SOMATOM Plus (80 kVp, 125 mAs). Note that the scan parameters are different for these two studies. This illustrates the fact that low dose values can be achieved via low kVp and low mAs. Changing from 120 kVp to 80 kVp with fixed mAs settings, Kalender [3] was able to measure a dose reduction of a factor of about 4. QCT BMD examinations are always accompanied by a lateral scout radiograph (tomogram) which is required to ensure reproducible positioning and involves an additional radiation dose. Kalender [3] measured an ESD of 0.46 mGy for lateral scout radiographs. He further estimated the ED from the tomogram and QCT to be 31.9 and 28.6 μSv, respectively. It is worth noting that his methodology was very crude. For instance, no depth dose was calculated. A QCT organ dose of 1.5 mSv estimated from the centre of a phantom was assumed for all the organs. This assumption is not valid since the centre of different organs/tissue varies and thus the organ dose also varies. However, he acknowledged the fact that his results were only accurate to within a factor of 2 to 3.

Huda and Morin [6] estimated the ED for SEQCT and DEQCT using a GE 9800Q CT scanner. The estimate, based on the methodology of Shrimpton [28], reported an effective dose of 10.8 mSv for a GE 9800Q scanner (120 kVp, 400 mAs, 22 slices). By assuming that the effective dose was linearly proportional to the number of slices and mAs, they calculated the ED as a function of these parameters. They also scaled the dose by a factor of 0.3 when 80 kVp was used instead of 120 kVp. When a tomogram (scout) and three CT images are acquired at 80 kVp, an effective dose equivalent in the range 200–360 μSv is estimated, depending on mAs used. Similarly, for a tomogram and six CT slices associated with the 80 and 120 kVp DEQCT, an ED of 700–1400 μSv is estimated. This also represents an over-simplification in the effective dose estimates. Their values are about eight times higher than those reported by Kalender [3]. The scaling method used may have overestimated the dose considering that the radio-sensitive organs are not located proportionally in the body. However, Karantanas *et al* [29] have also reported an effective dose equivalent of 370 µSv (no details of how EDE was calculated were given) for four slices and a scout view, using a Siemens Somaton DR2-G. They used 125 kVp, 60 mAs and 240° tube rotation in the data acquisition. They reported a computed tomographic dosimetric index (CTDI) of 5 mGy. The QCT BMD doses are similar in magnitude to other normal radiological examinations such as abdominal AP X-ray of 700 µSv, but lower than normal CT examinations, such as 3300 µSv for the lumbar spine [13].

The high doses encountered during normal imaging CT examination is necessary in order to obtain the low contrast resolution required to separate lesions from normal tissue and to use CT to its optimal diagnostic potential. However, when measuring high Z material such as bone, high doses such as those used for imaging are generally not necessary. Also the use of a large region of interest (ROI) to determine mean CT numbers for these studies allows much lower X-ray doses. Because bone mineral sensitivity goes up and dose goes down at lower kVp, the lowest possible setting is recommended for BMD measurements [30].

3.2.7. Patient dose from other techniques

For other techniques such as RA, PQCT and SXA no effective dose calculations have been reported, only entrance surface doses. Muller *et al* [31] reported an ESD of 100 µGy for a low dose pQCT. For a SXA DTX-100 (Osteometer A/s Rodovre, Denmark), an ESD of 16.6 µGy (40 kVp, 0.2 mA) has been reported [32]. Since very few radio-sensitive organs are irradiated using peripheral techniques, the ED is bound to be very low. For example, only a small fraction of these radio-sensitive tissues are exposed: skin, red bone marrow and bone surfaces.

For superseded techniques such as SPA and DPA, at the most only absorbed doses have been reported. These doses are dependent upon the strength of the source and the beam characteristics. The peak skin entrance dose for DPA ranged from 50 to 150 µGy, while for SPA of the radius it was about 50 µGy [33].

3.3. STAFF DOSE FROM DXA

In the United Kingdom, the maximum annual dose for a non-classified worker is 15 mSv, corresponding to a time-averaged dose rate in the workplace of 7.5 µSv h^{-1} [34]. If the dose rates approach this limit then the working area is defined as a Controlled Area and requires environmental monitoring and a written system of work. This includes monitoring the radiation exposure of staff members entering this area by ways such as TLD batches. At lower dose rates of up to $5\,\mathrm{mSv\ year^{-1}}$ ($2.5\,\mathrm{\mu Sv\,h^{-1}}$) the lesser Supervised Area requirement applies. In 1990 the International Commission on Radiological Protection (ICRP) recommended an annual dose limit of $1\,\mathrm{mSv\ year^{-1}}$ (equivalent to $0.5\,\mathrm{\mu Sv\ h^{-1}}$ in the work place) for members of the public [1].

For pencil beam systems such as the Lunar DPX and the Hologic QPR 1000, a time-averaged dose to staff from scatter is very low, even with the operator sitting as close as $1\,\mathrm{m}$ from the patient without shielding during scanning [5, 35]. For the Lunar DPX-L scanner, the maximum scattered dose-rate to an operator at $1\,\mathrm{m}$ from the patient for continuous exposure would be less than $1\,\mathrm{\mu Sv\,h^{-1}}$ [5]. The typical maximum workload for this type of scanner is 16 patients per day with two views per patient, with the X-ray beam on for $4\,\mathrm{min}$ per view. Thus, the maximum annual work-load would be approximately $533\,\mathrm{h}$ per year. The annual dose to the operator at $1\,\mathrm{m}$ would therefore be less than $0.4\,\mathrm{mSv}$ per year, which is similar to the National Radiological Protection Board (UK) suggested constraints of $0.3\,\mathrm{mSv}$ per year to members of the public from a single source [26]. Therefore, no protective screen for the operator, or additional protection for the public in adjoining areas, is necessary.

Associated with the increase in patient dose and higher patient throughput with fan beam DXA is a greater occupational hazard to staff from scattered radiation [5, 35]. The scattered dose from the fan beam systems such as the QDR 4500 and Lunar Expert XL is considerably higher and approaches limits set by the regulatory authorities for occupational exposure. The scattered dose-rate for the Lunar Expert scanner at $1\,\mathrm{m}$ from the patient is $64\,\mathrm{\mu Sv\,h^{-1}}$ (scan width $180\,\mathrm{mm}$) [5]. If one allows for a nominal scan width of $120\,\mathrm{mm}$, the scattered dose rate reduces to $42\,\mathrm{\mu Sv\,h^{-1}}$. Considering a maximum daily beam on time of $60\,\mathrm{min}$, the time average dose rate at $1\,\mathrm{m}$ would be approximately $5\,\mathrm{\mu Sv\,h^{-1}}$ (using an $8\,\mathrm{h}$ day). On this basis, the Controlled Area, as required by IRR 1985 [34], would not need to extend beyond $1\,\mathrm{m}$ around the radiation beam. However, in order to ensure that the operator receives an annual dose of less than $1\,\mathrm{mSv}$ (the new ICRP proposed limit for members of the public) [1], the operator would need to have a protective screen or be at least $3.5\,\mathrm{m}$ from the beam. Thus, in a relatively small room, to achieve doses as low as reasonably achievable, a protective screen offering a small degree of attenuation would be required. On the other hand, with a reduced scan time and workload, the daily beam-on time could be as low as $20\,\mathrm{min}$, in which case a protective screen would not be necessary, provided the operator is at least $2\,\mathrm{m}$ away from the beam. A similar staff dose analysis has been made for Hologic QDR 2000plus and QDR 4500 by Blake *et al* [35] and found to be 2.1 and $2.4\,\mathrm{\mu Sv\,h^{-1}}$ respectively. This occupational dose was found to be similar to that from a $^{99}\mathrm{Tc^m}$ MDP radionuclide bone scan (table 3.9).

3.4. STAFF DOSE FROM OTHER TECHNIQUES

Similar or lower scattered radiation dose have been reported for peripheral scanners such as SXA. Kelly *et al* [32] reported the scattered radiation to be less than 1 mSv at a distance of 10 cm from all sides of the scanner. Scattered dose assessment is important in the examination of the radius because of the proximity of sensitive organs such as the breast to the scattered radiation.

3.5. REDUCTION OF OCCUPATIONAL DOSE

Reduction of occupational exposure can be achieved by making appropriate use of shielding, distance, time and radiographic method as follows:

- *Shielding*: Shielding is the most important for protection of operators and members of the general public from unnecessary radiation exposure. When installing a fan beam DXA system, active precautions to reduce dose to staff such as the use of a radiation barrier should be considered
- *Distance*: Increasing the distance from the source of radiation can drastically reduce the radiation exposure. So, make use of the inverse square law by having a scanning room large enough to place the operator at least 3 m from the patient.
- *Time*: As expected, the less time one is exposed to radiation, the less dose is acquired. Use of scan modes with shorter scan times has been observed to approximately halve the dose for QDR systems [4].
- *Method*: Patel *et al* [35] have observed that scanning the patient's right hip in preference to the left can give a threefold reduction in potential occupational dose.

3.6. DOSE REDUCTION TECHNIQUES IN DXA APPLICATIONS

To minimize patient doses, as for instance required by the UK Government Ionizing Radiation Regulation (IRR) of 1988 [36], the smallest possible scan width and length should be used. For the Lunar Expert a range of collimation widths are available. Njeh *et al* [5] have been able to show that a 60% reduction in effective dose can be achieved by using smaller field sizes. The Lunar DPX-L scanners have an 'auto scan width' option which automatically narrows the width of the scan path when it locates the patient's bone mass. Using this facility will reduce the area exposed and thus the effective dose.

For the femur, the greatest contributor to the ED are the ovaries, but since a large number of female patients are post-menopausal the radiation hazard of concern is carcinogenesis, the genetic risk to future offspring no longer being a factor. Nevertheless, excluding the ovaries from the calculation results in a greater than 70% reduction in the ED and a good working protocol should exclude the ovaries from the scanning field for both femur and AP lumbar spine. This may not be possible for the fan beam femur scans.

The operator should be adequately trained in order to be able to locate the landmark for proper positioning. This will avoid repeated starting and terminating of a scan. Repeat measurements should be justified in terms of fast bone losses, borderline bone mineral density and monitoring for treatment efficacy.

3.7.   PROBLEMS WITH MEASURING PATIENT AND STAFF DOSE FROM ABSORPTIOMETRIC TECHNIQUES

Various detectors have been used to measure skin entrance and depth doses from DXA. These include ionization chambers [4, 18, 37], proportional counters [38], GM tubes [7], thermoluminescent dosimeters (such as LiF and LiBO) [5, 18] and film [9]. Lithium borate ($Li_2B_4O_7$:Mn) TLDs have an improved energy response compared with lithium fluoride (LiF). It also has better tissue equivalence than LiF, with an effective atomic number of 7.4 compared with 7.42 for tissue and 8.14 for LiF. For the dosimetry of X-rays with energies below 100 keV as used in DXA, the difference in effective atomic number between TLDs and tissue leads to a difference in mass attenuation coefficient and thus errors in measured dose [39].

The physical properties of the DXA machine pose a number of practical problems with regards to dosimetry. First, the dose rate during scanning is very low and multiple scans have been used to increase the radiation flux [5] above the detection limit. This is achievable for ESD but may be problematic for depth dose as the dose rate drops exponentially. For instance, Njeh *et al* [18] exposed TLDs to 60 and 167 scans for the PA spine and total body respectively using paediatric mode on a Lunar DPX-L in order to achieve a measurable dose. For first-generation DXA, the beam size is small and precise positioning of the detector in relation to the small DXA beam is required.

The accuracy of the effective dose estimates depends not only on the accuracy of the measurement of the dose rate at the surface and at depths but also on the correct determination of the relative positions, fractions and depths of the individual radio-sensitive organs/tissue exposed. It is apparent from the above that computation of effective doses is generally time consuming and cumbersome. Current computer programs compute EDs for common radiological examination but not for BMD measurements. Huda and Morin [6] have suggested a method of estimating ED using ESD, kVp and filtration information.

3.8.   CONCLUSION

Osteoporosis is a consequence of decrease in bone mass and leads to mechanical weakness of the bone. Management of this can be improved by assessment of bone mass using absorptiometric techniques such as RA, SXA, DXA and QCT. Availability and ease of use have made DXA the most widely used technique for measurements of bone density in clinical trials and epidemiological studies. Studies of radiation dose to patients from DXA confirms that patient dose is small compared with that given by many other diagnostic investigations involving ionizing radiation. Despite the increase in dose associated with the development of fan beam technology, patient dose is still relatively small. QCT gives rise to doses which are comparable with simple radiological examination such as the chest X-ray but lower than imaging CT. Radiation dose from other techniques such as RA and SXA are of the same order of magnitude as pencil beam DXA.

REFERENCES

[1] International Commission on Radiological Protection 1991 *1990 Recommendations of the International Commission on Radiological Protection* (Oxford: Pergamon Press)

[2] International Commission on Radiological Protection 1977 *Recommendations of the International Commission on Radiological Protection* ICRP publication 26 Report No 3

[3] Kalender W A 1992 Effective dose values in bone mineral measurements by photon absorptiometry and computed tomography *Osteoporos. Intl.* **2**(2) 82–7

[4] Lewis M K, Blake G M and Fogelman I 1994 Patient dose in dual x-ray absorptiometry *Osteoporos. Intl.* **4**(1) 11–15

[5] Njeh C F, Apple K, Temperton D H and Boivin C M 1996 Radiological assessment of a new bone densitometer—the Lunar EXPERT *Br. J. Radiol.* **69**(820) 335–40

[6] Huda W and Morin R L 1996 Patient doses in bone mineral densitometry *Br. J. Radiol.* **69**(821) 422–5

[7] Pye D W, Hannan W J and Hesp R 1990 Effect dose equivalent in dual X-ray absorptiometry [letter] *Br. J. Radiol.* **63**(746) 149

[8] Rawling D J, Faulkner K and Chapple CL 1992 The influence of scan time on patient dose and precision in bone mineral densitometry. In *Current Research in Osteoporosis and Bone Mineral Measurement II* ed E F G Ring (London: British Institute of Radiology) pp 23–4

[9] Bezakova E, Collins P J and Beddoe A H 1997 Absorbed dose measurements in dual energy X-ray absorptiometry (DXA) *Br. J. Radiol.* **70** 172–9

[10] Richardson R B 1990 Past and revised risk estimates for cancer induced by irradiation and their influence on dose limits *Br. J. Radiol.* **63**(748) 235–45

[11] Stewart SP, Milner D, Moore A C, Emery P and Smith M A 1996 Preliminary report on the Lunar Expert-XL imaging densitometer: dosimetry, precision and cross calibration. In *Current Research in Osteoporosis and Bone Mineral Measurement IV*

eds E F G Ring, D M Elvins and A K Bhalla (London: British Institute of Radiology) pp 101–2

[12] Blake G M and Fogelman I 1997 Technical principles of dual energy X-ray absorptiometry *Semin. Nucl. Med.* **27**(3) 210–28

[13] Shrimpton P C, Wall B F and Hart D 1998 Diagnostic medical exposures in the UK *Applied Radiation and Isotopes*

[14] Blake G M, Rea J A and Fogelman I 1997 Vertebral morphometry studies using dual-energy x-ray absorptiometry *Semin. Nucl. Med.* **27**(3) 276–90

[15] Falcini F, Trapani S, Civinini R, Capone A, Ermini M and Bartolozzi G 1996 The primary role of steroids on the osteoporosis in juvenile rheumatoid patients evaluated by dual energy X-ray absorptiometry *J. Endocrinol. Invest.* **19**(3) 165–9

[16] Seeman E, Young N, Szmukler G, Tsalamandris C and Hopper J L 1993 Risk factors for osteoporosis *Osteoporos. Intl.* **3** Suppl 1 40–3

[17] Laskey M A 1996 Dual-energy X-ray absorptiometry and body composition *Nutrition* **12**(1) 45–51

[18] Njeh C F, Samat S B, Nightingale A, McNeil E A and Boivin C M 1997 Radiation dose and in vitro precision in paediatric bone mineral density measurement using dual X-ray absorptiometry *Br. J. Radiol.* **70**(835) 719–27

[19] Koo W W, Walters J and Bush A J 1995 Technical considerations of dual-energy X-ray absorptiometry-based bone mineral measurements for pediatric studies *J. Bone Miner. Res.* **10**(12) 1998–2004

[20] Faulkner R A, Bailey D A, Drinkwater D T, Wilkinson A A, Houston C S and McKay H A 1993 Regional and total body bone mineral content, bone mineral density, and total body tissue composition in children 8–16 years of age *Calcif. Tissue Intl.* **53**(1) 7–12

[21] Martin C J, Farquhar B, Stockdale E and MacDonald S 1994 A study of the relationship between patient dose and size in paediatric radiology *Br. J. Radiol.* **67**(801) 864–71

[22] Kyriou J C, Fitzgerald M, Pettett A, Cook J V and Pablot S M 1996 A comparison of doses and techniques between specialist and non-specialist centres in the diagnostic X-ray imaging of children *Br. J. Radiol.* **69**(821) 437–50

[23] Chapple C L, Faulkner K, Lee R E and Hunter E W 1993 Radiation doses to paediatric patients undergoing less common radiological procedures involving fluoroscopy [see comments] *Br. J. Radiol.* **66**(789) 823–7

[24] Ruiz M J, Gonzalez L, Vano E and Martinez A 1991 Measurement of radiation doses in the most frequent simple examinations in paediatric radiology and its dependence on patient age *Br. J. Radiol.* **64**(766) 929–33

[25] Hart D, Jones D G and Wall B F 1996 Coefficient for estimating effective doses from pediatric X-ray examinations (London: NRPB) HMSO Report No R279

[26] NRPB 1992 *1993 Documents of the NRPB* (vol 4) *Occupational, Public and Medical Exposure* (Chilton: National Radiological Protection Board)

[27] Banks L M and Stevenson J C 1986 Modified method of spinal computed tomography for trabecular bone mineral measurements *J. Comput. Assist. Tomogr.* **10**(3) 463–7

[28] Shrimpton P C, Jones D G, Hillier M C, Wall B F, Le Heron J C and Faulkner K 1991 *1991 Survey of CT Practice in the UK.* Part 2. *Dosimetric Aspects* (Chilton, London: NRPB) Report No R249

[29] Karantanas A H, Kalef-Ezra J A and Glaros D C 1991 Quantitative computed tomography for bone mineral measurement: technical aspects, dosimetry, normal data and clinical applications *Br. J. Radiol.* **64**(760) 298–304

[30] Cann CE 1988 Quantitative CT for determination of bone mineral density: a review *Radiology* **166** 509–22

[31] Muller A, Ruegsegger E and Ruegsegger P 1989 Peripheral QCT: a low-risk procedure to identify women predisposed to osteoporosis *Phys. Med. Biol.* **34**(6) 741–9

[32] Kelly T L, Crane G and Baran D T Single 1994 X-ray absorptiometry of the forearm: precision correlation and reference data *Calcif. Tissue Intl.* **54**(3) 212–8

[33] Huddleston A L 1988 *Quantitative Methods in Bone Densitometry* (Boston: Kluwer Academic)

[34] IRR 1985 The Ionising Radiation Regulations (London: HMSO) 1985 Report No Statutory Instruments 1333

[35] Patel R, Blake G M, Batchelor S and Fogelman I 1996 Occupational dose to the radiographer in dual X-ray absorptiometry: a comparison of pencil-beam and fan-beam systems *Br. J. Radiol.* **69**(822) 539–43

[36] IRR 1988 The Ionising Radiation (Protection of Persons Undergoing Medical Examination) Regulations (London: HMSO) Report No Statutory Instruments 778

[37] Sorenson J A 1991 Relationship between patient exposure and measurement precision in dual-photon absorptiometry of the spine *Phys. Med. Biol.* **36**(2) 169–76

[38] Waker A J, Oldroyd B and Marco M 1992 The application of microdosimetry in clinical bone densitometry using a dual-photon absorptiometer *Br. J. Radiol.* **65**(774) 523–7

[39] Kron T 1994 Thermoluminescence dosimetry and its applications in medicine. Part 1. Physics, materials and equipment *Australas. Phys. Eng. Sci. Med.* **17**(4) 175–99

Chapter 4

Instrument evaluation

Christopher F Njeh and Didier Hans

4.1. INTRODUCTION

Physical measurement is a fundamental activity of everyday life and an essential component of science and medicine. In medicine, measurement plays a vital role in patient management. Although simple procedures like blood pressure and blood cholesterol level are taken for granted, the reliability and validity of these measurements are very important for the practice of medicine. In this book, various physical techniques for bone measurements will be discussed. These include *in vivo*, *in vitro*, invasive and non-invasive techniques. Any physical measurement has associated with it some degree of uncertainty (measurement error). These measurement errors affect the accuracy (validity) and precision (reliability) of a measurement. So, for the results measured with any technique to be believable, the level of uncertainty has to be reported or established.

This chapter will introduce the concepts of precision and accuracy as needed in instrument evaluation. Another crucial step in instrument validation is to check if the device is conforming with its clinical expectation and also if practical guideline is provided for immediate clinical use. Also elements of quality assurance (QA) as applied to these techniques will be discussed. It must be noted that while most of the discussion in this chapter relates to dual X-ray absorptiometry (DXA) and quantitative ultrasound (QUS), the concepts are generalizable to most physical measurement techniques.

4.2. MEASUREMENT ERRORS

It is familiar knowledge that the specification of a physical measurable quantity requires at least two items: a number and a unit. Frequently

91

neglected is a third term of virtually equal importance: an indication of the reliability or degree to which we can place confidence in the value stated. To understand and appreciate the methods of measurement analysis, it is important to realize that the numerical value of a measurement is not a unique number like the integer 3275. On the contrary, whether it results from a single trial or from a repeated observations, a measurement is only a sample from the population of all possible observations. It is subject to statistical fluctuations due to environmental and other agencies; is obtained by the use of instruments that cannot be made completely free of error and involves the observer, a fallible human. Even the choice of unit may introduce small errors due to uncertainty associated with the very specification or definition of the unit itself.

Measurements can simply be categorized into two groups, namely direct or indirect measurements. Direct measurements are the results of direct comparison, usually with the aid of instruments, of an unknown amount of a physical entity x with a known or standardized amount of the same entity s. Indirect or derived measurements result from calculation of a value as a function of one or more direct measurements. A very simple example of an indirect measurement is the determination of the volume of a sphere from a direct measurement of its radius. Similarly bone mineral density (BMD) is derived from the measurement of the attenuation of X-rays.

4.2.1.   Types of measurement error

Measurement errors can be broadly placed into two classes, systematic errors and random errors.

4.2.1.1.   Systematic errors

These are due to assignable causes and are, at least in principle, determinable or correctable if enough is known about the physics of the process. They are called systematic because they result in a consistent effect, e.g. values that are consistently too high or too low. Mainly three types of effect can be observed:

- *Errors without drift (on/off)*: These errors cause a complete malfunction of the device and are permanent. They are easy and immediate to detect because the system does not work at all.
- *Errors with drift*: The entity of the error increases over time and is not reversible. For example, as components age and equipment undergoes changes in temperature or sustains mechanical stress, critical performance gradually degrades. This is called drift. When this happens your test results become unreliable. Whilst drift cannot be eliminated, it can be detected and contained through the process of quality control (QC).

These types of error need to be detected as soon as they occur and possibly corrected *a posteriori*.

- *Error with shift*: These errors could be characterized as a 'by step' error, where systematic jump could be observed in the device performance usually because of a known condition, e.g specific malfunction, software change etc. They are easy to detect through the process of QC and possibly corrected *a posteriori*.

Most systematic errors fit neatly into one of four main categories: theoretical, instrumental, environmental and observational.

- *Theoretical errors*: These errors are concerned with the equations or relationships used in designing or used in the determination of indirect measurements. Generally, theoretical errors are allowed and tolerated (or corrected for) to enable the use of approximately correct equations or relations which may replace unattainable or very complex formulations. Science makes most advances by approximations—study the most important factors first; correct for secondary factors to improve matters later.

- *Instrumental and observational errors*: Instrumental errors are self-explanatory and are those due to the technique used. Instrument manufacturers will more often than not specify the instrument errors of their devices. By the design of instruments and the instructions supplied, manufacturers attempt to make it difficult for the observer to make errors, by reducing the need for judgement by automating as much as possible the user software. Still, such errors occur when an observer consistently reads too high, too low, too early or too late. Adequate training, careful attention, team checking and comparison of one observer with another are standard means for reducing or eliminating observational error.

- *Environmental errors*: A well-designed and beautifully engineered instrument, together with well thought out experimental procedure, may be drastically affected if the environmental impact is not evaluated. Certainly any experimenter must face the fact that changes in the properties of the environment will affect not only the behaviour of his/her instruments and the relations between his/her measured quantities but even that which he/she measures. The major environmental factors include temperature, pressure and humidity. To eradicate the environmental errors, two simple solutions are to isolate the experiment or control the environment. If these are not options, one could calculate the error from known and measured behaviour and apply a correction factor. Usually, to prevent such error the manufacturer will give the normal working range in the operating manual in terms of temperature, pressure and humidity. These conditions would need to be respected.

4.2.1.2.   Random errors

These are due to the summation of large numbers of small, fluctuating individual disturbances which combine to give results, which are too high at one time and too low at another. These errors, without drift, cause a random malfunction of the device and are often reversible. They can be difficult to detect as they are not permanent; when the cause of error stops, the system returns to normal functioning. Only a stringent and accurate quality control (QC) procedure and possible preventive maintenance enable the detection of such errors.

4.3.   EQUIPMENT VALIDATION

Equipment evaluation could be split into two main steps.

- Basic evaluation, characterizing the measurement errors associated with the measurement technique or device. The errors associated with any technique or device are the same as those discussed in the preceding section. However, these errors are sometimes discussed under two categories—accuracy and precision errors.
- Clinical evaluation aimed at determining the clinical utility of the technique. This more in-depth validation is not always required but is definitely of great importance in the case of the introduction of new technologies into your medical specialty. It will require the know-how of the right clinical question and also of which already-established method you may want to use as a gold standard. Of course, normal reference curves would not have to be left aside.

4.3.1.   Precision

Precision errors (also termed reliability) reflect the reproducibility of the technique. They measure the ability of a method to measure a parameter consistently over multiple measurements, so a precise measurement has nearly the same value each time it is measured. However, there is no quantitative measurement technique in clinical medicine that is perfectly reproducible. Precision error is usually expressed as a coefficient of variation. Imprecise measurement methods have a large coefficient of variation.

For bone densitometry, good precision is clinically crucial in the evaluation of serial measurements for longitudinal change (with time and/or therapy) to guarantee that any observed bone mineral density changes are real and not due to machine and/or operator variability. In the clinical context precision is often more important than accuracy.

Precision is also relevant to the method-comparison study (see section 4.4.1) because the repeatability of two methods of measurements limits the

Table 4.1. *Example of factors contributing to precision errors.*

Instrument variability
    Nonlinearity of measurement system response
    Electronic drift or shift
    Noise from environment factors such as temperature

Observer variability
    Positioning errors
    Analysis errors
    Parallax error reading

Subject variability
    Such as important fat or weight changes between visits for DXA
    Water content change
    Presence of artefacts
    Timing variability between paired measurements

---

amount of agreement that is possible. If one method has a poor precision— i.e. there is a considerable variation in repeated measurements on the same subject—the agreement between the two methods is also bound to be poor. Precision also has an important influence on the powers of a clinical study that uses that technique. The more precise a measurement, the greater is the statistical power at a given sample size to estimate mean values and to test hypotheses. There are three major sources of precision error: instrument related, technologist or observer and patient related. The global error accrued from these three precision errors can be estimated by the root mean square of the square of each individual error.

Instrument variability refers to variability in the measurement due to fluctuating environmental factors such as temperature and background noise.

Observer variability refers to variability in measurement that is due to the observer, including such things as patient position or hand–eye coordination in using a mechanical instrument.

Lastly *subject variability* refers to intrinsic biological variability in the study subjects due to such things as fluctuations in mood, change in body fat etc. (see table 4.1).

4.3.1.1. *Short-term precision*

Precision errors can be further separated into short-term and long-term. Short-term precision errors characterize the reproducibility of a technique and in bone densitometry are useful in describing the limitations of measuring changes in skeletal status. Generally short-term precision errors are assessed from measurements performed either on the same day or extending over a period of no more than two weeks. Precision can be expressed as follows.

- Absolute (standard deviation): the absolute precision errors are very useful in comparing the same parameter. However, because the absolute precision error depends on the unit of measurement, it is not adequate for comparison across techniques or measurements.
- Relative (%): we are often interested in the relative precision of a technique in discriminating osteoporotic patients or monitoring bone structure changes.

Short-term precision is determined by calculating the mean and standard deviation (SD) of two or more repeated measurements on each subject. The individual means and SDs for all the subjects are then averaged. However, since precision was measured for different subjects, there is the question of how to pool them. Gluer *et al* [1] reported that the correct estimate is not given by the arithmetic mean of the individual subjects' precision errors, but by root-mean-square r.m.s. SD. However, it is a popular convention to express precision data as the coefficient of variation (CV, relative) by dividing the SD by the mean for all subjects and expressing the result as a percentage [1, 2]. In the special case where two measurements are made on each subject, the CV can be expressed as

$$\mathrm{CV_{r.m.s.}} = \frac{\sqrt{\sum_{j=1}^{m} \dfrac{d_j^2}{2m}}}{\sum_{j=1}^{m} \dfrac{\bar{x}_j}{m}} \times 100\% \tag{4.1}$$

where d_j is the difference between the first and the second measurements for the duplicate measurements, m is the number of paired measurements and $\bar{x}$ is the mean of the paired measurements.

4.3.1.2. Long-term precision

Long-term precision errors are used to evaluate instrument stability. The assessment of long-term precision is complex since the variability of the data may be due to imprecision of the technique as well as to true biological changes in the tested variables. For repeated measurements taken on the same subject over time, the variability about the regression curve (i.e. the standard error of the estimate (SEE)) could be taken as an estimate of the person's long-term precision error. This is expressed as [3]

$$\mathrm{CV_L} = \frac{\sqrt{\sum_{j=1}^{m} \dfrac{\mathrm{SEE}_j^2}{m}}}{\sum_{j=1}^{m} \dfrac{\bar{x}_j}{m}} \times 100\% \tag{4.2}$$

where SEE results were predicted from the regression model.

Even when the underlying changes in tested parameter are truly linear, long-term precision would be expected to be larger than short-term precision due to small drifts in instrumental calibration, variation in the patient positioning and changes in soft tissue composition [3].

4.3.1.3. *Standardized precision*

Precision expressed as a ratio of standard deviation and mean does not take into account the different biological range or responsiveness of the different systems [4] not always allowing inter-device comparison. A technique with poor precision but with the ability to demonstrate larger changes (responsiveness) over time is preferable to a competing high precision and low responsiveness approach. Responsiveness depends on many factors including the measurement site and the technique. Different approaches have been suggested to standardize CV for easy comparison of different systems, although no consensus has been reached [4]. One suggested approach is to divide the CV by the annual percentage change in the parameter due to age (responsiveness) or treatment as derived from normative data [5]. The weakness of this approach is that the annual rate of change provided by manufacturer has been obtained on different populations and is cross-sectional and in addition the annual bone loss will change depending on the age range chosen (e.g. early post-menopausal versus peak bone mass period). A true comparison could be derived if this change were obtained on the same population and over the same age range. On the other hand, Miller *et al* [6] suggested standardizing precision errors by calculating the ratio of percentage precision to percentage range of results (5th–95th percentile).

$$sCV = \frac{CV\%}{4SD/Mean_{pop}} \tag{4.3}$$

where SD is the population standard deviation and mean is the average for all subjects.

One criticism of this approach is that it depends on the spread of the subject group and so the results are difficult to compare with other reports. Recently, Gluer *et al* [4] suggested using:

$$sCV = CV \frac{\text{response rate (reference technique)}}{\text{response rate (technique studies)}} \tag{4.4}$$

where CV is the uncorrected precision error and sCV the standardized precision error of the technique being studied. Gluer [4] suggested that for bone densitometry the ideal reference technique is the posterior–anterior DXA of the lumbar spine, since it has the lowest reported CV.

4.3.1.4   Methods to improve precision

The methods used to increase the measurement precision will depend on the technique. Some of the methods discussed here relate mainly to bone absorptiometry.

1. *Standardize the measurement.* All study protocol, *in vivo* or *in vitro*, should include operational definitions—specific instructions for making the measurements. There is a need to write out the directions on how to prepare the environment and the subject or sample, how to calibrate the instrument and so forth. For example, in QUS measurement, specific instructions like cleaning the foot for *in vivo* measurement or degassing the sample for *in vitro* measurement. This set of materials termed the 'operations manual' is essential. Even when there is only a single observer, specific written guidelines for making each measurement will help the observer/operator performance to be uniform over the duration of the study.
2. *Train and certify the operators/observers.* Training will improve the consistency of measurement techniques, especially when several observers are involved. It is often important to test the mastery of the techniques specified in the operations manual and to certify that observers have achieved the prescribed level of performance.
3. *Refine the instruments.* Mechanical and electronic instruments can be engineered to reduce variability.
4   *Automate the instrument.* Variation in the way human observers make measurements can be eliminated with automatic mechanical devices and self-running software. This strategy will improve precision only if the automatic device is itself relatively precise, so that its measurements have less variability than those made by human observers.
5. *Repetition.* The impact of random error of any source can be reduced by repeating the measurement and using the mean of the two or more readings. The precision can be substantially increased by this strategy, the primary limitations being the added radiation exposure in case of DXA and cost. Such an approach will not be possible for destructive *in vitro* measurements.

4.3.1.5.   How to carry out a precision study

When performing a precision study, one has to determine the minimal number of subjects required to characterize the technique. Secondly one has to decide on the characteristics of the subjects. Before proceeding to describe sample size and subjects, it is worth identifying two more subsets of precision, *in vivo* and *in vitro*.

•   *In vitro* precision studies are those carried out on a phantom. This characterizes the intrinsic reproducibility of the technique and hence the inherent/random errors caused by the technique. So, to carry out *in vitro* precision studies, a phantom is measured at least 10 times in a row without repositioning. Precision is calculated as discussed above by dividing the standard deviation over the sample by the mean of this sample. Normally *in vitro* precision is always better than *in vivo* because observer and subject sources of error are no longer present.

• *In vivo* precision studies are those carried out on live subjects. For individual patients, the precision error can help in comparing two measurements to determine if there is a true change in the tested parameters or whether the difference is due to the precision errors. Precision errors can also be used to design clinical studies or to evaluate changes in patient bone status.

To estimate the *in vivo* precision in clinical use of a specific parameter, we need to measure a representative set of individuals and combine their individual precision errors. However, since the individual precision may vary between subjects because of many factors such as age, anthropometric characteristics and so forth, a well designed study starts with a solid understanding of its goals and the selection of the right sample population to be tested. For example, if one wants to establish the minimum significant difference to detect in bone density between two visits for women under HRT, the choice of the population to estimate the *in vivo* precision error would need to be identical to the patients monitored in your facilities, i.e. post-menopausal women. Using a Monte Carlo simulation, Glüer *et al* [1] suggest estimating precision at a minimum 90% of confidence. To achieve this result the sample size needs to be based on either 14 or 27 subjects having triplicate or duplicate measurements respectively with repositioning. (See www.iscd.org for precision calculator.)

4.3.2 Accuracy

Accuracy (also termed validity) is the degree to which the measured result reflects the true value of the quantity being measured. In order to evaluate accuracy errors we need to know the true values of the measured parameters. Accuracy in the strictest sense should only be used when a new method is compared with its true value. For example, comparing BMD measured by one densitometer (such as QDR 4500) with another densitometer (such as QDR 2000) would not render an accuracy value, because varying degrees of measurement errors are inherent in both methods. However, such a comparison can be used to determine how much difference exists between the BMD measured using QDR 4500 and that measured using QDR 2000.

For DXA, accuracy has generally been defined as the degree to which bone densitometry techniques are able to estimate the calcium content. The measured bone mineral content (BMC) is compared directly with the

Table 4.2. *Comparison between precision and accuracy of measurement (adapted from Hulley and Cummings [8]).*

	Precision	Accuracy
Definition	The degree to which a variable has a nearly the same value when measured several times	The degree to which a variable actually represents what it is supposed to represent
How to measure	Comparison among repeated measurements	Comparison with a reference standard
Significant to study	Increase power to detect effects	Increase validity of conclusions or diagnosis
Types of errors	Random errors (caused or contributed by the observer, the subject, the instrument)	Systematic errors (contributed by the observer, the subject and the instrument)

ash weight of the measured bone. One should note that for clinical applications, only the fraction of the accuracy error that varies from patient to patient in an unknown fashion is of relevance. The other fraction, i.e. the one that is constant, can be averaged across subjects and is not important for two reasons. First, for diagnostic use of the technique, the reference data will be affected by the same error and thus the difference between healthy and osteoporotic bone is constant. Secondly for monitoring patient bone change, such an error is present in both baseline and follow-up measurement and does not contribute to the measured change [7]. For these reasons, smaller accuracy errors are of little clinical significance provided they remain constant, which is not always the case. Indeed, for example, body composition accuracy assessed by DXA will largely depend on the patient thickness. Under-estimation of the soft tissue content will be seen in obese subjects, while it is correctly estimated in normal range subjects.

Accuracy is different from precision as listed in table 4.2 and illustrated in figure 4.1, and the two are not necessarily linked. For example, a bone densitometer with a wrong calibration would be inaccurate but could still be precise, i.e. consistently off by a certain factor. Accuracy errors are caused by the systematic errors, so the greater the error, the less accurate the variable. The three main sources of measurement errors discussed for precision are also applicable for accuracy. These are observer bias, subject bias and instrument bias. Instrument bias can result from a faulty function of a mechanical instrument.

The major approaches to increasing accuracy include the first four of the strategies listed earlier for precision, and in addition:

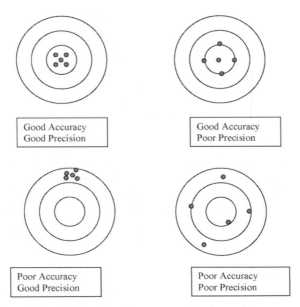

Figure 4.1. *Illustration of the difference in precision and accuracy. The centre of the circle represents the true value and the circles represent the measured values.*

- *Blinding*: This classic strategy does not prevent an overall bias in the measurements, but it can eliminate differential bias that affects one study group more than another. In a classic double-blind experiment, neither the observer nor the subject knows whether active medicine or identical-looking placebo has been assigned.
- *Calibrating the instrument*: The accuracy of many instruments, especially those that are mechanical or electrical, can be increased by periodic calibration using a gold standard.

4.3.3. When are two measurements significantly different?

When performing a patient follow-up over time (under treatment or not), it is important to determine whether the observed differences between several measurements are due to a real evolution or just due to the precision error of the techniques used. Classical statistical approaches could be used such as the Student's *t*-test and so forth, but their major limitation is that they do not take into account the clinical relevance. As an alternative, several authors proposed different methods to achieve the clinical relevance with a certain confidence in the interpretation. These methods are described below.

4.3.3.1   Least significant change

When assessing the suitability of a technique to use to follow up patients, the most important parameter is precision. This is because the least significant change (LSC) between two measurements on a single subject is related to the precision error [9]. This can be expressed as

$$\text{LSC} = Z \times \text{CV}\sqrt{\frac{1}{n_1} + \frac{1}{n_2}} \qquad (4.5)$$

where Z is the value for the desired level of confidence, CV is the precision (could be absolute or relative) and n_1 and n_2 are the number of measurements at each time point. It has been suggested that long-term precision should be used instead of short-term. So, for 95% confidence ($Z = 1.96$) and a single measurement at baseline and follow up, the LSC is given by $1.96\sqrt{2} \times \text{CV}$.

4.3.3.2.   Trend assessment margin

The least significant change is highly dependent on the power (Z) set to detect the change. Gluer [4] has proposed giving the name trend assessment margin (TAM) for a 80% confidence level ($Z = 1.26$) for a two-sided test or 90% confidence for a single-sided test. $\text{TAM} = 1.8 \times \text{CV}$.

4.3.3.3.   Monitoring time interval

The monitoring interval is the time required to observe a significant change in the majority of the patients. This is defined as

$$\text{MTI} = \frac{\text{LSC}}{\text{response rate}}. \qquad (4.6)$$

For a technique with a precision error of 1%, for example, the LSC (95% confidence) will be 2.8%. If the bone loss is 3% per year, the MTI will be 1 year. The concept of LSC and MTI can be useful for the identification of fast losers of bone. The inverse of the MTI (i.e. ratio of the response rate to the precision error) has been coined the term 'longitudinal sensitivity'. However, the response rate has to be based on the same class of age if aimed at comparing different techniques.

4.3.3.4.   Regression to the mean

In interpreting data due to treatment effect, clinicians should be aware of the statistical phenomenon known as regression to the mean. Regression towards the mean occurs whenever we select an extreme group based on one variable and then measure another variable for that group. The second group mean will be closer to the mean for all subjects than the first, and the weaker the correlation between the two variables the bigger the effect will be [10]. This phenomenon is due in part to measurement error and biological variation [11]. The clinical consequence is that a subject that has demonstrated a particularly high BMD gain in the first year after treatment is likely to show a reduced increase or perhaps even a decrease in BMD in the second year. Similarly, a subject who has lost an unexpectedly large amount of bone after the first year of treatment is likely to show less of a loss, even a gain, in the second year. Over a longer period of time, the extremes will be less deviant from the mean because some of the errors of repeated measurements will smooth the results [12]. Cummings *et al* [13] demonstrated this effect using data from two randomized, double-blinded, placebo-control trials of alendronate and raloxifene treatment. They concluded that effective treatment for osteoporosis should not be changed because of loss of BMD during the first year of use.

4.4.   STATISTICAL METHODS IN EQUIPMENT VALIDATION

4.4.1   Method-comparison studies

When new technologies are introduced, it is always desirable to establish their accuracy and precision as previously discussed. Also there is the tendency to determine how the new technique compares with current well-established methodology. In bone densitometry, the literature is full of such studies. For example, comparison between DPA and DXA and between QUS and DXA [14, 15]. These types of study are termed 'method-comparison' studies [16]. The goals of method-comparison studies are to determine

- how well the two methods agree and are repeatable with each other
- what degree of confidence can be placed in the new method.

Method-comparison studies are used to determine the level of agreement between two methods since the true value of the measured variable remains unknown in such studies because of inherent measurement errors of both the examined methods [17, 18]. Poor agreement between comparison values indicates that a large degree of systematic and/or random error is present in either one or both of the measurement methods. Systematic errors (bias) of a measurement method result in consistent under- or over-estimation of values with the comparison value. Systematic errors may arise from different sources such as those listed in table 4.3.

Random errors result in instability and unpredictability of values and are associated with the concept of precision (see section 4.2.1.2). Interaction can occur between precision and accuracy in method-comparison studies. This is because poor precision can limit the degree of agreement methods.

Table 4.3. *Factors contributing to inaccuracy in method-comparison studies: adapted from Szaflarski [16].*

Instrument bias of either method
    Constant (offset) error
    Drift
    Range error
    Response time
    Hysteresis

Observer bias
    Between-observer bias
    Within-observer bias in performing and reporting measurements

Ideally estimates of both accuracy and precision of a new measurement technique should be assessed.

There are various statistical approaches that can be used in the process of evaluating new equipment for clinical use

4.4.2. Bland and Altman plot

This is a statistical approach used to examine the nature and extent of agreement between two methods of clinical measurement [17, 18]. This approach provides estimates of both systematic and random errors of new techniques. The analyses are easy to compute and the associated plots provide an easy way to assess the degree of agreement and to identify outliers and differences between techniques across the observed physiological range.

Bland and Altman [17–19] recommended that paired data from two measurement methods first be plotted against each other with the clinical standard method on the x axis and the new method on the y axis. It is helpful if the scales of the x and y axes are the same so that the line of equality will make an angle of $45°$ to both axes. Regression lines should not be drawn on the plot. The relationship of the data to the line of equality should be closely examined across the measurement range for variability (see figures 4.2 and 4.3).

A second data plot, a Bland–Altman plot, is then created with the mean of both methods on the x axis and the difference of the two methods on the y axis (figure 4.2). It is often helpful to use the same scale for both axes. The mean of both methods is used to represent the best estimates of the true value of the measured variable, because the true value is not known in method-comparison studies. The difference should not be plotted against either value separately because the difference will be related to each, a well known statistical artefact [17].

The differences between methods (y axis) are likely to possess a normal distribution only if there is a random component to the error. The

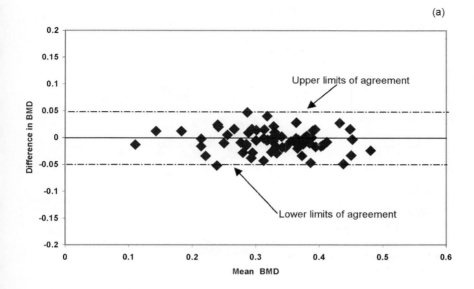

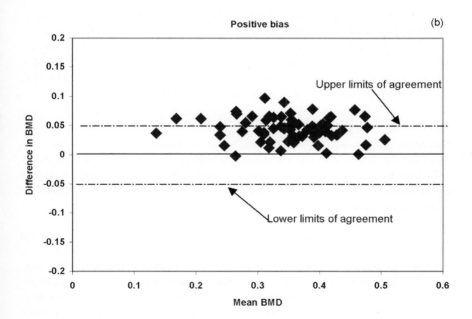

Figure 4.2. *Bland–Altman plot using data from techniques of measuring spine BMD in paediatrics: (a) real data are plotted showing good agreement, (b) one technique artificially added a bias of 0.05, (c) one technique added both positive and negative 0.07 bias, thus creating a poor agreement.*

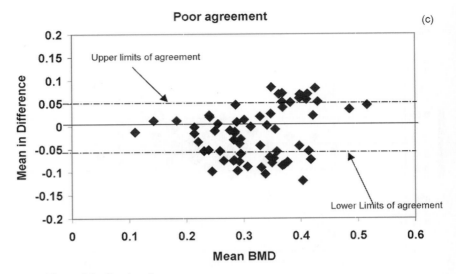

Figure 4.2. *Continued.*

distribution of differences can be checked for normality by constructing a simple histogram (or by using a statistical package). A Bland–Altman plot should then be examined visually for any patterns that may exist in the differences of the two methods (figure 4.2). The possible relationship between the differences and the mean can be investigated formally by calculating the Spearman's rank correlation coefficient (see section 4.4.3)

4.4.2.1.   Limits of agreement using Bland–Altman analysis

The difference between measurements by two methods on the same subject is called bias and is estimated by the mean difference ($\bar{d}$). There will also be variation about this mean, which can be estimated by the standard deviation of the difference (SD_d). If the differences are normally distributed, the 95% limit of agreement is given by

$$\bar{d} \pm 1.96SD_d \tag{4.7}$$

These estimates are meaningful only if we assume that bias and variability are normally distributed throughout the range of measurement, an assumption which can be checked graphically [18]. Provided differences within the observed limits of agreement would not be clinically important, the two measurement methods can be used interchangeably.

The limits of agreement are only estimates of the values that apply to the whole population. A second sample would give different limits. Provided the differences follow a distribution that is approximately normal, then the

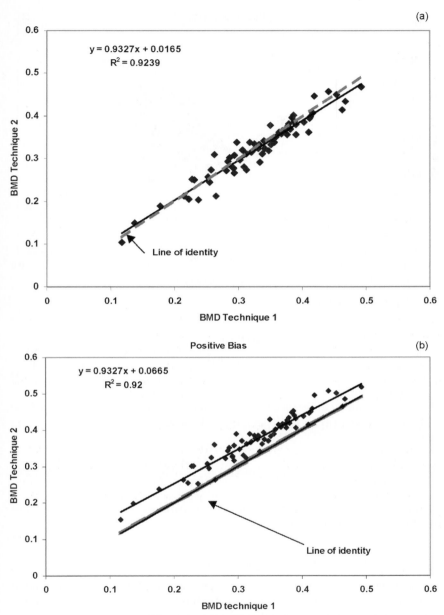

Figure 4.3. *Regression analysis of the data presented in Figure 4.2. (a) Excellent agreement. Note how close the line of identity is to the regression line. (b) Note the same R^2 as in (a). The only differences are in the intercept and displacement from the line of identity. This bias was easily seen in the Bland–Altman plot in figure 4.2(b). (c) Poor agreement between techniques.*

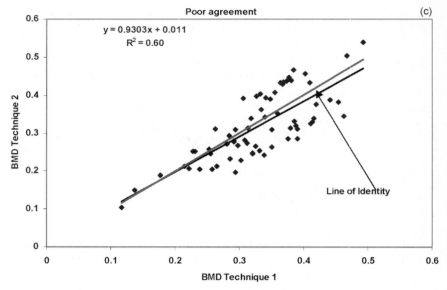

Figure 4.3. *Continued.*

confidence interval (CI) can be given as

$$\bar{d} \pm t\left(\frac{\text{SD}_d}{\sqrt{n}}\right) \tag{4.8}$$

where $\bar{d}$ is the mean difference (bias) and SD_d is the standard deviation of the differences. The formula for calculating CI for limits of agreement ($d - 2\text{SD}$ and $d + 2\text{SD}$) is

$$\text{CI} = (\bar{d} \pm 1.96\text{SD}) \pm t\left(\frac{1.71\text{SD}_d}{\sqrt{n}}\right) \tag{4.9}$$

where t is the appropriate point on the t distribution with $n - 1$ degrees of freedom and n is the sample size.

4.4.3.   *Regression analysis and correlations*

The reader is referred to statistics textbooks for detailed mathematical discussion of the topic [20]. This section will give a basic discussion of correlation and regression in method-comparison studies.

Correlation analysis gives a unitless number that summarizes the strength between two variables, while regression analysis gives a mathematical equation that can be used to estimate or predict values of one variable based on the known values of another variable. The statistic that summarizes the strength of a linear relationship between the two variables is called the correlation coefficient (R). Correlation coefficients range from -1 to $+1$ (R^2 is known as the coefficient of determination). There are various statistical formulas to determine the correlation between two variables but the most common is the Pearson product–moment correlation, which is based on the assumption that the variation comes from a bivariate normal distribution (i.e. each variable is normally distributed) [20, 21]. However, there are other approaches like Spearman's correlation coefficient which is non-parametric and calculates the correlation of two variables according to their rank. Correlation coefficient treats the two variables equally, or in other words the two variables are interchangeable.

Regression analysis considers the ability of one or more variables to predict a particular outcome of interest. The relationship between predictor variables and the outcome of interest may be linear, or one of any number of other mathematical relationships, e.g. logarithmic or quadratic. So simple linear regression is characterized by the effect of one variable predicting the effect on one contributing variable. Multiple regression analysis allows the investigator to assess the simultaneous effect of multiple variables on one continuous variable. Logistic regression assesses the effect of one or more variables on one dichotomous variable.

A linear regression analysis begins with the assumption that the relationship between any two variables may be summarized as a straight line

$$Y = \alpha + \beta X + \varepsilon \qquad (4.10)$$

where ε is the random error around the regression line and β is the regression coefficient or slope. When $\beta = 0$ there is no relationship between X and Y. The best fit line is drawn such that the distances from the data points to the line is minimized. These distances are squared and then summed to generate the least squares regression line. The underlying assumption of regression analysis is that X is given and has no measurement error. Usually, measurement errors exist for both X and Y. Special methods are needed to fit the linear relationship when both X and Y are measured imprecisely [22].

A high correlation between two techniques does not necessarily mean high agreement. Therefore correlation is not the best statistical method to assess the level of agreement between two techniques. Bland and Altman [17, 18] have put forward the following arguments against correlations:

- The correlation coefficient R measures the strength of a relationship between two variables, not the agreement between them. Correlation tests if the two methods are linearly related. Perfect correlation occurs when the points lie along a straight line on a graph. This is in contrast to the notion of agreement where exact agreement occurs only at the line of equality, defined as an intercept of zero and a slope of one.

- A change in the scale of measurement does not affect the correlation but certainly affects the agreement.
- Correlation depends on the range of the true quantity in the sample. If this is wide, the correlation will be greater than if it is narrow.

4.4.4.   Clinical evaluation of a new device

A successful clinical evaluation starts with a solid understanding of its goals and the establishment of appropriate hypotheses to be tested. Then, an appropriate study design is chosen to produce results that will test these study hypotheses and calculate the right sample size for a given power. The two main categories of study design are longitudinal studies which investigate changes over time or cross-sectional studies which mostly focus on describing a state or phenomenon at a fixed or an indefinite time. The latter are usually relatively inexpensive and are easier to conduct for rare events and are the most frequently used in osteoporosis research. For detailed discussion refer to Lu and Gluer [7, 20]

4.4.4.1.   Basic statistical method for baseline characteristic of a new device

When population distributions of the parameters of interest follow normal distributions, comparisons between means and standard deviations of two groups of subjects can use the t-test, paired t-test, and analysis of variance (ANOVA) depending on the study design. In case of a non-normal distribution, you may either try to transform the data by using, for example, a log transformation or you may simply use nonparametric statistical methods, such as the Wilcoxon rank test, the paired Wilcoxon signed rank test, or the Kruskal–Wallis test, etc.

4.4.4.2.   Statistical method for evaluating the diagnostic utility of a new device

The diagnostic utility of a technique depends more on its accuracy than precision. A possible means of diagnosing a specific pathology using a new device is to study the relative position of an individual's value in the population distribution. With a clear definition of a disease or an outcome, the accuracy of a diagnostic technique can be characterized by the probability of correct diagnosis, sensitivity and specificity.

- The sensitivity (or true-positive rate) of a binary test is defined as the probability that a truly diseased subject has a positive diagnosis.
- The specificity (or one minus the false-positive rate), on the other hand, is defined as the probability that a non-diseased subject has a negative diagnosis.

A method with good diagnostic ability should be high in both sensitivity and specificity. However, it is difficult to achieve both. In many cases, the cost for an increase in sensitivity is a decrease in specificity. However, the use of sensitivity and specificity does not allow us to compare two diagnostic tests because the differences may depend on the choices of cut-off values. To obtain the global assessment of diagnostic ability, receiver-operating characteristic (ROC) curves have been adopted in radiology literature. ROC analysis is a procedure derived from statistical decision theory. It plots the true-positive rate (sensitivity) as a function of the false-positive rate (one minus specificity). Thus, it shows the trade off between the sensitivity and specificity of a test. Because a ROC curve removes the effect of different choices of decision criteria and represents them as different points on the same curve, ROC analysis allows direct comparisons between two diagnostic techniques. Estimation of the area under the curve and comparison of it from two other ROC curves can be based on a non-parametric or a parametric approach. The advantages of the area under the curve are that they reflect the diagnostic accuracy over the whole range of possible operating points and have convenient and well-studied statistical properties. A disadvantage, however, is that such an area covers useless ranges of both sensitivity and specificity. The area under the curve becomes unreasonable when two ROC curves cross each other and the gain in the area of one ROC curve is due to the differences in very low specificities or sensitivities. Another limitation of ROC is that it cannot be adjusted for confounding factors such as weight and age, so ROC should only be used on similar population.

Two other diagnostic indices are the predictive values of a positive (PV+) and negative (PV−) diagnosis. A PV+ is the probability that a positively diagnosed subject actually has the disease. A PV− is the probability that a negatively diagnosed subject is actually normal. Both PV+ and PV− can be influenced by the prevalence of a disease and, therefore, may not be objective measurements for assessing techniques by themselves. However, they reflect the utility of a technique in clinical use and are very important to clinicians and epidemiologists.

4.4.4.3. *Statistical method for evaluating the risk assessment of disease*

Discriminant analysis.   One of the classic methods is discriminant analysis based on multivariate normal distribution theory. Discrimination analysis is a technique designed to create a classification rule that minimizes the probability of misclassification between two or more groups. However, discriminant analysis does not provide a summary measure of risk for each individual risk factor, such as the odds ratio from a logistic regression or relative risk from a Cox regression.

Logistic regression.   In such a model, the risk factors are directly linked to the event of the disease or the clinical outcome. The odds ratio of one unit change in risk factor, say bone density in this situation, is simply the exponential of the maximum likelihood estimate normally calculated using statistical packages. Like all the regression analyses, the units of risk factors affect the magnitude of regression coefficients and, therefore, the estimated odds ratios in logistic regression. To make the odds ratio comparable for different techniques or different measurements for the same technique, a unit of one standard deviation of the risk factor has been adopted in bone densitometry literature.

Both logistic regression and discriminant analysis provide an estimate of probability for osteoporotic fractures for a subject, given the values of his or her risk factors. In many applications, logistic regression and discriminant analysis give similar results. When the risk factors are normally distributed in both groups with equal variances and equal between-variable correlation, discriminant analysis is more efficient (but can only be used in cross-sectional studies) and has a greater statistical power than logistic regression. In the presence of discrete covariates and departures from normality assumptions, logistic regression is more efficient.

Proportional hazards model. Because of the nature of many studies, not all patients will have the same length of follow-up in a study. Furthermore, logistic regression or discriminant analysis does not distinguish between two subjects when one of them develops a fracture (or other outcome) many years earlier than the other. To resolve these difficulties, we can use the Cox proportional hazards model, also called the Cox regression model. Different from a logistic regression model for studying the probability of an event of interest, a Cox proportional hazards model studies the relationship between the risk factors and hazard rate of fractures or osteoporosis. Thus, a Cox regression model can only be applied to a prospective study.

Building models with multiple risk factors. The discriminant analysis, logistic regression and Cox regression models can study several risk factors in one model. It is always advisable to include age in the analysis, because it is a risk factor for osteoporosis and osteoporotic fractures and costs nothing to obtain. As a rule, the more variables included in a regression model, the better the agreement of the regression model to the observed data. However, there are some limitations in adding as many variables as one can. First, many risk factors are associated with each other. Therefore, information from one risk factor may already be covered by other risk factors. Second, as more risk factors are included, the regression model becomes more dependent on the observed data, and therefore becomes less generalizable. Thus, it is always important to select a model that contains a minimum of risk factors while maintaining statistical efficiency.

Remark: Nonparametric statistical procedures have been proposed to assess the risk of osteoporosis and osteoporotic fractures. Among them are the recursive partitioning methods, such as the classification and regression tree analysis (CART) and tree structured survival analysis (TSSA).

4.5. QUALITY ASSURANCE (QA)

4.5.1. Introduction

Quality assurance (QA) in medical applications is any method, procedure or approach for collecting, processing or analysing data aimed at maintaining or improving their reliability or validity [23]. Quality control (QC), on the other hand, is an aggregate of sampling and testing procedures based on statistical theory and analysis, and is designed to ensure adequate quality of the finished product. In short, QC is the process by which actual quality performance is measured. QA and QC are not new concepts in either the manufacturing industry or in the practice of medicine. Regular QA checks of equipment in diagnostic radiology and radiation oncology are very common and in some cases are mandatory by federal or state regulations.

The general goal of a good QA programme is to maximize the performance of the operator and reader as well as the reliability of the equipment. Indeed, positioning error is one of the major sources of imprecision in many quantitative techniques, which differs from many radiographic procedures whose only interpretation is based on the expert evaluation of a trained radiologist. As such, strict attention to the performance of the scanners is required. For example, this is important in bone status assessment, partly because changes in the measured variables (BMD, SOS and BUA) due to disease or treatment are relatively small (1–5%). Therefore, measurements of bone status changes have to be very precise because procedural errors, malfunctioning equipment or erroneous data analysis may cause substantial interference, even if the data are erroneous by only a few percentage points. QA of DXA or QUS includes longitudinal machine QC, monitoring of densitometer and software upgrades and review of patient data acquisition and analysis (table 4.4). Monitoring refers to observing an activity in relation to a defined specification, standard or targets.

The complexity of the QA program will depend on whether it is for an individual site or for a multi-centre clinical trial. At an individual site, for example, three types of measurements may be carried out: acceptance testing, routine testing and cross calibration.

- *Acceptance testings* are measurement carried out at the initial installation of the device. These may include accuracy, *in vivo* and *in vitro* precision, power output measurement and radiation safety evaluation. As previously described, *in vivo* measurement should either cover a wide age and body status range (precision has been reported to vary between normal and osteoporotic subjects) or a very specific range depending on the objectives. The acceptance testing has two objectives: firstly to establish baselines for future comparison, secondly to establish that the performance of the machine meets the vendor's initial specifications.

Table 4.4. *Examples of some of the quality assurance requirements for absorptiometry and QUS techniques.*

---

1. QA of DXA/QUS devices
   Machine QC
      Periodic graphical analysis of phantom data
      Periodic status reports on equipment stability: monitoring for trends
      Notification of centres in the events of problematic performance
      Descriptive statistics, tabular cumulative sum (CUSUM), regression analysis
         techniques and visual inspection
   Machine upgrades
      Phantom and patient cross-calibration
   Machine software upgrades
      Cross-calibration

2. QA of data acquisition
   Training of operators
      Patient positioning
      Data input
      Error sources

3. QA of data analysis
      Central review (for cases of multi-centre study)
      Central analysis
      Outlier check

- *Routine testing* includes regular measurement of a QC phantom to detect any drift in the machine, and periodic graphical analysis of QC data. Indeed, over the course of time, a given scanner can 'drift' upward or downward, creating the false impression on an *in vivo* analysis that a patient is gaining or losing bone mineral density. Careful attention to daily QC scan review and scanner performance over time reduces errors and problems which are time-consuming or impossible to correct afterwards.

- *Cross-calibration testing* is simply the comparison of instrument performance to a standard of known accuracy. It may simply involve the determination of deviation from nominal or include correction (adjustment) to minimize the errors. Properly calibrated equipment provides confidence that your products/services meet their specifications. In summary, the cross-calibration will increase production yields, optimize resources, assure consistency and ensure measurements (and perhaps products) are compatible with those made elsewhere. By making sure that your measurements are based on international standards, you promote customer acceptance of your products around the world. For example, in the case of a multi-centre clinical trial, by design, one would observe diverse populations, scanning equipment, and operators, leading to inevitable differences in patient data among centres. To overcome this problem, a cross-calibration of scanner data would be required to allow for pooling of the data and characterization of the demographics of the studied population.

4.5.2. Tools for QA

4.5.2.1. Training and standardization

One of the most simple but powerful tools in QA is the proper training of the technician who will perform the measurements. Indeed, as a foundation and guidepost for the quality of your centre, detailed and standardized procedure manuals would be necessary to acquire accurate, precise and reproducible quantitative measurements, as well as to ensure consistency across technician and/or centres. Such procedure should at least include the following.

- Specific instructions for scan settings.
- Patient positioning.
- Procedures and requirements for repeat scans.
- Instruction on the proper image/data analysis.
- Listing of potential artefacts which may influence the reading and interpretation.
- Instructions for the daily QC.
- Instructions for archiving the data.
- Instructions for scanner maintenance or upgrade.
- A set of tolerance levels—for example the example spine *in vivo* BMD precision should be less than 1%.

The QC protocol should be written for each device and should specify the following.

- The test method.
- The specific equipment to be used.
- Parameters to be tested. This is easy for DXA where only BMD is recorded, but there could be incidences where there are lots of parameters that can be measured. A good example is the QC of a linear accelerator, where mechanical and output measurements are important.
- Frequency of measurement, e.g. daily, weekly etc.
- Responsibilities of different groups of staff, e.g. technologist acquires the data and physicist analyses.
- Expected references values.
- Tolerances, for example spine BMD cv of less than 1%, or 2%.
- Action to be taken, e.g. when to repeat the measurement and when to call the service engineer.

- Rules for documenting all the QC process. For example, a file should be kept to record the results of these investigations, as well as copies of engineers' reports following all maintenance visits.

4.5.2.2. Phantoms

An accepted method of monitoring equipment performance is to measure the parameters of interest using a test object which may be either a standard or a phantom. A standard is an object of known physical, X-ray or acoustic properties, which does not attempt to resemble the anatomy of interest. It is usually a simple geometric form and can be used to test one or more specific aspects of the scanner performance. On the other hand, a phantom should be anthropomorphic, i.e. it should reproduce as close as possible the interaction between the *in vivo* measured skeletal site and the measuring device. As such it is called an 'anthropomorphic phantom'. It is hoped that changes in the scanner that affect the *in vivo* measurement will also affect the phantom measurement and thus be detected by the QC program, and that the variations in the *in vivo* results may be predicted from the phantom-based device calibration. As a result, in theory, one can correct patient measurements for the effect of an unstable device based on the drift or shift of the phantom measurements by applying a correction factor to the patient data. This procedure would guarantee that the patient data reflect the 'biological or therapeutic' reality and are not affected by device malfunctions.

Unfortunately *in vivo* is not always easy to mimic. While anthropomorphic phantoms used for DXA are relatively easy to construct and the interaction between X-rays and a bone-equivalent phantom is well known (e.g. European Spine phantom), this is not the case for quantitative ultrasound for which the theory of propagation through cancellous bone is quite complex and not well understood. As a result, the manufacture of an anthropomorphic phantom specific for ultrasounds is very difficult and expensive. Consequently, today, there are no universally accepted 'ultrasound' phantoms, but only 'manufacturer specific' phantoms which are not anthropomorphic (even if some are getting close, such as the Leeds phantom) and whose daily changes may not reflect what happens *in vivo*.

Nevertheless, phantoms are usually accepted to monitor equipment drift or detect electro-acoustic or electronic malfunction, although the reliability of these phantoms may be affected by external factors such as temperature, ageing and storage. The confidence in the results will then depend highly on the statistical method and alarm thresholds one may apply as well as on the stringent approach of the QC.

4.5.2.3. Statistical methods in QA

All statistical methods related to the acceptance test including precision, accuracy, comparability and so forth have been largely described previously.

However, specific QC statistics will be needed to evaluate the data acquired from routine testing. Generally the scanner software has limits of performance against which each QC test result is compared. In addition, some applications may want to conduct more detailed analysis of the trends in performance using various control chart tests that are available in many standard statistical packages.

The standard QC procedures mostly available are visual inspection, the Shewhart chart with sensitizing rules, the Shewhart chart with sensitizing rules and a filter for clinically insignificant mean changes, a moving average chart and standard deviation, and cumulative sum chart (CUSUM) [24–26]. Many researchers compared these different QC approaches and currently, based on their respective advantages and disadvantages, it is accepted that control charts with visual inspection, the Shewhart chart and CUSUM method are the preferred QC statistical tools. The other tests could be used to evaluate the statistical significance of observed changes and provide estimates of the date and magnitude of change allowing the development of correction factors, if desired.

Visual inspection.   Control charts with visual inspection are useful in determining whether observed variations are within normal scanner performance, although it is relatively subjective and depends on the operator's experience and attention. This can be done by plotting longitudinal BMD data over time and using visual judgement to identify the potential change points created by drifts or sudden jumps.

Shewhart chart.   The Shewhart chart with sensitizing rules and an additional filter for clinically insignificant mean changes, represents one class of test which has been applied in DXA and has the highest specificity but a relatively low sensitivity [27, 28]. Nevertheless it is acceptable at the individual sites provided their scanner performance is followed up by CUSUM analysis (see figure 4.4).

For example, in DXA, a Shewhart chart is a graphic display of a phantom BMD that has been measured over time. The chart contains a centreline that represents the mean reference BMD value (to be chosen over a stable period, usually a month, with x observations). The reference value changes whenever the Shewhart chart indicates an out-of-control signal. The new reference value will then be the mean of the x observations after the date of the signal. The interpretation of Shewhart charts is based on a set of rules which can be used to detect systematic deviation from normal performance. One set of rules which has been applied to DXA scanner QC data is listed in table 4.5 (adapted from Faulkner *et al* [28]).

Once a change point is identified by any of the tests given above, we use the next x observations to generate new reference values and apply the tests on the subsequent data according to the new reference value. They can be applied prospectively or retrospectively and provide an objective, reproducible

Table 4.5. *Shewhart chart criteria for the evaluation of QC charts.*

One phantom measurement more than 1.5% of standard deviation from the established centreline.

Two consecutive phantom measurements either above or below 1% of the established centreline.

Four consecutive phantom measurements more than 0.5% above or below 1% of the established centreline.

Ten consecutive measurements either above or below the established centreline.

evaluation of scanner performance. However, the sensitizing rules increase the sensitivity of the Shewhart chart, but also increase the number of alarms that are clinically insignificant, which is not desirable. To overcome this, a threshold based on the magnitude of mean shifting is proposed as an alternative to reduce alarms due to changes with smaller magnitudes.

CUSUM method. An alternative to the Shewhart method is the CUSUM method. It allows an objective and analytic evaluation of the QC data collected and should be used for monitoring DXA scanner performance at a central quality assurance centre. Indeed, it offers good sensitivity and specificity values. In addition, this method provides an accurate estimate of the time of failure, which facilitates the investigation of cause and the application of correction factors to be applied to the patients as necessary.

Practically, the tabular CUSUM charts which have been used in DXA represent a running sum of the deviations of each day's BMD measurement result from the established mean BMD as mathematically represented below. If we consider X_i as the BMD value of the ith scan, upper one-sided tabular CUSUM $S_H(i)$ and lower one-sided tabular CUSUM $S_L(i)$ can be defined as

$$S_H(i) = \max \left[0, \frac{X_i - \mu_0}{\sigma} - k + S_H(i-1) \right]$$

$$S_L(i) = \max \left[0, \frac{\mu_0 - X_i}{\sigma} - k + S_L(i-1) \right].$$

(4.11)

Here, μ_0 is the reference BMD value, σ is the standard deviation, and k is selected as 0.5. The initial values of $S_H(0)$ and $S_L(0)$ are 0. For an instrument that is in control, normal deviation above and below the mean keeps the sum near zero. However, systematic departure from the mean BMD or a single large deviation will result in a large sum. An alarm is raised when the cumulative sum exceeds a predefined threshold (usually if $S_L(i)$ or $S_H(i)$ is greater than 5). After each alarm, a new mean is established from the first points after the discontinuity and the process continues (see figure 4.4).

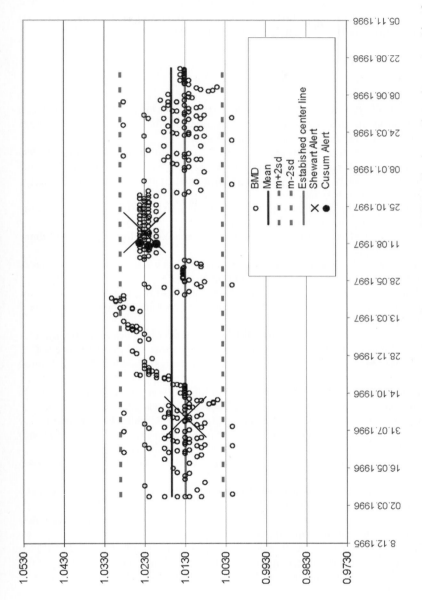

Figure 4.4. *Results of analysis of longitudinal QC data from a densitometer using CUSUM, Shewhart and visual inspection. Identified change points by each technique are indicated by the large cross and filled circle marks on the graph.*

The QC tests discussed above have been compared with one another by different researchers [24–26]. Lu *et al* [25] found CUSUM charts to have the best performance, with high sensitivity and specificity. These charts gave estimates of the time of failure with near zero bias and were second only to Shewart charts in time taken to detect a change.

REFERENCES

[1] Glüer C C, Blake G, Lu Y, Blunt B A, Jergas M and Genant H K 1995 Accurate assessment of precision errors: how to measure the reproducibility of bone densitometry techniques *Osteoporos. Intl.* **5**(4) 262–70
[2] Ravn P, Overgaard K, Huang C, Ross P D, Green D and McClung M 1996 Comparison of bone densitometry of the phalanges, distal forearm and axial skeleton in early postmenopausal women participating in the EPIC Study *Osteoporos. Intl.* **6**(4) 308–13
[3] Blake G M and Fogelman I 1997 Technical principles of dual energy X-ray absorptiometry *Semin. Nucl. Med.* **27**(3) 210–28
[4] Glüer C C 1999 Monitoring skeletal changes by radiological techniques 1999 *J. Bone Miner. Res.* **14**(11) 1952–62
[5] Glüer C C 1997 How to characterize the ability of diagnostic technique to monitor skeletal changes *J. Bone Miner. Res.* **12** (Suppl 1) S378
[6] Miller C G, Herd R J, Ramalingam T, Fogelman I and Blake G M 1993 Ultrasonic velocity measurements through the calcaneus: which velocity should be measured? *Osteoporos. Intl.* **3**(1) 31–5
[7] Lu Y and Glüer C C 1999 Statistical tools in quantitative ultrasound applications. In *Quantitative Ultrasound: Assessment of Osteoporosis and Bone Status* eds C F Njeh, D Hans, T Fuerst, C C Glüer and H K Genant (London: Martin Dunitz) pp 77–100
[8] Hulley S B, Cummings S R and Browner W S 1988 *Designing Clinical Research: An Epidemiologic Approach* (Baltimore: Williams & Wilkins)
[9] Bonnick S L, Johnston C C, Kleerekoper M, Lindsay R, Miller P, Sherwood L *et al* 2001 Importance of precision in bone density measurements *J. Clin. Densitom.* **4**(2) 105–10
[10] Bland J M and Altman D G 1994 Regression towards the mean *Br. Med. J.* **308**(6942) 1499
[11] Johnson J and Dawson-Hughes B 1991 Precision and stability of dual-energy X-ray absorptiometry measurements *Calcif. Tissue Intl.* **49**(3) 174–8
[12] Glüer C C 2000 The use of bone densitometry in clinical practice *Bailliere's Best Pract. Res. Clin. Endocrinol. Metab.* **14**(2) 195–211
[13] Cummings S R, Palermo L, Browner W, Marcus R, Wallace R, Pearson J *et al* 2000 Monitoring osteoporosis therapy with bone densitometry: misleading changes and regression to the mean. Fracture Intervention Trial Research Group *J. Am. Med. Assoc.* **283**(10) 1318–21
[14] Holbrook T L, Barrett-Connor E, Klauber M and Sartoris D 1991 A population-based comparison of quantitative dual-energy X-ray absorptiometry with dual-photon absorptiometry of the spine and hip *Calcif. Tissue Intl.* **49**(5) 305–7

[15] Njeh C F, Boivin C M and Langton C M 1997 The role of ultrasound in the assessment of osteoporosis: a review *Osteoporos. Intl.* **7**(1) 7–22

[16] Szaflarski N L and Slaughter R E 1996 Technology assessment in critical care: understanding statistical analyses used to assess agreement between methods of clinical measurement [see comments] *Am. J. Crit. Care* **5**(3) 207–16

[17] Bland J M and Altman D G 1986 Statistical methods for assessing agreement between two methods of clinical measurement *Lancet* **1**(8476) 307–10

[18] Bland J M and Altman D G 1999 Measuring agreement in method comparison studies [see comments] *Stat. Methods Med. Res.* **8**(2) 135–60

[19] Altman D G and Bland J M 1983 Measurement in medicine: the analysis of method comparison studies *Statistician* **32**(3) 307–17

[20] Armitage P and Berry G 1988 *Statistical Methods in Medical Research* (Oxford: Blackwell Scientific Publications)

[21] Greenfield M L, Kuhn J E and Wojtys E M 1998 A statistics primer. Correlation and regression analysis *Am. J. Sports Med.* **26**(2) 338–43

[22] Mandel J 1984 Fitting straight lines when both variables are subject to error *J. Quality Technol.* **16**(1) 1–14

[23] Meinert C L 1996 *Clinical Trials Dictionary* (Baltimore)

[24] Garland S W, Lees B and Stevenson J C 1997 DXA longitudinal quality control: a comparison of inbuilt quality assurance, visual inspection, multi-rule Shewhart charts and Cusum analysis *Osteoporos. Intl.* **7**(3) 231–7

[25] Lu Y, Mathur A K, Blunt B A, Gluer C C, Will A S, Fuerst T P *et al* 1996 Dual X-ray absorptiometry quality control: comparison of visual examination and process-control charts *J. Bone Miner. Res.* **11**(5) 626–37

[26] Pearson D and Cawte SA 1997 Long-term quality control of DXA: a comparison of Shewhart rules and Cusum charts *Osteoporos. Intl.* **7**(4) 338–43

[27] Orwoll E S, Oviatt S K and Biddle J A 1993 Precision of dual-energy X-ray absorptiometry: development of quality control rules and their application in longitudinal studies *J. Bone Miner. Res.* **8**(6) 693-9

[28] Faulkner K G and McClung M R 1995 Quality control of DXA instruments in multicenter trials *Osteoporos. Intl.* **5**(4) 218–27

SECTION 2

INVASIVE TECHNIQUES

Chapter 5

Mechanical testing

Christopher F Njeh, Patrick H Nicholson
and Jae-Young Rho

5.1. INTRODUCTION

5.1.1. Bone

The biological aspects of bone have been introduced in chapter 1. However, in this chapter we will summarize aspects of bone pertinent to our discussion on the mechanical measurement. Bone is a very specialized form of connective tissue composed of organic and inorganic phases. The organic matrix of bone consists mainly of type 1 collagen fibrils (90%), the remaining 10% corresponding to non-collagenous proteins, proteoglycans and phospholipids. The mineral (inorganic) substance of bone is a calcium phosphate hydroxyapatite.

Whole bone, i.e. bone as an organ, consists not only of calcified bone matrix and bone cells but also of non-osseous cells, blood vessels, nerve fibres and bone marrow. Depending on the skeletal sites, bones may appear long tubular segments (long bones), bilaminar plates (flat bones) or short irregular prismatic structures (short bones). In long bones, three different regions can be distinguished (figure 5.1): the diaphysis (central shaft) which represents the longest part; the epiphysis, which are present at the two extremities; and the metaphysis, which lie between. A cartilaginous layer, the so-called growth plate, separates the metaphysis from the epiphysis in the growing skeleton, but tends to ossify as the skeleton matures. The long bones are present in the peripheral or appendicular skeleton, i.e. limbs, ribs and clavicles; the flat bones are typically found in the skull, scapula and pelvis and the short bones in the axial skeleton (vertebrae, sternum), carpus and tarsus. Independently of their macroscopic anatomy, all skeletal segments consists of an outer layer of compact bone and an inner zone (the medulla) which contains bone marrow (figure 5.1).

Bones provide important functions to the body including mechanical support and site of muscle attachment for locomotion, protection of various

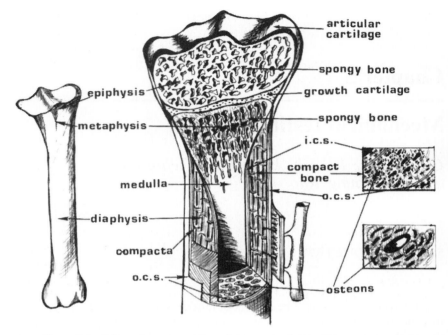

Figure 5.1. *Schematic representation of the proximal tibia, illustrating cortical and cancellous bone (reprinted with permission from An and Draughn [179]).*

vital organs such as the brain, spinal cord, heart, lungs and bone marrow, and metabolic pathways associated with mineral homoeostasis, representing calcium and phosphate reserves [1].

5.1.1.1.   Cortical versus cancellous bone

Anatomically, two forms of bone are distinguishable: the cortical (compact) and trabecular (spongy or cancellous) bone. Cortical bone appears as a solid continuous mass in which spaces can be seen only with the aid of a microscope. Cancellous bone on the other hand consists of a three-dimensional network of trabeculae, with the interspaces occupied by bone marrow. Both cortical and cancellous bones have a very similar basic composition, although the true density of fully calcified cancellous bone is a little lower (3%), and its proteoglycan content a little greater than those of the fully calcified compact bone. Although still contentious, the real difference between compact and cancellous bone depends on its porosity: that of compact bone, mainly due to the voids provided by osteon canals, Volkmann's canals, osteocytes and their canaliculi and resorption lacunae, varies from 5 to 30% (apparent density about $1.8\,\mathrm{g/cm^3}$); the porosity of cancellous bone, chiefly due to the wide vascular and bone marrow intertrabecular spaces, ranges from 30 to more than 90% (apparent density 0.1 to 0.9 g/cm^3) [2]. It has also been argued that since cancellous bone is more metabolically active compared with cortical bone, this could create newer and more mechanically competent bone [3]. Trabeculae are the unit components of the cancellous bone.

5.1.2. Bone structure

Cancellous bone is made up of an interconnected network of rods or plates or both [4]. Cancellous bone structure therefore can be viewed as a three-dimensional representation of trabeculae (an individual structural element of cancellous bone) in space (figure 5.1). On a broader scale, the architecture of cancellous bone includes its porosity, connectivity (the extent to which structural elements are connected together) and anisotropy. Traditionally, from a two-dimensional histological section, four primary measurements can be made: area, length (usually of a perimeter or boundary), distance between points or between lines and number [5, 6] (see chapter 6). Stereological theorems are then used to convert these two-dimensional parameters into three dimensions such as: bone volume (BV/TV), bone surface (BS/TV), bone surface/bone volume (BS/BV), trabecular thickness (Tb.Th), trabecular separation (Tb.Sp) and trabecular number (Tb.N) [6]. All stereological theorems require that sampling be random and unbiased, a condition only rarely fulfilled in bone histomorphometry [7]. With the exception of the conversion of area fractions to volume fractions, most stereological theorems also require that the structure be isotropic, meaning that a perpendicular to any element of surface has an equal likelihood of pointing in any direction in space [6]. Hence an assumption of the structure such as parallel plate or cylindrical rod model is required for the conversion. For example, according to the parallel plate model, trabecular number and trabecular separation can be calculated as (BV/TV)/Tb.Th and Tb.Th ∗ (TV/BV-1) respectively [6].

A problem with histomorphometry is that most of the structure parameters are derived from area and perimeter measurements using either plate or rod models. However, most of the structure in cancellous bone is not distinctively isotropic. Hildebrand and Ruegsegger [8] developed direct techniques of measuring Tb.N, Tb.Th and Tb.Sp based on the distance transformation of the binary object. They found that using this direct method Tb.N was less dependent on BV/TV than the derived Tb.N [9, 10]. However, with the advent of high resolution three-dimensional measurements there is the possibility of assessing model-independent architecture parameters.

Other methods have been described for characterizing cancellous bone's two- and three-dimensional architectural features, utilizing direct three-dimensional digitization, microcomputed tomography [10–12] and magnetic resonance imaging [13–15]. This has been reviewed recently by Cortet *et al* [16] and Odgaard [17]. With these three-dimensional approaches, other aspects of bone structure not easily accessible via histological measurements can be addressed. These include anisotropy and connectivity.

5.1.2.1. *Bone anisotropy*

Materials that have different properties in different directions are termed anisotropic. Bone structural and mechanical anisotropy refers to the variation in orientation of trabeculae and consequently mechanical properties and is an important architectural property of cancellous bone. As many as 21 independent elastic constants are required to characterize the mechanical behaviour of bone completely. Most materials have planes of symmetry that reduce the level of anisotropy and hence the number of material constants required to fully characterize the material. Depending on the variation with direction the material could be isotropic (no variation), transversely isotropic or orthotropic [18]. Orthotropic materials have properties that differ in each of the three mutually perpendicular directions, and nine elastic constants are required to characterize their mechanical properties fully. Bovine femur is an example of a tissue with orthotropic material symmetry. Materials that have properties that are constant within a given plane are termed transversely isotropic. Human osteonal bone is an example of a transversely isotropic material, because it has the same Young's modulus in all transverse directions, but higher Young's modulus in the longitudinal directions. Materials that have the same elastic properties in all directions have the highest order of symmetry and are termed isotropic. Complete characterization of the mechanical behaviour of anisotropic material requires mechanical testing to be performed in several different orientations. Ideally, mechanical testing of a specimen should be oriented relative to the axes of material symmetry.

Anisotropy can be defined using mean intercept length (MIL) measurements [19]. MIL denotes the average distance between bone/marrow interfaces and can be evaluated by a stepwise rotated analysis grid. When MIL are plotted in a polar diagram, an equation of an ellipse (two-dimensional) or ellipsoid (three-dimensional) can be fitted to the data according to the method of Harrigan and Mann [20]. The ellipsoid may be expressed by the quadratic form of a second-rank tensor, known as the MIL fabric tensor. The radii of the ellipsoid expressed as eigenvalues can be used to define the degree of anisotropy [11, 20–23]. MIL results depend entirely on the interface between bone and marrow [24]. This has the effect that anisotropic structures may appear isotropic when examined with the MIL method [25]. Because MIL is unable to detect some forms of architectural anisotropy, volume-based measures such as volume orientation (VO) [25], star volume distribution (SVD) [26] and star length distribution (SLD) [27] have been introduced. See the review by Odgaard [17] for details of these new approaches, where it was emphasized that a structure may be isotropic with respect to one architectural variable and anisotropic with respect to another architectural variable.

5.1.2.2. *Connectivity*

Connectivity relates to the branching nature of the trabeculae. Many parameters have been used to quantify connectivity including trabecular bone pattern factor (TBPf) [28], star volume [29], trabecular number and trabecular thickness [30, 31]. However, these approaches have limitations that can be overcome by using a topological approach [32, 33]. In the topological approach the Euler number is the measure of connectivity, which can be determined using a technique described by Felkamp *et al* [11].

5.1.3. *Why study the mechanical properties of bone?*

There are many reasons why the study of the mechanical properties of bone is important, and the breadth of reasons is evidenced by the varied backgrounds of those interested and active in this study of these properties. These include orthopaedic surgeons, trauma surgeons, orthodontists, prothodontists, radiologists, practitioners of general medicine, medical physicists and biomedical engineers.

- A good knowledge of these properties can help predict how bones can be expected to behave in the body. For example, 'the loads they can and cannot bear or the amount of energy they will absorb before fracturing' [34]. This knowledge can also be used as a predictor of the effects of ageing and disease on bone behaviour.
- Mechanical tests provide input for computational models of bone mechanics, adaptation and repairs.
- Knowledge of the material properties of cancellous bone has a twofold implication for bone implantation. Firstly, if other materials are to be substituted for bone, their mechanical properties must be compatible with those of bone to ensure a viable system. Secondly, the whole process of fracture fixation-implant design and implant insertion depend upon the internal distribution of the material properties [35]. This is more so when the prostheses are predominantly surrounded by cancellous bone and accordingly rely on cancellous bone for fixation. An example is the total hip-joint replacement, in which one part of the artificial joint occupies the medullary canal of the femur, the other part occupying the acetabular region of the pelvis [36].
- Data on the strength characteristics and other mechanical properties of bone might also be useful in the selection of sites for obtaining bone grafts.
- The ability of the human body to withstand acceleration and deceleration forces of various magnitude without severe trauma is of major importance to the safety engineer who must design the automobile,

airplane cockpit or space vehicle to protect the occupant as much as possible from the effect of these forces.

- Some knowledge of these limits of tolerance is likewise of practical significance to the designers and manufacturers of protective clothing and equipment used in sports such as football, skiing and motor vehicle racing. Other researchers like anatomists, zoologists, physiologists, anthropologists, physicists, crystallographers and bio-engineers are also interested in the mechanical properties of bone.

The importance of the mechanical properties has been discussed. We will now look in greater detail at the mechanical properties of bone. Before looking at the experimental methods and the values obtained, we shall first establish the basic definitions used in discussing bone mechanical properties.

5.1.4. Basic concepts in bone mechanics and definition of terms

5.1.4.1. Structure versus material

Mechanics is a physical science that assesses the effect of forces on objects. Mechanical properties of bone are basic parameters, which reflect the structure and function of bone. In studying bone mechanics, a distinction is usually made between the behaviour of a whole bone as a *structure* and the intrinsic mechanical properties of the bone tissue as a *material*. At the structural level a range of factors contribute to the mechanical behaviour including the direction and magnitude of the imposed loads, the size and architecture of the bone being loaded and the material properties of the bone tissues. The structural behaviour of bone is determined by conducting mechanical tests on whole-bone specimens subjected to physiological or traumatic loading conditions. The material behaviour of a specimen is not influenced by its geometry, but reflects the intrinsic properties of the material itself. Studies at the material level are generally performed on specimens of standardized shape under controlled conditions, enabling comparisons to be made with other materials and potentially allowing the basic phenomena governing the mechanical behaviour to be revealed. One of the great ongoing challenges in bone mechanics is that of integrating knowledge gained at the material level back into models of whole bone structural behaviour without introducing overwhelming complexity.

5.1.4.2. Hooke's Law, strain and stress

Force is a measurable vector (the SI unit is the Newton), which has a magnitude, direction and point of application. Forces act on a body and tend to change the velocity of the body (external effect) or shape of the body (internal effect). There are basically three types of force: tensile, compressive and shear, which are determined by the direction and effect of the forces acting on the body. In normal day-to-day activities a complex and constantly-changing pattern of forces is imposed on the skeletal system. Like any other material, bone deforms under the action of these imposed forces. When the forces are removed a solid, such as bone, tends to completely recover its original form provided the deformation induced by the applied forces was not above a certain level known as the *elastic limit*. Within this elastic range the deformation is linearly proportional to the magnitude and direction of the applied force, a relationship known as *Hooke's law*.

Strain is a dimensionless measure of relative deformation or percentage change in length. Stretching a sample to 101% of its original length corresponds to a strain of 0.01, or 1%. When a load is applied axially to a bar or rod of uniform cross-section, the proportional changes in length $\Delta l/l$ and width $\Delta b/b$ are called normal strains, denoted by ε (figure 5.2). A

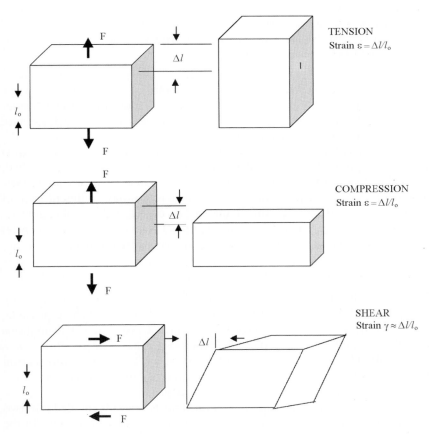

Figure 5.2. *Schematic representation of tension, compression and shear loading where F is the applied load, l_o is the original length, l is the final length, Δl is the change in dimension $(l - l_o)$ due to the applied load.*

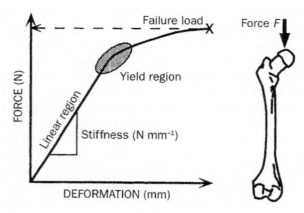

Figure 5.3. *Schematic diagram of a typical load–deformation curve. The slope of the curve in the linear region is known as the stiffness and the maximum load applied is termed the ultimate or failure load (reprinted with permission from Njeh et al [49]).*

shear load produces a shear strain, usually quantified as the change in angle undergone by two lines originally at right angles. Strain rate is the deformation per unit time, which is an important parameter for viscoelastic (see section 1.4.6) materials such as bone.

Stress is defined as the force per unit area and may be classified as tensile, compressive or shear depending upon how the load is applied. Normal axial stresses (σ) can be either *tensile* or *compressive*. Shear stresses, denoted by τ, occur when equal and opposite forces have different lines of action which tend to alter the shape of the object without changing its volume. Stress has units of pascals ($1\,\mathrm{Pa} = 1\,\mathrm{N\,m^{-2}}$). The physiological stress levels for bone are generally below the megapascal range.

5.1.4.3.   The stress–strain curve

The relationship between the load applied to a structure and the resulting deformation is called a *load-deformation curve* (figure 5.3). Where possible, it is generally preferable to convert loads into stresses, and deformations into strains, and thereby replot the relationship as a *stress–strain curve* (figure 5.4). The advantage is that the stress–strain curve gives information directly relating to the intrinsic material properties of the specimen, and is independent of the specimen size and geometry.

The stress–strain curve (figure 5.4) for bone is typically divided into two regions, the elastic and plastic, which are divided by the *yield point*. In the first region, behaviour is linearly elastic (for cancellous bone the cell walls bend or compress axially). The second phase is known as the *plastic region* (cellular collapse for cancellous bone) where an increase in strain results in

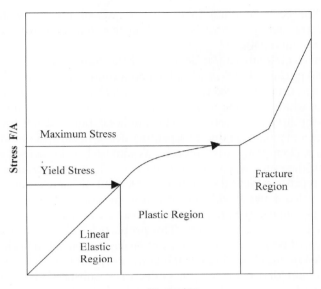

Figure 5.4. *Schematic diagram of a typical stress–strain curve for a compressive test of a cancellous bone. The linear, plastic and fracture region are shown. Note that stress is computed as the applied force F divided by the cross-sectional area (A) and the strain is computed as the deformation (Δl) divided by the original length l_o.*

little or no increased stress. For cancellous bone this so-called plastic behaviour can be caused by a great variety of phenomena at the microscopic level, including collapse of the cells by elastic buckling, plastic yielding and crack growth or brittle fracture of the cell walls [37]. The point/region where the deformation changes from being elastic to at least partially plastic is the yield point. Although bone reaching this point may still have far to go before it actually breaks, it is permanently damaged to some extent once it enters the plastic region [38]. The amount of post yield strain that occurs in a material before fracture is a measure of the *ductility* of the material. A material showing a large post-yield strain is referred to as ductile, whereas a material showing little post-yield strain is described as *brittle*. Brittleness can be estimated from the reciprocal of the width of the displacement to fracture.

The yield strength is the stress at the yield point. In practice the yield point can be difficult to define in bone since there may be a very gradual transition from linear elastic behaviour into plastic nonlinear behaviour. Researchers such as Hvid and Jensen [39] have attempted to define it as the point where the stress–strain curve begins to become nonlinear. Another approach, known as the offset approach, defines the yield point as the point where a line corresponding to the linear part of the stress–strain curve translated through a specified strain offset (e.g. 0.2% strain) intersects the stress–strain curve [40].

Mechanical failure can be defined as the degradation of a material property beyond its elastic limits or loss of material continuity. The maximum stress the bone can sustain is called the *ultimate strength*, and the *breaking strength* is the stress at which the bone actually breaks catastrophically. Ultimate strength and breaking strength are often identical in bone, but need not necessarily be so. *Fatigue* is the damage due to repetitive stresses below the ultimate stress. Fatigue is a slow progressive process, as opposed to an acute catastrophic process which results when the ultimate strength of a material is surpassed. Typically repetitive cyclical loading (smaller than ultimate strength) causes a crack through a material with subsequent separation of the object into pieces.

The area under the stress–strain curve is a measure of the amount of *energy* needed to cause a fracture. This property is called energy absorption or *toughness* of bone and is an important property from a biomechanics point of view. A low toughness makes the bone more liable to fracture, and it is widely thought that bone toughness decreases substantially with age [41]. On the other hand, measuring fracture toughness is not easy because samples with dimensions which meet the test requirements can be cut only from large bones [4].

5.1.4.4.   *Modulus of elasticity*

For axial loading the slope of the stress–strain curve within the elastic region is called the *modulus of elasticity*, or *Young's modulus*, E (expressed in pascals, $N\,m^{-2}$), the reciprocal of which is known as *compliance*. The modulus of elasticity is defined as

$$E = \frac{\sigma}{\varepsilon} = \frac{F/A}{d/L} \tag{5.1}$$

where F is the force, A is the cross-sectional area, d is the change in length and L is the original length. Similarly, the *shear modulus* G characterizes the relationship between shear stress and shear strain.

If the material has the same properties in all directions it is said to *isotropic*, and it can be characterized using only two moduli, Young's modulus E and the shear modulus G [42]. Young's modulus, as we know from above, is defined as the ratio of the axial stress to axial strain in simple elongation. Where simple elongation is the deformation produced in a long prismatic rod by forces acting parallel to its axis, the resultant force acts at the centroid of the cross section [43]. Young's modulus is the modulus often referred to in the literature on mechanical properties of bone because it is perhaps the most straightforward in terms of experimental determination. The *bulk modulus* (K) is defined as the modulus of volume expansion; that is, the ratio of the isotropic stress to the relative change in volume. The isotropic stress simply implies that the specimen is being stressed equally in all three planes or $P_x = P_y = P_z$ [42].

5.1.4.5. *Poisson's ratio*

Balanced axial forces applied to a prismatic bar not only produce an axial strain but also give rise to a lateral or transverse strain (ε_l) with a consequent change in cross-sectional area. The ratio of the lateral strain to the axial strain is quantified by another material constant known as *Poisson's ratio* (ν) of the material [44]. For an isotropic linear elastic material, relationships among the elastic constants can be expressed as

$$E = 3K(1 - 2\nu) \tag{5.2}$$

$$E = 2G(1 + \nu) \tag{5.3}$$

where E is Young's modulus, G is the shear modulus and K is the bulk modulus. Since E and K are both positive, equation (5.2) indicates that ν cannot have a value greater than 0.5 [44]. Poisson's ratio of human cancellous bone is 0.2 to 0.3.

5.1.4.6. *Viscoelasticity*

The elastic theory assumes perfect elasticity, but for most materials, including plastics, wood and bone, the mechanical behaviour depends upon both the rate and duration of application of the stress. This phenomenon is known as *viscoelasticity*, and materials that exhibit it are termed viscoelastic. Viscoelasticity is the result of internal energy losses due to friction in the structure (intrinsic viscoelasticity) or fluid flow (fluid-dependent visco-elasticity) during deformation. Bone is viscoelastic [2, 45], but in many cases it is acceptable to make the simplifying assumption of linear elastic behaviour [46]. Modelling a viscoelastic material as a Kelvin body, the stress (σ) can be expressed as

$$\sigma = E\varepsilon + n\dot{\varepsilon} \tag{5.4}$$

where E is the elastic modulus, ε is the strain and $\dot{\varepsilon}$ is the strain rate ($d\varepsilon/dt$). Equation (5.4) implies that the true elastic modulus can only obtained by letting $\dot{\varepsilon} \rightarrow 0$. This suggests that an infinitely low deformation rate (quasi-static testing) should be used to determine E [47]. However, it should be remembered that the *in vivo* strains in bone are dynamic, and that quasi-static tests may therefore not be physiologically representative. Strains in human bone range from 0 to 0.12, and strain rates are typically 0.001–$0.03\,\mathrm{s}^{-1}$. Carter and Hayes [48] suggested that the longitudinal strength and stiffness of mineralized bone tissue are approximately proportional to the strain rate raised to the power 0.06. The viscoelasticity of bone implies that strain rate should be specified whenever mechanical properties are being reported in order to facilitate cross-comparison of results.

The behaviour of viscoelastic materials which are important in quantifying the mechanical properties of bone are *stress relaxation* and *creep*. Stress relaxation is the decay of stress within a material subjected to a constant strain. Creep is the gradual increase in strain of a material subjected to a constant load (or deformation under a constant load). In a linearly viscoelastic material, energy is dissipated by plastic or viscous flow within the material so the loading and unloading curves do not overlap, instead forming a closed hysteresis.

Bone, like wood and other natural materials, is *heterogeneous* (i.e. its elastic properties vary from point to point) and *anisotropic* (i.e. its elastic properties and strength depend upon the orientation of the microstructure with respect to the direction of loading) [38]. For cancellous bone the mechanical anisotropy arises from the orientation of the trabecular struts and plates. This anisotropy was demonstrated by the work of Brown and Ferguson [35] who reported detailed quantitative information about the spatial and directional variations of the material properties of cancellous bone from the proximal femur.

The basic theory and terminology having been established, the next section will review the experimental methods used to determine the mechanical properties of bone, the values obtained, and the problems associated with the various measuring modalities.

5.2. EQUIPMENT AND SPECIMEN CONSIDERATION

5.2.1. *Equipment*

A brief introduction of the equipment used for mechanical testing will be introduced. The equipment to be used for mechanical measurement will depend on the technique used. The typical mechanical testing machine has several key components including the hydraulic power supply (HPS), actuator, controller, load unit, force and deformation transducers, gripping or coupling devices and a system for data recording [49]. The testing machines can be screw-driven or pneumatic, but the most versatile machines are servo-hydraulically controlled. The HPS provides the fluid pressure necessary to drive the actuator piston. A servo-valve is used to transform electrical energy into hydraulic fluid pressure. The pressure is then applied as load to the specimen, at a variety of different rates and magnitudes.

The load unit of the system employs actuator rods to apply load to the specimen. In servo-hydraulic systems, the actuator operates under control of a servo-valve and contains a linear variable differential transducer which provides rod displacement information to the controller.

Load applied to the specimen during the testing procedure is measured with a force transducer, called a load cell, which is mounted on the fixed side opposite the load unit. The load cell measures the output displacement of the loading platen. Transducers are available in a variety of load ranges and should be selected based both on the operating limitations of the actuator and on the intended use of the testing system. Most modern machines provide a computer interface for data recording and control of the testing procedure (see figure 5.5).

Displacement measurements are very sensitive to machine compliance that can be caused by compliance of the load cell, deformation within the coupling or gripping devices or even deformation of the frame of the testing machine itself [50]. Therefore, an accurate length measurement requires externally applied strain gauges or extensometers. These devices are directly bonded to the specimen surface by glue or clipping tools or they are attached to the testing columns close to the specimen [49, 51]. Extensometers and strain gauges change electrical resistance when strains are applied. Therefore, a Wheatstone bridge amplifier must be used in concert with these transducers to give a voltage output. Recording of a stress–strain curve can be done by connecting the load cell and strain measurement device to an x–y recorder.

Proper preparation of the specimen for mechanical testing requires several key tools. A band saw with a 6.5 mm ($\frac{1}{4}$ in) fine tooth blade is an invaluable general resource for preparing gross bone sample for testing. For making small, cylindrical bone specimens for compression testing, a tabletop drill press with a trephine bit is sufficient. A wheel grinder/polisher is also a useful tool in final preparation of bone specimens in which specific dimensions must be identical.

5.2.2. Specimen handling

Bone samples used for mechanical testing must be carefully prepared and stored to avoid damage to the specimen and degradation of the mechanical properties [52]. The conditions required for accurate maintenance of mechanical integrity such as specimen preservation, hydration and temperature are discussed in section 5.5. However, it is worth mentioning here that tissue autolysis begins within hours following removal of bone from the body and it is likely that this adversely affects the mechanical properties of bone [53]. The preferred method for storage is freezing in an airtight container at −20 °C or lower [54–56]. The following steps are recommended for specimen handling:

1. Prior to mechanical testing specimens must be completely thawed at room temperature.
2. Samples should be kept wet throughout specimen preparation (machining) and testing.

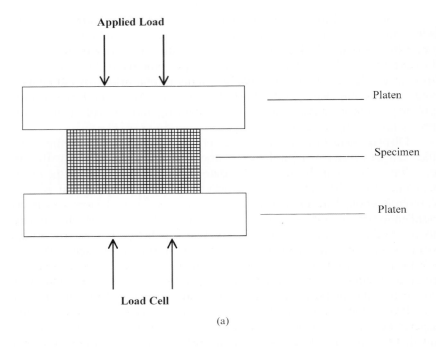

(a)

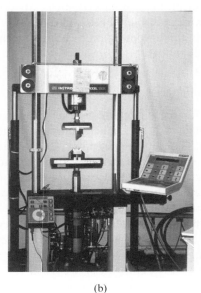

(b)

Figure 5.5. *(a) Schematic diagram of the mechanical testing setup, (b) a photograph of an instron machine used for mechanical testing.*

3.   When samples are not being tested, and in between tests, they should be kept at 4 °C in isotonic saline.
4.   Samples should be harvested by using a band saw driven at a low speed and under constant water irrigation to prevent heat build up.
5.   Testing should ideally be performed under physiological conditions such as at a temperature of 37 °C. However, errors introduced by measurements at room temperature are acceptable for most cases.

5.3.   METHODS OF MEASURING THE MECHANICAL PROPERTIES OF BONE TISSUE

There are three classical biomechanical tests commonly used for assessing the biomechanical properties of bone: tension, compression and torsion. Bending or flexure tests that combine compression and tension are also routinely used [51, 57]. The American Society for Testing and Material (ASTM) has developed standards to which these measurements should be carried out [58]. These standards can be generally applied to bone, although modifications to specimen size and method of gripping the test specimen may be necessary. As previously discussed, the biomechanical test used will also depend on whether material (intrinsic) or structural (extrinsic) properties are being assessed. For example, when dealing with cancellous bone, the mechanical properties can be investigated at the structural level (i.e. that of the trabecular framework) or at the material level (i.e. that of the material making up individual trabeculae). A compression test on a cancellous core measures properties at the structural level, whereas a nano-indentation test on individual trabeculae measures material level properties.

   The subsequent sections will attempt to address the methodology and problems associated with the different types of measurement techniques.

5.3.1.   *Uniaxial compressive test*

Compressive testing is the mostly widely used method for characterizing the mechanical properties of cancellous bone, but it has also been used for cortical bone and whole bone specimens. Whilst it is conceptually straightforward, obtaining accurate results in practice requires great attention to detail and a thorough appreciation of the potential sources of error. Test specimens, generally in the form of cylinders or rectangular prisms, are machined from the anatomical region of interest with the desired orientation. Care must be taken not to damage the specimen during preparation and machining.

   In the compression test, load is applied by squeezing the specimen between two flat rigid surfaces (or platens) (see figure 5.5(a)). The deformation of the specimen is measured either from the movement of the platens or,

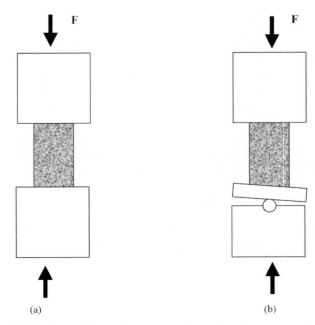

Figure 5.6. *(a) Compressive testing specimen setup, (b) use of pivoting platens to reduce the effect of irregularities.*

optimally, using an extensometer or extensometers attached to the central section of the specimen. *Preconditioning* may be used, whereby relatively small loads are applied cyclically before performing the main test. This leads to a 'bedding-in' of the specimen, flattening out any small irregularities at the specimen faces in contact with the platens.

End-effects are a major problem in compression testing of cancellous bone, and arise from non-ideal loading conditions where the specimen meets the platens [59, 60]. Stress concentration occurs at the ends of the specimen so that proportionally larger deformations tend to occur there. This tends to decrease the measured elastic modulus.

The validity of the test also depends on the two ends of the samples being plane and parallel to each other. If the faces are not parallel, a non-uniform distribution of stress occurs, again causing an underestimation of both Young's modulus and strength. To some extent, the problem of alignment can be reduced by including a ball-joint to account for slightly non-parallel specimen faces (see figure 5.6). The specimens are not usually given shoulders, but rather have a uniform cross-section over their entire length.

Another problem arises from the frictional interaction between specimen and platens. As the Poisson's ratio effect tries to make the specimen expand radially outwards, frictional forces arise at the specimen/platen interface and add to the stress distribution pattern within the specimen. However, the effect of this on the results of compression testing is believed to be small [61]. Recent work has shown that many of the problems experienced in compressive testing of cancellous bone can be avoided by embedding the bone specimens into rigid caps which are then gripped in the testing machine [60]. Using this approach, initial highly nonlinear 'toe-in' region of the stress–strain curve is absent [40]. Finally, it should be noted that whilst most compression testing is performed using impermeable platens and an unconfined specimen, in special cases the specimen can be confined in a rigid cylindrical cavity or loaded using porous compression platens that permit fluid flow into and out of the bone [48].

The stress–strain curve obtained from mechanical compression testing reveals three distinct regimes of behaviour (see figure 5.4), linear elastic, plastic region and fracture region. The second phase of *collapse* progresses at roughly constant load until the cell walls meet and touch. Once this happens the resistance to load increases, giving rise to a final increasing steep portion of the stress–strain curve. This is usually called the *densification phase*.

5.3.2. *Uniaxial tensile test*

Tensile testing can be one of the most accurate methods for measuring bone properties, but specimens must be relatively large and carefully machined from the region of interest. For accurate results, specimens should have a reduced and uniform cross-sectional area over the central length, widening out at the ends where the specimen will be gripped. The most common geometry used is the dumbell shape (figures 5.7 and 5.8), although rectangular paralleliped specimens have also been used [62].

Tensile test specimens for cortical bone and cancellous bone are illustrated in figure 5.8. The inner length (parallel length) is also known as the gauge region. Dimensions are derived from ASTM standards [58]. The ratio b/a should be around 2. The inner length (l) should ideally be as large as possible but realistically at least 20 mm. A specimen thickness of 4–8 mm is required. The grip length m should be one quarter of the whole specimen length L. Because of the relatively homogeneous microstructure of cortical bone, its bone specimens can be made comparatively small in size (gauge diameter $= a = 3$ mm). In principle the same geometry used for cortical bone tensile test specimens applies to trabecular bone. However, because of its structure a minimum gauge diameter of 5 mm is required to ensure continuum scale criteria are met. Furthermore, the specimen ends must be embedded in resin or acrylic cement or glued to the specimen holder to provide sufficient clamping support (figure 5.7(b)) [40].

In theory, the applied tensile force is distributed uniformly over the cross-sectional area within the central region and, for best results, deformation should be measured using an extensometer or extensometers in this region.

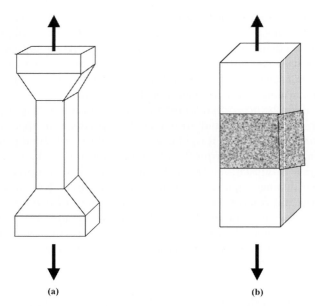

Figure 5.7. *(a) Tensile testing cortical bone specimen setup, (b) tensile testing cancellous bone specimen setup.*

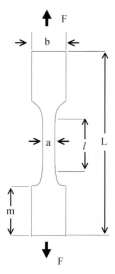

Figure 5.8. *Illustration of a tensile test specimen for cortical bone. This design will consistently produce the majority of the strain and final failure in the midsection of the specimen.*

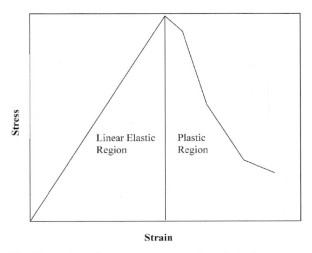

Linear Elastic Region

Plastic Region

Strain

Figure 5.9. *Illustration of stress–strain curve of tensile testing.*

Stress is calculated as the applied force divided by the bone cross-sectional area measured in the specimen gauge region. The stress–strain curve recorded for tensile testing has a slight variation from compression. There is the initial linear region followed by a region that is nonlinear until a maximum stress is achieved. Beyond this point progressive trabecular fracture occurs, resulting in a gradual reduction of stress with increasing strain until total separation of the fracture surfaces is observed (figure 5.9) [63]. The gradual decrease in stress after the maximum is accompanied by a tearing phenomenon wherein the trabeculae of the cancellous bone progressively tear free from adjacent bone.

Studies of the mechanical properties of cancellous bone have concentrated on compressive properties, both for reasons of experimental expediency and because compressive failure of trabecular bone (e.g. in vertebra fracture) is a common problem clinically. Difficulties in tensile measurement arise principally from the problem of gripping cancellous bone without crushing it [63].

The tensile behaviour of most porous materials differs significantly from the compressive behaviour. The modes of failure in the two types of loading could account for some of the differences. The struts in porous materials do not buckle under tensile loading, whereas buckling is almost certainly a major phenomenon under compressive loading. In addition, since tensile fracture is associated with distraction of two fracture surfaces, no further energy is absorbed with continued loading after fracture. On the other hand, in compression cancellous bone continues to support loads after fracture. Porous materials in general absorb much less energy in tension than in compression. The above arguments support the finding of Stone *et al* [64]

Table 5.1. *Examples of studies comparing the tensile and compressive strength of cancellous bone.*

Reference	Specimen	Tensile strength (MPa)	Compressive strength (MPa)
Evans [67]	Human vertebrae	1.18	6.27–8.62
Carter *et al* [63]	Human femora	NS*	NS
Stone *et al* [64]	Bovine humerus	2.63	8.29
Neil *et al* [68]	Human vertebrae	NS	NS
Kaplan *et al* [65]	Bovine humerus	7.6 ± 2.2	12.4 ± 3.2
Rohl *et al* [66]	Human	2.42	1.90
Keaveny *et al* [40]	Bovine tibia	15.6	21.3

* NS, no statistical difference.

and Kaplan *et al* [65], that the compressive strength of cancellous bone is greater than the tensile strength. The average ultimate strength recorded by Kaplan *et al* [65] was 7.6 ± 2.2 MPa in tension and 12.4 ± 3.2 MPa in compression for bovine cancellous bone from the proximal humerus. More recently, Keaveney *et al* [40] reported yield strength to be 30% lower in tension than compression, again in bovine cancellous bone. However, these results are contradicted by other studies (see table 5.1). Carter *et al* [63] found no significant difference between tensile and compressive strength for human trabecular bone (femora), although they noted that the energy absorption capacity was significantly lower in tension. Rohl *et al* [66] reported that the tensile strength was actually higher than the compressive strength. Reasons for the discrepancies amongst these studies include differences in anatomical site and species, and inaccuracies in the test methods.

5.3.3. Bending test

An *in situ* bone in the living body is subjected to a variety of force systems— gravitational, muscular activity, static and dynamic support and impact— which tend to bend and twist the bone. Logic will dictate that the strength and other mechanical properties of bone be determined from bending and torsion tests. Since bone is weaker in tension than in compression [69], failure in bending tests usually occurs on the tensile side. Furthermore, there is always some deformation due to induced shear stresses. To minimize shear stresses the span length of the specimen has to be sufficiently long. Ideally, the span length between the outer supports (figure 5.10) should be 16 times larger than the thickness of the specimen [58]. However, this requirement is hardly ever realized in whole-bone testing [49]. Experimentally, bending tests are generally performed using either a three- or four-point bending configuration on relatively long, thin specimens (figure 5.10). In three-point bending,

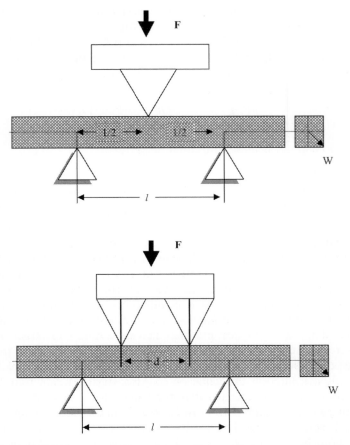

Figure 5.10. *Three- and four-point bending applied to rectangular bone specimens (adapted from Bouxsein and Augat [49]). Bending moments are determined by the inner (d) and outer (l) span lengths and the cross-section moments of inertia around the bending axis. Stress and strain are determined by the distance of the surface from the neutral axis (W).*

unwanted shear stresses are generated near the mid-section of the bone, but this can be avoided by using four-point bending. This, however, requires equal forces to be applied at each loading point, which can be difficult to accomplish with bones that are not uniformly shaped and have irregular geometries. Thus the three-point bending test is used more often to measure the biomechanical properties of whole bone.

To derive the elastic modulus (E) from bending tests the specimen geometry must be known, and it is often assumed that E is the same in tension as in compression and that the material is homogeneous and isotropic. Determination of the elastic modulus by bending also requires knowledge of the moment of inertia about the axis of the deformation. If the moment of inertia (I) cannot be determined for the given loading direction, only the specimen's extrinsic bending stiffness (or flexural rigidity) EI can be assessed. The stress, strain and Young's modulus can be calculated from the force and displacement of the loader in figure 5.10. For three-point loading the equation is

$$E = \frac{Fl^3}{\Delta l 48I}, \qquad \sigma = \frac{Flw}{4I}, \qquad \varepsilon = \frac{12\Delta lw}{l^2} \tag{5.5}$$

and for a four-point bending the equation is

$$E = \frac{F(l-d)^2(l+2d)}{\Delta l 48I}, \qquad \sigma = \frac{F(l-d)w}{4I}, \qquad \varepsilon = \frac{12\Delta lw}{(l-d)(l+2d)} \tag{5.6}$$

where σ is stress, ε is strain, w is the distance from the centre of mass, F is the applied force, Δl is deflection, and d and l are the inner and outer span lengths (figure 5.10).

Several authors consider bending to be a rather unsatisfactory test for determining the mechanical properties of a material [41, 67]. On the other hand, Currey [34] and Simkin and Robin [70] were in favour of the method, their argument being that clamping or supporting cortical bone specimens is simpler when this method is used than with tension or compression tests. They also pointed out that the deflections are larger and the method is relatively insensitive to inaccuracies in the centricity of the load. Bending tests also allow testing to be done on samples which are slightly curved, without altering the value of E. Also, drying which may happen during the test will not greatly alter the value of E obtained, whereas in the compression tests the shrinkage is of the same order of magnitude as the strain. There are few data in the literature on bending measurements on cancellous bone, and it seems much more suited to cortical bone characterization. The nature of cancellous bone as a porous material makes preparation of suitable test specimens for bending difficult.

5.3.4. *Torsion test*

Torsional testing assesses the shear properties of the specimen. When a specimen is loaded in torsion (twisting moment), shear stress varies from zero at the centre of the specimen to a maximum at the surface (figure 5.11). For any cross-section, the maximum shear stress in torsion can be calculated using the formula

$$\tau = Tr/J \tag{5.7}$$

where T is the applied torque, r is the maximum radial distance from the centre of the cross-section to its surface and J is the polar moment of inertia of the cross-sectional area.

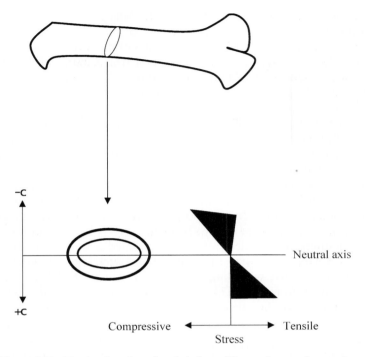

Figure 5.11. *Torsional testing of a whole bone. The two bone ends are gripped and twisted relative to each other under a twisting moment T.*

To carry out this test, the two ends of the bone sample are usually embedded in plastic blocks for easy gripping during testing, and the two blocks are then twisted (applying a torque T) relative to each other. The angular deformation θ is then measured. The shear modulus, G, is given simply by

$$G = TL/\theta K \tag{5.8}$$

where L is the length of the unembedded portion of the specimen and K is the torsional constant. For regularly machined specimens, K is equal to the polar moment of inertia. However, in practice, tensile stresses tend to be produced in addition to pure shear stresses, and this can affect the measurement. It is possible to devise pure shear tests but generally these require a complex non-standard specimen geometry, and have rarely been implemented with bone.

5.3.5. Fatigue

Fatigue is a well-known phenomenon in engineering structures which are subject to prolonged cyclical loading. If the induced cyclical stresses are large enough, or the number of loading cycles high enough, a *fatigue failure* may occur [67].

During the daily activities of life, the bones of the skeleton are repeatedly subjected to stress by gravity and muscle action. The response of the bones to this repetitive loading is important in understanding the nature of the so-called 'fatigue', 'stress' or 'march' fracture. This type of fracture is commonly encountered in the metatarsal.

Fatigue testing measures the ability of a material to resist degradation of mechanical properties due to cyclically applied load. Bone fatigue can occur in two ways: either creep by slippage of the cement lines or crack accumulation [71, 72]. Fatigue testing in the laboratory can use tensile, compressive, bending or torsional loading. Good specimen preparation is again a vital prerequisite. The key feature is the application of cyclic loading through a suitable testing machine. In the most common approach, the maximum stress level is kept constant and the specimen is cyclically loaded until failure occurs. The number of cycles to failure is counted and, using data from a number of different specimens, a graph of the stress against the cycles to failure can be plotted providing characteristic curves with which fatigue behaviour can be compared [73, 74]. Fatigue tests are considerably more time-consuming than the other methods discussed.

5.3.6. *Indentation/hardness tests*

Hardness is usually defined as the resistance of the material to indentation by another solid body under static or dynamic loading. Hardness or indentation tests measure hardness by driving an indenter with a specified geometry into the polished surface of the material with a known loading for a specified time. The tests can be categorized based on the geometry and/or the size of the indenter employed. Different geometries such as the Brinell, Rockwell, Vickers and Knoop indenters have been used [75]. Based on the size of the indenter, different sizes of the sample will be evaluated and hence macroindentation, microhardness and nano-indentation are defined. Each of these methods assesses bone properties at different scales. Macroindentation testing measures bone mechanical properties at the macrostructural level.

The first indentation test was reported in 1900 by Brinell and colleagues using a spherical steel ball indenter. Another indentation test by Rockwell also uses a spherical indenter. An osteopenetrometer measures the hardness of bone at the macrostructural level. This is a semi-quantitative approach based on the principle of recording the force necessary to advance a needle into bone and also the depth of penetration [76]. The strength data are reported as the average penetration strength (Pa), given by the force averaged over a 1 mm interval divided by the projected area of the measuring profile [77, 78]. Penetration tests do not reflect any well defined mechanical property

Figure 5.12. *Laboratory setup for ultrasound measurement, showing a Perspex water tank containing the two transducers. The received signal is digitized by the oscilloscope. The computer controls the excitation of the transducers.*

of trabecular bone, but Hvid *et al* [76] have shown that the relationship to the ultimate strength of bone cylinder could be expressed as average penetration strength given by $1.74\sigma\mu^{1.06}$. This method has a fundamental flaw, however, which is the lack of an explicit relationship between the measured values and the mechanical properties of bone.

5.3.7. Ultrasound

The use of ultrasonic waves in the measurement of the elastic properties of bone is well known [62, 79–86] (figure 5.12) (see chapter 14). The elastic properties can be deduced from velocity measurements of shear and longitudinal waves propagating in particular directions in the bone specimens if the density and the elastic anisotropy of bone are specified. The relationships between velocity and elastic properties follow from the theory of small-amplitude elastic wave propagation in anisotropic solids [87]. There are two modes of wave propagation characterized by overall specimen geometry. The first case, as overall specimen geometry tends to infinity, is referred to as bulk wave propagation. Since the cross-sectional dimension is large the wave does not perceive the solid boundaries.

$$E = \frac{(1+\nu)(1-2\nu)}{(1-\nu)}\rho\nu^2 \qquad (5.9)$$

where E is the bulk modulus, ν is Poisson's ratio and ρ is the apparent density.

The second case, where overall specimen geometry tends to 0, is called bar wave propagation. In this case the entire cross-section is excited by the passing wave. The bar wave velocity can be written in terms of Young's modulus directly, whereas the bulk wave velocity is related to the bulk modulus [62].

$$E = \rho\nu^2. \tag{5.10}$$

Several investigators have assumed cortical bone specimens to be orthotropic with nine independent elastic constants [62, 79, 81–83]. Consideration must be given to the different structures of cortical and cancellous bone. Techniques for the ultrasonic velocity measurement for cortical bone have utilized frequencies between 2 and 10 MHz. These relatively high frequencies allow accurate determination of the time delay due to propagation through cortical specimens with dimensions as small as 5 mm [62]. Both cylindrical and cubic specimen shapes have been used, but cubic specimens offer the possibility of velocity measurement in several different directions. The use of ultrasound to measure the elastic properties of the porous structure of cancellous bone is a much more complex problem. Lower frequency waves (50–100 kHz) or longer wavelength are necessary to determine the elastic properties of the cancellous bone structure due to the high porosity [84, 85]. Porosity can be quantified by measuring the void fraction, defined as the ratio of void volume to total volume. Void fractions range from 0.5 to 0.8 for cancellous bone. A direct consequence of the large void fraction is that ultrasonic waves are strongly attenuated in cancellous bone, much more strongly than in cortical bone. In other words, the velocity of propagation of an ultrasonic wave is strongly dependent on the frequency of oscillation. The dispersive nature of cancellous bone must be analysed carefully in order to obtain meaningful values of the elastic properties.

A total of 18 different velocities can be measured, three in each of six directions using the specimen shape developed by Van Buskirk *et al* [83]. However, only nine velocities could be measured from human cortical specimens due to thin cortex [79]. Two of the nine independent orthotropic elastic constants (c_{12} and c_{13}) cannot be obtained using this specimen shape, but could be obtained by relationships assuming transverse isotropy: $c_{13} = c_{23}$ and $c_{66} = \frac{1}{2}(c_{11} - c_{12})$. However, seven elastic coefficients are sufficient to determine whether the elastic symmetry of the material is orthotropic, transversely isotropic or isotropic.

Cancellous bone is also considered linearly elastic and anisotropic [88]. The most straightforward relationships between velocities and elastic properties are derived for waves of long wavelength propagating in bar-shaped specimens of narrow cross-section [87]. Longitudinal velocities, v_i, of waves propagating in the i direction, and shear wave velocities, v_{ij}, for propagation in the i direction with particle motion in the j direction, are related to Young's moduli and the shear moduli by

$$E_i = \rho v_i^2 \qquad (5.11)$$

and

$$G_{ij} = \rho v_{ij}^2 = \rho v_{ji}^2 \qquad (5.12)$$

where ρ is the apparent density of the structure and E_i represents the elastic moduli and G_{ij} the shear moduli in the i direction. Two requirements must be satisfied for correct application of equations (5.10) and (5.11). First, the wavelength must be greater than the characteristic dimensions of the structure, if the structure is to be considered a continuum. Second, the wavelength must be larger than the cross-sectional dimensions of the specimen. Wave propagation fulfilling the requirements is often referred to as bar wave propagation. In order for an ultrasonic wave to propagate at a velocity governed by the elastic properties of the gross structure, the first requirement must be satisfied. Harrigan *et al* [89] presented arguments that 5 mm characteristic dimension is the lower limit at which cancellous bone can be considered a continuum. The 50 kHz frequency of the transmitted ultrasonic wave, and measured velocities between 1000 m/s and 1600 m/s, predict a wavelength (velocity/frequency) of the order of 20 mm, well above the 5 mm characteristic dimension of the cancellous structure and the 10 mm cross-sectional dimension of the specimen.

Ultrasonic techniques have a significant advantage over mechanical testing methods in the determination of the elastic properties of bone [62, 79, 85], being able to use smaller, more simply shaped specimens. Also, several anisotropic properties can be measured from a single specimen. Due to inhomogeneity, anisotropy and limited size of bone, both advantages are significant.

The shear moduli and Poisson's ratio for the cancellous bone are rarely measured. Bar wave propagation techniques are not adequate to measure Poisson's ratio. Ashman *et al* [62] showed that the bulk mode worked well for the measurement of the orthotropic elastic properties of cortical bone. Bulk wave propagation in cancellous bone might provide an adequate method for measuring Poisson's ratio. Bulk wave propagation in cancellous bone specimens of approximately 10 mm^3 might be achieved by raising the frequency of the ultrasonic waves from 50 kHz to approximately 500 kHz. This would reduce the wavelength from approximately 40 mm to about 4 mm [89]. To achieve this task, one should be careful not to increase the frequency so much as to cause continuum breakdown, where the waves begin to propagate along individual trabeculae. Ultrasonic velocity measurements should be made at intermediate frequencies between 50 and 500 kHz to determine at what frequency a transition between bulk and bar waves occurs.

Table 5.2. *Survey of compressive Young's modulus reported for cancellous bone.*

Reference	Region	Modulus (MPa)
McElhaney *et al* [187]	Vertebral	151.7
Pugh *et al* [46]	Distal femur	413–1516
Schoenfield *et al* [45]	Proximal femur	344.7
Townsend *et al* [91]	Patella	121.3–580
Lindahl [92]	Proximal tibia	1.4–79.2
	Vertebral	1.1–139
Carter and Hayes [88]	Proximal tibia	20–200
Ducheyne *et al* [93]	Distal femur	58.8–2942
Brown and Ferguson [35]	Proximal femur	1000–9800
William and Lewis [94]	Proximal tibia	100–500
Goldstein *et al* [95]	Proximal tibia	4.2–430
Martens *et al* [96]	Proximal femur	58–2248
Ashman *et al* [97]	Vertebral	158–378
Ashman and Rho [84]	Distal femur	1636
Hodgskinson and Currey [98]	Tibia, femur	4–1830
Jensen *et al* [77]	Calcaneus	54–861
Langton *et al* [209]	Calcaneus	35.6 ± 23.18

Many researchers have considered that elastic properties obtained from ultrasonic techniques are dynamic properties (high strain rate), because the ultrasonically determined moduli would appear to be relatively higher than moduli obtained from mechanical tests. Williams and Johnson [90] found that the ultrasonically determined value of the modulus was about twice the 'static' value. They attributed this result to the difference in strain rate. However, Ashman *et al* [85] showed a correlation coefficient of 0.96 between moduli measured with ultrasonic and tensile techniques. This issue may be addressed by performing non-destructive mechanical tests with several different strain rates and ultrasonic tests for same specimen.

5.3.8. *Conclusion*

We have described above the main methods used in the literature for the determination of the mechanical properties of bone. Some of the reported compressive moduli for cancellous bone are presented in table 5.2. Table 5.3 reviews some of the reported data for cortical bone. Each approach has inherent problems and advantages, but it seems likely that, given its long history and the large quantity of published data, uniaxial compressive testing will remain the method of choice in many applications.

Table 5.3. *Survey of compressive and tensile Young's modulus reported for cortical bone.*

Reference	Species	Type of bone	Type of test	Strength (MPa)	Young's modulus (GPa)
Reilly *et al* [99]	Human	Femur	Compression	167–215	14.7–19.7
Burstein *et al* [100]	Human	Femur	Compression	179–209	15.4–18.6
Cezayirlioglu *et al* [101]	Human	Femur	Compression	205–206	–
Ascenzi and Bonucci [102]	Human	Femur	Compression	90–167	4.9–9.5
Burstein *et al* [100]	Human	Tibia	Compression	183–213	24.5–34.3
McElhaney *et al* [103]	Bovine	Femur	Compression	133	24.1–27.6
Reilly *et al* [99]	Bovine	Femur	Compression	240–295	21.9–31.4
Simkin and Robin [70]	Bovine	Tibia	Compression	165	23.8 ± 2.2
Reilly *et al* [99]	Bovine	Tibia	Compression	228 ± 31	20.9 ± 3.26
Reilly *et al* [99]	Human	Femur	Tensile	107–140	11.4–19.7
Burstein *et al* [100]	Human	Femur	Tensile	120–140	15.6–17.7
Cezayirlioglu *et al* [101]	Human	Femur	Tensile	133–136	
Burstein *et al* [100]	Human	Tibia	Tensile	145–170	18.9–29.2
Vincetelli and Grigorov [104]	Human	Tibia	Tensile	162 ± 15	19.7 ± 2.4
Reilly *et al* [99]	Bovine	Femur	Tensile	129–182	23.1–30.4
Burstein *et al* [100]	Bovine	Tibia	Tensile	188 ± 9	28.2 ± 6.4

5.4. METHODS OF MEASURING THE MECHANICAL PROPERTIES OF THE TRABECULAE

The trabeculae can be considered to be the underlying unit structure of cancellous bone. The field of biomechanics has acquired a good understanding of trabecular bone behaviour on the continuum level (i.e. the macroscopic or structural level). Investigators have now begun to study trabecular bone behaviour at the level of individual trabeculae (i.e. the microstructural or material level). As mentioned elsewhere, the density of cancellous bone can be assessed in two different ways, corresponding to the two levels outlined above—the apparent or bulk density (typically 100–1000 kg m^{-3}) and the real or material density (typically 1600–2000 kg m^{-3}).

Characterization of the mechanical properties of human cancellous bone is a very important aspect of musculoskeletal research. The progress has been hampered to some extent by controversy over whether the properties of the trabeculae are the same as that of cortical bone, i.e. whether they are the same material or not. It all started with the speculation by Wolff [210] that cancellous bone possesses similar properties to cortical bone. This was supported by the much-cited empirical model of Carter and Hayes [48], where they indicated that the strength and Young's modulus were proportional to the square and cube of the apparent density respectively. In their model, they proposed that all bone should be viewed as a single material. The smallness of the struts (typically 100–200 µm in thickness) makes direct mechanical property measurements difficult to implement. Also to harvest the samples for measurement can be problematic. However, several different methods have been tried, including microhardness testing, nano-indentation, ultrasonic testing, buckling of single struts, three-point bending and finite element analysis. The results of some of the investigations are summarized in tables 5.2 and 5.3.

5.4.1. Microhardness

This section will discuss briefly some of the methods used to characterize trabecular material properties.

When indentations ranging from 20 to 150 mm in length are made with a load of less than 200 g, then the hardness should be included within the microhardness range. Vickers and Knoop are the two main measures of microhardness used today. If a Vickers hardness tester is used then the Vickers hardness number (VHN) is calculated from the expression

$$\text{VHN} = 1854.4 \frac{P}{d^2} \tag{5.13}$$

where P is the applied force (g) and d is the mean of two measured diagonals (µm). Microhardness is an attractive method of quantifying the physical effects of small-scale spatial variation in the composition of bone. This is more so for material (trabeculae) measurement of cancellous bone because bone specimens can be as small as 150 µm across. Weaver [105] showed that microhardness is an accurate and reliable measure of the degree of mineralization. Microhardness data provide a way of estimating the mechanical properties of specimens that are not easily accessible by conventional testing methods. However, this method cannot measure directly elastic moduli. The advantage of hardness tests is that they can be done on very small specimens. Cortical bone specimens can be plates 0.5 mm thick for a Vickers hardness test.

5.4.2. Nano-indentation

Nano-indentation is an evolution of the conventional hardness test for the assessment of the mechanical properties of thin films, small volumes and small microstructural features. This development is due to the availability of depth-sensing nano-indentation instruments capable of measuring displacements of the order of nanometres [106, 107]. From analysis of nano-indentation load-displacement data, it is possible to derive values of the elastic modulus [106, 108], hardness [106] and viscoelastic properties such as storage and loss moduli [109]. Nano-indentation can be used to probe a surface and map its properties on a spatially-resolved basis, often with a resolution of better than 1 μm, and to measure the properties of small microstructural features such as individual osteons and trabeculae [110–112].

Conventional mechanical testing has a lower limit of a few hundreds of microns at best. Micro-indentation measures a dimension of a few tenths to hundreds of microns [113, 114], while nano-indentation can study on a micron or smaller length scale. Since many important microstructural components of bone have dimensions of only a few microns or less, the nano-indentation technique can be used to investigate the mechanical properties of osteonal, interstitial and trabecular lamellar bone at the largely unexplored micron and submicron level. As an example, the nano-indentation method has been used to examine variations in the individual lamellar properties within osteons, as a function of distance from the osteonal centre. The technique also potentially allows one to examine the properties in different directions and therefore explore elastic anisotropy, even in very small specimens such as individual trabeculae. Characterization of lamellar bone properties by nano-indentation methods would thus aid in further theoretical developments of the mechanisms of fracture in bone.

Because of experimental complications associated with testing in liquid environments and keeping specimens wet during testing, most nano-indentation work has focused on dried bone [110]. It is well documented, however, that the mechanical properties of bone show notable changes after dehydration [67, 115, 116]. In general, drying increases the Young's modulus of bone, decreases its toughness and reduces the strain to fracture. An important question thus arises as to the degree to which nano-indentation mechanical property measurements are affected by drying. To this end, a series of tests was recently conducted on wet and dry bovine femur using a special testing fixture in which specimens could be tested fully immersed in a liquid [117]. Drying was found to increase the elastic modulus by 9.7% for interstitial lamellae and 15.4% for osteons. The hardness was also found to increase by 12.2% for interstitial lamellae and 17.6% for osteons.

The nano-indentation techniques elucidated in this chapter can be used to study mechanical properties at bone–implant interfaces, pathologic metabolic bone diseases, the properties of woven bone and calcified cartilage in fracture callus, the properties of the growth plate and biological aspects of implant loosening. Such results would fill a gap in our present knowledge by providing a better understanding of the mechanical function of bone at the microstructural and submicrostructural level.

The most common method for analysing nano-indentation load-displacement data is that of Oliver and Pharr [106], which expands on ideas developed by Doerner and Nix [118], but is not constrained by the assumption of a flat punch indenter geometry. A typical load–displacement

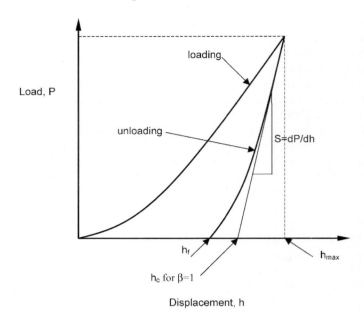

Figure 5.13. *A schematic representation of load versus indenter displacement data for an indenation experiment. showing quantities used in data analysis as well as a graphical interpretation of the contact depth. h_f is the final depth of the contact impression after unloading; h_{max} is the indenter displacement at peak load; h_c is the vertical distance along which contact is made; S is the initial unloading stiffness.*

indentation curve is shown in figure 5.13. During loading, both elastic and plastic deformation occur under the indenter as the contacts change with increasing depth. The unloading part of the curve is dominated by elastic displacements. The hardness is found by computing the mean pressure under the indenter at the point of maximum load. This requires a knowledge of the contact area at that point. Because the unloading curve is dominated by elastic displacement, it is possible to determine the elastic modulus of the material being indented from the slope of the unloading curve [119].

Figure 5.14 shows a cross-section of an indentation and identifies the parameters used in the analysis. At any time during loading, the total displacement h is written as

$$h = h_c + h_s, \qquad (5.14)$$

where h_c is the vertical distance along which contact is made (called contact depth) and h_s is the displacement of the surface at the perimeter of the contact. At peak load, the load and displacement are P_{max} and h_{max}, respectively. Upon unloading, the elastic displacements are recovered and, when

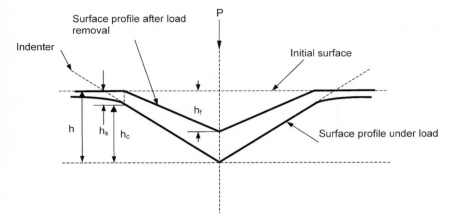

Figure 5.14. *A schematic representation of a section through an indentation showing various quantities used in the analysis.*

the indenter is fully withdrawn, the final depth of the residual hardness impression is h_f.

The analysis begins by fitting the unloading curve to the power-law relation

$$P = b(h - h_f)^m \qquad (5.15)$$

where P is the indentation load, h is the displacement, b and m are empirically determined fitting parameters and h_f is the final displacement after complete unloading, determined by curve fitting. The unloading stiffness, S, that is the slope of the unloading curve during the initial stages of unloading, is then established by differentiating equation (5.15) at the maximum depth of penetration, $h = h_{max}$, giving

$$S = \frac{dP}{dh}\bigg|_{h=h_{max}} = mb(h_{max} - h_f)^{m-1}. \qquad (5.16)$$

The depth along which contact is made between the indenter and the specimen, h_c, can also be estimated from the load–displacement data using

$$h_c = h_{max} - \varepsilon \frac{P_{max}}{S} \qquad (5.17)$$

where P_{max} is the peak indentation load and ε is a constant which depends on the geometry of the indenter [$\varepsilon = 1$ for the flat punch; $\dot{\varepsilon} = 0.75$ for the paraboloid of revolution (Berkovich indenter); $\varepsilon = 0.72$ for the cornical indenter]. With these basic measurements, the projected contact area of the hardness impression, A, is derived by evaluating an empirically determined indenter shape function at the contact depth, h_c; that is, $A = f(h_c)$.

Finally, the hardness, H, and the effective elastic modulus, E_{eff}, are derived from

$$H = \frac{P_{\max}}{A} \qquad (5.18)$$

and

$$E_{\text{eff}} = \frac{1}{\beta} \frac{\sqrt{\pi}}{2} \frac{S}{\sqrt{A}} \qquad (5.19)$$

where β is 1 for a circular (flat punch), 1.034 for a triangular (Berkovich indenter) and 1.012 for a square (Vickers indenter) indentation. Note that the values of b for the triangular and square geometries deviate from the circular one by only 3.4% and 1.2% respectively [120].

An effective modulus is used in the analysis to account for the fact that elastic deformation occurs in both the specimen and the indenter. The effective modulus is related to the specimen modulus through

$$\frac{1}{E_{\text{eff}}} = \frac{1 - \nu^2}{E} + \frac{1 - \nu_i^2}{E_i} \qquad (5.20)$$

where E and ν are indentation modulus and Poisson's ratio for the specimen, and E_i and ν_i are the same quantities for the indenter. The elastic properties of the diamond indenter, ν_i and E_i, are 0.07 and 1140 GPa, respectively. We assume that Poisson's ratio for bone is $\nu = 0.3$. A sensitivity study showed that varying ν in the range 0.2 to 0.4 changed the measured values of E by no more than 8%.

The attractiveness of this approach is that direct observation and measurement of the contact area is not needed for the evaluation of H and E, thus facilitating property measurement from very small indentations. Clearly, however, the accuracy with which H and E can be measured depends on how well equations (5.16)–(5.20) describe the indentation deformation behaviour. In this regard, it is important to note that these equations were derived by Sneddon [121], assuming elastic deformation only. One important way in which an elastic solution fails to properly describe the elastic/plastic behaviour observed in indentation, concerns the pile-up or sink-in of material around the indenter. In the purely elastic contact solution, material always sinks in, while for elastic/plastic contact, material may either sink in or pile up. Since this has important effects on the indentation contact area, it is not entirely surprising that the Oliver–Pharr method has been found to work well for hard ceramics, in which sink-in predominates, but significant errors can be encountered when the method is applied to soft metals that exhibit extensive pile-up [122]. Pile-up leads to contact areas that are greater than the cross-sectional area of the indenter at a given depth. These effects lead to errors in the absolute measurement of mechanical properties by nano-indenta-indentation [107]. However, the elastic and plastic properties of bone are such that pile-up is not an important issue [123], and can largely be ignored.

5.4.3. Buckling

It has been suggested that bending and buckling are highly probable modes of deformation of trabeculae at most cancellous bone sites [46]. This has encouraged the direct use of mechanical buckling to investigate the trabeculae properties. The elastic buckling theory is based on the Euler's equation given by

$$P_c = \frac{\pi^2 EI}{k^2 l^2} \tag{5.21}$$

where P_c is the critical applied load, E is Young's modulus, I is the area moment of inertia, k is the end condition constant and l is the free length of the beam. Researchers like Runkle and Pugh [124] and Townsend *et al* [115] have used the above method to obtain 8.75 and 11.46 GPa (wet) and 14.1 GPa (dry) respectively for Young's modulus of trabeculae.

5.4.4. Ultrasound technique

Ultrasound presents another interesting modality for the measurement of the trabeculae. As previously indicated, the propagation is dependent on the sample size and on the frequency/wavelength of the ultrasound waves [125]. For higher frequencies such as 2 MHz, the wave will travel along the individual trabeculae and allow calculation of Young's modulus of the trabecular material [84]. Inhomogeneity and anisotropy of cancellous bone may contribute to velocity measurement error. Acoustic waves travel faster through denser regions of bone; therefore, in an inhomogeneous specimen, acoustic waves arrive at slightly different times. The variation in anisotropy of cancellous bone may also be taken into account in the calculation of velocity. The true pathway of ultrasound is not well defined. Rho *et al* [126] reported that the actual pathway of cancellous bone measurement was slightly longer (6.5%) than the average length of the specimen. However, the material elastic modulus predicted by high-frequency ultrasonic velocity [84] was recaluated to be 14.7 GPa by taking account of the actual pathway of cancellous bone, which is a considerably higher value than that reported by others. This value is threefold greater than the most commonly used value in analytical models (5.4 GPa; Choi *et al* [127]). However, there is still controversy as to the value of the material elastic modulus for trabecular bone (table 5.4).

5.4.5. Other techniques

Acoustic microscopy is another ultrasound modality that has proven useful in measuring the material properties of bone [128, 129]. Acoustic microscopy provides a direct means of measuring the material properties of tissues at a

Table 5.4. *Estimates and determinations of the elastic modulus of individual trabeculae of cancellous bone.*

Reference	Type of bone	Analytical/test method	Estimate of trabecular moduli (GPa)
Wolff [210]	Human	*	17–20 (wet)
	Bovine		18–22 (wet)
Pugh et al [46]	Human distal femur	Finite element	Concluded that $E_{trab} < E_{compact}$
Townsend et al [91]	Human proximal tibia	Inelastic buckling	11.38 (wet) 14.13 (dry)
Runkle and Pugh [124]	Human distal femur	Buckling	8.69 ± 3.17 (dry)
Williams and Lewis [94]	Human proximal tibia	Experiment with two-dimensional, finite element	1.3
Ryan and Williams [132]	Fresh bovine femur	Tension test, single trabeculae	0.76 ± 0.39
Mente and Lewis [133]	Dried human femur	Experiment with finite element	5.3 ± 2.6
Ku et al [130]	Fresh frozen human tibia	Three-point bending of ultra-small machined	3.17 ± 1.5
Ashman and Rho [84]	Bovine femur	Microtensile testing	10.90 ± 1.60 (dry)
		Ultrasound test	14.8 ± 1.4 (wet)
Rice et al [131]	Human	Statistical data	0.61
	Bovine	Statistical data	1.17
Hodgskinson et al [134]	Bovine proximal femur	Microhardness	Concluded no statistical difference
Rho et al [135]	Human	Nano-indentation	19.6 ± 3.5 (dry) longitudinal direction 15.0 ± 3.0 (dry) transverse direction

microscopic scale. Very high frequencies are required starting at 50 MHz with a resolution of about 60 µm. However, increases in frequency are accompanied by rapidly increasing ultrasonic attenuation, which limits the testing range.

Other techniques, as shown in table 5.4, have been implemented in the study of trabeculae. The discrepancy and controversy still exist as to the value of the elastic modulus of trabeculae as compared with cortical bone. While Ku *et al* [130] showed no difference (3.8 GPa (cortical) and 3.17 GPa (trabeculae)), William and Lewis [94] predicted an order of 10 difference (1.3 GPa (trabeculae) 16 GPa (cortical)). The small value for Young's modulus of cortical bone obtained by Ku *et al* [130] is rather puzzling in comparison with published data for cortical bone. Rice *et al* [131] speculated that the very small specimen size may enhance surface imperfections' effects on the measured value.

5.4.6. *Conclusions*

- Dissection of single trabeculae from cancellous bone is a problem of its own.
- The various measuring procedures have their own inherent problems and assumptions which contribute to uncertainty in the results.
- The turnover rate in cancellous tissue is eight times that in cortical bone tissue, thus the calcium content of trabeculae bone cannot be assumed to be homogeneous [131].

The above factors may be wholly or partly responsible for the discrepancies in the values obtained by different researchers. On the other hand, the question of the similarity of the elasticity of trabeculae and cortical bone remains in dispute on two accounts. Firstly it has been shown that the bone mineral content (BMC) and material density of trabeculae are less than those of cortical bone (BMC = 63% (cortical), 51% (trabeculae)) [136]. The density of ash cortical bone is consistently higher than in cancellous bone tissue [137]. Secondly, small changes to mineral content and density have disproportionately large effects on bone stiffness and strength [38, 88, 138]. Ashman and Rho [84], using the equation quoted by Currey [38] for the dependence on BMC, estimated that the elastic modulus of trabecular bone (at the material level) would be expected to be 25% lower than cortical bone. This argument, coupled with the fact that most investigations find Young's modulus of the trabeculae to be less than that of cortical bone, call for a rethink of the assumption made by Wolff [210] and later by Carter and Hayes [48]. Hopefully, in the future, more refined modes of measurement will help resolve categorically the controversy.

5.5. FACTORS INFLUENCING THE MECHANICAL PROPERTIES OF BONE

Young's modulus of bone (at the structural level) reported in the literature ranges spans an extraordinary five orders of magnitude, ranging from approximately 0.2 to 10 000 MPa in compression. A survey of the published data is given in tables 5.2 and 5.3. In principle, mechanical testing of bone is a straightforward process. Experimental results, however, can be affected by specimen preparation, test methods and environmental conditions. In particular loading rate, deformation rate, specimen size, specimen shape, mode of loading (tensile, compressive, bending etc.) and the method of gripping the test specimen can influence the mechanical response of bone and engineering materials. The mechanical properties are also influenced by factors such as age of the bone, disease state of the specimen, species and anatomical location.

The following sections will discuss how some of these conditions affect the mechanical properties of bone.

5.5.1. *Specimen configuration*

The impact of specimen configuration depends on the type of testing and the type of bone. For cancellous bone, high porosity makes complicated specimen shapes difficult to machine since the bone may be fragile and prone to damage during machining. Human cancellous bone from the vertebra and calcaneus of elderly subjects can have porosity of up to 95% and poses particular problems with regard to machining. Different specimen configurations have been used for different methods of testing, with cubic and cylindrical specimen geometries commonly used in compressive testing. Generally, the specimen must be large enough for a representative section of pores and struts to be characterized, yet small enough to resolve the heterogeneity of the tissue. It has been shown that a $5\,mm^3$ test sample is the minimum size for which the continuum characterization is realistic [35, 89]. This minimum size is, however, disputed, with Linde *et al* [139] advocating $7\,mm^3$.

Surface irregularities, induced bending and difficulties associated with accurately measuring strains all present problems in compressive testing. The problems can be minimized by width to length ratio, by making the measuring surfaces as parallel as possible, and by measuring strain with extensometers directly attached to the specimen. The use of a confining ring has been reported to result in the data being higher by a factor of about one third than that which would be measured in an unconfined uni-axial test [37]. This may be because side bending is prevented. On the other hand, loading in tension, though less susceptible to non-uniform loading, requires larger and more complicated shapes.

Cancellous bone is anisotropic but the degree of anisotropy varies, for example, with anatomical site and with age. Characterizing the mechanical anisotropy requires measurements to be made in different directions. One approach has been to measure in three orthogonal directions corresponding to the anatomical axes [140]. If a non-destructive testing methodology is adopted, in which applied strains do not exceed the yield strain, individual specimens can be tested in each of the three directions in turn [140].

5.5.2. Specimen preservation

A problem of great importance in studies on mechanical properties of bone is the preservation of a specimen during the time between harvest and testing. The main concerns are to prevent the specimen from drying and degradation of the mechanical properties. The degradation process starts soon after life-to-death transition and leads to dramatic changes in bone mineral properties [49]. Degradation can be minimized by harvesting the specimen as early as possible after death and transferring it to some form of preservation. This can be done by keeping it in water, physiological saline solution, embalming fluid, alcohol or by freezing it. The possible effect on mechanical properties will depend on the method and length of preservation. The effects of the various methods of preservation have been investigated by different researchers including Sedlin and Hirsch [141], Evans [67] and Linde and Sorensen [54].

Freezing at $-20\,^{\circ}\text{C}$ has been shown not to affect the physical properties of bone so long as the specimen is adequately hydrated and thawed before testing [67, 141]. Most researchers have adopted freezing as the mode of storage, although there is variation in the reported temperatures used. For example, Evans *et al* [142] used $-20\,^{\circ}\text{C}$, Brown and Ferguson [35] used $-10\,^{\circ}\text{C}$ and Hvid *et al* [143] used $-30\,^{\circ}\text{C}$. There is a consensus that the best method of long-term preservation prior to testing is to freeze the specimen at $-20\,^{\circ}\text{C}$ or lower in saline-soaked gauze [51, 55, 56]. Freeze-drying, on the other hand, has disastrous effects on strength [144]. Sedlin and Hirsh [141] showed that embalmed bone exhibits different values of strength and elasticity compared with fresh tissue.

5.5.3. Bone hydration

The process of drying has a marked effect on the mechanical properties of bone. Kraus concluded that the ultimate tensile strength, elasticity and hardness were increased, whereas the ultimate shear strength and elongation were reduced in bone in the process of drying. Complete drying of bone has been shown to increase its tensile strength compared with that of wet bone [67, 141, 145, 146]. Townsend *et al* [115] reported an increase of about 25–30% in the elastic modulus of both cortical and cancellous bone. Therefore, all bones used throughout the course of this study were maintained wet and from unembalmed adults with no history of bone disease.

5.5.4. Sterilization

There is an increasing awareness of the risk of human samples being contaminated with a variety of pathogens, therefore there is an increasing need for bone tissue to be sterilized. Sterilization is also important for bone samples used in bone grafts. Sterilization can be achieved using chemical fixation, although there is only indirect evidence that fixation destroys potential pathogens in tissues [147]. However, chemical fixation may have an effect on the mechanical properties of bone. Chemical fixation using aldehydes has a direct effect on the primary amine groups of the polypeptide collagen chains by forming an increased number of inter- and intra-fibrillar cross-links [148, 149]. The effects of fixation on the mechanical properties of bone have been reported in the literature [67, 103, 141, 150]. McElhaney *et al* [103] reported a significant reduction in bone's ultimate compressive strength, but an insignificant reduction of ultimate tensile strength. Sedlin [141] also reported an insignificant increase (4.3%) in stiffness after treatment with 10% formalin (4% formaldehyde). However, Evans [67] found a 68% increase in Young's modulus and ultimate strength following chemical fixation. Recently, Currey *et al* [150] reported that fixing with formaldehydes has a small effect on Young's modulus and that there was no compelling evidence that other properties measured in static tests are much affected. However, they found that impact energy absorption was reduced by quite a large amount.

Another mode of sterilization is the use of ionizing radiation. The effects of radiation on the biomechanical properties of bone remain a matter of debate [151–157]. Zhang *et al* [157] found that there was no significant difference in mechanical or material properties of the iliac crest wedges exposed to dosages of 2–2.5 kGy when compared with the non-irradiated wedges. They suggested that 2–2.5 kGy is an acceptable dose for gamma irradiation sterilization of the iliac crest. Komender [151] found that the bending, compression and torsion strength of human femoral diaphysis to be compromised significantly when irradiated with doses of 60 kGy but not when doses of 30 kGy were used. Another study by Hamer *et al* [156] showed a significant reduction (64%) in bending strength of human femur after irradiation with 28 kGy. Currey *et al* [155] found that Young's modulus was unchanged by any level of radiation (17, 29.5 and 94.7 kGy). However, radiation significantly reduced bending strength, work to fracture and impact energy absorption. The study of Currey *et al* [155] and those of other have consistently demonstrated that the reduction in strength is dependent on the dose [155, 156]. Radiation, even at relatively low doses, makes the bone more brittle and thereby reduces its energy-absorbing capacity. To limit the impact of radiation on the mechanical properties of bone, low levels of radiation sufficient to produce bacterial safety should be used in conjunction with biological tests.

5.5.5.   *Strain rate*

The viscoelasticity of bone poses problems of reproducibility associated with respect to variation in strain rate used across different studies. The resistance of bone to load increases with increasing rate of application of the load, so

Table 5.5. *Effect of strain rate on Young's modulus of bovine cortical bone [159]*

Strain rate (s^{-1})	Young's modulus (GPa)
0.001	18.6
0.01	20.0
0.1	24.4
1	27.6
300	33.1
1500	42.0

the mechanical properties of bone vary slightly with strain rate [158]. The effect of strain rate was originally demonstrated in compression by McElhaney [159], as shown in table 5.5. This fact was confirmed by many workers [88, 160–162]; with the comprehensive study of Wright and Hayes [163] in tension, it was established that the elastic modulus increased by a factor of 2 (i.e. ~17–40 GPa), from a 'quasi-static' strain rate range to a very rapid strain rate. Carter and Hayes [88] found that both strength and elastic modulus of cancellous bone were approximately proportional to strain rate raised to the power 0.06. The strain rate to which bone is normally subjected ranges from $0.001 \, s^{-1}$ for slow walking to $0.01 \, s^{-1}$ for vigorous activity. Using the results of Carter and Hayes [88], the elastic modulus might vary as much as 15% during normal activity [88].

Measurement of actual strain in the gauge section of the specimen, rather than the travel of the grips or testing machine platens [50, 85, 164] has been found to be problematic. For small deformations there are no theoretical reasons for Young's modulus to be different in the two strain measurement configurations. However, deformation measurement without an extensometer according to Linde and Hvid [50] resulted in a 30% error. Allard [165] showed differences of up to 50% when elastic modulus was measured with an extensometer versus load platens.

5.5.6. *Age and disease*

The mechanical properties of cancellous bone have shown a significant age-related decrease [166–169]. Mosekilde *et al* [167] observed a 75–80% decrease in stiffness from age 20 to 80 years. Metabolic bone disease is known to lower values of the mechanical properties below those of healthy bone [170–172]. For example, Katz and Yoon [171] and Ashman *et al* [173] showed that the ultrasound velocities in osteopetrotic bone are significantly lower than those in normal bone of essentially the same density, while the osteoporotic bone values are consistent with their reduced densities.

5.5.7. Temperature

The elastic properties of many materials are temperature dependent [55, 56]. The majority of tests on cancellous bone have been carried out at room temperature (21 °C) and it is likely that the mechanical properties will have different values at body temperature (37 °C). Smith and Walmsley [174] demonstrated that Young's modulus was approximately linearly but inversely proportional to the temperature in the temperature range used (5–37 °C). Brear *et al* [175] demonstrated that loading at 37 °C rather than room temperature results in lower values for all the mechanical properties they tested, but the decrease was not large (2–4%).

5.5.8. Miscellaneous

Cancellous bone, as explained earlier, has the porous cells filled with marrow. Carter and Hayes [48] reported that the presence of marrow during testing did not generally alter the elastic properties of cancellous bone. While some researchers carry out their investigations on defatted specimens [97], others either work on specimens with fat *in situ* or do not bother to report on whether or not marrow is present [176]. It seems safe to assume that, provided very high strain rate testing is not anticipated, removal of bone marrow will not affect measured mechanical properties.

Carter *et al* [63] compared two methods for marrow removal—warm water jetting and solvents (methanol and chloroform)—and found that the method of marrow extraction did not affect the elastic properties. They also showed that samples which were kept wet throughout the preparation procedure were generally more compliant than samples which were dried and rewetted. Sharp *et al* [177] showed that only immersion in trichloroethylene in an ultrasound bath for 4 h removed all the fat.

5.6. MECHANICAL PROPERTIES OF BONE

5.6.1. Introduction

Bone is a very complex organ and has been proposed to have a hierarchy [3, 178]. In order to understand its mechanical properties, it is important to understand the mechanical properties of its component phases, and the structural relationship between them at various levels of hierarchical organization. The following hierarchy has been proposed: macrostructure (whole bone), architecture (compact or cancellous bone), microstructure (osteon, trabeculae), submicrostructure (lamella, large collagen fibres) and ultrastructure (collagen fibres and molecules) [3]. The mechanical properties will depend on what level the assessment is carried out. The detailed analysis is beyond the scope of this chapter. The reader is referred to more detailed books [179, 180].

5.6.2.   *Mechanical properties of cancellous bone*

5.6.2.1.   *Introduction*

There is now a general consensus that the mechanical properties of cancellous bone, like any other cellular material, are affected by three intrinsic factors:

1.   apparent density (volume fraction of solids)
2.   architecture or fabric of the trabeculae
3.   properties of the cell wall material, notably the mineral content of the bone material itself [21, 37, 48, 63, 88, 134, 181].

Additional variables include the temperature at which the properties are being determined, whether the bone is living or dead, embalmed or fresh, and the age, sex, race and species of animal from which the bone is obtained (see previous section). There may be other microstructural factors such as amorphous-to-crystalline calcium phosphate ratio, and mineral to non-mineral bonding [67, 182]. Over the years many researchers have tried to relate the above physical properties to the mechanical properties (compressive strength, elastic modulus, tensile strength). This has stemmed from the societal need to be able to predict fracture risk from physical factors that may readily be measured *in vivo*. The following sections will summarize the various attempts to correlate these physical factors to mechanical properties.

5.6.2.2.   *Density*

Two types of density are usually quoted in the literature: apparent density (bulk or structural density) ρ_a and real density (material density) ρ_m. The porosity of a porous material is of course linearly related to the apparent density, assuming that material density is constant. Therefore either porosity or apparent density is useful and, in many cases, equivalent measures of the amount of mineralized tissue present.

   Different methods of measuring density have been used in the literature, sometimes making comparison of results difficult. Sharp *et al* [177] reviewed some of the methods used in the literature. They went on to suggest a protocol for measuring the density of trabecular bone, as follows. The total specimen volume and total specimen mass are measured prior to mechanical testing, and the sample is then defatted in trichlorethylene in an ultrasound bath for 4 h. The defatted specimen is then rehydrated in distilled water for 4 h under vacuum, and then suspended on a hook, submerged in distilled water, and the submerged weight measured using Archimedes' principle. The specimen is then centrifuged for 15 min at 17g on blotting paper, and finally weighed in air to obtain the hydrated tissue mass. The density of water multiplied by the difference between the hydrated tissue mass and the submerged mass gives the bone tissue volume. The densities may then be calculated.

There is a wealth of data attempting to relate mechanical properties (Young's modulus, compressive strength and tensile strength) of cancellous bone to the apparent density. The attractiveness of the approach lies in the fact that these measures are relatively easy and may relate to similar non-invasive techniques, enabling the characterization of bone properties to be performed *in vivo* and reviewed by Goldstein [183]. For example, apparent density is directly related to bone mineral density measured *in vivo* using X-ray absorptiometry. The need to be able to predict fracture risk, for example in the context of osteoporosis, has driven much of the research on the determinants of bone mechanical properties and specifically on the role of bone density as an explanatory factor. Both linear and power law functions have been used to relate mechanical properties to density (see table 5.6). Power law models are generally found to offer better fits to the data, especially when a wide density range is studied, and are expected from a theoretical standpoint to provide a better description of the characteristic behaviour of a cellular material [184].

Table 5.6. *The relationship between mechanical properties and density of trabecular bone.*

Reference	Region	Relationship
Weaver and Chalmers [186]	Human vertebral	Linear
	Human calcaneus	Linear
Galante *et al* [137]	Vertebral	Linear
McElhaney *et al* [187]	Cranial bone	$E_c \propto \rho^3$
		$\sigma_c \propto \rho^2$
Behrens *et al* [182]	Distal femur	Linear
	Proximal tibia	Linear
Lindahl [92]	Vertebral	Linear
Carter and Hayes [88]	Human and bovine tibia	$\sigma_c = 68\dot{\varepsilon}^{0.06}\rho^2$
		$E_c = 3790\dot{\varepsilon}^{0.06}\rho^2$
Carter *et al* [63]	Proximal femur	As above
Marten *et al* [96]	Proximal femur	Linear
Stone *et al* [64]	Bovine humeri	$\sigma_s = 21.6\rho^{1.65}$
Kaplan *et al* [65]	Bovine humeri	$\sigma_T = 14.5\rho^{1.71}$
		$\sigma_c = 32.4\rho^{1.85}$
Brear *et al* [175]	Bovine femur	$E_c \propto \rho^{1.87}$
		$\sigma_c \propto \rho^{1.77}$

E is Young's modulus, σ is the strength, ρ is the density and subscripts c, s and T are compressive, shear and tensile respectively.

Carter and Hayes [48, 88] reported a power relationship between apparent density and compressive strength and Young's modulus for tests on human and bovine cancellous bone. Employing data from the work of McElhaney [159] and Galante *et al* [137], Carter and Hayes [88] concluded that the axial Young's modulus E depended upon the apparent density to the third power and strain rate $\dot{\varepsilon}$ to the power 0.06, thus arriving at E in MPa as

$$E = 3790\dot{\varepsilon}^{0.06}\rho^3 \tag{5.22}$$

where ρ is the apparent density in $g\,cm^{-3}$ and $\dot{\varepsilon}$ is the strain rate in s^{-1}. They also concluded that the axial compressive stress σ depended upon strain rate to the 0.06 power and upon the square of apparent density, thus

$$\sigma = 68\dot{\varepsilon}^{0.06}\rho^2 \tag{5.23}$$

where σ is in MPa. Two assumptions were made before empirically arriving at the above results, namely:

1. The Young's modulus and strength of both human and bovine cancellous bone have exactly the same dependence on apparent density.
2. Based on the suggestion of Wolff's law [210], cortical bone is simply more dense than cancellous.

The validity of both of the above assumptions is debatable.

On the other hand, other researchers such as Brear *et al* [175] and Hodgskinson and Currey [185] could only demonstrate a quadratic relationship for both strength and Young's modulus. Rice *et al* [131] conducted a comprehensive survey of the literature. By taking into consideration the factors of species (bovine/human), direction (longitudinal, anterior–posterior and medial–lateral), axial stress condition (compressive or tensile) and method of testing (confined, unconfined and indentation), they carried out a rigorous statistical analysis and came to the following conclusions.

*   Elastic modulus is proportional to the square of apparent density.
*   Strength is proportional to the square of apparent density.
*   Elastic modulus is directly proportional to strength.
*   The data on mechanical properties of cancellous bone tissue from different species cannot be combined.
*   The mechanical properties of compact bone cannot be obtained by extrapolation of the mechanical properties of cancellous.

The results obtained by Carter and Hayes [48, 88] were derived empirically. Theoretical calculation of elastic modulus and strength of very highly porous materials will depend on the state of the pores of the material, that is (a) whether the cells are closed or open, and (b) whether the loads applied to the cell walls deform them primarily in simple axial loading (tension or compression) or in bending.

Cancellous bone structure ranges from open cell symmetric to closed cell columnar [4, 37], so it will depend on the specimen in question. Most of the evidence of cancellous bone mechanics suggests that bending is the prime mode of deformation of the cell walls [37, 46], although William and Lewis [94] identified situations where deformation was primarily axial loading.

Gibson [37] has attempted to tie together the power relationships (quadratic and cubic) obtained by different researchers, using theoretical models. He argued that, at low apparent densities, the structure of cancellous bone is one of connecting rods to form open cells. At higher apparent densities the cells fill in more, and the structure becomes one of plates forming closed cells. It was further argued that the principal mechanism of the linear elastic deformation is cell wall bending and that elastic collapse is caused by elastic bending and buckling of cell walls. The model and analysis predict (for both compressive strength and Young's modulus) a quadratic relation for an open cell and a cubic relation for a closed cell structure. The limitation of the above theoretical treatise is that it is based on regular repeating cells. The cell size of trabecular bone, however, is not uniform. Also, in some locations, the pores in the trabecular bone have a distinct orientation that induces a mechanical anisotropy.

The bone mineral content (BMC) or bone mineral density (BMD) has the advantage that it can be measured *in vivo* by X-ray absorption techniques (see chapters 8–11). Weaver and Chalmers [186] as well as other investigators have sought a correlation between mineral content and mechanical properties. Behrens *et al* [182] reported a statistically significant, although not strong, relationship between the linear photon absorption coefficient and the compressive strength of cancellous bone. Hvid *et al* [78] reported a linear relationship between BMC and mechanical properties. Currey [184], using the data of Hvid *et al* [78], demonstrated that a power relationship should be favoured. Recent studies have also confirmed the high association between BMD and mechanical properties [188].

The problem and cause of the scatter in the relationship between mechanical properties and density is that the mechanical properties are influenced by many factors, some of which it may not be possible to control in a given experiment [189]. For example, it must be remembered that bone is anisotropic, and that consequently different correlations with density may exist for different material directions. Furthermore, the degree of anisotropy varies significantly throughout most cancellous structure. Therefore one should use caution when predicting *tensorial* properties (e.g. moduli) from a *scalar* property (e.g. density) [97].

5.6.2.3. Architecture

It is generally known that the physiology of bone function directly influences the structure and strength of bone, a relationship known as Wolff's law [210].

It follows that the arrangement of trabeculae in space follows the lines of stress and the architecture of trabecular bone then represents a physiological optimization aimed at maintaining mechanical integrity while minimizing bone mass. These variations in mechanical properties have been shown to be a function of anatomical position and loading direction. This has been supported by work done on cancellous bone at knee [182], proximal femur [35], ankle joint [143, 190] and proximal tibia [39].

Early evidence of the role of architecture was reported by Bell *et al* [166], who pointed out that, in ageing human vertebrae, cancellous bone loses strength more rapidly than the reduction in the amount of bone tissue. They attributed their findings to the fact that the loss is largely in transverse struts. The function of these struts is probably to act principally as lateral braces so that the longitudinal struts do not buckle. Reducing the number of points at which sideways support is provided for a slender column considerably reduces the load it can bear even though the cross-sectional area remains unchanged [34]. Galante *et al* [137] showed that specimens (vertebral bone) loaded in the superior–inferior direction were more than twice as strong as those loaded in the lateral–medial direction. Behrens *et al* [182] also observed that the compressive strength of the knee varies with the location. Brown and Ferguson [35] produced detailed quantitative information about the spatial and directional variations of the material properties of cancellous bone (proximal femur).

The continuous process of remodelling that occurs throughout life changes the internal architecture of bone. The modelling and remodelling activity of bone is influenced by the complete loading history to which the tissue is exposed over some period of time [191]. Behren *et al* [182] were able to show a strength difference between loaded and unloaded condyles of specimens taken from patients suffering from vagus. Vajjhala *et al* [192] found that modulus and strength of trabecular bone are reduced more dramatically by density losses from resorption of trabeculae than by those from uniform thinning of trabeculae.

Researchers including Brown and Ferguson [35] have tried to link the stress pattern to the actual orientation of trabeculae. This has been thanks to advanced stereology, digital analysis algorithms and refined imaging processes. Pugh *et al* [46], Townsend *et al* [91] and Hodgskinson and Currey [98, 185] have attempted to correlate the architectural properties (anisotropy, fabric and connectivity) and morphologic measures (trabecular plate thickness and trabecular plate separation) to the structural level mechanical properties of cancellous bone. The problem of multi-collinearity makes it difficult to establish the true underlying pattern of relationships. However, there is convincing evidence for a significant role for fabric, independent of apparent density, in determining mechanical properties [98, 185].

Clinically Ciarelli *et al* [193] found that there was an architectural difference between hip fracture and control subjects in terms of the three-dimensional spatial arrangement of the trabeculae. This emphasizes the importance of structure in predicting fracture risk.

5.6.3.   Mechanical properties of cortical bone

5.6.3.1.   Introduction

The mechanical properties of cortical bone depend on various factors including the type of mechanical testing [164]. The compressive strength and elastic modulus for cortical bone range from 133 to 295 MPa and from 14.7 to 34.3 GPa respectively, while tensile strength and elastic modulus range from 92 to 188 MPa and from 7.1 to 28.2 GPa respectively (see table 5.3). Bending strength and elastic modulus have also been reported in the literature and vary from 35 to 283 MPa and 5 to 23 GPa respectively. It has to be noted that the value will depend on whether the measurement is on whole bone or on cortical cut beams. An *et al* [179] reported from the literature that both strength and elastic modulus of whole bone are about 60% of the cortical bone beams.

5.6.3.2.   Bone density

The apparent density of cortical bone is the dry wet divided by the specimen volume. It is a function of both the porosity and mineralization of the bone materials. Cortical bone has an average apparent density of approximately 1.9 g/cm^3. For cortical bone, apparent density and material density are approximately the same, as there are no marrow spaces in cortical bone. There is a positive correlation between apparent density of cortical bone and its mechanical properties [194]. Density as measured using DXA is also positively correlated with the strength and stiffness of various bones [195–197].

The effects of porosity of cortical bone on mechanical properties have been well studied [180]. In cortical bone the porosity is determined by Haversian canals, absorption cavities and vascular channel. The variation in the elastic modulus of cortical bone can be explained by a power-law relationship with porosity. Different authors have reported different values for the power; for example, Schaffler and Burr [198] proposed the following relationship for bovine cortical bone using tensile testing:

$$E = 33.9(1 - p)^{10.9}. \tag{5.24}$$

5.6.3.3.   Anisotropy and heterogeneity

Cortical bone is considered to be anisotropic; in particular, human cortical bone is transversely isotropic [69, 82, 171, 199, 200]. This is because in cortical bone the collagen fibres and osteons are longitudinally oriented.

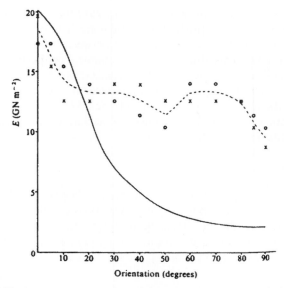

Figure 5.15. *Variation in Young's modulus (E) of bovine femur specimens with the orientation of the specimen axis to the long axis of the bone, for wet (○) and dry (×) conditions, compared with the theoretical curve predicted from a fibre-reinforced composite model by Bonfield and Grynpas [199].*

The longitudinal direction is considered to be approximately parallel to the long axis of the diaphysis. Thus the elastic modulus of cortical bone tested in the longitudinal direction is approximately 50% greater than its modulus in the transverse direction. The elastic modulus of cortical bone is similar in tension and compression, but lower in shear [99]. The variation of elastic modulus (E) with orientation for bovine compact bone shows a non-uniform decrease from the longitudinal (0°) to the transverse orientation (90°), as illustrated in figure 5.15. These data do not follow exactly the predictions of a model of bone as a fibre-reinforced composite [199]. The relevant data points in figure 5.15 agree reasonably with the values obtained by Reilly and Burstein [69] for human compact bone with values of elastic modulus at 0°, 30°, 60° and 90° of 23.1, 16.7, 12.8 and 10.4 GPa, respectively.

Similar to the elastic properties, the strength properties of cortical bone also depend on the loading direction [69, 82, 201]. For example, both the tensile and compressive strength of specimen loaded in the longitudinal direction are approximately 2.5 and 1.5 times greater, respectively, than those loaded in the transverse direction. The strength properties are higher in compression than in either tension or shear. For longitudinally oriented specimens, the ultimate compressive strength is about 50% greater than the tensile strength, whereas in transversely oriented specimens the difference between compressive and tensile strength is over twofold.

Values for mechanical properties of bone vary from one bone to another as well as within different regions of the same bone. There is great variation in mechanical properties between animals and between bones [38, 67, 95, 202]. It has be proposed that the differences in mechanical properties are produced mainly by differences in the amount of mineralization [38, 203, 204]. Property differences may also be explained by differences in mechanical function or differences in the microstructure between bones. Moreover, even for similar bones, or within a given bone, differences in mechanical properties can exist, which reflect local structural variations [62, 85].

5.6.3.4. *Miscellaneous*

Other factors that may influence the mechanical behaviour of cortical bone include its histological structure (primary versus osteonal bone), the collagen and orientation of collagen fibres, the number and composition of cement lines and the presence of fatigue-induced microdamage [205–208].

REFERENCES

[1] Njeh C F, Cheng X G, Elliot J M and Meunier P J 1999 Bone, bone diseases and bone quality. In *Quantitative Ultrasound: Assessment of Osteoporosis and Bone Status* eds C F Njeh, D Hans, T Fuerst, C-C Gluer, H K Genant (London: Martin Dunitz) pp 1–20

[2] Carter D R and Spengler D M 1978 Mechanical properties and composition of cortical bone *Clin. Orthop.* **135** 192–217

[3] Rho J Y, Kuhn-Spearing L and Zioupos P 1998 Mechanical properties and the hierarchical structure of bone *Med. Eng. Phys.* **20**(2) 92–102

[4] Gibson L J and Ashby M F 1997 *Cellular Solids: Structure and Properties* 2nd edn (Cambridge, New York: Cambridge University Press)

[5] Revell P A 1983 Histomorphometry of bone *J. Clin. Pathol.* **36**(12) 1323–31

[6] Parfitt A M, Drezner M K, Glorieux F H, Kanis J A, Malluche H, Meunier P J *et al* 1987 Bone histomorphometry: standardization of nomenclature, symbols, and units. Report of the ASBMR Histomorphometry Nomenclature Committee *J. Bone Min. Res.* **2**(6) 595–610

[7] Vesterby A, Gundersen H J and Melsen F 1987 Unbiased stereological estimation of osteoid and resorption fractional surfaces in trabecular bone using vertical sections: sampling efficiency and biological variation *Bone* **8**(6) 333–7

[8] Hildebrand T and Ruegsegger P 1997 A new method for the model independent assessment of thickness in three-dimensional images *J. Microsc.* **185** 67–75

[9] Laib A and Ruegsegger P 1999 Calibration of trabecular bone structure measurements of *in vivo* three-dimensional peripheral quantitative computed tomography with 28-micron resolution microcomputed tomography *Bone* **24**(1) 35–9

[10] Hildebrand T, Laib A, Muller R, Dequeker J and Ruegsegger P 1999 Direct three-dimensional morphometric analysis of human cancellous bone: microstructural data from spine, femur, iliac crest, and calcaneus *J. Bone Miner. Res.* **14**(7) 1167–74

[11] Feldkamp L A, Goldstein S A, Parfitt A M, Jesion G and Kleerekoper M 1989 The direct examination of three-dimensional bone architecture in vitro by computed tomography *J. Bone Miner. Res.* **4**(1) 3–11

[12] Muller R and Ruegsegger P 1997 Micro-tomographic imaging for the non-destructive evaluation of trabecular bone architecture *Stud. Health. Technol. Inform.* **40** 61–79

[13] Majumdar S, Kothari M, Augat P, Newitt D C, Link T M, Lin J C *et al* 1998 High-resolution magnetic resonance imaging: three-dimensional trabecular bone architecture and biomechanical properties *Bone* **22**(5) 445–54

[14] Wehrli F W, Hwang S N and Song H K 1998 New architectural parameters derived from micro-MRI for the prediction of trabecular bone strength *Technol. Health Care* **6**(5–6) 307–20

[15] Genant H K, Engelke K, Fuerst T, Glüer C C, Grampp S, Harris S T *et al* 1996 Noninvasive assessment of bone mineral and structure: state of the art *J. Bone Miner. Res.* **11**(6) 707–30

[16] Cortet B, Colin D, Dubois P, Delcambre B and Marchandise X 1995 Methods for quantitative analysis of trabecular bone structure *Rev. Rhum. Engl. Ed.* **62**(11) 781–93

[17] Odgaard A 1997 Three-dimensional methods for quantification of cancellous bone architecture *Bone* **20**(4) 315–28

[18] Cowin S C and Mehrabadi M M 1989 Identification of the elastic symmetry of bone and other materials *J. Biomech.* **22**(6–7) 503–15

[19] Underwood E E 1981 *Quantitative Stereology* (Reading, MA: Addison-Wesley)

[20] Harrigan T P and Mann R W 1984 Characterization of microstructural anisotropy in orthotropic materials using a second rank tensor *J. Mater. Sci.* **19** 761–7

[21] Goulet R W, Goldstein S A, Ciarelli M J, Kuhn J L, Brown M B and Feldkamp L A 1994 The relationship between the structural and orthogonal compressive properties of trabecular bone *J. Biomech.* **27**(4) 375–89

[22] Kuhn J L, Goldstein S A, Feldkamp L A, Goulet R W and Jesion G 1990 Evaluation of a microcomputed tomography system to study trabecular bone structure *J. Orthop. Res.* **8**(6) 833–42

[23] Turner C H, Cowin S C, Rho J Y, Ashman R B and Rice J C 1990 The fabric dependence of the orthotropic elastic constants of cancellous bone *J. Biomech.* **23**(6) 549–61

[24] Kanatani K I 1984 Stereological determination of structural anisotropy *Int J. Eng. Sci.* **22** 531–546

[25] Odgaard A, Jensen E B and Gundersen H J 1990 Estimation of structural anisotropy based on volume orientation. A new concept *J. Microsc.* **157**(2) 149–62

[26] Odgaard A, Kabel J, van Rietbergen B, Dalstra M and Huiskes R 1997 Fabric and elastic principal directions of cancellous bone are closely related [see comments] *J. Biomech.* **30**(5) 487–95

[27] Smit T H, Schneider E and Odgaard A 1998 Star length distribution: a volume-based concept for the characterization of structural anisotropy *J. Microsc.* **191**(3) 249–57

[28] Hahn M, Vogel M, Pompesius-Kempa M and Delling G 1992 Trabecular bone pattern factor—a new parameter for simple quantification of bone microarchitecture *Bone* **13**(4) 327–30

[29] Vesterby A 1993 Star volume in bone research. A histomorphometric analysis of trabecular bone structure using vertical sections *Anat. Rec.* **235**(2) 325–34

[30] Parfitt A M, Mathews C H, Villanueva A R, Kleerekoper M, Frame B and Rao D S 1983 Relationships between surface, volume, and thickness of iliac trabecular bone in aging and in osteoporosis. Implications for the microanatomic and cellular mechanisms of bone loss *J. Clin. Invest.* **72**(4) 1396–409

[31] Croucher P I, Garrahan N J and Compston J E 1996 Assessment of cancellous bone structure: comparison of strut analysis, trabecular bone pattern factor, and marrow space star volume *J. Bone Miner. Res.* **11**(7) 955–61

[32] Compston J E 1994 Connectivity of cancellous bone: assessment and mechanical implications [editorial] *Bone* **15**(5) 463–6

[33] Odgaard A and Gundersen H J 1993 Quantification of connectivity in cancellous bone, with special emphasis on 3-D reconstructions *Bone* **14**(2) 173–82

[34] Currey J D 1970 The mechanical properties of bone *Clin. Orthop.* **73** 209–31

[35] Brown T D and Ferguson A B 1980 Mechanical property distributions in the cancellous bone of the human proximal femur *Acta Orthop. Scand.* **51**(3) 429–37

[36] Charnley J 1970 *Acrylic Cement in Orthopaedic Surgery* (Baltimore: Williams and Wilkins)

[37] Gibson L J 1985 The mechanical behaviour of cancellous bone *J. Biomech.* **18**(5) 317–28

[38] Currey J D 1984 *The Mechanical Adaptation of Bones* (Princeton, NJ: Princeton University Press)

[39] Hvid I and Jensen J 1984 Cancellous bone strength at the proximal human tibia *Eng. Med.* **13**(1) 21–5

[40] Keaveny T M, Guo X E, Wachtel E F, McMahon T A and Hayes W C 1994 Trabecular bone exhibits fully linear elastic behavior and yields at low strains *J. Biomech.* **27**(9) 1127–36

[41] Burstein A H and Frankel V H 1971 A standard test for laboratory animal bone *J. Biomech.* **4**(2) 155–8

[42] Blair S G W 1969 *Elementary Rheology* (London: Academic Press)

[43] Hall I H 1968 *Deformation of Solids* (London: Thomas Nelson)

[44] Sprackling M T 1985 *Liquids and Solids* (London, Boston: Routledge & Kegan Paul)

[45] Schoenfeld C M, Lautenschlager E P and Meyer P R Jr 1974 Mechanical properties of human cancellous bone in the femoral head *Med. Biol. Eng.* **12**(3) 313–7

[46] Pugh J W, Rose R M and Radin E L 1973 Elastic and viscoelastic properties of trabecular bone: dependence on structure *J. Biomech.* **6**(5) 475–85

[47] Linde F and Hvid I 1987 Stiffness behaviour of trabecular bone specimens *J. Biomech.* **20**(1) 83–9

[48] Carter D R and Hayes W C 1976 Bone compressive strength: the influence of density and strain rate *Science* **194**(4270) 1174–6

[49] Bouxsein M L and Augat P 1999 Biomechanics of Bone. In *Quantitative Ultrasound Assessment of Osteoporosis and Bone Status* eds C F Njeh, D Hans, T Fuerst, C C Glüer and H K Genant (London: Martin Dunitz) pp 21–46

[50] Linde F and Hvid I 1989 The effect of constraint on the mechanical behaviour of trabecular bone specimens *J. Biomech.* **22**(5) 485–90

[51] Turner C H and Burr D B 1993 Basic biomechanical measurements of bone: a tutorial *Bone* **14**(4) 595–608

[52] Linde F, Hvid I and Madsen F 1992 The effect of specimen geometry on the mechanical behaviour of trabecular bone specimens *J. Biomech.* **25**(4) 359–68

[53] Burstein A H, Currey J D, Frankel V H and Reilly D T 1972 The ultimate properties of bone tissue: the effects of yielding *J. Biomech.* **5**(1) 35–44

[54] Linde F and Sorensen H C 1993 The effect of different storage methods on the mechanical properties of trabecular bone *J. Biomech.* **26**(10) 1249–52

[55] Borchers R E, Gibson L J, Burchardt H and Hayes W C 1995 Effects of selected thermal variables on the mechanical properties of trabecular bone *Biomaterials* **16**(7) 545–51

[56] Matter H P, Garrel T V, Bilderbeek U and Mittelmeier W 2001 Biomechanical examinations of cancellous bone concerning the influence of duration and temperature of cryopreservation *J. Biomed. Mater. Res.* **55**(1) 40–4

[57] Athanasiou K A, Zhu C, Lanctot D R, Agrawal C M and Wang X 2000 Fundamentals of biomechanics in tissue engineering of bone *Tissue Eng.* **6**(4) 361–81

[58] ASTM 1994 Metal test methods and analytical procedures. In *1994 Annual Book of ASTM Standards* (Philadelphia: ASTM) section 3

[59] Odgaard A and Linde F 1991 The underestimation of Young's modulus in compressive testing of cancellous bone specimens *J. Biomech.* **24**(8) 691–8

[60] Keaveny T M, Pinilla T P, Crawford R P, Kopperdahl D L and Lou A 1997 Systematic and random errors in compression testing of trabecular bone *J. Orthop. Res.* **15**(1) 101–10 [published erratum appears in 1999 *J. Orthop. Res.* **17**(1) 151]

[61] Keaveny T M, Borchers R E, Gibson L J and Hayes W C 1993 Theoretical analysis of the experimental artifact in trabecular bone compressive modulus *J. Biomech.* **26**(4–5) 599–607 [published erratum appears in 1993 *J. Biomech.* **26**(9) 1143]

[62] Ashman R B, Cowin S C, Van Buskirk W C and Rice J C 1984 A continuous wave technique for the measurement of the elastic properties of cortical bone *J. Biomech.* (*UK*) **17**(5) 349–61

[63] Carter D R, Schwab G H and Spengler D M 1980 Tensile fracture of cancellous bone *Acta Orthop. Scand.* **51**(5) 733–41

[64] Stone J L, Beaupre G S and Hayes W C 1983 Multiaxial strength characteristics of trabecular bone *J. Biomech.* **16**(9) 743–52

[65] Kaplan S J, Hayes W C, Stone J L and Beaupre G S 1985 Tensile strength of bovine trabecular bone *J. Biomech.* **18**(9) 723–7

[66] Rohl L, Larsen E, Linde F, Odgaard A and Jorgensen J 1991 Tensile and compressive properties of cancellous bone *J. Biomech.* **24**(12) 1143–9

[67] Evans F G 1973 *Mechanical Properties of Bone* (Springfield, IL: Thomas)

[68] Neil J L, Demas T C, Stone J L and Hayes C W 1983 Tensile and compressive properties of vertebral trabecular bone *Trans. Orthop. Res Soc.* 344

[69] Reilly D T and Burstein A H 1975 The elastic and ultimate properties of compact bone tissue *J. Biomech.* **8**(6) 393–405

[70] Simkin A and Robin G 1973 The mechanical testing of bone in bending *J. Biomech.* **6**(1) 31–9

[71] Lakes R and Saha S 1979 Cement line motion in bone *Science* **204** 501–3

[72] Forwood M R and Parker A W 1989 Microdamage in response to repetitive torsional loading in the rat tibia *Calcif. Tissue Intl.* **45**(1) 47–53

[73] Choi K and Goldstein S A 1992 A comparison of the fatigue behavior of human trabecular and cortical bone tissue *J. Biomech.* **25**(12) 1371–81

[74] Zioupos P and Casinos A 1998 Cumulative damage and the response of human bone in two-step loading fatigue *J. Biomech.* **31**(9) 825–33

[75] McKoy B E, Kang Q and An Y H 2000 Indentation testing of bone. In *Mechanical Testing of Bone and the Bone–Implant Interface* eds Y H An and R A Draughn (New York: CRC Press) pp 233–40

[76] Hvid I, Andersen K and Olesen S 1984 Cancellous bone strength measurements with the osteopenetrometer *Eng. Med.* **13**(2) 73–8

[77] Jensen N C, Madsen L P and Linde F 1991 Topographical distribution of trabecular bone strength in the human os calcanei *J. Biomech.* **24**(1) 49–55

[78] Hvid I, Jensen N C, Bunger C, Solund K and Djurhuus J C 1985 Bone mineral assay: its relation to the mechanical strength of cancellous bone *Eng. Med.* **14**(2) 79–83

[79] Rho J Y 1996 An ultrasonic method for measuring the elastic properties of human tibial cortical and cancellous bone *Ultrasonics* **34**(8) 777–83

[80] Lang S B 1970 Ultrasonic method for measuring elastic coefficients of bone and results on fresh and dried bovine bones *IEEE Trans. Biomed. Eng.* **17**(2) 101–5

[81] Yoon H S and Katz J L 1976 Ultrasonic wave propagation in human cortical bone—I. Theoretical considerations for hexagonal symmetry *J. Biomech.* **9**(6) 407–12

[82] Yoon H S and Katz J L 1976 Ultrasonic wave propagation in human cortical bone—II. Measurements of elastic properties and microhardness *J. Biomech.* **9**(7) 459–64

[83] Van Buskirk W C, Cowin S C and Ward R N 1981 Ultrasonic measurement of orthotropic elastic constants of bovine femoral bone *J. Biomech. Eng.* **103**(2) 67–72

[84] Ashman R B and Rho J Y 1988 Elastic modulus of trabecular bone material *J. Biomech.* **21**(3) 177–81

[85] Ashman R B, Rho J Y and Turner C H 1989 Anatomical variation of orthotropic elastic moduli of the proximal human tibia *J. Biomech.* (*UK*) **22**(8–9) 895–900

[86] Njeh C F, Boivin C M and Langton C M 1997 The role of ultrasound in the assessment of osteoporosis: a review *Osteoporos. Intl.* **7**(1) 7–22

[87] Kolsky H 1963 *Stress Waves in Solid* (Oxford: Clarendon Press)

[88] Carter D R and Hayes W C 1977 The compressive behavior of bone as a two-phase porous structure *J. Bone Joint Surg.* [*Am*] **59**(7) 954–62

[89] Harrigan T P, Jasty M, Mann R W and Harris W H 1988 Limitations of the continuum assumption in cancellous bone *J. Biomech.* **21**(4) 269–75

[90] Williams J L and Johnson W J 1989 Elastic constants of composites formed from PMMA bone cement and anisotropic bovine tibial cancellous bone *J. Biomech.* **22**(6–7) 673–82

[91] Townsend P R, Raux P, Rose R M, Miegel R E and Radin E L 1975 The distribution and anisotropy of the stiffness of cancellous bone in the human patella *J. Biomech.* **8**(6) 363–7

[92] Lindahl O 1976 Mechanical properties of dried defatted spongy bone *Acta Orthop. Scand.* **47**(1) 11–9

[93] Ducheyne P, Heymans L, Martens M, Aernoudt E, de Meester P and Mulier J C 1977 The mechanical behaviour of intracondylar cancellous bone of the femur at different loading rates *J. Biomech.* **10**(11/12) 747–62

[94] Williams J L and Lewis J L 1982 Properties and an anisotropic model of cancellous bone from the proximal tibial epiphysis *J. Biomech. Eng.* **104**(1) 50–6

[95] Goldstein S A, Wilson D L, Sonstegard D A and Matthews L S 1983 The mechanical properties of human tibial trabecular bone as a function of metaphyseal location *J. Biomech.* **16**(12) 965–9

[96] Martens M, Van Audekercke R, Delport P, De Meester P and Mulier J C 1983 The mechanical characteristics of cancellous bone at the upper femoral region *J. Biomech.* **16**(12) 971–83

[97] Ashman R B, Corin J D and Turner C H 1987 Elastic properties of cancellous bone: measurement by an ultrasonic technique *J. Biomech.* **20**(10) 979–86

[98] Hodgskinson R and Currey J D 1990 Effects of structural variation on Young's modulus of non-human cancellous bone *Proc. Inst. Mech. Eng.* [H] **204**(1) 43–52

[99] Reilly D T, Burstein A H and Frankel V H 1974 The elastic modulus for bone *J. Biomech.* **7**(3) 271–5

[100] Burstein A H, Reilly D T and Martens M 1976 Aging of bone tissue: mechanical properties *J. Bone Joint Surg. Am.* **58**(1) 82–6

[101] Cezayirlioglu H, Bahniuk E, Davy D T and Heiple K G 1985 Anisotropic yield behavior of bone under combined axial force and torque *J. Biomech.* **18**(1) 61–9

[102] Ascenzi A and Bonucci E 1967 The tensile properties of single osteons *Anat. Rec.* **158**(4) 375–86

[103] McElhaney J H, Fogle J L, E B and Weaver G 1964 Effect of embalming on the mechanical properties of beef bone *J. Appl. Physiol.* **19** 1234–6

[104] Vincentelli R and Grigorov M 1985 The effect of Haversian remodeling on the tensile properties of human cortical bone *J. Biomech.* **18**(3) 201–7

[105] Weaver J K 1966 The microscopic hardness of bone *J. Bone Joint Surg. Am.* **48**(2) 273–88

[106] Oliver W C and Pharr G M 1992 An improved technique for determining hardness and elastic modulus using load and displacment sensing indentation experiments *J. Mater. Sci.* **7** 1564–83

[107] McElhaney K W, Vlassak J J and Nix W D 1998 Determination of indenter tip geometry and indentation contact area for depth-sensing indentation experiments *J. Mater. Res.* **13**(5) 1300–6

[108] Vlassak J J and Nix W D 1994 Measuring the elastic properties of anisotropic materials by means of indentation experiments *J. Mech. Phys. Solids* **42**(8) 1223–45

[109] Asif S A S, Wahl K J, Colton R J and Warren O L 2001 Quantitative imaging of nanoscale mechanical properties using hybrid nanoindentation and force modula-tion. *J. Appl. Phys.* **90**(3) 1192–200

[110] Rho J Y, Tsui T Y and Pharr G M 1997 Elastic properties of human cortical and trabecular lamellar bone measured by nanoindentation. *Biomaterials* **18**(20) 1325–30

[111] Rho J Y, Roy M E II, Tsui T Y and Pharr G M 1999 Elastic properties of microstructural components of human bone tissue as measured by nanoindentation *J. Biomed. Mater. Res.* **45**(1) 48–54

[112] Roy M E, Rho JY, Tsui T Y, Evans N D and Pharr G M 1999 Mechanical and morphological variation of the human lumbar vertebral cortical and trabecular bone *J. Biomed. Mater. Res.* **44**(2) 191–7

[113] Blackburn J, Hodgskinson R, Currey J D and Mason J E 1992 Mechanical proper-ties of microcallus in human cancellous bone *J. Orthop. Res.* **10**(2) 237–46

[114] Ziv V, Wagner H D and Weiner S 1996 Microstructure–microhardness relations in parallel-fibered and lamellar bone *Bone* **18**(5) 417–28

[115] Townsend P R, Rose R M and Radin E L 1975 Buckling studies of single human trabeculae *J. Biomech.* **8**(3–4) 199–201

[116] Currey J D 1988 The effects of drying and re-wetting on some mechanical properties of cortical bone *J. Biomech.* **21**(5) 439–41

[117] Rho J Y and Pharr G M 1999 Effects of drying on the mechanical properties of bovine femur measured by nanoindentation *J. Mater. Sci.—Mater. Med.* **10**(8) 485–8

[118] Doerner M F and Nix WD 1986 A method for interpreting the data from depth-sensing indentation instruments *J. Mater. Res.* **1**(4) 601–9

[119] Nix W D 1997 Elastic and plastic properties of thin films on substrates: nano-indentation techniques. *Mater. Sci. Eng.—Struct. Mater. Properties. Microstructure and Processing* **234** 37–44

[120] Pharr G M, Oliver W C and Brotzen F R 1992 On the generality of the relationship among contact stiffness, contact area, and elastic-modulus during indentation *J. Mater. Res.* **7**(3) 613–7

[121] Sneddon I N 1965 The relation between load and penetration in the axisymmetric Boussinesq problem for a punch of arbitrary profile. *Int J. Eng. Science* **3** 47–57

[122] Bolshakov A and Pharr G M 1998 Influences of pileup on the measurement of mechanical properties by load and depth sensing indentation techniques *J. Mater. Res.* **13**(4) 1049–58

[123] Rho J Y and Pharr G M 2000 Nanoindentation testing of bone. In *Mechanical Testing of Bone and the Bone–Implant Interface* eds Y H An and R A Draughn (New York: CRC Press) pp 257–69

[124] Runkle J C and Pugh J 1975 The micro-mechanics of cancellous bone. II. Determination of the elastic modulus of individual trabeculae by a buckling analysis *Bull. Hosp. Joint Dis.* **36**(1) 2–10

[125] Rho J Y 1998 Ultrasonic characterisation in determining elastic modulus of trabecular bone material *Med. Biol. Eng. Comput.* **36**(1) 57–9

[126] Rho J Y 1998 Characterization of ultrasonic pathway and wavelength dependence in determining elastic properties using ultrasound method. *Med. Biol. Eng. Comp.* **36**(1) 57–9

[127] Choi K, Kuhn J L, Ciarelli M J and Goldstein S A 1990 The elastic moduli of human subchondral, trabecular, and cortical bone tissue and the size-dependency of cortical bone modulus *J. Biomech.* **23**(11) 1103–13

[128] Takano Y, Turner C H and Burr D B 1996 Mineral anisotropy in mineralized tissues is similar among species and mineral growth occurs independently of collagen orientation in rats: results from acoustic velocity measurements *J. Bone Miner. Res.* **11**(9) 1292–301

[129] Turner C H, Rho J, Takano Y, Tsui T Y and Pharr G M 1999 The elastic properties of trabecular and cortical bone tissues are similar: results from two microscopic measurement techniques *J. Biomech.* **32**(4) 437–41

[130] Ku J L, Goldstein S A, Choi K W, London M and Herzig L S 1987 *The Mechanical Properties of Single Trabeculae* (Orthopedic Research Society)

[131] Rice J C, Cowin S C and Bowman J A 1988 On the dependence of the elasticity and strength of cancellous bone on apparent density *J. Biomech.* **21**(2) 155–68

[132] Ryan S D and Williams J L 1989 Tensile testing of rodlike trabeculae excised from bovine femoral bone *J. Biomech.* **22**(4) 351–5

[133] Mente P L and Lewis J L 1989 Experimental method for the measurement of the elastic modulus of trabecular bone tissue *J. Orthop. Res.* **7**(3) 456–61

[134] Hodgskinson R, Currey J D and Evans G P 1989 Hardness, an indicator of the mechanical competence of cancellous bone *J. Orthop. Res.* **7**(5) 754–8

[135] Rho J Y, Flaitz D, Swarnakar V and Acharya R S 1997 The characterization of broadband ultrasound attenuation and fractal analysis by biomechanical properties *Bone* **20**(5) 497–504

[136] Gong J K, Arnold J S and Cohn S H 1964 Composition of trabecular and cortical bone. *Anat. Rec.* **149** 325–32

[137] Galante J, Rostoker W and Ray R D 1970 Physical properties of trabecular bone. *Calcif. Tissue Res.* **5**(3) 236–46

[138] Currey J D 1969 The mechanical consequences of variation in the mineral content of bone *J. Biomechanics* **2** 1–11

[139] Linde F, Gothgen C B, Hvid I and Pongsoipetch B 1988 Mechanical properties of trabecular bone by a non-destructive compression testing approach. *Eng. Med.* **17**(1) 23–9

[140] Nicholson P H, Cheng X G, Lowet G, Boonen S, Davie M W, Dequeker J *et al* 1997 Structural and material mechanical properties of human vertebral cancellous bone. *Med. Eng. Phys.* **19**(8) 729–37

[141] Sedlin E D and Hirsch C 1966 Factors affecting the determination of the physical properties of femoral cortical bone. *Acta Orthop. Scand.* **37**(1) 29–48

[142] Evans J A and Tavakoli M B 1990 Ultrasonic attenuation and velocity in bone. *Phys. Med. Biol.* **35**(10) 1387–96

[143] Hvid I, Rasmussen O, Jensen N C and Nielsen S 1985 Trabecular bone strength profiles at the ankle joint *Clin. Orthop.* **199** 306–12

[144] Pelker R R, Friedlaender G E, Markham T C, Panjabi M M and Moen C J 1984 Effects of freezing and freeze-drying on the biomechanical properties of rat bone *J. Orthop. Res.* **1**(4) 405–11

[145] Sedlin E D 1965 A rheologic model for cortical bone. A study of the physical properties of human femoral samples. *Acta Orthop. Scand. Suppl.* **83** 1–77

[146] Evans F G and Lebow M 1951 Regional differences in some of the physical properties of human femur *J. Appl. Physiol.* **3** 563–72

[147] Leong A S and Gilham P N 1989 The effects of progressive formaldehyde fixation on the preservation of tissue antigens *Pathology* **21**(4) 266–8

[148] Nimni M E, Cheung D, Strates B, Kodama M and Sheikh K 1987 Chemically modified collagen: a natural biomaterial for tissue replacement *J. Biomed. Mater. Res.* **21**(6) 741–71

[149] Boskey A L, Cohen M L and Bullough P G 1982 Hard tissue biochemistry: a comparison of fresh-frozen and formalin-fixed tissue samples *Calcif. Tissue Intl.* **34**(4) 328–31

[150] Currey J D, Brear K, Zioupos P and Reilly G C 1995 Effect of formaldehyde fixation on some mechanical properties of bovine bone *Biomaterials* **16**(16) 1267–71

[151] Komender A 1976 Influence of preservation on some mechanical properties of human haversian bone *Mater. Med. Pol.* **8**(1) 13–7

[152] Anderson M J, Keyak J H and Skinner H B 1992 Compressive mechanical properties of human cancellous bone after gamma irradiation *J. Bone Joint Surg. Am.* **74**(5) 747–52

[153] Godette G A, Kopta J A and Egle D M 1996 Biomechanical effects of gamma irradiation on fresh frozen allografts in vivo *Orthopedics* **19**(8) 649–53

[154] Jinno T, Miric A, Feighan J, Kirk S K, Davy D T and Stevenson S 2000 The effects of processing and low dose irradiation on cortical bone grafts *Clin. Orthop.* **375** 275–85

[155] Currey J D, Foreman J, Laketic I, Mitchell J, Pegg D E and Reilly G C 1997 Effects of ionizing radiation on the mechanical properties of human bone *J. Orthop. Res.* **15**(1) 111–7

[156] Hamer A J, Strachan J R, Black M M, Ibbotson C J, Stockley I and Elson R A 1996 Biochemical properties of cortical allograft bone using a new method of bone strength measurement. A comparison of fresh, fresh-frozen and irradiated bone *J. Bone Joint Surg. Br.* **78**(3) 363–8

[157] Zhang Y, Homsi D, Gates K, Oakes K, Sutherland V and Wolfinbarger L Jr 1994 A comprehensive study of physical parameters, biomechanical properties, and statistical correlations of iliac crest bone wedges used in spinal fusion surgery. IV. Effect of gamma irradiation on mechanical and material properties. *Spine* **19**(3) 304–8

[158] Linde F, Norgaard P, Hvid I, Odgaard A and Soballe K 1991 Mechanical properties of trabecular bone. Dependency on strain rate *J. Biomech.* **24**(9) 803–9

[159] McElhaney J H 1966 Dynamic response of bone and muscle tissue *J. Appl. Physiol.* **21**(4) 1231–6

[160] Currey J D 1975 The effects of strain rate, reconstruction and mineral content on some mechanical properties of bovine bone *J. Biomech.* **8**(1) 81–6

[161] Currey J D 1989 Strain rate dependence of the mechanical properties of reindeer antler and the cumulative damage model of bone fracture *J. Biomech.* **22**(5) 469–75

[162] Currey J D 1988 Strain rate and mineral content in fracture models of bone *J. Orthop. Res.* **6**(1) 32–8

[163] Wright T M and Hayes W C 1976 Tensile testing of bone over a wide range of strain rates: effects of strain rate, microstructure and density *Med. Biol. Eng.* **14**(6) 671–80

[164] Reilly D T and Burstein A H 1974 Review article. The mechanical properties of cortical bone *J. Bone Joint Surg. Am.* **56**(5) 1001–22

[165] Allard R N 1990 *Mechanical Compression Testing to Determine the Elastic Modulus of Cancellous Bone* (Arlington: University of Texas at Arlington)

[166] Bell G H, Dunbar O, Beck J S and Gibb A 1967 Variations in strength of vertebrae with age and their relation to osteoporosis. *Calcif. Tissue Res.* **1**(1) 75–86

[167] Mosekilde L and Danielsen C C 1987 Biomechanical competence of vertebral trabecular bone in relation to ash density and age in normal individuals *Bone* **8**(2) 79–85

[168] Mosekilde L 1989 Sex differences in age-related loss of vertebral trabecular bone mass and structure–biomechanical consequences *Bone* **10**(6) 425–32

[169] Rüegsegger P, Durand E P and Dambacher M A 1991 Differential effects of aging and disease on trabecular and compact bone density of the radius *Bone* **12** 99–105

[170] Bartley M H Jr, Arnold J S, Haslam R K and Jee W S 1966 The relationship of bone strength and bone quantity in health, disease, and aging *J. Gerontol.* **21**(4) 517–21

[171] Katz J L and Yoon H S 1984 The structure and anisotropic mechanical properties of bone. *IEEE Trans. Biomed. Eng.* **31**(12) 878–84

[172] Keller T S, Hansson T H, Abram A C, Spengler D M and Panjabi M M 1989 Regional variations in the compressive properties of lumbar vertebral trabeculae. Effects of disc degeneration. *Spine* **14**(9) 1012–9

[173] Ashman R B, Van Buskirk W C, Cowin S C, Sandborn P M, Wells M K and Rice J C 1985 The mechanical properties of immature osteopetrotic bone. *Calcif. Tissue Intl.* **37**(1) 73–6

[174] Smith J W and Walmsley R 1959 Factors affecting the elasticity of bone *J. Anatomy* **93** 503–23

[175] Brear K, Currey J D, Raines S and Smith K J 1988 Density and temperature effects on some mechanical properties of cancellous bone. *Eng. Med.* **17**(4) 163–7

[176] Vahey J W, Lewis J L and Vanderby R Jr 1987 Elastic moduli, yield stress, and ultimate stress of cancellous bone in the canine proximal femur *J. Biomech.* **20**(1) 29–33

[177] Sharp D J, Tanner K E and Bonfield W 1990 Measurement of the density of trabecular bone *J. Biomech.* **23**(8) 853–7

[178] Ascenzi A 1988 The micromechanics versus the macromechanics of cortical bone—a comprehensive presentation *J. Biomech. Eng.* **110**(4) 357–63

[179] An Y H and Draughn R A editors 2000 *Mechanical Testing of Bone and the Bone–Implant Interface* (New York: CRC)

[180] Martin R B, Burr D B and Sharkey N A 1998 *Skeletal Tissue Mechanics* (Princeton, NJ: Princeton University Press)

[181] van der Linden J C, Birkenhager-Frenkel D H, Verhaar J A and Weinans H 2001 Trabecular bone's mechanical properties are affected by its non-uniform mineral distribution *J. Biomech.* **34**(12) 1573–80

[182] Behrens J C, Walker P S and Shoji H 1974 Variations in strength and structure of cancellous bone at the knee *J. Biomech.* **7**(3) 201–7

[183] Goldstein S A 1987 The mechanical properties of trabecular bone: dependence on anatomic location and function *J. Biomech.* **20**(11–12) 1055–61

[184] Currey J D 1986 Power law models for the mechanical properties of cancellous bone *Eng. Med.* **15**(3) 153–4

[185] Hodgskinson R and Currey J D 1990 The effect of variation in structure on the Young's modulus of cancellous bone: a comparison of human and non-human material *Proc. Inst. Mech. Eng.* [H] **204**(2) 115–21

[186] Weaver J K and Chalmers J 1966 Cancellous bone: its strength and changes with aging and an evaluation of some methods for measuring its mineral content *J. Bone Joint Surg. Am.* **48**(2) 289–98

[187] McElhaney J H, Fogle J L, Melvin J W, Haynes R R, Roberts V L and Alem N M 1970 Mechanical properties of cranial bone *J. Biomech.* **3**(5) 495–511

[188] Augat P, Link T, Lang T F, Lin J C, Majumdar S and Genant H K 1998 Anisotropy of the elastic modulus of trabecular bone specimens from different anatomical locations *Med. Eng. Phys.* **20**(2) 124–31

[189] Keaveny T M, Morgan E F, Niebur G L and Yeh O C 2001 Biomechanics of trabecular bone *Ann. Rev. Biomed. Eng.* **3** 307–33

[190] Jensen N C, Hvid I and Kroner K 1988 Strength pattern of cancellous bone at the ankle joint *Eng. Med.* **17**(2) 71–6

[191] Cowin S C 1986 Wolff's law of trabecular architecture at remodeling equilibrium *J. Biomech. Eng.* **108**(1) 83–8

[192] Vajjhala S, Kraynik A M and Gibson L J 2000 A cellular solid model for modulus reduction due to resorption of trabeculae in bone *J. Biomech. Eng.* **122**(5) 511–5

[193] Ciarelli T E, Fyhrie D P, Schaffler M B and Goldstein S A 2000 Variations in three-dimensional cancellous bone architecture of the proximal femur in female hip fractures and in controls *J. Bone Miner. Res.* **15**(1) 32–40

[194] Yeni Y N, Brown C U and Norman T L 1998 Influence of bone composition and apparent density on fracture toughness of the human femur and tibia *Bone* **22**(1) 79–84

[195] Jurist J M and Foltz A S 1977 Human ulnar bending stiffness, mineral content, geometry and strength *J. Biomech.* **10**(8) 455–9

[196] Stromsoe K, Hoiseth A, Alho A and Kok W L 1995 Bending strength of the femur in relation to non-invasive bone mineral assessment *J. Biomech.* **28**(7) 857–61

[197] Currey J D 1990 Physical characteristics affecting the tensile failure properties of compact bone *J. Biomech.* **23**(8) 837–44

[198] Schaffler M B and Burr D B 1988 Stiffness of compact bone: effects of porosity and density *J. Biomech.* **21**(1) 13–6

[199] Bonfield W and Grynpas M D 1977 Anisotropy of the Young's modulus of bone. *Nature* **270**(5636) 453–4

[200] Katz J L 1980 The structure and biomechanics of bone. *Symp. Soc. Exp. Biol.* **34** 137–68

[201] Pope M H and Outwater J O 1974 Mechanical properties of bone as a function of position and orientation *J. Biomech.* **7**(1) 61–6

[202] Dempster D T and Liddicoat R T 1952 Compact bone as a non-isotropic material. *Am J. Anatomy* **91** 331–62

[203] Catanese J III, Iverson E P, Ng R K and Keaveny T M 1999 Heterogeneity of the mechanical properties of demineralized bone *J. Biomech.* **32**(12) 1365–9

[204] Bowman S M, Zeind J, Gibson L J, Hayes W C and McMahon T A 1996 The tensile behavior of demineralized bovine cortical bone *J. Biomech.* **29**(11) 1497–501

[205] Carter D R and Hayes W C 1977 Compact bone fatigue damage—I. Residual strength and stiffness *J. Biomech.* **10**(5–6) 325–37

[206] Martin R B and Ishida J 1989 The relative effects of collagen fiber orientation, porosity, density, and mineralization on bone strength *J. Biomech.* **22**(5) 419–26

[207] Burr D B, Schaffler M B and Frederickson R G 1988 Composition of the cement line and its possible mechanical role as a local interface in human compact bone *J. Biomech.* **21**(11) 939–45

[208] Burstein A H, Zika J M, Heiple K G and Klein L 1975 Contribution of collagen and mineral to the elastic-plastic properties of bone *J. Bone Joint Surg. Am.* **57**(7) 956–61

[209] Langton C M, Njeh C F, Hodgskinson R and Currey J D 1996 Prediction of mechanical properties of the human calcaneus by broadband ultrasonic attenuation *Bone* **18**(6) 495–503

[210] Wolff J 1892 *Des Gesetz der Transformation der Knochen* (Berlin: Hirchwald)

Chapter 6

Histomorphometry

Jean E Aaron and Patricia A Shore

6.1. INTRODUCTION

The contemplation of bone as an organized tissue seems to have commenced with descriptions by van Leeuvenhoek (1632–1723) of its penetration by various kinds of 'tubuli'. These provide an extensive internal surface area for exchange which ranges from the Haversian systems and Volkmann canals (about $3\,m^2$ in a normal male skeleton) to the trabecular network (about $16\,m^2$), osteocyte lacunae (about $90\,m^2$) and canaliculi (about $1000\,m^2$; see Aaron [1] for references). To this might also be added the surface area presented by the nanometre world of the enshrouded calcium phosphate interface. Embryologically bone may arise directly within a fibrous membrane (intramembranous ossification) or indirectly within a cartilage model (endochondral ossification). Thereafter it is governed by three cell systems, the osteoblasts, the osteocytes and the osteoclasts, which act anabolically or catabolically upon the many square metres of intrinsic surfaces above. The perpetual modulation of the microanatomy externally and internally ensures that the skeleton is environmentally responsive and at the various levels this may be instantaneous or almost imperceptibly slow. The resultant change may be either a beneficial adaptation that accommodates the skeleton to unaccustomed stress, or a detrimental and progressive attrition such as seems to accompany the endemic physical under-activity characteristic of modern societies, the penalty of which is an increased predisposition to minimum trauma fractures in later life. At the same time, bony tissue does not function in isolation from the soft tissues; it is bonded to the musculature by Sharpey's fibres and is influenced by blood biochemistry and by the metabolism of the gut and kidney, such that there exists a fine balance between the hard skeleton and the other tissues it supports, anchors, protects and supplies with calcium and phosphate. If this balance is chronically disturbed, for example by hormonal changes, metabolic bone disease may follow.

The extent to which bone 'quality' contributes to the biomechanical performance of the skeleton independent of its mass is still debated and there is a natural tendency to underestimate a tissue whose most characteristic feature seems superficially akin to that of a geological rock. However, this is to overlook not only the structural complexity but also the considerable histo-chemical heterogeneity. In addition, it is to ignore increasing evidence of a mineral metamorphosis in response to manipulation that may limit observa-tional reliability to such a degree that the Greek name for apatite, meaning 'the deceiver', seems presciently apt. The histomorphometric and microscopy methods for the analysis of bone described below and in chapter 7 are directed at the determination of the 'quality' of the bony tissue rather than its quantity since the latter can be measured by non-invasive and less time-consuming/labour-intensive means. In rounding up some of the qualitative sheep for counting, there are presented micro-anatomical and histological variables of some maturity that have already entered the scientific/clinical application fold, together with new ultrastructural and macromolecular lambs now coming through the gate. Methods range from cryopreservation to tissue incineration, from thick slices to fractionated fine particles, from bulk-stained microcracks to immuno-stained molecules, and from manual measurement to computer-driven automation and back again.

SECTION A: MICROARCHITECTURE USING COMPUTERIZED AND MANUAL TECHNIQUES

In vivo imaging, thin sections, serial sections, thick slices

Although the adult skeleton consists mainly of cortical bone the research attention surrounding the remaining 20% that is cancellous tissue stems from its greater gross structural complexity (figure 6.1) and its prominence at the major sites of osteoporotic fracture. The outer cortical envelope transfers strain to the inner spongiosa and where it is thin (and in some places it almost disappears) it may deflect up to 500 μm upon loading [2]. There is disagreement about the relative contribution to structural strength made by the two types of bone, particularly in vulnerable regions such as the vertebral body. For example, there is evidence to suggest that more than 50% of the strength of a vertebral body is in its cortex [3–5] which thins with age [6]. Overaker *et al* [7], using finite element modelling, demonstrated that doubling the cortical width of the vertebral body apparently doubled its strength (see also [8]). In contrast, McBroom *et al* [9] observed that the removal of the cortex produced a reduction in compressive strength of only 10% and subsequently a number of other authors attributed from 67 to 90% of the

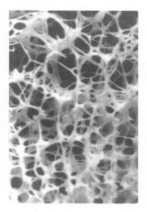

Figure 6.1. *Typical trabecular architecture in a defatted sagittal slice, 4 mm thick, of the vertebral body of a middle-aged man. Optical microscope, ×7.*

compression strength of this site to the cancellous bone [7, 10–12], creating an uncertain state of affairs at the outset.

It has been suggested that a loss of trabecular bone mass is accompanied by a disproportionate weakening of the region and that the relationship between bone density and strength is nonlinear with 20–40% of the variance in strength being unaccountable [9, 13–15]. Consistent with this is the observation that the age-related fall in vertebral bone mass is insufficient to explain the accompanying reduction in strength [16]. At the same time, there is the familiar significant overlap in bone mass when age-matched patients with and without fractures are compared [17–20]. A literature survey led Ott [21] to the conclusion that even in those subjects with a particularly low bone mass the chance of fracturing during the following year was only 10%, while a similar literature appraisal by Marshall *et al* [22] indicated that although the bone mineral density can predict fracture risk it cannot identify those individuals who will actually have a fracture. Obversely a pharmacologically stimulated restoration of bone density is not invariably accompanied by a reduction in fractures either [23, 24]. A solution was offered by Legrand *et al* [25], who reported that only the trabecular architectural parameters were significant predictors of the presence of vertebral fractures.

The performance of a vertebral body depends upon the direction of the load, compressive strength being anisotropic, such that it is greater in the vertical than the horizontal direction irrespective of the overall density [26–28] (see review of bone anisotropy by Odgaard [29]). In accordance with Wolff in 1899 [30] there is a correlation between lines of stress and trabecular orientation. By combining the measurement of bone density with that of anisotropy (preferred orientation, or fabric) Kabel *et al* [31] claimed to explain 90–96% of the variance for elastic modulus (see also [32–34]) thereby placing anisotropy high on the list of architectural factors influencing the mechanical properties of cancellous bone [35]. The loss of whole trabeculae apparently effects the elastic modulus more than does their generalized thinning [36] and there are many recent histomorphometric investigations of the structural significance of reduced trabecular interconnection [37–42]. By modelling cycles of cancellous thinning and thickening in which connectivity was allowed to vary in some cycles and not in others, Kinney and Ladd [43] reported a direct linear relationship between connectivity and the elastic modulus. Using similar models Ulrich *et al* [44] found that the prediction of elasticity constants was improved by adding trabecular number, width and spacing to the bone density and anisotropy recommended above. Finally, while key microarchitectural variables may correlate closely with the bone mass in young healthy tissue, the interrelationship is likely to be substantially altered by age and disease.

6.2.   TRABECULAR ARCHITECTURE—NON-INVASIVE, NON-DESTRUCTIVE

The derivation of trabecular patterns from radiographs to avoid invasive biopsy procedures and gain access to the most vulnerable skeletal sites is not new [45]. Following in this tradition is the recent introduction of computerized texture analysis of radiographs of cancellous bone [28, 46, 47], though radiological summation divorces the processed image from structural reality. To surmount this problem, high-resolution magnetic resonance imaging (MRI) has been applied to the appendicular skeleton [48], while clinical computed tomography (CT) has been targeted at the axial skeleton [49, 50]. This has resulted in factors such as the ridge number density (a measure of trabecular number by Laib *et al* [51]), the 'trabecular fragmentation index' whereby the length of the trabecular network is divided by the number of discontinuities [52], the T-texture (trabecular width) and I-texture (trabecular separation [53]), to which Gordon *et al* [54] have added the 'hole area' (similar to the histological star volume described below). Using 1 mm thick CT slices it was found possible to separate matched patients with and without fractures, so improving upon the sensitivity of bone mineral density measurement alone (see also [50]). Similarly, using a slice thickness of 0.7 mm and with a resolution of 150 μm, a relationship between the tomography and certain histomorphometric variables has been reported [41,55], particularly with respect to the trabecular strut and node numbers (A Mortimer and R Soames, unpublished data). However, the radiation exposure to achieve this remains a major limitation clinically [50, 56]. The next step in validating the tomography above with the histology below has been hastened by the increasing accessibility of micro-CT technology, with a reported resolution of about 10 μm [57, 58]. This enables the laboratory assessment of structure in a non-destructive manner [44, 59, 60] (see also the NMR micro-imaging of Chung *et al* [61] and Hipp *et al* [62]), thereby conserving the samples for secondary investigation of the associated mechanical properties.

6.3. TRABECULAR ARCHITECTURE—TWO-DIMENSIONAL HISTOLOGY

Following the introduction of the technology that enabled undecalcified bone to be routinely prepared for optical microscopy, attention was immediately directed at establishing the basis for remodelling imbalance and the trabecular microarchitecture was largely overlooked (with a few notable exceptions [63–65]). The widespread availability of computers brought image analysis to all, expanding the range of structural variables that had been previously constrained by manual methods employing eyepiece graticules [66]. The indirect character of many of the micro-anatomical variables in present usage has generated misgivings among the stereologists about their mathematical veracity. For example, Odgaard [29] dismisses 'surrogate measures' with the firm declamation that the only acceptable variables are ConnEuler [67] and star volume [68] where there is no doubt about precisely what is being assessed. While this may be so, other variables are often more convenient in a comparative clinical context where they are determined either individually or by taking several in parallel for a more reliable interpretation [69]. In particular the so-called 'parallel plate' parameters continue to be extensively applied and include the trabecular number, thickness and separation. Derived from area/perimeter measurements (table 6.1), these were originally known as the mean trabecular plate density (MTPD), thickness (MTPT) and separation (MTPS) respectively [70] before conforming to the recommended ASBMR nomenclature [71] to which authors should now adhere.

Cancellous atrophy may take place as a result of generalized trabecular thinning, as the total removal of individual trabeculae, or as a combination of the two. As well as influencing biomechanical properties the pattern of bone loss also determines therapeutic outcome since while thinned bars can be thickened, the replacement of lost bars is problematic [66, 72] (figure 6.2). For example, the atrophied spongiosa of osteoporotic women responds positively to fluoride therapy. Nevertheless, the restoration of the bone mass in these subjects does not prevent continuing fractures apparently because the thickening of the trabecular remnants, which provides increased resistance to compression forces, is not accompanied by an improved capacity to withstand bending forces, since the latter is dependent upon more regular trabecular interconnection and the restitution of cross struts [24]. There is the possibility that even the most successful treatment regimens will never be capable of reconnecting a disconnected trabecular system [73], as reviewed by Dempster [74]. However, recent observations [75, 76] suggest

Table 6.1. *The equations for the derivation of trabecular architectural variables based upon area and perimeter measurements (parallel plate parameters).*

Trabecular bone volume (bone volume/total volume)%

$$BV/TV = 100 \times BAr/TAr$$

where Ar is area

Trabecular width mcm (µm)

$$TbWi = 2000 \times BAr \; (mm^2)/BPm \; (mm)$$

where Pm is perimeter

$$TbTh = TbWi/1.199$$

where Th is thickness and 1.199 is the correction factor for section obliquity for iliac trabecular bone [70]

Trabecular number (mm)

$$TbN = BV/TV \; (\%) \times 10/TbWi \; (\mu m)$$

Trabecular separation

$$TbSp = 1000/TbN \; (mm^{-1}) - TbWi \; (\mu m)$$

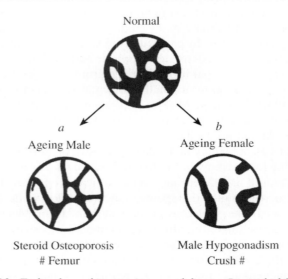

Normal

a   *b*

Ageing Male   Ageing Female

Steroid Osteoporosis   Male Hypogonadism
\# Femur   Crush \#

Figure 6.2. *Trabecular architecture in age and disease. Diminished formation leads to trabecular attenuation (a) as seems to be the case in ageing men, in femoral fracture patients and in patients treated with corticosteroid drugs. Increased resorption leads to trabecular atrophy (b) and disconnection—a weaker arrangement that probably contributes to the sex-related difference in fracture since it occurs in ageing women, in crush fracture patients and also in hypogonadal men.*

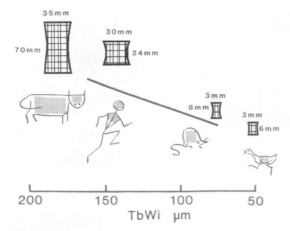

Figure 6.3. *The trabecular width in man and mouse. The trabecular width is normally biologically controlled for optimum efficiency and interspecies variation is small. It is in accordance with Ham's constant [77] that no osteocyte is more than 100 μm from the nearest bone surface such that any trabecula exceeding 200 μm must be perforated by a central blood vessel. In the graph, while the stylized vertebral body alters by an order of magnitude between the largest (including the dinosaur) and smallest vertebrates, the width of the trabeculae alters relatively little.*

a possible avenue whereby this might be achieved through exactly the same sequence of events that ensures that bone trabeculation remains one of Nature's constants (figure 6.3). That is accomplished by combining trabecular thickening with angiogenesis (figure 6.4).

6.4. THE TRABECULAR ANALYSIS SYSTEM (TAS)

As indicated in table 6.2, a wide range of computer-related systems have been applied to histological preparations of cancellous bone [80] which vary from small iliac crest biopsies (table 6.3) to large vertebral bodies. The region of the iliac crest is the internationally accepted standard skeletal sampling site because of its accessibility, cancellous character and freedom from weight bearing. On the other hand the location has the disadvantage that, as trabecular structure is determined by function, any micro-anatomical changes at the iliac crest may not be representative of the common clinical fracture areas. However, as all bones apparently undergo the changes that are characteristic of systemic bone disease or ageing (see Aaron [1] for references) albeit at different rates, it remains acceptable. For example, the correlation between architectural variables at the iliac crest and lumbar spine has been recently calculated to be about 0.7 [81]. To analyse the material there was

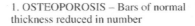

1. OSTEOPOROSIS – Bars of normal
thickness reduced in number

2. TREATMENT – Bars increase
in thickness beyond 200 microns

3. TREATMENT – Vascular resorption
cavities appear in thickness bars

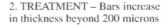

4. TREATMENT – Resorption cavities
expand into new marrow spaces
as trabeculae continue to thicken
and occasionally fuse

5. RESTITUTION – A complex network
of new trabeculae is generated

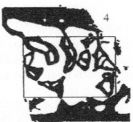

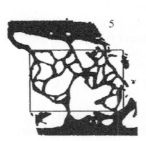

Figure 6.4. *The restitution of osteoporotic bone—a proposed microanatomical sequence. A new spongiosa may develop from the remnants of the old (1) when trabecular thickening (2) is followed by intra-trabecular resorption (3) combined with angiogenesis, stimulated as a sequel to the inadequate nutrition of osteocytes progressively distanced from the bone surface [77]. The proliferating blood vessels occupy the newly created intra-osseous channels which widen internally (4) into marrow spaces as more bone is added externally (with drug treatment or exercise) to the trabecular surface, transforming each expanding bony bar into a new network (5). Bony bridges will interconnect the progressively juxtaposed bars. Diagram created by the stepwise modification of the binary image of an osteoporotic transilial bone biopsy.*

Table 6.2. *Some of the computerized systems reported in the literature for measuring the trabecular architecture.*

System	Variable	Reference
Summagraphic digitizer plus PDP11/10 computer	Trabecular perimeter Trabecular width (direct) Trabecular separation (indirect) Trabecular number (indirect)	Birkenhager-Frenkel *et al* [78]
Videoplan/Osteoplan (Kontron, Germany)	Trabecular width (direct) Trabecular separation (indirect) Trabecular number (indirect)	Schnitzler *et al* [79]
OsteoMeasure (OsteoMetrics Inc., Atlanta) Charlie@osteometrics.com	Trabecular width (indirect) Trabecular number (indirect) Trabecular separation (indirect)	Hordon *et al* [80]
Optomax V Analytical Measurement System (Optomax Inc, Hollos, NH)	Trabecular strut analysis	Mellish *et al* [81] Parisien *et al* [83]
Zeiss MOP 3 digitizer (Carl Zeiss Inc, New York)	Trabecular thickness (indirect) Trabecular number (indirect) Trabecular separation (indirect) (Parallel plate parameters from area and perimeter)	Parfitt *et al* [70]
IBAS II image analyser (Kontron, Germany)	Trabecular width (direct) (Derived from expanding circles around the median axis of each trabecula until opposing boundaries reached when diameter of circle is trabecular width)	Garrahan *et al* [84]
IBAS II image analyser	Trabecular strut analysis	Garrahan *et al* [85]
IBAS 2000 image analyser (Kontron, Germany)	Trabecular pattern factor	Hahn *et al* [86]
Personal computer system (e.g. TAS, see below))	Marrow space star volume	Vesterby [68] Croucher *et al* [87]
Quantimet 570 image analyser (Leica, Malmaison, France))	Interconnectivity index	Legrand *et al* [25]
Quantimet 520 image analyser (Leica, Cambridge, UK)	Fractal analysis	Weinstein *et al* [88] Fazzalari and Parkinson [89])

Table 6.2. *(Continued)*

Personal computer system	Fabric/anisotropy	Hodgskinson and Currey [33] Whitehouse [90])
Trabecular Analysis System (TAS) (Using a personal computer). Supplied by S.K.Paxton@leeds.ac.uk Charlie@osteometrics.com	Comprehensive range	Aaron *et al* [82] Hordon *et al* [80]
Personal computer analysis	Comprehensive range	Croucher *et al* [87]

a pressing need for a dedicated system based upon the personal computer. This was propelled by a rising interest in trabecular micro-architecture as indicated by the proliferating literature on this subject, combined with the prohibitively expensive hardware available which often necessitated high-contrast stains unsuitable for other histomorphometric purposes. The solution for our laboratory was the in-house development of our automated trabecular analysis system (TAS) [82] (software available on request as

Table 6.3. *The bone biopsy procedure [1, 91–93].*

Site
2 cm behind the anterior superior spine and 2 cm below the iliac crest (transilial; originally specimens were taken perpendicular to the iliac crest, 2 cm behind the anterior spine).

Anaesthesia
Sedation, e.g. with pethidine. After cleaning the skin a local anaesthetic, e.g. xylocaine, is infiltrated, with more injected following insertion of the trephine. The operation is performed as a 1 day case (with a full blood count and clotting screen beforehand).

Operation
A small incision enables the muscles to be separated by blunt dissection to expose the periosteum. The serrated cylindrical sleeve of the biopsy trephine (e.g. Bordier trephine) grips the periosteum, while the inner sharp-toothed cylinder (about 8 mm in diameter; smaller specimens are less representative and proportionally more fragmented) cuts a core of bone as it is rotated under pressure (generally manually, although a mechanical drill may be used).

Post-operation
After withdrawal of the trephine its special pin is used to expel the biopsy and the wound is closed with two or three sutures. The patient rests for 2 h to ensure satisfactory haemastasis, and avoids heavy physical activity for several days.

NB. Tetracycline labelling is a tissue time marker that aids interpretation; its administration commences about three weeks in advance of a proposed biopsy (see p 218).

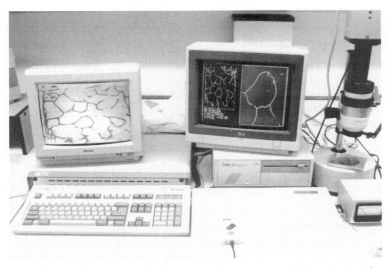

Figure 6.5. *The automated trabecular analysis system (TAS) showing the low-power optical microscope with closed-circuit TV camera, image analyser and microcomputer. On the PC monitor is a typical bone section. (Computers & Biomed. Res.* **25***, p. 4, Fig 1b, permission of Academic Press Inc).*

indicated in table 6.2 and figure 6.5). It encompasses most of the micro-architectural variables described in the literature and serves as a rapid route to comprehensive shape analysis.

Undecalcified bone sections are processed routinely by embedding in methylmethacrylate (table 6.4), sectioning on a heavy-duty microtome such as the Jung K or Polycut (Reichert-Jung, Heidelberg, now Leica, Germany) and staining in toluidine blue stain (table 6.5) or the Goldner method (table 6.6). The section for analysis is placed upon the microscope stage of a

Table 6.4. *General processing technique for the embedding of undecalcified bone in methyl-methacrylate.*

1. 70% ethanol or 10% buffered formalin (for frozen material)	1 day
2. 96% ethanol	1 day
3. 100% ethanol	1 day
4. 100% ethanol/chloroform 1:1	1 day
5. Methylmethacrylate	3 days
6. Methylmethacrylate/0.1% benzoyl peroxide	3 days
7. Embed in methylmethacrylate with 25% dibutylphthalate and 2.5% benzoyl peroxide. Polymerize in a water bath at 30 °C	

Table 6.5. *Toluidine blue staining procedure.*

1. Distilled water	3 min
2. 1% toluidine blue	30 min
3. Distilled water	5 min
4. Dehydrate in 100% ethanol	3×2 min
5. Clear in methylcyclohexane	2×3 min
6. Mount in neutral mounting medium, e.g. XAM (BDH Chemicals, Poole, UK)	

Mineralized bone, blue/purple. Osteoid, light blue.

Table 6.6. *Modified Goldner trichrome stain for methylmethacrylate sections.*

Solutions	Method
(i) Weigert's haematoxylin Solution A: 1 g haematoxylin in 100 ml 96% ethanol Solution B: 1.1 g $FeCl_3 \cdot 6H_2O$ + 1 ml 25% HCl Made up to 100 ml with distilled water Before use, mix A + B in 1 : 1 ratio	1. Distilled water 3 min 2. Weigert's haematoxylin 20 min 3. Rinse in running tap water 20 min 4. Distilled water 5 min 5. Panceau/fuchsin/azophloxine 5 min 6. Acetic acid 1% 15 sec 7. PTA/Orange II 20 min
(ii) Masson-Panceau de xylidine 0.75 g Panceau de xylidine 0.25 g Acid fuchsin 1 ml glacial acetic acid Made up to 100 ml with distilled water	8. Acetic acid 1% 15 sec 9. Light green 5 min 10. Acetic acid 1% 3 min 11. Distilled water 5 min 12. Dehydrate in absolute alcohol 3×2 min 13. Clear in methylcyclohexane 2×3 min 14. Mount in neutral mounting medium e.g. XAM (BDH Chemicals, Poole, UK)
(iii) Azophloxine 0.5 g azophloxine (a.k.a. acid red) 0.6 ml glacial acetic acid Made up to 100 ml with distilled water	
(iv) Panceau/fuschin/azophloxine 7.5 ml Panceau de xylidine 2 ml azophloxine 88 ml 0.2% glacial acetic acid	
(v) Light green 1 g light green 1 ml glacial acetic acid Made up to 500 ml with distilled water	
(vi) Phosphotungstic acid (PTA)/Orange II 3 g phosphotungstic acid 2 g Orange II Made up to 100 ml with distilled water	
(vii) 1% Acetic acid	

Cytoplasm and muscle, red. Mineralized bone, green. Osteoid, red. Nuclei, blue/black.

Table 6.7. *Procedure for the analysis of trabecular architecture using the TAS system [82].*

(i)	**Calibration**
	Performed microscopically with a millimetre ruler aligned along the x and y axes.
(ii)	**Image loading (grabbing)**
	A frame grabber retains the image of the bone section on the microscope stage.
(iii)	**Image thresholding (segmenting)**
	Separates the trabecular network from background features (e.g. marrow) in the original image which is composed of individual pixels with a grey-level value between 0 (black) and 63 (white). The optimum threshold value is selected manually; all pixels below this become white and all those above become black creating a binary 'bitmap' (figure 6.6) which may be compressed and saved on disk.
(iv)	**Defining the area of interest (AOI)**
	A coloured box is superimposed upon the binary image (figure 6.6) and its size and position adjusted manually.
(v)	**Image editing**
	An algorithm removes 'noise' in the form of single dissociated pixels, while an editing facility that includes erasure or addition allows minor artefacts (e.g. sectioning cracks; see figure 6.7) to be removed. Zoom magnification enables close inspection for accuracy; reference is also made to the original histological slide for clarification.
(vi)	**Image processing**
	All features outside the AOI (e.g. the eortices and peripheral trephine damage) are automatically removed to minimize processing. The binary image is thinned to its medial axis ('skeletonized') by a modification of Hilditch algorithm H [94] reducing the trabeculae to strings of single pixels, i.e. struts (figure 6.6).
(vii)	**Image analysis**
	The repeated passes and deletions above generate BV/TV (% or mm^2) and BS (mm^2/mm^3) or BPm (mm). Their ratio (table 6.1) provides TbWi (μm) followed by TbN (/mm) and TbSp (μm). The routine also produces the node number (NNd, joint points of three or more strings of pixels), terminus number (NTm, end points of a pixel string) and the NNd:NTm ratio, together with the strut number TbN (direct) and their character (e.g. Nd-Nd, Nd-Tm, Tm-Tm) and isolated bone profile number (NPf [71]). Also derived from the BS and BPm measures is the trabecular pattern factor TbPF (figure 6.6).
(viii)	**Other functions**
	• Point to point distance using a mouse cursor, e.g. enabling TbWi to be measured directly by tipping at intersections of the perpendicular to the long axis of the trabecula, and avoiding junctions [78].
	• A 'radial' scan function of path lengths through marrow spaces [95] using either a superimposed regular grid pattern [87] or a random point selector instructed to perform as many passes as required to indicate the star volume.
	• A program to produce a fractal analysis [96].
	• A facility for determining preferred trabecular orientation based upon the Mean Intercept Length (MIL) method, an intersection counting technique where MIL is the total line length divided by the number of intersections (see [29] for references).

TAS v2.09 January 2001   Trabecular Analysis Report

b2258 preG2
Report Generated on 7 Feb 2001 17:21:11
File TimeStamp is undefined

Thresholds: 0,160
Pixels are 0.0120 x 0.0126mm
BV   13.6297%
BS   1.7582mm/mm2
TbTh 129.3094mcm
NNd 6
NTm 20
NNd:NTm ratio   0.3000

BAr 2.5826mm2
BPm 33.3154mm
TbWi 155.0420 mcm
TbN 1.0540
TbSp 819.4266 mcm
TbPF 1.2724
Tar 18.9487mm
BS3d 1.4664 mm2/mm3
SDn 10.7587mm
Total Strut Number 21

Cuts 0
Singles 0
Nd-Nd 4
Tm-Nd 5
Tm-Tm 6
Tm-Cut 1
Nd-Cut 5
Mean Nd-Nd 0.9509mm
Mean Tm-Tm 0.6147mm
Mean Tm-Nd 0.6916mm
Mean Tm-Cut 2.0763mm
Mean Nd-Cut 0.9694mm
TbLe 17.8724mm

AOI used.
AOI Status
AOI information for new-42.tif
AOI area is 18.9487 mm2
AOI is Rectangular
Size: 4.5783 x 4.1388 mm
Pos: 2.1687, 0.6435 mm

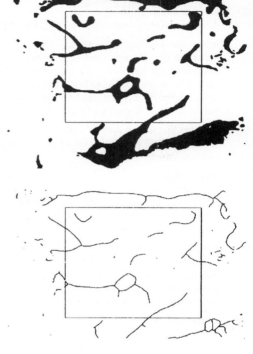

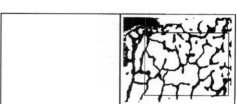

Figure 6.6. *A typical image and data print-out from the automated trabecular image analysis system (TAS) showing the binary intact and thinned images from an osteoporotic subject; the area of interest is defined by the rectangle. Bottom (inset) shows a typical binary image from a young healthy subject for comparison.*

transmitted light dissecting microscope (for example a Wild M7A) with a zoom lens. A low magnification of about ×15 is adequate and the area of interest is defined by the user within a deformable rectangular window of variable size. Attached to the viewing tube of the microscope is a

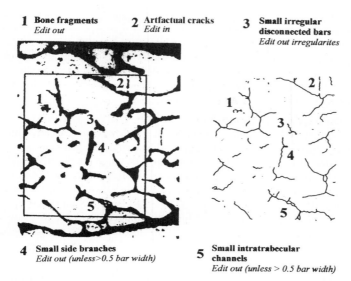

1 Bone fragments
Edit out

2 Artfactual cracks
Edit in

3 Small irregular disconnected bars
Edit out irregularites

4 Small side branches
Edit out (unless>0.5 bar width)

5 Small intratrabecular channels
Edit out (unless > 0.5 bar width)

Figure 6.7. *Histological features for attention when editing the binary image prior to an automated TAS analysis to avoid disproportionate complexity produced by artefacts, small channels and minor surface irregularities when the image is 'skeletonized'.*

closed-circuit black and white television camera that transmits the image for capture in a 256×256 pixel, 64 grey-level format by a **VIP** image analyser (Sight Systems, Newbury, Berkshire, UK), connected in turn to an IBM-compatible PC (512k or better), programmed in Turbo Pascal v5, with specific sections in Assembler for enhanced speed of computation. The steps involved in analysis are outlined in table 6.7.

For any network, the topological property of connectivity indicates the maximal number of branches that can be broken before the network is separated into two parts; it has been described as the number of struts minus one [97]. If a single path links the junctions the structure is simply connected and vulnerable to fragmentation when one branch is broken, while a multiple connected structure is more stable. [98]. In the cancellous network, removal of cross-bracing struts weakens the bone to an extent that is disproportionate to the mass of tissue lost (doubling the distance between the cross-ties reduces the resistance to bending forces fourfold). This is illustrated by the apparent location of the majority of microcallus found in the vertebral body and indicative of microfracture repair on the vertical trabeculae [99], since a reduction in horizontal cross-ties with age leads to a lengthening of these unsupported struts causing them to buckle [38]. It is also worth bearing in mind that topological measures of connectivity cannot discriminate between rod-like connections and fenestrated plates, and that

Table 6.8. *Histological indices of trabecular interconnection in cancellous bone.*

(i)   **Node : terminus ratio NNd : NTm**
The computer-generated thinned image of struts varies with age and pathology [81]. Following introduction by Garrahan *et al* [85] the ratio has been applied by others (e.g. [83, 100, 101]) and has usefully differentiated between groups despite the majority of termini being planar artefacts.

(ii)   **Trabecular pattern factor (TbPF)**
Proposed by Hahn *et al* [86] on the basis that a highly connected structure presents concave surfaces while a poorly connected one presents convex surfaces. It is calculated by determining the trabecular area (A1) and perimeter (P1) before and after (A2, P2) dilatation (e.g. corresponding to one pixel) of the binary image, a process in which concave perimeters decrease and convex perimeters increase. The quotient $(P1 - P2)/(A1 - A2)$ is the TbPF. The results are influenced by magnification and computer smoothing procedures [102]. Dilatation should be small enough to avoid the confluence of juxtaposed surfaces and large enough to accommodate minimal curvature [103].

(iii)   **Marrow space star volume (V^*)**
This indicates the mean volume of marrow space that can be seen unobstructed in all directions from a random point within it, and is derived from the distance cubed between the point and its intercepting radius with the bone surface [68, 104]

$$V^* = \pi/3 \times l_0^3.$$

In an interconnected structure the value is low, in a disconnected structure it is high.

(iv)   **Profile number (NPf)**
Isolated bone profiles in cancellous tissue increase with age as connectivity falls, followed by their decrease later as some structures disappear [71].

(v)   **ConnEuler**
This is a stereological index [67, 97] calculated from the number of holes (marrow cavities) and bridged components (bone particles) in two parallel sections separated by a distance of 10–40 μm [98]. The edge problem created when a specimen is extracted introduces an error that is inverse to specimen size [29]. It has been reported that when connectivity measured in this way increases, the mean trabecular plate density (i.e. TbN indirect) paradoxically may decrease [105].

(vi)   **Fractal analysis**
This is an index of structural complexity and was developed from similarities observed between nature (natural fractals) and certain types of mathematical objects (ideal fractals). While ideal fractals have a profile that is unchanged by low or high magnification (self-similarity [106]), natural fractals possess self-similarity only within a limited range. The fractal dimension may be used to describe structural changes in disease. It has been applied to cancellous bone (see [89, 103] for references) since an accurate measurement of the bone surface is central to reliable spatial description. However, bone surface increases with the magnification. Fractal analysis provides a dimensionless constant derived at a range of magnifications and has been used on histological sections and radiographs to identify disease- or age-related change.

Table 6.8. *(Continued)*

(vii)   **Interconnectivity index (ICI)**
First used on porous biomaterials [107], it has been applied to the connectivity of the marrow cavities of cancellous bone [25] by creating the 'skeleton' of their profiles and determining the number of nodes (N), node-to-node branches (NN) and node to free end branches (NF), together with the number of 'trees' (T), these being the individual structures composed of interconnected nodes. Then

$$ICI = (N \times NN)/(T \times (NF + 1))$$

the component ($NF + 1$) being a device to avoid computational problems due to division by zero in some cases. The greater the connectivity of the marrow spaces (high N, low T) the higher the ICI and the greater the fragmentation of the trabecular network.

while severing a trabecular rod will decrease connectivity, the fenestration of a trabecular plate will apparently increase connectivity, both events producing the same end result of reduced structural strength. Histological methods relating to aspects of trabecular interconnection are shown in table 6.8.

6.5.   TRABECULAR ARCHITECTURE—THREE-DIMENSIONAL IMAGE

Trabecular reconstruction presents certain difficulties to be overcome if mechanical strength and histological structure are to be more precisely related. A recent critical review of methods adopted to measure the third dimension of cancellous architecture by Odgaard [29] cautions against over-confidence in relating the results to mechanical properties derived from the standard, simple and popular compressive test. This he finds almost invariably incomplete in the elastic properties ascribed to a particular orientation (there being up to 21 elastic constants involved) and to the range of orientations considered. In addition are other inherent problems that include friction at end plates, structural end phenomena, specimen geometry, storage methods, temperature effects, continuum assumptions and visco-elasticity, all culminating in an accuracy that may be 40% in error. That is without taking account of the suspected changes that may take place in a tissue when the rest of the animal is removed from around it (to borrow the comment of Nobel laureate Albert Szent Gyorgy about extracted muscle). Histologically, reconstruction may be accomplished by sequential thin sectioning on the one hand, or by taking thick slices on the other; both procedures remove the uncertainty presented by the two-dimensional section concerning trabecular inteconnection.

6.5.1. Serial section techniques

Amstutz and Sissons [108] reconstructed vertebral spongiosa from carefully aligned serial sections. Despite the passage of time and technological advances [109] the process has never been placed on a routine footing. This is because the spacing between consecutive sections should be between one third and one tenth of the length of the feature of interest, i.e. 15–50 μm distant for trabeculae 150 μm long [97, 110] making tissue preparation time-consuming, section alignment a challenge, image processing excessively demanding of computer memory space and data storage capacity, while in addition the sections are prone to shrinkage. These difficulties can be largely avoided by means of an unusual and relatively rapid approach (M F Wilson and J E Aaron, unpublished) which is to repeatedly photograph the two-dimensional image of the surface of the bone embedded in plastic as it is systematically sectioned on a heavy-duty microtome such as the Jung K or the Reichert-Jung Polycut E (Leica, Germany; table 6.9). Not only is the block surface of value in this way (figure 6.8), but also the discarded sections may be collected and used for other histomorphometric purposes. This is because the black dye added to the colourless embedding medium to provide better contrast with the white bone is an unobtrusive translucent pale grey in a typical microscope section 5–20 μm thick. After independently devising a similar procedure, Odgaard *et al* [111, 112] recorded alternate sections with their automated system and reported the digitization of 170 images per hour, later rising to 600 per hour, producing substantial data sets from which a vertebral body can be reconstructed in about 2 h with the assistance of powerful computers.

In accomplishing this impressive technical achievement, Odgaard *et al* [111, 112] removed the marrow tissue before commencement to gain better image quality. This tacitly assumes that the trabecular network is entirely integrated, as is probably the case in young healthy material where the number of bone particles is 1. The same may not apply to unhealthy material where part of the network may be detached from the main framework. No matter how efficient the automation, or how fine the final product, there remains a fundamental error in extracting the marrow tissue from the spongiosa for the sake of image enhancement without first ensuring that isolated trabeculae and possibly larger trabecular profiles are not removed at the same time. The unanswered question is not whether isolated fragments of the cancellous network are lost, for the evidence indicates that they will be [60, 113, 114], but how much is lost in this way. For example, if the result from a video reconstruction of the block surface of a large ilial autopsy specimen from a normal 80-year-old man is typical and representative, it indicated by extrapolation the presence within his skeleton of not only 80 000 real trabecular termini but also 8000 real islands (M F Wilson and J E Aaron, unpublished data). Finally with respect to the marrow tissue

Table 6.9. *The procedure for serial section reconstruction of cancellous bone by repeated photography of the block surface as the tissue is sectioned.*

Preparation of the tissue block for good image contrast

Either[1]: Remove the marrow tissue by air jet and immersing in ethanol/acetone for 48 h or by heating to 55 °C in 0.75 M NaOH for 8 h. Bleach the bone white in 3% H_2O_2 for up to 5 h.

Or[2]: Leave the marrow tissue in place (so avoiding the loss of isolated bone islands and producing sections that remain useful for histomorphometry).

Either[1]: Add black Araldite Colouring Paste (Ciba-Geigy, Basle, Switzerland; 30 g in 210 ml Epofix resin, Struers, Copenhagen; with 30 ml hardener, 30 ml acetone), warming the resin to 37 °C; penetration into the alcohol-dehydrated bone is aided by a few seconds under vacuum.

Or[2]: Add Trylon opaque black carbon-based colour paste (0.1 g in 3 ml resin; stocked by model making retailers) to methylmethacrylate monomer and refrigerate at 4 °C for 3 days (longer for large specimens). Transfer to the final embedding mixture (2.5 g benzoyl peroxide, 25 ml di-*n*-butylphthalate plasticizer and 3.3 g Trylon paste in 100 ml methylmethacrylate). To prevent separation of the Trylon paste, place on an automatic shaker at room temperature until polymerization is complete.

Image capture

Either[1]: Use a PC attached to a CCD camera (Videk Megaplus 1400, Eastman Technology Inc, Canandaigua, NY) using an 8 bit resolution 3020×1035 frame grabber with 256 grey levels (e.g. VS-100-AT, Imaging Technology Inc, Woburn, MA) to digitize and segment into a binary image for storage first on the PC hard disk until sectioning is complete, then transfer to a network of SUN SparcStations.

Or[2]: Use a good quality videorecorder attached to a CCTV camera fitted with a C-mount to a Wild M7A dissecting microscope with zoom lens suspended over the microtome and record grey-level images on VHS videotape. (These may subsequently be converted into a format acceptable to the TAS system.) Alternatively a PC with an 8 bit resolution 768×512 frame grabber with 256 grey levels may be used (Data Translation Vision-EZ or PCVision Plus).

Modifications to the heavy duty microtome for optimum results[1]

A controller module monitors microtome movement by optoelectronic interruption to ensure precise block alignment after each cutting cycle. (Without this facility, reference features, e.g. three bristles, may be added to the embedding medium [118].)

A ring light (Intralux 5000, Volpi AG, Schlieren, Switzerland) ensures uniform illumination of the block, while a layer of low-viscosity mineral oil on the specimen surface eliminates reflections.

---

[1] References [111, 112].
[2] M F Wilson and J E Aaron, unpublished.

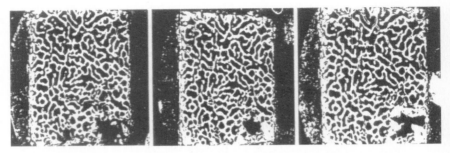

Figure 6.8. *Three serial images of bovine cancellous bone taken from a VHS videorecording of the block surface during microtomy. Magnification ×3.*

itself, there also arises the possibility that it too contributes biomechanically. Like the growing plant shoot or root tip, it has turgor pressure which when biologically channelled can lift concrete paving stones.

6.5.2.   Thick slice technique

For those with little hardware and even less aptitude for statistical and stereological fireworks, there may be a simpler route. As indicated above, the separation of elderly fracture and non-fracture subjects on the basis of either their densitometry or their histology has proved unsatisfactory. Factors have been proposed to improve discrimination that range from predisposition to falls in some subjects to fundamental changes in the macro-molecular composition of the bone matrix in others, with associated changes in the genetic code for collagen, vitamin D and oestrogen receptors also reported. In view of the perceived significance of trabecular interconnection and its apparent independence from the bone mass in some circumstances (for example, [8, 25, 80, 102, 115, 116]) knowledge about the number of trabecular termini may assist in the determination of fracture predisposition. Because bone histomorphometry traditionally is performed using sections that are 5–10 µm thick, the image is two-dimensional. Consequently any evaluation of trabecular disconnection based upon simply counting trabecular termini is inherently unreliable as the majority of apparent termini will be inevitably artefacts of the plane of section (figure 6.9).

There are descriptions in the literature of thick slices of cancellous bone used by some authors to study trabecular micro-architecture [6, 63, 117, 118]. Such methods almost invariably recommend the removal of the obscuring marrow tissue (see Hahn *et al* [119] for a possible exception and also for useful methodological advice about preparing large undecalcified bone specimens). This has the consequence for osteopenic tissue described above (p. 202) that portions of detached trabeculae may be lost with the extracted marrow, an eventuality that would cause real termini to be underestimated. A solution

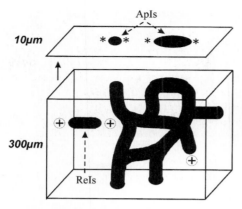

Figure 6.9. *Diagram showing apparent (∗) and real (+) trabecular termini and also apparent (ApIs) and real (ReIs) islands, i.e. isolated trabeculae, in a thin (two-dimensional) histological section (top) and in a thick (three-dimensional) slice (bottom) of cancellous bone. (Courtesy of Dr R C Shore.)*

is offered in the form of a novel, rapid and inexpensive method for identifying real trabecular termini that uses intact thick slices of plastic-embedded bone (300 μm) combined with a surface staining procedure [113]. A single stain on upper and lower surfaces, such as alizarin red, toluidine blue, light green or the von Kossa silver nitrate method, is effective (figure 6.10(a,b); table 6.10); alternatively two different stains may be combined, with for example alizarin red stain applied to one surface and light green to the other (figure 6.10(c,d); table 6.11). Either procedure displays the stained two-dimensional trabecular image (upper and lower stained surfaces) while sandwiched within the depth of the slice, is the unstained three-dimensional trabecular image which may be enhanced by viewing in partially polarized light. In these preparations the apparent termini are stained red or green (or brown in the case of the von Kossa stain) while real termini (and also real islands of bone) are unstained and appear white.

The thick slices of undecalcified embedded bone (see table 6.4 for the embedding procedure) are cut using a Microslice 2 (Malvern Instruments, Ultra Tech Manufacturing Inc., Santa Anna, California, USA) which has a rotating diamond-impregnated disc, cooled by a jet of water (see also the alternative grinding/polishing procedure of Hahn *et al* [119] using an automatic grinding machine). Slice thickness may be subsequently confirmed by a micrometer. It was found that a width of 300–400 μm gave optimum results with osteopenic bone viewed under the optical microscope, the clarity deteriorating in thicker slices, especially of young healthy bone where the image tends to be completely interconnected and the marrow tissue obscures the complexity. Using a hand counter and standard slice thickness, real (i.e. unstained) trabecular termini and real isolated islands of bone (unstained

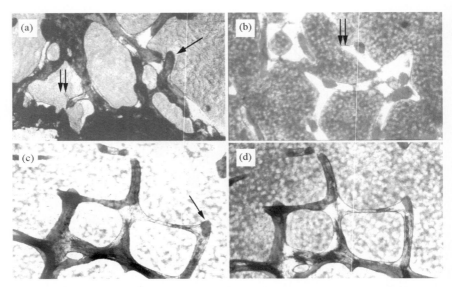

Figure 6.10. *Photomicrographs of cancellous bone in slices, 300 μm thick, showing the two- and three-dimensional architecture and real and apparent termini. (a) Von Kossa stain on both surfaces, partially polarized light. (b) Alizarin red on both surfaces, partially polarized light. (c) Alizarin red on one surface and light green on the other, viewed in plain light for comparison with (d) in polarized light for added clarity. The single arrows indicate apparent termini (i.e. stained); the double arrows indicate real termini (i.e. unstained) present within the depth of the slice and relatively rare ×20.*

throughout) may be counted accurately within the depth of the slice as numbers are not great (rarely exceeding two in a bone biopsy section). The region of interest may be delineated by placing a loose masked coverslip on top of the slice, the dimension of the window selected to exclude cortical bone and peripheral damage. Recent application of this technique to two

Table 6.10. *The modified von Kossa staining method.*

1. Place the sections in distilled water.	5 min
2. Immerse in 1% silver nitrate and illuminate 20 cm from a 100 W bulb until the bone appears dark brown.	Approx. 30 min
3. Wash in distilled water.	2 × 3 min
4. Dehydrate in 100% ethanol.	3 × 2 min
5. Clear in methylcyclohexane.	2 × 3 min
6. Mount in neutral mounting medium, e.g. XAM (BDH Chemicals, Poole, UK).	

Mineralized bone, dark brown.

Table 6.11. *Dual staining procedure for 300 μm slices using Alizarin red and light green.*

1. Coat one side of the slice with a thin layer of petroleum jelly.
2. Immerse the slice in 1% Alizarin red S, pH 4.2 until maximum staining is achieved (approximately 5 min).
3. Wash the slice in tap water and blot dry.
4. Remove the petroleum jelly with methylcyclohexane (approximately 5 min) and blot dry.
5. Immerse the slice in 1% light green, pH 2.7 for 5 min.
6. Wash in distilled water.
7. Dehydrate rapidly in three changes of absolute ethanol.
8. Clear in methylcyclohexane for 5 min.
9. Blot excess liquid and observe unmounted or, if necessary, mount in neutral mounting medium, e.g. XAM (BDH Chemicals, Poole, UK) which will obscure any surface imperfections.

Mineralized bone, red or green.

groups of patients, with and without vertebral fracture but with the same bone mass (as measured by dual-energy X-ray absorptiometry), showed a significant difference in the number of trabecular termini when all other histomorphometric methods failed to do so [114]. It was estimated from the analysis that the elderly female skeleton contains 300 000 real trabecular termini and 20 000 real islands. This estimate is greater than that in the elderly man (80 000 real trabecular termini and 8000 real islands) described after serial section reconstruction above (see p. 202), the difference possibly relating to the better maintenance of the trabecular framework in elderly men than women [66]. At the same time, it suggests that not all discontinuous (and therefore stress-free) trabeculae are subject to the immediate aggressive osteoclastic resorption observed by Mosekilde [120].

Another application of the thick slice method may be to determine more reliably the number of trabecular nodes (i.e. junctions) since they may be underestimated in the two-dimensional image. It has also been suggested that the method might usefully be applied in association with the ConnEuler principle [67].

SECTION B: MICROFRACTURE AND MICROCALLUS

Bulk-staining techniques

While the trabecular termini above may be generated by the complete transgression of cancellous bony bars by regular, well documented remodelling events involving osteoclast-related imbalance, they may also be produced by another route. For some time there has been evidence of a cyclic process of trabecular destruction and callus repair that is independent of regular

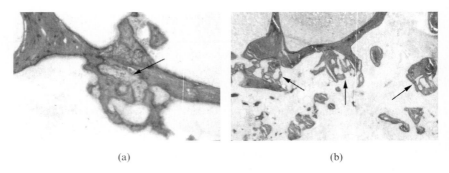

(a)   (b)

Figure 6.11. *Trabecular microcallus. (a) A single typical microcallus surrounding a fracture (arrowed) across the trabecula. Goldner stain, ×50. (b) An area of extensive trabecular fragmentation with several sites of micro-callus repair (arrowed) in the subchondral femoral head of an osteoarthritic subject. Goldner stain, ×20.*

bone remodelling but which may be no less significant in some conditions [121] (figure 6.11(a)). Its sporadic incidence, however, means that quantification using small standardized biopsies is inappropriate and in consequence any contribution to skeletal status remains poorly understood [122]. Nevertheless, illustrations of discrete sites of microcallus repair have been regularly presented for many years in the spine and femur [33, 123–128]. In particular they are a frequent feature of subchondral bone [129] especially in the arthritic hip (figure 6.11(b)). Sites occur without any external damage to the bone and may indicate mechanical insufficiency even in the static state [117]. The trabecular stiffness of microcallus was estimated by Blackburn *et al* [130] in terms of microhardness. This differed little from that of the adjacent bone, though the presence of extensive thick callus is likely to stiffen a bar more than its fracture-free neighbour, with particular implications for subchondral bone and joint degeneration. There remains also considerable uncertainty about how closely the mechanical properties of isolated, dead, dried, fixed or frozen bone resemble those of its living counterpart *in situ*; for example, isolated dry bone is harder than wet bone [131], while storage in formalin produces a 20% increase in hardness [132].

Stages in microcallus production commence with a bridge of woven bone [133] and conclude with a locally thickened trabecula, the irregularity of which relates to the degree of remodelling that has subsequently taken place [125]. Coarse parallel uncalcified collagenous fibres (rich in collagen Type III) are instrumental in bonding the trabecular disconnection and create a framework for fracture repair; there is also the apparent role of the bone fragments in promoting their own adhesion into mosaic-like trabecular arrangements [76] (figure 6.12(a)). It was reported by Vernon-Roberts and Pirie [123] and by Hansson and Roos [125] that the number of sites of

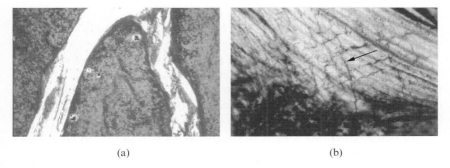

(a)   (b)

Figure 6.12. *Before and after trabecular microfracture. (a) Post-failure, showing two trabeculae, one (left) pristine in polarized light with regular lamellation; the other (right) a repaired composite of irregular fragments; in plain light the two appear identical, ×55. (b) Pre-failure, showing cross-hatched microcracks (arrowed) apparent only in polarized light. Goldner stain, ×270.*

microcallus in the vertebral body relates inversely to the amount of bone such that they are commonplace in osteopenia. This is consistent with early descriptions by Freeman *et al* [134] of accumulated sites of microcallus in the proximal femur of osteoporotic patients. The majority of sites of microcallus in the vertebrae, which may constitute up to 10% of the trabecular bone volume, are reported close to the end plate in association with vertical rather than horizontal trabeculae [38, 99, 128], the latter being preferentially lost [63]. This may relate to descriptions using acoustic emission of a susceptibility to microdamage near the endplates of osteopenic vertebral bodies resulting in the characteristic biconcave deformity and weakening the bone to later insult [135]. Normally signals from adjacent osteocytes or from the disturbed matrix itself may activate remodelling, removing the potential problem early [136]. It has also been suggested that the cement lines are weak interfaces [137] and act as preferred locations for microfissuring, having evolved as a protective device to channel damage away from less readily repaired locations. In some instances repair seems to fail to take place at all due to changes in the ability of the tissue to perceive and react to damage, producing isolated islands of bone, detached fragments and the disconnected struts described above (see p. 205).

The microdamage that precedes the formation of microcallus probably occurs as a result of normal daily physical activity with a regularity that is physiologically tolerable [138, 139]. It is only when pre-failure planes and microcracks accumulate (figure 6.12(b)), such that a proportion of the skeleton cannot bear its usual load, that clinically evident fractures may result in response to minimal trauma [140]. An exponential correlation between age and trabecular microfracture [138] has been attributed to the combination of an increased incidence and decreased repair [141–144]. The architecture of bone is such that typical peak strains are well below the threshold of its breaking strain [122] but repeated cyclic loads may disrupt intermolecular bonds in the mineralized matrix causing a loss of stiffness and leading to fatigue micro-fissures [140, 145, 146]. It has been calculated [134] that normal stress at the capito-cervical junction of the proximal femur at a bone density of $0.5\,g$ per cm^3 is sufficient to cause fatigue damage as a physico-pathological precursor to clinical osteoporotic fracture [126, 147].

In vivo microcracks were first described by Frost in 1960 [148]. Despite the passage of time and a renewed interest (for example, [128, 137, 149]) there persists the hindrance to progress of their uncertain histological identification in the absence of the repair response. Without this, the separation of *in vivo* microdamage from *in vitro* artefactual damage arising during specimen excision and microtomy is unreliable [150]. A distinguishing marker is essential. To provide this the technique of bulk staining was adopted by Frost [148] on the premise that much of the intact bone matrix is dense and impervious to stains. It follows that only in those regions where cracks are present in bone *before* sectioning would the entry of the stain be permitted, while those cracks introduced later would be unstained. The bulk-staining method of Frost received little further attention until almost 30 years later, when its veracity was confirmed with the addendum that the technique did not cause additional artefactual cracking through its associated dehydration stage [149]. Subsequently a growing number of authors have applied this 'en bloc' staining method to the demonstration of micro-fissures under various biomechanical and pathological circumstances. However, confidence in the results may be misplaced if due consideration is not given to the substantial pressure generally exerted in sampling from the skeleton in the first place, the specimens invariably being excised from relatively large bones (for example, long bone, rib, femoral head). This might be avoided by using small bones that can be removed intact from cadavers (A Lenehan, D Bridges, P Shore and J Aaron, unpublished). The removal of the phalangeal bones of the feet requires minimal manipulation and moreover they are small enough for histological preparation in their entirety. At the same time, although they are not a prime site for osteoporotic fracture, they may serve as a general monitor of bone quality as they are normally subjected to considerable stress during locomotion, as is well illustrated by the development of 'march fractures' in inexperienced young soldiers following long and unaccustomed treks. Some stains used to demonstrate microcracks and the results are shown in table 6.12 and figure 6.13. When trying the method for the first time and because the number of micro-fissures is likely to be low in normal bone, they may be artificially generated under control by using, for example, a Holden Compressing Machine (Department of Mechanical Engineering, University of Leeds) and progressive loads of up

Table 6.12. *Bulk-staining histological method for the demonstration of microcracks.*

The bulk-staining properties of five stains were compared as 1% neutral solutions in absolute alcohol:

- Xylenol orange (most rapid, 24 h sufficient for human phalanges)
- Calcein (neutralized stain deteriorates within 24 h, needing replenishing)
- Tetracycline (neutralized stain deteriorates within 24 h, needing replenishing)
- Basic fuchsin (tends to leach into alcohol lubricant during microtomy; use water)
- Gentian violet (best results)

Procedure for small intact specimens the size of human middle/distal phalanges. (Staining may be accelerated if the tip of the bone is gently removed, allowing direct access to the marrow space.)

- Preserve in 70% alcohol for 72 h.
- Place in about 30 cm^3 (or ten times the specimen volume) of 1% gentian violet and leave for 7 days at room temperature with the vessel screw-cap lightly secured to prevent evaporation; a vacuum is optional.
- Remove the screw-cap and allow the staining solution to evaporate (about 10 days).
- Wash in absolute alcohol for 1 h to remove excess stain.
- Embed in methylmethacrylate as in table 6.4 (the bone may be cut transversely in half at this stage to enable penetration of the embedding medium); cut sections 15 μm thick on a Jung K microtome using minimal lubrication of the knife with water.
- Mount sections in DePeX and observe under the plain, polarized or ultraviolet light microscope.

Optimum staining for microcracks is indicated by well stained cells, including osteocytes, and a poorly stained/unstained mineralized matrix.

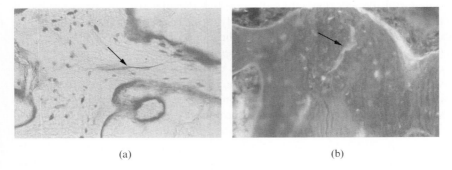

(a)   (b)

Figure 6.13. *Bulk-stained microcracks (arrowed) produced in vivo. (a) Gentian violet stain, plain light. (b) Tetracycline stain, ultraviolet light. (Basic fuchsin and other fluorochromes give similar results). ×100.*

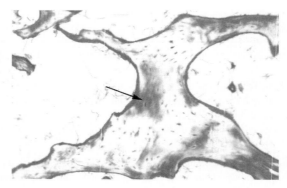

Figure 6.14. *Increased bone matrix permeability to basic fuchsin in the stained areas (arrowed), resulting from applied stress and possibly indicative of the appearance of ultra-small fissures below the resolution of the optical microscope. The surrounding matrix remains impermeable and unstained, ×85.*

to 2.0 kN (200 N) and 2.5 kN (250 N). This produces fissures indistinguishable from the 'physiological' microcracks defined by Frost [148] as labelled by stain that has penetrated them by diffusion. They differ from the small parallel cracks that regularly appear during microtomy and which are characterized by sharp unstained borders and an empty intervening space. While all stained microcracks originate *in vivo*, there are those that may remain inaccessible to the stain such that counts may be an underestimate.

Using the above method it has been observed that while sections of normal unstressed bone contain the occasional single discrete microcrack, those from bone subjected to controlled stress contain groups called 'microfracture episodes' [136]. At the same time, a compressive load of 2.0 kN not only visibly creates many microfissures but also increases the permeability of the matrix such that it stains apparently intact regions, separating them from unstained structurally sound tissue. These compression-induced 'leaky' areas (figure 6.14) resemble a naturally occurring tetracycline staining phenomenon previously noted in the mature matrix of osteoporotic subjects [121] which seemed to have become similarly 'leaky', this time *in situ*. In both instances these diffusely stained regions may indicate the presence of ultra-small micro-fissures below the resolution of the optical microscope, and witness to an unrecorded facet of bone biodynamics. As such they may provide insight into a normal hydraulic cycle within the matrix whereby transitory ultrastructural fissures are judicially opened and closed in response to the application and release of stress, only becoming permanent as a result of fatigue or other failure. Exposure to the higher compression force described above rendered permeable to stain not only most of the

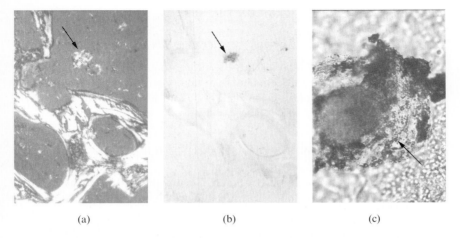

(a)   (b)   (c)

Figure 6.15. *Multiple microfissures and autoclasis. (a) A fragmented trabecula (arrowed) is evident in an unstained section viewed in polarized light, and (b) is the specific location of extracellular acid phosphatase activity (arrowed; lead sulphide method), ×70. (c) The hydrolytic enzyme is apparently associated with the separation and dispersal of the particulate mineral phase (arrowed), ×270.*

mineralized matrix, but also the associated soft tissues as membrane and tissue boundaries were ruptured.

Finally, there are instances in cancellous bone when the damage to a trabecula seems excessive, with no evidence of callus repair despite the presence of a compact cluster of trabecular fragments apparently in an advanced state of disintegration. It is tempting to dismiss this as an artefact, yet the location sometimes suggests that this cannot be the case. For example, they can be found isolated in the middle of large specimens (whole vertebral bodies or heads of femur) where they are surrounded by an extensive and intact trabecular network and undisturbed marrow tissue. As osteoclasis is absent from such regions, a spontaneous self-destruction of the living bone seems to be the only explanation, hence the name autoclasis was proposed [121]. Associated with the extracellular matrix in these areas is the hydrolytic lysosomal enzyme acid phosphatase (figure 6.15) the distribution of which seems integral to the micron-sized calcified particles apparently dissociating from the inorganic phase (see p. 255). In this way autoclasis seems to be the lysis of bone at some distance from a phagocytic cell, when multiple fissured and fragmented regions become detached from the trabecular network before final dissolution and dispersal. The contribution made to trabecular disconnection with age and osteoporosis remains to be established.

SECTION C: MATRIX REMODELLING

Semi-automated histomorphometry, tetracycline labelling and staining

The extracellular matrix with its translucent homogeneity in plain light is not the structural continuum it seems but is in reality a heterogeneous mosaic of remodelled units of chronological, metabolic and histochemical diversity.

6.6.   COMPUTER-ASSISTED HISTOMORPHOMETRY

The visual nine-point scale of Beck and Nordin in 1960 [91] was an early approach to quantitative bone histology in which samples were compared with nine photographs with calculated trabecular bone areas ranging from 6 to 27%. From this there developed more extensive histomorphometry using integrating eyepieces (manufactured, for example, by Zeiss, Oberkochen) engraved for point sampling (based upon the principle of the geologist Delesse in 1848 for rock porosity) and with parallel lines for intercept measurement. These were applied to each of four to eight fields within groups of about 16 sections per specimen. Computers were confined at first to only a few laboratories since the hardware was prohibitively expensive and the software limited, requiring particularly high contrast preparations. Computer-assisted methods are now in common usage, generating a wide range of direct and indirect remodelling variables including the bone volume, osteoid borders, resorption cavities and associated cell populations (as well as certain trabecular architectural variables described in section A, such as the trabecular width, number and separation based upon surface/perimeter measurement). Accurate analysis continues to require considerable expertise and remains time-consuming despite the reduction in the number of sections examined to 2–4 per specimen (see Birkenhager-Frenkel *et al* [151] concerning the need for adequate numbers of sections from different levels; there is also the dictum that it is better to measure more less well than a few precisely). The Goldner tetrachrome stain [152] is widely used for histomorphometry (table 6.6) and differentiates cellular and extracellular features well (figure 6.16), although some authors have criticized its reliability for osteoid tissue (when in doubt this can be monitored by comparison with a toluidine blue stained section). The Goldner stain is effective for computer-assisted systems such as the semi-automated OsteoMeasure (OsteoMetrics Inc, Atlanta, GA: E-mail Charlie@osteometrics.com) or the Morphomat (Zeiss, Germany).

6.6.1.   *The OsteoMeasure system*

This is attached to an optical microscope with a side-arm and a digitizing pad upon which the features to be measured are outlined. A sequence of square

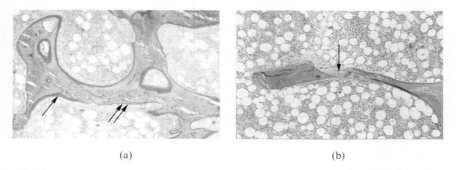

(a)   (b)

Figure 6.16. *Bone remodelling features. (a) A typical Goldner-stained section showing sites of formation (single arrow) and resorption (double arrow), plain light, ×30. (b) A similar preparation showing an osteoclatic resorption cavity (arrowed) that has almost perforated the trabecula, ×30.*

fields is systematically scanned and analysed under the optical microscope, commencing in the top left-hand corner beneath the cortex and traversing the section to produce an exactly-matched composite image of the whole cancellous area of interest presented as a colour print-out (figure 6.17). On

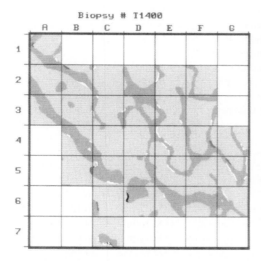

Figure 6.17. *A typical OsteoMeasure print-out (normally in colour) of cancellous bone. Analysis commences in the top left-hand corner and proceeds horizontally, each of the sequential squares representing a single field. Blank fields contain artefactual damage and are excluded. The mineralized bone may be colour-coded blue, the dark osteoid tissue red (shown as black borders), while a white outline marks the extent of resorption cavities and yellow (not shown here) indicates those resorption cavities containing osteoclasts. The surface extent of osteoblasts can be similarly mapped.*

Table 6.13. *Histomorphometric variables as listed in a typical OsteoMeasure-assisted analysis; those remodelling parameters in most frequent use are italic.*

Bone Biopsy Report	Abbreviation	Value	Units	Normals
Static indices				
Bone volume	BV/TV	23.4069	(%)	23.19 (4.37)
Osteoid volume (/TV)	OV/TV	0.0643	(%)	0.32 (0.19)
Osteoid volume (/BV)	OV/BV	0.2746	(%)	1.48 (0.93)
Mineralized volume	Md.V/TV	23.3427	(%)	
Bone surf/tissue volume index	BS/TV	3.5735	(mm^2/mm^3)	3.50 (0.46)
Bone surface/volume index	BS/BV	15.2668	(mm^2/mm^3)	
Total osteoid surface	OS/BS	2.0569	(%)	12.10 (4.64)
Bone interface (total)	BI/BS	2.1210	(%)	
Osteoblast surface	Ob.S/BS	0.2687	(%)	3.90 (1.94)
Eroded surface	ES/BS	9.1788	(%)	4.09 (2.33)
Osteoclast surface	Oc.S/BS	0.1967	(%)	0.69 (0.61)
Quiescent surface	QS/BS	88.7643	(%)	
Reversal surface	Rv.S/BS	8.9821	(%)	2.70 (2.96)
Remodelling surface	Rm.S/BS	11.2357	(%)	
Labelled surface	LS/BS	*	(%)	9.68 (4.95)
Single labelled surface	sL.S/BS	*	(%)	
Double labelled surface	dL.S/BS	*	(%)	5.30 (0.70)
Single labelled surface (Rel)	sL.S/LS	*	(%)	
Double labelled surface (Rel)	dL.S/LS	*	(%)	
Double/single label ratio	dL.S/sL.S	*	(%)	
Active eroded surface (/BS)	S1.S/BS	0.3446	(%)	
Trabecular thickness	Tb.Th	131.0029	(μm)	133.00 (22.0)
Trabecular profile thickness	Tb.Pf.Th	0.0000	(μm)	
Osteoid thickness	O.Th	8.2351	(μm)	10.34 (2.05)
Wall thickness	W.Th	*	(μm)	34.16 (2.23)
Inter-label thickness	Ir.L.Th	*	(μm)	
Osteoblast volume density	N.Ob/T.Ar	0.6111	$(/mm^2)$	
Osteoblast surface density	N.Ob/B.Pm	0.2052	(/mm)	
Osteoblast index	N.Ob/Ob.Pm	76.3838	(/mm)	
Osteoclast volume density	N.Oc.T.Ar	0.1222	$(/mm^2)$	
Osteoclast bone surface density	N.Oc/B.Pm	0.0410	(/mm)	
Osteoclast eroded surface density	N.Oc/E.Pm	0.4472	(/mm)	
Osteoclast index	N.Oc/Oc.Pm	20.8699	(/mm)	
Structural indices				
Trabecular separation	Tb.Sp	*	(μm)	570.60 (98.9)
Trabecular number	Tb.N	*	(/mm)	1.75 (0.23)
Kinetic indices				
Mineralizing surface	MS/BS	*	(%)	76.40 (11.3)
Mineralizing surface (osteoid)	MS/OS	*	(%)	≥60.00

Table 6.13 *(Continued)*

Mineral apposition rate	MAR	*	(μm/d)	0.51 (0.04)
Adjusted apposition rate	Aj.AR	*	(μm/d)	
Bone formation rate (surface)	BFR/BS	*	($μm^3/μm^2/y$)	35.80 (8.9)
Bone formation rate (Rel)	BFR/BV	*	(%/y)	
Bone formation rate	BFR/TV	*	(%/y)	
Mineralization lag time	Mlt	*	(d)	
Osteoid maturation time	Omt	*	(d)	
Total period	Tt.P	*	(d)	
Formation period	FP	*	(d)	
Active formation period	FP (a+)	*	(d)	
Resorption period	Rs.P	*	(d)	
Reversal period	Rv.P	*	(d)	
Remodelling period	Rm.P	*	(d)	
Quiescent period	QP	*	(d)	
Activation frequency	Ac.f	*	(/y)	

* Value not calculated.

completion a list of variables is automatically calculated, some of which are primary while many are secondary derivations from these (the recommended nomenclature is used throughout [71]). The variables are shown in table 6.13 and most are measured under the plain light optical microscope (figure 6.16) at a magnification sufficient to resolve cellular detail, i.e. about ×150. For an analysis that includes the dynamic variables an incident light fluorescence (epi-fluorescence) attachment to the microscope enables measurement of the distance between the fluorescent bands in material that has been labelled at recorded times with tetracycline (inter-label thickness, Ir.L.Th). From this information and the time interval, the mineral apposition rate (MAR) is calculated together with other related parameters (for example, the mineralization lag time [153]) and the extent of the calcification front (mineralizing surface, i.e. all double-labelled sites plus half single-labelled sites expressed relative to either the bone surface or more commonly the osteoid surface). For the formation variable known as the mean wall thickness (W.Th; a measure of the depth of new bone apposed to the surface by a team of osteoblasts before 'switching off' into a resting phase; figure 6.18) polarized light displays the defining cement line and the distinguishing sheaf of parallel lamellae [66, 154]. There is also a thionin staining method that avoids the need for polarized light [155, 156]. Alternatively the mineralization rate (μm or mcm/day) and mean wall thickness (μm or mcm) can be determined separately by simple manual methods using a calibrated eyepiece and taking four (or more) equidistant readings at each site, a procedure that is convenient and no more time-consuming than the above for a range of width measurements, including the cortical thickness. Essential reading is Parfitt *et al* [71] for the comprehensive tabulation of the histomorphometric variables and their definition.

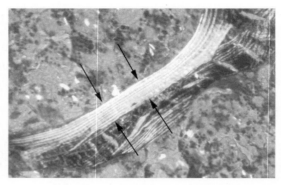

Figure 6.18. *Mean wall thickness. Individual remodelling units are evident in polarized light demarcated by the cement line (arrowed) and sheafs of parallel lamellae. The paired arrows indicate the mean wall thickness. Goldner stain, ×80.*

6.6.2.   Tetracycline labelling and staining of the calcification front

The antibiotic tetracycline forms a stable complex with calcium phosphate/ calcium carbonate as a result of which it has a long history in association with bone formation [157–159]. Tetracycline labelling *in vivo* prior to taking a bone biopsy is common practice (figure 6.19), providing dynamic information to aid interpretation of the static histological image. Less well known but with a history that is equally long is tetracycline staining (see

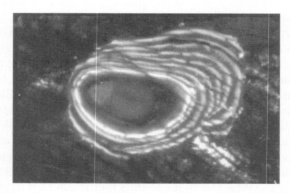

Figure 6.19. *Fluorochrome labelling and staining. Undecalcified iliac bone multiply labelled in vivo with tetracycline (double labelling is more usual) and stained in vitro with xylenol orange (innermost band, contrasting in colour). The successive position of the calcification front is indicated by the concentric fluorescent bands. The differential 'switching on' and 'switching off' taking place during the formation phase of this remodelling unit is indicated by the variation in the number of bands distributed around the central canal. Epi-fluorescence microscope, UV light, ×65.*

Table 6.14. *Tetracycline labelling and fluorochrome staining of the calcification front [160].*

Tetracycline labelling. Markers applied *in vivo*.

Day 1   Label 1 *in vivo* 300 mg of demethylchlortetracycline orally twice daily for 3 days.

Day 10   Label 2 *in vivo* 300 mg of demethylchlortetracycline orally twice daily for 3 days. (The tetracycline specification or labelling duration may be varied to give bands contrasting slightly in colour or width.)

Day 14   Biopsy (table 6.3). Immerse the fresh biopsy in 70% EtOH prior to embedding (table 6.4). Examine without staining using a fluorescent microscope (Zeiss exciter filter BG 12, peak transmission $\lambda = 4000$ A and barrier filters 53 and 44, transmission $>\lambda = 5300$ A).

MAR is the distance between the band midpoints or between the corresponding edges measured in four or more equidistant places.

Fluorochrome (including tetracycline) staining. Marker applied *in vitro*.

Day 1   Label 1 *in vivo* 300 mg of tetracycline (as above).

Day 10   Biopsy + stain 2 *in vitro*. Immerse the fresh biopsy in 1% neutral tetracycline hydrochloride (achromycin) in 10% buffered formalin or 70% EtOH. Since a toxicity problem does not arise, 1% xylenol orange or calcein may be used for contrasting colours. Wash prior to embedding (table 6.4). Examine without further staining under a fluorescence microscope and measure the inter-marker distance as above.

Fluorochrome stain may be used alone to indicate the extent of the calcification front.

NB. Store all sections in the dark.

Aaron *et al* [160] for references). This is a useful adjunct for analysis as the fluorochrome marker is applied not *in vivo* as is customary, but *in vitro*, i.e. the fresh *intact* biopsy is immersed in a tetracycline solution before further processing (table 6.14). The fluorescent image after *in vitro* staining closely resembles that after *in vivo* labelling (figure 6.19), since only the mineral at sites of bone formation is accessible to the marker, the remainder being packaged, compressed and generally inaccessible [161]. As there are no toxic considerations for the subject with the staining method a diversity of fluorochromes may also be used for colour separation of the bands.

Reference books on the above topic of remodelling and histomorphometry include Anderson [162], Malluche and Faugere [163], Recker [164], Eriksen *et al* [165], Cowin [166] and An and Martin [167].

6.7.   ACKNOWLEDGMENTS

The valuable contribution of a number of undergraduate students in Human Biology at Leeds is gratefully recognized, especially that of Annette Lenehan,

Alexander Mortimer, David Bridges, Fiona McClelland, Martin Wilson, Anthony Aimes, Verity Burt, Mark Brown, Hannah Welbourn, Matthew Thomas and Linda Freidrich. We are also indebted to Dr Roger C Shore, Division of Oral Biology, Leeds Dental Institute, for his generous support in processing the illustrations.

REFERENCES

[1] Aaron J E 1976 *Calcium, Phosphate and Magnesium Metabolism* (Edinburgh, London, New York: Churchill Livingstone) p 298

[2] Brickmann P, Frobin W, Hierholzer E and Horst M 1983 *Spine* **8** 851–6

[3] Rockoff J D, Sweet E and Bleustein J 1969 *Calc. Tissue Res.* **3** 163–75

[4] Moselkilde L I and Moselkilde L 1986 *Bone* **7** 207–12

[5] Yogandon N, Nyklebust J B, Cusick J F, Wilson C R and Sances A 1988 *Clin. Biomech.* **3** 11–8

[6] Ritzel H, Amberg M, Posl M, Hahn M and Delling G 1997 *J. Bone Min. Res.* **12** 89–95

[7] Overaker D W, Langrana N A and Cuitino A M 1999 *J. Biomech. Eng.* **121** 542–9

[8] Oleksik A, Ott S, Vedi S, Bravanboer N, Compston J and Lips P 2000 *J. Bone Min. Res.* **15** 1368–75

[9] McBroom R J, Hayes W C, Edwards W T, Golderg R P and White A A 1985 *J. Bone Joint Surg.* **67A** 1206–13

[10] Faulkner K G, Cann C E and Hasegawa B H 1991 *Radiology* **179** 660–74

[11] Mizrahi J, Silva M J, Keaveny T M, Edwards W T and Hayes W C 1993 *Spine* **18** 2088–96

[12] Silva M J, Keaveney T M and Hayes W C 1997 *Spine* **22** 140–50

[13] Dalen N, Hellstrom L and Jacobson B 1976 *Acta Orthoped. Scand.* **47** 503–8

[14] Carter D R and Hayes W C 1976 *Science* **194** 1174–6

[15] Rice J C, Cowin S C and Bowman J A 1988 *J. Biomech.* **21** 155–68

[16] Mosekilde L I, Mosekilde L and Danielsen C C 1987 *Bone* **8** 79–85; 1986 *Bone* **7** 207–12

[17] Kaplan F S, Dalinka M, Karp J S, Fallon M D, Katz M, Boden S, Simpson E, Attie M and Hadden J G 1989 *Orthopedics* **12** 949–55

[18] Ross P D, Heibrun L K, Wasnich R D, Davis J W and Vodel J M 1989 *J. Bone Min. Res.* **4** 649–56

[19] Melton L J, Kan S H, Frye M A and Vahner H W 1989 *Ann. J. Epidemiol.* **129** 1000–11

[20] Kimmel D, Recker R, Gallagher J, Vaswani A and Aloia J L 1990 *Bone* **11** 217–35

[21] Ott S M 1993 *Calcif. Tissue Intl.* **53** S7–13

[22] Marshall D, Johnell O and Widel H 1996 *Br. Med. J.* **312** 254–9

[23] Kleerekoper M, Peterson E L, Nelson D A, Philips E, Shork M A, Tilley B C and Parfitt A M 1991 *Osteoporos. Intl.* **1** 155–61; 1989 *J. Bone Min. Res.* **4** S1 376

[24] Aaron J E, de Vernejoul M-C and Kanis J A 1991 *Bone* **12** 307–10

[25] Legrand E, Chappard D, Pascaretti C, Duquenne M, Krebs S, Rohmer V, Basle M and Audran M 2000 *J. Bone Min. Res.* **15** 13–9

[26] Galante J, Rostoker W and Ray R D 1970 *Calc. Tissue Res.* **5** 236–46

[27] Moselkilde L 1989 *Bone* **10** 425–32
[28] Ouyang X, Majumdar S, Link T, Lu Y, Augat P, Lin J, Newitt D and Genant H K 1998 *Med. Phys.* **25** 2037–41
[29] Odgaard A 1997 *Bone* **20** 315–28
[30] Wolff J 1899 *Virchow's Arch. Pathol. Anat. Physiol.* 155 156; 1870 **50** 389–450
[31] Kabel J, van Rietbergen B, Odgaard A and Huiskes R 1999 *Bone* **25** 481–6
[32] Hodgskinson R and Currey J D 1990 *Proc. Instn. Mech. Engrs.* **204** 43–51
[33] Hodgskinson R and Currey J D 1990 *Proc. Instn. Mech. Engrs.* **204** 115–21
[34] Goldstein S A, Goulet R and McCubbrey D 1993 *Calcif. Tissue Intl.* **53** S127–31
[35] Keaveny T M and Hayes W C 1993 *J. Biomech. Eng.* **115** 534–42
[36] Silva M J and Gibson L J 1997 *Bone* **21** 191–9
[37] Dempster D W, Fergusson-Pell M W, Mellish R W E, Cochran G V B, Xie F, Fey C, Horbert W, Parisien M and Lindsay R 1993 *Osteoporos. Intl.* **3** 90–6
[38] Snyder B, Piazza S and Edwards W 1993 *Calcif. Tissue Intl.* **53** S14–22
[39] Grote H J, Amling M, Vogel M, Posl M and Delling G 1995 *Bone* **16** 301–8
[40] Amling M, Posl M, Ritzel H, Hahn M, Vogel M, Wenig V J and Delling G 1996 *Archiv. Orthop. Trauma Surg.* **115** 262–9.
[41] Cendre E, Mitton D, Roux J P, Arlot M E, Duboeuf F, Burt-Pichat B, Rumelhart C, Peix G and Meunier P J 1999 *Osteoporos. Intl.* **10** 353–60
[42] Thompsen J S, Ebbesen E N and Mosenkilde L 2000 *Bone* **27** 129–38
[43] Kinney J H and Ladd A J C 1998 *J. Bone Min. Res.* **13** 839–45
[44] Ulrich D, van Rietbergen B, Laib A and Ruegsegger P 1999 *Bone* **25** 55–60
[45] Singh M, Riggs B L, Beabout J W and Jowsey J 1972 *Ann. Intern. Med.* **77** 63–7
[46] Geraets W G, van der Stelt P F, Netelenbos C J and Elders PM 1990 *J. Bone Min. Res.* **5** 227–33
[47] Caligiuri P, Giger M L, Favus M J, Jia H, Doi K and Dixon L B 1993 *Radiol.* **186** 471–4
[48] Majumdar S, Genant H K, Grampp S, Jaegas M D, Newitt D C and Gies A A 1994 *Europ. Radiol.* **4** 517–24
[49] Durand E and Ruegsegger P 1991 *J. Comput. Assisted Tomog.* **15** 133–9
[50] Link T M, Majumdar S, Lin J C, Augat P, Gould G G, Newitt D, Ouyang X, Lang T F, Mathur A and Genant H K 1998 *J. Comput. Assisted Tomog.* **22** 15–24
[51] Laib A, Hildebrandt T, Hauselmann H J and Reugsegger P 1997 *Bone* **21** 541–6
[52] Chavalier F, Laval-Jentet A M, Laval-Jentet M and Bergot C 1992 *Calcif. Tissue Intl.* **51** 8–13
[53] Ito M, Ohki M, Hayashi K, Yamanda M, Uetani U and Nakamuru T 1995 *Radiology* 194 55–9
[54] Gordon C L, Lang T F, Augat P and Genant K 1998 *Osteoporos. Intl.* **8** 317–25
[55] Mitton D, Cendre E, Roux J P, Arlot M E, Peix G, Rumelhart C, Babot D and Meunier P J 1998 *Bone* **22** 651–8
[56] Engelke K, Graeff W, Meiss L, Hahn M and Delling G 1993 *Invest. Radiol.* **28** 341–9
[57] Muller R 1998 *Bone* **23** 59–66
[58] Hildebrand T, Laib A, Muller R, Dequecker J and Ruegsegger P 1999 *J. Bone Min. Res.* **14** 1167–74
[59] Feldkamp L, Goldstein S, Parfitt M, Jesion G and Kleerekoper M 1989 *J. Bone Min. Res.* **4** 3–11
[60] Kinney J H, Lane N E and Haupt D L 1995 *J. Bone Min. Res.* **10** 264–70

[61] Chung H W, Wehrli F W, Williams J L and Wehrli S I 1995 *J. Bone Min. Res.* **10** 1452–61

[62] Hipp J A, Janujwicz A, Simmons C A and Snyder B 1996 *J. Bone Min. Res.* **11** 286–92

[63] Atkinson P J 1967 *Calc. Tissue Res.* **1** 24–32

[64] Watamatsu E and Sissons H A 1969 *Calc. Tissue Res.* **4** 147–61

[65] Dyson E D, Jackson C K and Whitehouse W J 1970 *Nature (Lond.)* **225** 957–9

[66] Aaron J A, Makins N B and Sagreiya K 1987 *Clin. Orthop. Rel. Res.* **215** 260–71

[67] Gunderson H J, Boyce R W, Nyengaard J R and Odgaard A 1993 *Bone* **14** 17–22

[68] Vesterby A 1990 *Bone* **11** 149–155

[69] Chappard D, Legrand E, Pascaretti C and Basle M F 1999 *Microsc. Res. Tech.* **45** 303–12

[70] Parfitt A, Mathews C H E, Villanueva A R, Kleerekoper M, Frame B and Rao D S 1983 *J. Clin. Invest.* **72** 1396–409

[71] Parfitt A, Drezner M K, Glorieux F H, Kanis J A, Malluche H, Meunier P J, Ott S M and Recker R R 1987 *J. Bone Min. Res.* **2** 595–610

[72] Aaron J E, Francis R M, Peacock M and Makins N B 1989 *Clin. Orthop. Rel. Res.* **243** 294–305

[73] Parfitt A M 1992 *Bone* **13** S41–7

[74] Dempster D W 2000 *J. Bone Min. Res.* **15** 20–3

[75] Aaron J E, de Vernejoul M-C and Kanis J A 1992 *Bone* **17** 399–413

[76] Aaron J E and Skerry T M 1994 *Bone* **25** 211–30

[77] Ham A W 1952 *J. Bone Joint Surg.* **34A** 701–28

[78] Birkenhager-Frenkel D H, Courpron P, Hupsher E A, Clermonts E, Coutinho M F, Schmitz P I M and Meunier P J 1988 *Bone* **4** 197–216

[79] Schnitzler C M, Pettifor J M, Mesquita J M, Bird M D, Schnaid E and Smyth A E 1990 *Bone* **10** 183–199

[80] Hordon L D, Raisi M, Aaron J E, Paxton S K, Beneton M and Kanis J A 2000 *Bone* **27** 271–6

[81] Mellish R W, Fergusen-Pell M W, Cochran G V, Lindsay R and Dempster D W 1991 *J. Bone Min. Res.* **6** 689–96

[82] Aaron J E, Johnson D R, Kanis J A, Oakley B A, O'Higgins P and Paxton S K 1992 *Comput. Biomed. Res.* **25** 1–16

[83] Parisien M, Mellish R W Silverberg S J, Shane E, Lindsay R, Bilezikian J and Dempster D 1992 *J. Bone Min. Res.* **7** 913–9

[84] Garrahan N J, Mellish R W, Vedi S and Compston J E 1987 *Bone* **8** 227–30

[85] Garrahan N J, Mellish R W, Vedi S and Compston J E 1986 *J. Microsc.* **142** 341–9

[86] Hahn M, Vogel M and Pompesius-Kempa M 1992 *Bone* **13** 327–30

[87] Croucher P I, Garrahan N J and Compston J E 1996 *J. Bone Min. Res.* **11** 955–61

[88] Weinstein R S, Majumdar S and Genant H K 1992 *Bone* **13** A38

[89] Fazzalari N L and Parkinson L H 1997 *J. Bone Min. Res.* **12** 632–9

[90] Whitehouse W J 1974 *J. Microsc.* **101** 153–68

[91] Beck J S and Nordin B E C 1960 *J. Path. Bact.* **80** 391–7

[92] Rao D S 1990 *Bone Histomorphometry: Techniques and Interpretation* (Boca Raton, FL: CRC Press) p 3

[93] Hordon L and Aaron J 1996 *Rheumatology in Practice* Summer Issue 8–11

[94] Naccache N J and Shinghal R 1984 *Pattern Recog.* **17** 279–89

[95] Spiers F W and Beddoe A H 1977 *Phys. Med. Biol.* **22** 670–80

[96] http://math.bu.edu/DYSYS/chaos-game/node6.html
[97] Odgaard A and Gunderson H J 1993 *Bone* **14** 173–82
[98] De Hoff R T, Aigeltinger E and Craig K 1972 *J. Microsc.* **95** 69–91
[99] Moselkilde L 1990 *Bone* **10** 13–35
[100] Shahtaheri S M, Aaron J E, Johnson D R and Purdie D W 1999 *Brit. J. Obstet. Gynaecol.* **106** 432–8
[101] Shahtaheri S M, Aaron J E, Johnson D R and Paxton S K 1999 *J. Anat.* **194** 407–21
[102] Flautre B and Hardouin P 1994 *Bone* **15** 477–81
[103] Compston J E 1994 *Bone* **15** 463–6
[104] Vesterby A, Gunderson H, Melsen F and Mosekilde L 1989 *Bone* **10** 7–13
[105] Boyce R W, Wronski T J, Ebert D C, Stevens M L, Paddock C L, Youngs T A and Gundersen H J G 1995 *Bone* **16** 209–13
[106] Mandelbrot B B 1977 *Fractals: Form, Chance and Dimension* (San Francisco: W H Freeman)
[107] Le H M, Holmes R E, Shors E and Rosenstein D A 1992 *Acta Stereol.* **11** S1 267–72.
[108] Amstutz H C and Sissons H A 1969 *J. Bone Joint Surg.* **51B** 540–50
[109] Kothari M, Keaveny T M, Lin J C, Newitt D C and Majumdar S 1999 *Bone* **25** 245–50
[110] De Hoff RT 1982 *J. Microsc.* **131** 25–63
[111] Odgaard A, Anderson K, Melson F and Gunderson H J 1990 *J. Microsc.* **159** 335–42
[112] Odgaard A, Andersen K, Ullerup R, Frich L H and Melsen F 1994 *Bone* **15** 335–42
[113] Shore P A, Shore R C and Aaron J E 2000 *Biotechnic. Histochem.* **75** 183–92
[114] Aaron J E, Shore P A, Shore R C, Beneton M and Kanis J A 2000 *Bone* **27** 277–82
[115] Kleerekoper M, Villanueva A R, Stanciu J, Rao D S and Parfitt A M 1985 *Calcif. Tissue Intl.* **37** 594–7
[116] Recker R R 1983 *Calcif. Tissue Intl.* **53** S139–42
[117] Amling M, Hahn M, Wening V J, Grote H J and Delling G 1994 *J. Bone Joint Surg.* **76** 1840–6; 1994 *Bone* **27** 193–208
[118] Yanuka M, Dullien FAL and Elrick D E 1984 *J. Microsc.* **135** 159–68
[119] Hahn M, Vogel M and Delling G 1991 *Virchow's Archiv. A. Pathol. Anat.* **418** 1–7
[120] Mosekilde L 1990 *Bone Min.* **10** 13–35
[121] Aaron J E 1977 *Calcif. Tissue Res.* **22** S247–54
[122] Frost H M 1991 *Calcif. Tissue Intl.* **49** 229–31
[123] Vernon-Roberts B and Pirie C J 1973 *Ann. Rheum. Dis.* **32** 406–12
[124] Watson M 1975 *J. Bone Joint Surg.* **57A** 696–8
[125] Hansson T and Roos B 1981 *Spine* **6** 375–80
[126] Ohtani T and Azuma H 1984 *Acta Orthop. Scand.* **55** 419–22
[127] Fazzalari N L, Vernon-Roberts B and Daracot J 1987 *Clin. Orthop. Rel. Res.* **216** 224–33
[128] Hahn M, Vogel M and Amling M 1995 *J. Bone Min. Res.* **10** 1410–16
[129] Atkinson P J, Walsh J A, and Haut R C 1998 *Biotechnic. Histochem.* **74** 27–33
[130] Blackburn J, Hodgskinson R, Currey J and Mason J E 1992 *J. Orthop. Res.* **10** 237–46
[131] Amprino A 1958 *Acta Anat.* **34** 161–86
[132] Weaver J K 1966 *J. Bone Joint Surg. (Am.)* **48** 273–88
[133] Wong S Y P, Kariks M D and Evans R A 1985 *J. Bone Joint Surg.* **67** 274–83
[134] Freeman M A R, Todd R C and Pirie C J 1974 *J. Bone Joint Surg.* **56B** 698–702
[135] Kopperdahl D L, Pearlman J L and Keaveny T M 1999 *J. Orthop. Res.* **17** 346–53

[136] Wenzel T E, Schaffler M B and Fyhrie D P 1996 *Bone* **19** 89–95
[137] Schaffler M B, Choi K and Milgrom C 1995 *Bone* **17** 521–5
[138] Fazzalari N L 1993 *Calcif. Tissue Intl.* **53** S143–7
[139] Mori S, Harruff R and Burr D B 1993 *Arch. Pathol.* **117** 196–8
[140] Frost H M 1989 *Calcif. Tissue Intl.* **44** 367–81
[141] Koszyca B, Fazzalari N I and Vernon-Roberts B 1989 *Clin. Orthop. Rel. Res.* **244** 208–126
[142] Sokoloff L 1993 *Arch. Pathol. Lab. Med.* **117** 191–95
[143] Villanueva A R, Longo J A and Weiner G 1994 *Biotechnic. Histochem.* **69** 81–8
[144] Schaffler M B, Choi K and Milgrom C 1995 *Bone* **17** 521–5
[145] Carter D R 1984 *Calcif. Tissue Intl.* **36** S19–24
[146] Burr D B, Forwood M R and Fyhrie D P 1997 *J. Bone Min. Res.* **12** 6–13
[147] Urovitz E P M, Fornasier V L and Risen M I 1977 *Clin. Orthop. Rel. Res.* **127** 275–80
[148] Frost H M 1960 *Henry Ford Hosp. Bull.* **8** 25–35
[149] Burr D B and Stafford T 1989 *Clin. Orthop. Rel. Res.* **260** 305–8
[150] Frost H M 1985 *Clin. Orthop. Rel. Res.* **200** 198–225
[151] Birkenhager-Frenkel D H, Schmitz P I M, Breuls P N W M, Lockefeer J H M and van der Heul R O 1977 *Bone Histomorphometry 1976* (Toulouse: Société de la Nouvelle Imprimerie Fournie) p 63
[152] Schenk R K, Merz W A and Muller J 1969 *Acta Anat. (Basel)* **74** 44–53
[153] Eriksen E F 1986 *Endocrine Rev.* **7** 379–408
[154] Lips P, Courpron P and Meunier P J 1978 *Calc. Tissue Res.* **26** 13–7
[155] Birkenhager-Frenkel D H, Nigg A L, Hens C J J and Birkenhager J C 1993 *Bone* **14** 211–6
[156] Derkx P and Birkenhager-Frenkel D H 1995 *Biotechnic. Histochem.* **70** 70–4
[157] Andre T 1956 *Acta Radiol. (Stockh.)* **142** S1–89
[158] Harris W H, Jackson R H and Jowsey J 1962 *J. Bone Joint Surg.* **44A** 1308–20
[159] Frost H M 1969 *Calc. Tissue Res.* **3** 211–37
[160] Aaron J E, Makins N B and Francis R M 1984 *J Histochem. Cytochem.* **32** 1251–61
[161] Tapp E, Kovacs K and Carroll R 1965 *Stain Technol.* **41** 57–60
[162] Anderson C 1982 *Manual for the Examination of Bone* (Boca Raton, FL: CRC Press)
[163] Malluche H H and Faugere M C 1986 *Atlas of Mineralized Bone Histology* (Basel: Karger)
[164] Recker R R 1990 *Bone Histomorphometry: Techniques and Interpretation* (Boca Raton, FL: CRC Press)
[165] Eriksen E F 1994 *Bone Histomorphometry* (New York: Raven Press)
[166] Cowin S C 2000 *Bone Mechanics Handbook* (Boca Raton, London, New York, Washington: CRC Press)
[167] An Y H and Martin K L 2003 *Handbook of Histology Methods for Bone and Cartilage* (Totawa, NJ: Humana Press)

Chapter 7

Microscopy and related techniques

Jean E Aaron and Patricia A Shore

with specialist contributions by Roger C Shore and Jennifer Kirkham

7.1. INTRODUCTION

As a consequence of developments in microsampling and the observations of authors such as Ascenzi based upon the mechanical testing of individual trabeculae, separate osteons and lamellae, increasing attention is directed at the properties of bone at a microstructural and ultrastructural level. Technological advances have resulted in an expanding range of general and specialist microscopes and accessories, many of which are useful for bone. These include light microscopes (fluorescence, polarization, interference contrast, laser confocal which may be combined with absorption spectrometry), electron microscopes (e.g. transmission, scanning, high voltage), composition analysis systems (electron backscatter diffraction, X-ray microanalysis, i.e. EDX and XRF, energy filtering transmission electron microscopy (EFTEM), together with scanning probe microscopes (tunnelling, atomic force). These are used with histological, histochemical and immunohistochemical stains and tissue markers which differentiate or label (a) structural features (e.g. osteoid tissue, cell types, organelles), (b) specific enzymes (e.g. acid or alkaline phosphatase, proteinases such as collagenase), (c) specific structural proteins in the cells or extracellular matrix (e.g. collagenous and non-collagenous proteins), or (d) general and specific genomic activity in health and disease (e.g. the detection of messenger ribonucleic acid (mRNA) for natural cytokines such as growth factors and for pathogenic infection such as the canine distemper virus in pagetic bone). There are also markers for (e) inorganic ions (e.g. radioisotopes and calcium-sensitive fluorescent dyes such as tetracycline). These latter act as molecular probes to facilitate time-lapse investigation of cells in culture. They may be combined with laser confocal cellular three-dimensional morphology which enables interaction with the matrix to be observed *in vitro* (for an earlier alternative method see Aaron

225

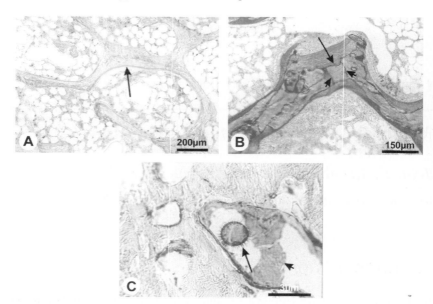

Figure 7.1. *Photomicrographs showing the identification of (a) iron from ethnic cooking vessels at the calcification front (arrowed) and within marrow cells as a blue stain using Perls' method, (b) aluminium at the calcification front (arrowed) and cement lines (arrowheads) following renal dialysis, stained purple using the solochrome azurine method, and (c) fungal hyphae (arrowhead) with a spore (arrowed) stained purple in 16th-century bone from an ancient burial ground, using Stoughton's method.*

and Pautard [1] who viewed living bone cells under the optical microscope *in situ* possibly for the first time, and advocated the necessity of identifying 'the right cell, in the right place, at the right time, using the right technique'). Added to stains for bone salt are (f) diverse special stains (e.g. to determine the matrix distribution of iron from ethnic cooking vessels (figure 7.1(a)), or excessive aluminium accumulation at the calcification front in the bone matrix after renal dialysis (figure 7.1(b)), where aluminium is used to chelate and control phosphate). Further afield in biology are (g) other stains with a more unusual skeletal application (e.g. to identify fungal spores and hyphae in palaeopathological specimens (figure 7.1(c)), crucial in separating the disturbances of putative past bone disease from post-interment changes [2]).

Material for analysis may be living tissue, or fresh-frozen for either cryomicrotomy or freeze-fracturing analysis (fracture patterns relate to differences in structural composition), or it may be chemically fixed before slicing with a diamond saw directly or after embedding for microtomy/ultra-microtomy. The more extensive the physical manipulation, dehydration and chemical intervention and the higher the magnification, the more divorced the final image may be from reality (table 7.1). This is especially true for

Table 7.1. *Summary of the routine histology processing steps in preparing undecalcified bone for light microscopy and transmission electron microscopy.*

Light microscopy (LM)	Process	Transmission electron microscopy (TEM)
Bone specimens as large as 50 cm^3 can be processed without apparent structural detriment at the LM level.	**Sample acquisition**	The bone is cut into cubes not exceeding 8 mm^3 using cutting devices such as diamond saws or diamond impregnated discs.
	↓	
A typical fixation regime is neutral buffered formaldehyde, followed by a wash in distilled water. Alternatively, 70% ethanol may be used.	**Chemical fixation** Preserves the structural. integrity of the tissue by preventing post-mortem degradation by microbial and autolytic factors and disruption by the osmotic effects of subsequent processing solutions.	A typical fixation regime is neutral buffered glutaraldehyde followed by buffered osmium tetroxide followed by washing in distilled water.
	↓	
Achieved with a graded series of ethanol solutions up to 100%.	**Dehydration** Tissue fluid must be removed in order to facilitate the complete impregnation by hydrophobic embedding media.	Achieved with a graded series of ethanol or acetone solutions up to 100%.
	↓	
Typically the acrylic resin, methyl methacrylate, is utilized. Firstly the methylmethacrylate monomer is allowed to infiltrate the tissue. This fluid is then replaced with monomer containing a plasticizer, dibutyl phthalate, and a catalyst, benzoyl peroxide. High polymerization temperatures should be avoided as bubbles will form within the resin.	**Embedding in resin** It is necessary to infiltrate the tissue with an embedding material which, when hardened, will support it during the sectioning procedure.	Typical embedding media are epoxy or acrylic resins. The mixture containing resin and catalyst is allowed to infiltrate the tissue. Polymerization is promoted by incubating at 60 °C overnight.
	↓	
Sections of a thickness between 10 and 20 μm are cut using a heavy-duty sledge microtome equipped with a hardened steel knife.	**Sectioning** Sections are obtained sufficiently thin to allow the transmission of the illuminating beam (light or electrons).	Sections of a thickness in the region of 100 nm are cut using an ultramicrotome equipped with a diamond knife. The sections are floated on to water at pH 7 and picked up on metal em grids.
	↓	
Sections are immersed in the chosen stain, followed by washing, drying and mounting on glass microscope slides. Some commonly used stains are those which stain the mineral (e.g. silver nitrate solutions or alizarin red) or the matrix (e.g. trichrome stains).	**Section staining** Staining increases specimen contrast and facilitates discrimination of tissue components.	TEM stains consist of solutions of heavy metals which impart electron contrast. The stains are taken up by cellular components and extracellular matrices differentially. Typical heavy metals used are uranium and lead.

bone, where for example particulate calcified structures (microspheres and nanoparticles, p. 255) rather than the textbook classical image of random sheets of uniform crystallites may be a closer approximation to the morphology of the inorganic phase *in situ* within the vital hydrated fibrous matrix. In imaging a dynamic biological 'colloid' such as this, the difficulties of alteration by preparation (fixation, staining and dehydration), particularly in electron microscopy, are only partly solved and may be misleading even with the best techniques of slam-freezing [3] and freeze-fracturing [4]. These uncertainties can only be reliably examined by recourse to such methods as short-term Brownian Dynamics through micro-laser optical analysis and simulation [5]. However, a reliable (?) observation and identification of biological compounds in a naturally anhydrous 'static' state has apparently been obtained in one hard tissue, calcified keratin [6], and may be used for comparison. Here unlike bone, the same (similar) mineral but a different fibrous component can be seen without processing at a resolution of less than 1 nm by enhanced digital methods (see p. 256). Selected aspects of bone microscopy are considered below.

SECTION A: MOLECULAR LABELLING

Autoradiography, immunohistochemistry, laser confocal imaging

7.2. RADIOISOTOPE-LABELLING OF BONE— AUTORADIOGRAPHY

This is a labelling technique that provides optical information about cell division and synthetic activity, bone turnover, transport and the localization of isotope-labelled macromolecular constituents. The isotopes may be bone matrix-seeking (for example ^{45}Ca, ^{90}Sr and ^{125}I-labelled albumin) or they may be soluble precursors of macromolecular constituents (for example tritiated thymidine). Embedded sections of labelled bone are placed underneath or on the surface of a layer of photographic emulsion in the darkroom and are left in light-sealed containers for a period that is dependent upon the type of ray emitted by the radio-active element, its half-life and its concentration. During incubation, radiation from any 'hot-spots' collides with and reduces silver halide particles in the emulsion to silver; after photographic development these appear under the light microscope as regular discrete black dot-like deposits (figure 7.2). Since after processing the emulsion is transparent and permeable, the section beneath can be stained to show the corresponding histology, or the section may be removed and stained and mounted separately. The resolution is determined by (i) the distance between the radiation source and emulsion, (ii) the path length of the radiation from

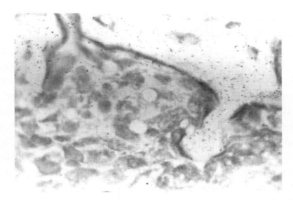

Figure 7.2. *Autoradiograph of a frozen section of a young (3 week-old) rabbit femur, only 15 min after intravenous injection with ^{45}Ca when the isotope is already extensively incorporated in the sub-periosteal bone matrix and bone cells as dark dots. (Courtesy of Dr M Owen.)*

the isotope and (iii) specific properties of the emulsion itself. Ideally for maximum resolution the emulsion would consist of a monolayer of small densely packed silver bromide grains. In practice the silver is then too sparse for optical detection and a low grain size emulsion about 5 µm thick is generally used. Emitters of soft, short-range β-rays such as tritium are preferred because there is less confusing 'cross-fire' between adjacent sources. On the other hand, high-energy particles such as ^{45}Ca can pass obliquely for some distance through both section and emulsion, producing misleading silver grains away from the source; to reduce this path length, the sections used are as thin as possible. An author long established in this field is Owen [7]. More recently autoradiography has been combined with *in situ* hybridization (p. 237) and for example, Chen *et al* [8] used ^{35}S-labelled RNA as a probe for bone sialoprotein and osteopontin to show the increased expression of the former and the suppression of the latter in osteoblastic cells in vitamin D deficiency.

Another silver-based method that has been applied to proliferating tumour cells [9] and might usefully be used to determine bone cell nuclear activity is the AgNOR process which targets the nucleoli. The colloidal silver technique (50% aqueous silver nitrate mixed 2:1 by volume in 2% gelatin in 1% formic acid) recognizes non-histone proteins associated with specific loops of DNA called nucleolar organizer regions (NORs) which encode ribosomal RNA. Staining takes 60 min for routine formalin-fixed paraffin-embedded sections and is performed in the darkroom by safelight, after which the sections are washed and may be counterstained. The AgNORs are optically visible dots within the nuclei which can be counted as an index of nuclear activity and protein synthesis.

7.3. CRYOMICROTOMY, BONE HISTOLOGY AND IMMUNOHISTOCHEMISTRY

Bone histomorphometry is generally performed on thin plastic-embedded, undecalcified sections (chapter 6). However, results are delayed since processing requires a minimum of 12 days. At the same time, there is an expanding demand for sections suitable for histochemistry and immuno-histochemistry and while cold-setting plastics have sometimes been used successfully (for example [10, 11]) as also have decalcified paraffin-embedded sections (for example Vidal *et al* [12] showed oestrogen receptors in osteo-blasts and osteocytes), fresh frozen material is more immediate and biochemically reliable, particularly for less stable proteins. On the other hand, the cellular morphology may be inadequate in frozen sections of hard tissues. The advantages of both techniques may be combined and the disadvantages avoided by cutting frozen sections and plastic-embedded sections from the same tissue block [13]. The procedure commences with the routine preparation of sections of fresh frozen undecalcified bone and ends with the routine preparation of plastic-embedded sections that are indistinguishable from those prepared directly without the intervention of freezing, as indicated in table 7.2.

7.3.1. *Immunohistochemistry*

This exploits the specific affinity between an antibody (serum immuno-globulins (IgG)) and its corresponding antigen (table 7.3). To see the antibody complex optically it is labelled. The marker may be a peroxidase–anti-peroxidase (PAP) enzyme reaction product, as is used widely in routine pathology, or the antibody may be biotinylated such that it will form an exceptionally strong covalent bond with the labelled glycoprotein avidin or with streptavidin (produced by *Streptomyces avidinii* and particularly specific in its reaction). Innovatory kits are regularly introduced by the manu-facturers to simplify processing. For example, the Dako EnVision kit (DakoCytomation Ltd, Cambridgeshire) whereby the marker enzyme is bound to the secondary antibody by dextran is claimed as a more rapid and sensitive supercedent to the above avidin–biotin complex (ABC) tech-nique. For research purposes, however, and because there are few processing steps, conjugation of the antibody with a fluorescent marker, fluoroscein isothiocyanate (FITC), is frequently the method of choice (table 7.4), with observation of the section by incident (i.e. epi-) fluorescence microscopy (transmitted fluorescence tends to be too weak). While the antibody itself may be labelled before application (direct method) it is more efficient for the often scarce and expensive primary antibody to be applied to the tissue in an unlabelled form (indirect method) where it binds to its specific epitope and in this bound form becomes the target of a less valuable labelled

Table 7.2. *The rapid preparation of fresh frozen bone sections, followed by plastic embedding as required [13, 14].*

Freezing	Submerge the undecalcified specimen in either cardice/liquid *n*-hexane ($-75\,°C$) or liquid nitrogen/isopentane ($-150°$) until frozen (quenched). Attach the specimen to the microtome chuck with carboxymethylcellulose (CMC) gel, 1.6% at $4\,°C$ (BDH Chemicals, Poole, UK). Support with more CMC contained by a removable rectangular frame that surrounds the specimen, until the entire block is frozen in more quenching fluid, when the gel functions as reinforced ice.
Microtome	LKB PMV 2258 or 450 MP heavy duty cryomicrotome (Leica, Germany) with a hardened steel D-profile ($35°$) knife set at angle $2°$.
Microtomy	During cryomicrotomy the section is supported by a film of polyvinylpyrrolidone (PVP). This is composed of equal volumes of 40% PVP_{10} and 20% PVP_{360} (combining strength and flexibility) and is spread in two thin consecutive layers over the block face; a square of cigarette paper attached to the fresh film provides additional support. Cut sections, $8\,\mu m$ thick, slowly and press on to cool glass slides coated with a pressure-sensitive adhesive (e.g. Durotak, Product No. 180-1197, National Adhesives and Resins Ltd, Slough, UK). Thaw the sections in 10% buffered formalin pH 7.2 for 5 min when the PVP dissolves and the paper detaches.
Staining	Morphology: (i) von Kossa/eosin method. Place the slide flat in 5% $AgNO_3$ and illuminate for 30 s at 20 cm distance from a 100 W bulb. Wash gently in distilled water before immersing in 3% sodium thiosulphate for 1 min. Counterstain in 1% aqueous eosin for 10 min. Wash gently and mount in Hydromount (National Diagnostics, Aylesbury, Bucks, UK). NB. An aqueous mountant is essential. (Colour: bone brown, osteoid red, cells yellow.) (ii) Haematoxylin/eosin. Place the slide flat and stain progressively with Harris's haematoxylin. Wash gently and blue in tap water. Counterstain in 1% aqueous eosin for 10 min, wash gently in distilled water and mount in Hydromount, as above. (Colour: bone purple, osteoid pink, nuclei dark blue.) (iii) Toluidine blue, 0.1%, pH 3.5. Place the slide flat and stain for 30 min. Wash gently in distilled water, dehydrate in absolute EtOH (3×3 min), clear in methycyclohexane (2×3 min) and mount in neutral mounting medium, e.g. XAM (BDH Chemicals, Poole, UK). (Colour: bone dark blue, osteoid light blue, calcification front black.)
Staining	Immunohistochemistry: Specific antibody of choice (tables 7.3, 7.4).
Thawing/embedding	Place overnight in 10% buffered formalin, dehydrate and embed in methylmethacrylate (chapter 6, table 6.4) or other resin and section on a Jung K or Polycut heavy duty microtome (Leica, Germany). NB. Tetracycline labelling is not eluted by the above processes.

Table 7.3. *The four main criteria defining a good specific antibody preparation for immuno-staining. (Derived from Bendayan [16].)*

Property	Determinant
Specificity	Purity of the antigen used in the immunizing protocol.
Affinity	The longer the amino acid chain of the epitope, the greater the affinity and the less cross-reaction with other proteins.
Avidity	Binding is increased when a variety of antibody molecules in the antiserum conjoin with different epitopes in the same antigen, multiplying the antibody–antigen interaction.
Titre	The number of antibody molecules per millilitre of antiserum determines the greatest dilution still producing adequate staining.

Table 7.4. *The procedure for immunohistochemical localization in frozen bone sections using either the peroxidase–antiperoxidase (PAP) label (conjugate) or the fluorescein isothiocyanate (FITC) label. (After Carter et al [17, 18].)*

Example antibodies for illustration
- Primary polyclonal antibodies raised in rabbit against human collagen Types I and III (affinity-purified; Institut Pateur, Lyons, France), tenascin (a gift of Dr R Chiquet-Ehrismann) and fibronectin (ICN Biomedicals, High Wycombe, UK).
- Partially purified monoclonal antibodies raised against rat bone sialoprotein (BSP; clone WVIDI[9C] and rat osteopontin (OP; clone MPIIIBIO[1]; Iowa Hybridoma Bank).

Immunoperoxidase method (PAP)
Pre-treatment
- Fix the 8 μm thick fresh-frozen sections attached to glass microscope slides (table 7.2) in cold buffered formalin (6.5 g sodium dihydrogen phosphate monohydrate and 4.0 g disodium hydrogen phosphate anhydrous dissolved in 1 litre of 4% formaldehyde, pH 7.0) for 5 min.
- Demineralize the sections in 20% tetrasodium ethylenediaminetetracetic acid (EDTA), pH 7.0 for 3 min at room temperature to expose any masked epitopes.
- Reduce endogenous peroxidase activity with 1% H_2O_2 in phosphate buffered saline (PBS), pH 7.2 for 1 min.
- To increase staining by exposing any additional isotopes, pre-treat with bovine testicular hyaluronidase (Sigma) 0.1 mg/1 ml PBS, pH 7.2 for 20 min at 22 °C.
- To reduce nonspecific staining, pre-treat with normal swine serum (Flow, Rickmansworth, UK) diluted 1 : 5 in PBS, pH 7.2 for 5 min at 37 °C.

Incubation in the primary antibody
Dilute the primary antibody in PBS, pH 7.2 (e.g. collagen Type I 1 : 100; collagen Type III 1 : 80; tenascin 1 : 100; fibronectin 1 : 100; BSP 1 : 200; OP 1 : 100) and add sparingly, sufficient to cover the section. Incubate at 37 °C for 60 min in a humid chamber.

Table 7.4. *(Continued)*

Incubation in the secondary antibody
Depending upon the primary antibody species:
Either: Incubate with swine anti-rabbit IgG (Dako) diluted 1 : 10 in PBS pH 7.2 for 30 min followed by PAP–rabbit complex (Dako, High Wycombe, UK) diluted 1 : 40 for 30 min, both at 37 °C in a humid chamber.
Or: Incubate with anti-mouse IgG (cross-absorbed against rat serum; Sigma) diluted 1 : 40 in PBS pH 7.2 for 30 min followed by PAP–mouse complex.

Visualization for optical microscopy
Add to section 3,3′-diaminobenzidine tetrahydrochloride [dissolve 10 mg DAB tablet (Sigma) in 16 ml Tris-saline 0.05 M, pH 7.6 and 0.2 ml H_2O_2 and incubate for 10 min. Mount without dehydration in DABCO (0.25 g 1,4-diazabicyclo(222)octane (Sigma) dissolved in 9 ml glycerol and 1 ml PBS, pH 7.2 corrected to pH 8.6 with NHCl]. Alternatively mount after air-drying in XAM (BDH, Poole, UK).

Between all the above stages wash the sections for 3×5 min in PBS, pH 7.2.

Controls
- Incubate without the primary antibody.
- Stain known positive soft tissue sections (e.g. normal oral mucosa) alongside the hard tissue.

The avidin–biotin immunoperoxidase method (ABC) is similar to the above, primary antibody binding being visualized by incubation with biotinylated goat antibody for 30 min, followed by streptavidin–biotin peroxidase complex (streptavidin ABC/hrp Duet kit; Dako, UK) for 30 min, followed finally by DAB above, and washing with Tris-buffered saline pH 7.6 for 5 min between each stage.

Indirect immunofluorescence method (FITC)
Pretreatment
Demineralize in EDTA as above. (Compare the final results with undemineralized sections.)

Incubation in the primary antibody
Incubate as above for 30 min at 37 °C.

Incubation in the secondary antibody
Depending upon the primary antibody species:
Either: Incubate in FITC-conjugated swine anti-rabbit IgG (Dako, High Wycombe, UK) diluted 1 : 40 in PBS pH 7.2 for 30 min at 37 °C in a dark humid chamber.
Or: Incubate in FITC-conjugated anti-mouse IgG (cross-absorbed against rat serum; Sigma) diluted 1 : 40 in PBS pH 7.2 for 30 min at 37 °C in a dark humid chamber.

Visualization
Mount in DABCO (above) and examine under UV illumination (excitation wavelength 490 nm), photographing with Ektachrome 160 tungsten colour film or Ilford HP5 black and white film.
Between all the above stages wash the sections for 3×5 min in PBS, pH 7.2.

Controls: As above.

---

secondary antibody with a non-specific high affinity for immunoglobulins. Specific antibodies may be polyclonal (i.e. raised in whole animals against complete antigen macromolecules) or monoclonal (i.e. raised in cell cultures against a specific part of the antigen macromolecule) [15]. Although monoclonal antibody probes may identify sequences as small as two or three amino acids, these are likely to occur in unrelated proteins, producing false positives. Conversely false negatives may arise if the epitope is altered by tissue fixation and associated protein cross-linking, emphasizing the need for appropriate controls (for technical details and references see Bendayan [16]).

7.3.2. Immunohistochemistry of the extracellular matrix

By locating osteocalcin in osteoblasts and young osteocytes, the immunohisto-chemical technique has validated the clinical measurement of osteocalcin in blood as a routine biochemical marker of bone formation, frequently eliminat-ing the need for a bone biopsy. Similarly it has demonstrated oestrogen receptors in osteoblastic cells [12, 19] and PTH/PTHrp receptors in hyper-trophic chondrocytes at the growth plate and osteoblasts in the primary spongiosa [20]. At the same time, the increased availability of specific bone-related antibodies is putting a structural face on to the biochemistry of the organic extracellular matrix and in particular on to that part (including osteo-calcin above) that is not collagen Type I. The procedure is most powerful when the protein of interest is a significant component of a discrete histological feature (for example, an osteon or coarse fibrous insertion) rather than a diffuse humoral agent or scarce commodity which is frequently difficult to dissociate with confidence from non-specific background staining.

Evidence is now accumulating that bone is sub-divided into sectors or domains that are heterogeneous in their macromolecular composition and also in their sensitivity to environmental stimuli, providing insight into why some parts of the skeleton are more responsive than others (for example the proximal and distal femur are dissimilar in their stress-related behaviour [21] as also are endochondral and intramembranous areas of bone [22]. Regional differences are well illustrated by labelled antibodies to collagen Type III. These show that the coarse (5–25 µm) uncalcified birefringent Sharpey fibres that anchor muscle to bone do not terminate in the superficial cortex as regular histology suggests (figure 7.3(a,b)), but penetrate the bone to a considerable degree (figure 7.3(c)), even transgressing the Haversian bone to the spongiosa in some areas [17, 23]. The extent and configuration of the fibrous ramifications (with their stabilizing sheath of collagen Type VI and helix of elastin) was confirmed when their distribution was mapped throughout the entire rat femur using numerous consecutive transverse and longitudinal immuno-stained sections, enabling their three-dimensional reconstruction [24, 25]. This indicated that the apparently random collagen

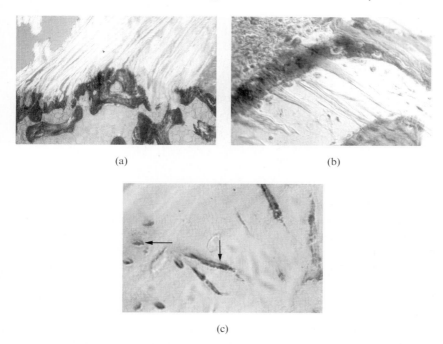

(a)   (b)

(c)

Figure 7.3. *Immunohistochemistry and the distant ramifications of Sharpey's fibres. (a) Low-power view of a tendinous insertion (T) terminating in Sharpey's fibres that penetrate the outer bone surface; the birefringent insertions are uncalcified. Human tissue, Alizarin red stain, polarized light. ×30. (b) A higher-power view of similar fibrous insertions, 2–25 μm wide. Goldner stain, Nomarski optics, ×120. (c) The same fibres in longitudinal (vertical arrow) and transverse (horizontal arrow) aspect, deep within the cortex and apparent only by immuno-techniques. Stained for collagen Type III by the PAP method, Nomarski optics, ×700. (Courtesy of Dr D H Carter.)*

Type III-stained fibres above were in reality marshalled into a discrete sub-periosteal domain that dominated the cortex of the proximal rat femur like a bony cap which tapered distally, terminating at the mid-shaft by attachment to small intracortical cartilage islands. In the form of such regional cohorts of complex uncalcified collagenous insertions, Sharpey's fibres may serve not only to inter-link muscle and bone but also may function as pro-active channels of musculoskeletal exchange, ensuring a synchrony between the two tissues. This exchange may be engaged in the retraction of the immuno-stained fibrous network of the rat following ovariectomy, when positively stained areas of the cortex become negative as a diminution of the fibrous insertions precedes the hypogonadal bone loss. Conversely in response to exercise and prior to bone gain the same fibres expand in number and extent and the positively stained area increases [26]. At the

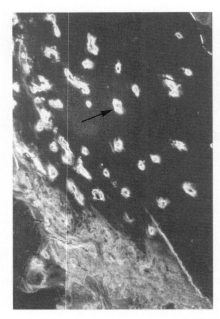

Figure 7.4. *Immunohistochemistry and osteopontin distribution. The bone domain (lower left) is extensively and heterogeneously stained for the matrix protein osteopontin and fluoresces apple green; it is separated by a sharply defined boundary from another domain (upper right) that is unstained and dark with the exception of bright remodelling osteons (arrowed). Rat femur, FITC immunofluorescence, ×50. (Courtesy of Dr D H Carter.)*

same time, another bone domain immediately adjacent is apparently stimulated into a synchronized remodelling episode with increased vascularity and a resultant accumulation of osteopontin-rich osteons (figure 7.4). In turn these latter may complete the sequence of events by their impact upon the inorganic phase and its biomechanical performance, since osteopontin apparently controls crystal growth [27] among other things. This example of an unsuspected chain reaction of macromolecular modulation that has been exposed by immunohistochemistry presents a formidable technological challenge if these basic aspects of bone quality are to be further unmasked and consolidated, and clinically and biomechanically translated.

7.3.3. *Immunohistochemistry and colloidal gold labelling*

Due to its high atomic number gold is especially electron dense. While the above procedures enable the localization of specific structural components under the optical microscope, the colloidal gold method uses the same underlying principle to locate them primarily in the electron microscope by affixing

Table 7.5. *The criteria defining a reliable immunogold staining technique.*
(Derived from Petrusz [29].)

Property	Determinant
Specificity	Reagent-dependent.
Sensitivity	Minimum concentration of the specific molecule detected by the reagent.
Efficiency	Maximum reagent dilution to generate a signal.
Accuracy	Number of processing steps required; direct labelling in one step produces the optimum resolution.
Precision	Consistent pattern when staining repeated.
Practicality	Technical skills required. Cost of consumables.

an electron-dense marker to the antibody of interest (for example, intracellular bone phosphoproteins have been traced in relation to cell organelles in this way [28]. Commonly known as the immunogold technique (table 7.5), it was developed by Faulk and Taylor [30] (see [16] for review). A number of physicochemical factors influence the adsorption of antibody proteins to the colloidal gold reagent including electrostatic van der Waals forces and the isoelectric point of the protein. Colloidal gold consists of hydrophobic negatively charged particles and its stability in water is maintained by electro-static repulsion; if this is reduced the colloidal form flocculates. The particles have a regular spherical shape and may be produced in different sizes to reduce their tissue-masking effect although the smallest particles can produce nonspecific background staining. The conjugation of gold particles to macro-molecules tends not to affect their biological activity; however, with low molecular weight molecules they may produce a complex that is unstable.

7.3.4. *In situ hybridization*

This is mentioned briefly because similar principles of intermolecular affinity to those above apply. *In situ* hybridization is used by molecular biologists as a research tool at the interface of biochemistry and histology for the location and expression of nucleic acids within histological sections, an advantage being that the isolation and purification of the genes from the tissue is not necessary. Labelled RNA or DNA probes (most commonly cDNA and oligonucleotides, 20–50 mers, the sequence of which is specified) are placed in solution upon fixed histological sections where they locate the complemen-tary nucleic acid chain by forming a specific hybrid molecule that may be DNA + RNA, DNA + DNA or RNA + RNA. Before commencement the hydrogen bonding between the *in situ* DNA strands are first broken to permit hybridization with the labelled complementary strands of either DNA or RNA. To enable visualization under the optical microscope there is a further (indirect) affinity reaction using a second labelling reagent

Table 7.6. *Brief outline of a typical in situ hybridization procedure. (Derived from Bendayan [16].)*

INCUBATE the sections for the first affinity reaction with primary specific labelled probes (1–10 µg/ml) in prehybridization buffer (0.6 M NaCl, 30% deionized formamide) containing 150 µg/ml of salmon sperm DNA overnight at 37 °C in a humid chamber.

VISUALIZE by means of a second affinity reaction either by incubation with labelled streptavidin if the first probe is biotinylated or with a labelled anti-digoxigenin antibody if the first probe is digoxigenin (radioisotopes, colloidal gold and FITC are common labels).

CONTROL by:
- Incubation of sections with excess unlabelled probe.
- Prior-treatment of sections with DNA-ase and/or RNA-ase.
- Omission of the specific first probe.
- Use of tissues known *not* to express the specific nucleic acid sequence.

(table 7.6). This may contain radioactive molecules (for example ^{35}S [8]) for detection by autoradiography (p. 228), or more recently biotin or digoxigenin followed by labelled streptavidin or anti-digoxygenin respectively, or an immunohistochemical reaction including immuno-gold if electron microscopy is proposed (see Bendayan [16] for further general details and Saito *et al* [31], Kaneko *et al* [32] and Sherwin *et al* [33] for application to calcified tissues).

7.4. LASER CONFOCAL MICROSCOPY

The confocal microscope (e.g. Biorad MRC-600 series; Meridian ACAS 570) scans with a laser beam (e.g. argon ion and krypton–argon) with visible or UV windows to generate a sharp image of a thin plane within the specimen that is free from extraneous light. The reflected light may be used to produce a high resolution (submicron) three-dimensional image of the surface of the specimen, or fluorescent molecular probes for specific features may be used (e.g. specific labelled antibodies, hydrophilic markers of cytoplasm, specific toxin labels for cell surface proteins, nucleic acid labels such as DAPI, lipophilic membrane labels for endoplasmic reticulum and mitochondria, GFP (Green Fluorescent Protein) stain for the Golgi apparatus, ion-sensitive dyes such as Fluo-3 AM or tetracycline for calcium; see the Molecular Probes Company brochure of Reagents for Cell Biology and Imaging for a more comprehensive range). Twin detectors enable dual labelling and stereo images and computer reconstruction of sequential sections provide three-dimensional information. The probes may simply diffuse into the cell, for example as membrane-permeable ester derivatives, or they may be injected from microelectrodes (with the assumption that the cell does not behave abnormally after the assault). Some dyes are dextran conjugates which when injected remain within the compartment of interest, being too large to readily escape. Other probes are introduced by transfection of the nucleus with a plasmid encoding. For example, cells may be nucleofected with a plasmid encoding GFP stain and analysed some days later by fluorescence for its expression as a Golgi protein. In biomedical research confocal microscopy has been most frequently applied to the morphology of the extensively branched neuronal network that is not best displayed in two-dimensional sections. A similar application is envisaged for the osteocyte network in bone since the 'young' osteocytes in newly formed bone package the calcium and phosphate before export to the calcification front [34, 35]. On this premise, living chondrocytes in 50 μm thick slices of growth plate cartilage have been observed by confocal microscopy for 1 h using an Indo-1 AM fluorescent probe for calcium [36]. Changes in staining during the observation period suggested a cellular loading and unloading of mineral-primed vesicles, confirming the process reported in osteocytes in regions of the young mouse calvarium [1, 35]. A similar event was observed by confocal microscopy in cultured osteosarcoma cells using tetracycline as a calcium probe [37]. Coupled with laser absorption spectrometry the composition of the areas examined can be better detailed.

SECTION B: MINERAL MICROANALYSIS AND MORPHOLOGY

Demineralization, ashing and volumetic displacement, density fractionation, microradiography, back-scattered image analysis, electron probe and EDX, scanning electron microscopy, slam-freezing, microspheres and nanoparticles, atomic force microscopy

The division of bone into a composite of domains, as witnessed immunohistochemically in the organic phase above, is repeated in the inorganic phase in density and also in mineral mobility. This was illustrated by the progressive demineralization of cylindrical transiliac autopsy specimens *in bulk* with various strong (1 N HCl) and weak (10% potassium citrate) agents [38] when a solution of 10% EDTA, for example, was found to remove about half the mineral in 7 h, but not in a uniform manner. Mineral was preferentially lost from some surfaces and some trabeculae but not from others and the sharp transition from dense bone to demineralized matrix resembled the appearance of natural well defined osteoid borders. In comparison, *sections* of bone (whether plastic-embedded or cryo-cut) when progressively treated in the same way lost about half their mineral in 5 s. However, the etching pattern was remarkably similar despite the exposure in the section of mineral not readily accessible in bulk. This non-homogeneity of mineral extraction *in vitro* suggests that *in vivo* the bone salt in some domains is considerably more mobile than in others. (A specifically formulated and commercially available decalcification agent is Decalc; Histolab, Gothenburg, Sweden.)

7.5. MINERAL DENSITY

7.5.1. *Ashing and volume displacement*

Two-thirds of the weight of bone and half its volume consists of calcium and phosphorus. For those laboratories lacking hard tissue histology, that nevertheless require some form of microanalysis of bone quality to complement bone mass measurements, two methods [39, 40] remain useful [41]. One determines the relative bone salt content as the percentage ash weight and may be applied, for example, as an index of osteomalacia. The other is based upon Archimedes' Principle and determines the trabecular bone tissue volume relative to the total sample volume (%BV/TV) and may be used as an index of osteopenia. Both procedures are outlined in table 7.7.

7.5.2. *Density gradient fractionation of powdered bone*

Microanalysis can be taken a step further by determining the distribution of mineral density by density gradient fractionation of powdered bone and comparing normal and pathological conditions [42–46]. For example, Grynpas *et al* [45] froze alcohol-fixed bone specimens in liquid nitrogen before lyophilizing and pulverizing in a percussion mill (Spex Freezer Mill, Metuchen, NJ), cooling in liquid nitrogen and sieving to isolate particles below 20 μm. Nusgens *et al* [44] ground dehydrated and fat-free bone into particles with a maximum size of 40 μm, while Richelle [42] used a specially designed grinder with carborundum wheels and a cooling jacket to produce particles 5–10 μm in size, which he considered small enough to belong to no more than one histological structure. All authors separated the particles into fractions of different density by successive centrifugation of the precipitate at progressively decreasing specific gravity in bromoform-toluene density gradient ranging from 1.7 (no material being found in the supernatant of a liquid of 1.6) to 2.3 g/cm^3. After washing in ethanol, drying in a desiccator at room temperature and weighing the precipitate, a mineralization profile is obtained. The method used by Fincham [43] is described in table 7.8 and included a 'Coulter Counter' electronic particle size analyser (lower size limit 2 μm), leading to the conclusion that the density spread of the particles increased with age. The procedure has been used to show, for example, a shift towards higher-density fractions with immobility [45] and more low-density fractions in response to parathyroid hormone treatment, possibly as a result of the preferential resorption of the high-density fraction [42], or perhaps

Table 7.7. *Two rapid and simple microanalytical methods to determine the relative volume of bony tissue (%BV/TV) and the % ash weight.*

% Bony tissue volume/total tissue volume (i.e. bone + marrow)

- The cortex and any adhering soft tissue is removed.
- The specimen is trimmed into a regular shape, e.g. cylinder or cube.
- Blood is eliminated by a water or air jet, after which fat is extracted either by immersion for 48 h in ethanol/acetone or by refluxing in petroleum ether at 35 °C for 3 h for a typical bone biopsy specimen 8 mm in diameter × 10 mm in length. Alternative to this is prolonged immersion (1–4 days) in an ultrasound bath with a detergent.
- Dry in an oven overnight at 105 °C, cool in a desiccator and weigh the marrow-free sample in air.
- Evacuate in a flask of carbon tetrachloride to eliminate air pockets among the trabeculae, before reweighing in carbon tetrachloride, specific gravity 1.592.
- Measure the specimen with a micrometer and calculate its total volume (TV).
- It follows that:

$$\text{density of bony tissue} = \frac{\text{Wt}_{\text{air}} \times 1.592}{\text{Wt}_{\text{air}} - \text{Wt}_{\text{CCl}_4}}$$

$$\text{relative volume bony tissue, i.e. BV/TV} = \frac{\text{Wt}_{\text{air}} - \text{Wt}_{\text{CCl}_4}}{1.592 \times \text{TV}}.$$

The result for cancellous bone from the iliac crest is generally BV/TV 15–35% for normal bone and may be as low as 5% in osteoporosis. NB. For accuracy, all the air spaces previously occupied by marrow tissue must be evacuated.

% Ash content

- Weigh the sample of bone, dried and defatted as above.
- Ash the dried sample in platinum dishes for 24 h at 600 °C.
- Reweigh and calculate the percentage ash content.

The result for normal bone ranges between 57 and 67% (Vogt [47] and Dequecker [48] described the normal range as 54–62%). In osteomalacia it is lower than this, for example 45% or less. NB. The procedure is not as sensitive as histology in detecting mild cases.

---

also due to more remodelling and the presence of immature less calcified bone in consequence. The disadvantage of such procedures is that there is no direct histological association.

7.6. MINERAL MICROANALYSIS

7.6.1. Microradiography

Microradiography is the preparation of soft X-ray images of undecalcified bone sections or slices to display the pattern of mineral density in the extracellular matrix. Once prevalent in microscopic studies of normal and

Table 7.8. *Preparation and density fractionation of bone powder. (After Fincham [43].)*

Preservation
Store specimens unfixed at $-10\,^{\circ}C$ prior to use.

Preparation
Saw the frozen material (e.g. trephined human vertebral bone cylinders, 1 cm wide) transversely into slices, 3–4 mm thick. Remove marrow tissue with a fine water jet and dissect away any adherent soft tissue (e.g. blood vessels) or unwanted hard tissue (e.g. cortical bone). Extract lipids by continuous treatment of the slices in chloroform/methanol (1 : 1 v/v) for 12 h. Dry at $105\,^{\circ}C$ for 48 h.

Fragmentation
Powder the prepared slices by grinding 200 mg aliquots in a few ml of carbon tetrachloride in a rotary disc mill (Tema Machinery Ltd) for about 3 min. Wash the powder (size range about 10–25 µm) from the mill barrel with the same solvent before wet-sieving in more solvent through micromesh sieves, mesh sizes 37, 28 and 10 µm (Endicotts Test Sieves Ltd).

Fractionation
Dried bone density is 1.9–$2.2\,g/cm^3$. Therefore mixtures of carbon tetrachloride (density $1.6\,g/cm^3$) and bromoform (density $2.9\,g/cm^3$) encompass this density range. Aliquots (1–2 mg) of bone powder are placed in capped centrifuge tubes with 2–3 ml of appropriate density mixture, and dispersed to a uniform suspension in an ultrasonic bath for 3 min. Centrifugation (3000 rpm for 20 min) separates 'floats' and 'sinks.'

Analysis
The density fractionated particles (i.e. floats and sinks) may be transferred to volumetric flasks (100 ml) and made up to volume with an electrolyte solution (lithium chloride in isopropyl alcohol) to avoid swelling and immiscibility. They may be subsequently analysed by a Coulter Counter particle size analyser (Coulter Counter Electronics Ltd).

---

pathological undecalcified bone (for example Jowsey [49]), it was eclipsed by histomorphometry because the cell populations and other soft tissues of interest are not shown. Nevertheless the reliability concerning mineral distribution is unsurpassed (figure 7.5) and the technique continues to be useful, for example Li *et al* [50]. The specimen is placed in direct contact with a fine grain high-resolution emulsion and exposed to a beam of X-rays which is differentially absorbed by the mineralized tissue. After development, radiopaque heavily mineralized areas appear white, less calcified areas are shades of grey and unmineralized regions are black when the image is viewed under the optical microscope. The resolution is influenced by the focusing or collimation of the beam and depends upon the wavelength of the X-rays, the size of the focal spot on the target, the distance between

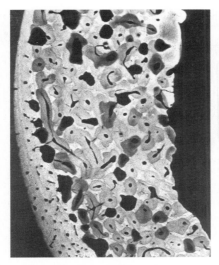

Figure 7.5. *Microradiograph of cortical bone in transverse section showing osteons of variable density; osteocyte lacunae are clearly resolved. ×30. (Courtesy of Dr J Jowsey.)*

the specimen and the emulsion and the distance between the target and the specimen, according to the equation

$$\text{resolution} = \frac{\text{focal spot size} \times \text{specimen-to-emulsion distance} + \text{emulsion thickness}}{\text{target-to-sample distance}}.$$

A resolution of approximately 1 µm, visualizing osteocyte lacunae but not their canaliculi, was achieved by Jowsey [51] using a focal spot size of 1 mm^2 and a distance between target and sample of 20 cm, and it was recommended that slices 100 µm thick be used, thinner sections lacking contrast and thicker ones producing poor anatomical definition. This is because the microradiograph is a summation of many levels, such that detail is lost in the superimposed shadows cast and by parallax distortions. To diminish this Bohatirchuk and Jeletzky [52] recommended for X-ray imaging the unusual step of taking sections 5–10 µm thick by means of which they too were able to resolve mineral deposits of 1 µm [53]. To complement the microdensity the section may be stained after microradiography when the corresponding soft tissues will become apparent. Especially fine examples of the technique have been presented by Heuck (figure 7.6) [54].

7.6.2. Backscattered electron image analysis

Where a quantitative index of mineral density as well as topography is required for comparative purposes, backscattered electron image analysis

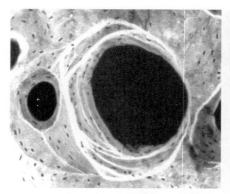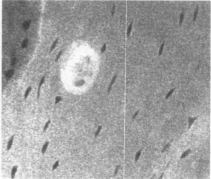

Figure 7.6. *Microradiographic details showing (left) high-density cement lines and (right) high-density osteocyte halos in marble bone disease in a 35-year-old woman, ×160. (Courtesy of Dr F Heuck 1971 Israel J. Med. Sci.* **7** *477.)*

is increasingly the method of choice. Backscattered electrons vary in number and direction with the surface topography of the specimen and its composition, crystallinity and magnetism. Backscattered electron (BSE) imaging is achieved using 2–3 mm thick bone slices cut with a water-cooled diamond saw, that are subsequently embedded in plastic and polished and carbon-coated before examination in a Zeiss digital scanning microscope (DSM 962). BSE imaging was applied to bone by Boyde and Jones [54] (see also [55, 56]). Digital images are produced at a magnification of about ×30 and analysed by means of a dedicated backscattered electron detector. Image histograms with a range of grey-level values ranging from 0 to 255 are produced with a peak for mineralized tissue in the range 150–200. The fraction of the total tissue occupied by low (65–151), medium (152–166), high (167–181) and very high (182–239) mineral density, is recorded. By this means it was observed that normal men had more low and medium density bone than women and that in osteoarthritis high mineral density featured widely, suggesting poorer quality due to increased stiffness. Osteogenesis imperfecta has been similarly examined [57].

7.6.3.   *Electron probe X-ray microanalysis (by specialist Dr Roger C Shore)*

The term electron probe X-ray microanalysis encompasses a number of techniques, often with slightly confusing names and acronyms. The terms energy dispersive X-ray (EDX) microanalysis and energy dispersive X-ray spectroscopy (EDS) are synonymous and may be associated with both transmission (TEM) and scanning (SEM) electron microscopes where in both cases the probe is formed from a focused beam of electrons. This may be used, for example, to target a small area of the polished surface of an embedded specimen block mounted on a metal stub in the SEM or an ultrathin section in the TEM. Two further acronyms which may be confused are WDS and WEDS. WDS refers to wavelength dispersive X-ray spectroscopy (as opposed to EDS, energy dispersive X-ray spectroscopy) where elements are discriminated by the characteristic wavelength of the X-ray photons emitted rather than their energy. Perhaps the major limitation of this type of spectrometer is that it is only able to detect X-rays from a single X-ray peak (i.e. one element) in one analytical run. This property generally results in EDS being the method of choice in biological systems and therefore the remainder of this topic will be concerned with this technique only. WEDS refers to *windowless* energy dispersive X-ray spectroscopy. In a normal EDS the solid state detector is isolated from the microscope vacuum by a thin protective window, usually made from a film of beryllium between 2 and 10 μm thick. An unfortunate property of this beryllium film is that it absorbs the low-energy X-ray photons characteristic of the light elements. In practice, a detector with a beryllium window is unable to detect elements with an atomic number below sodium. In a biological tissue such as bone the elements carbon, nitrogen and oxygen obviously constitute a significant proportion of the analysed sample but would be undetected by a conventional detector. Windowless detectors remove or reduce this problem by incorporating a removable window which exposes the detector crystal during the analytical run or by replacing the beryllium window with an ultrathin film of, for example, hydrocarbon. In the latter case this may also be termed 'windowless', even though a form of window is present. However, these typically will permit the passage of X-rays emitted from elements as far down as boron.

The use of EDS (or WEDS) for the study of the elemental composition of bone samples has many common features in the TEM and SEM but some important differences in terms of signal generation and peak quantification. The nature of the material, a biological composite with a very high inorganic ion content, also presents problems in terms of interpretation and quantification. One of the main problems of spectrum interpretation and *true* quantification (i.e. determination of absolute elemental mass per volume) is that the true extent of the excitation volume, particularly in bulk samples in the SEM, is difficult to measure as the penetration of the beam into the specimen and its lateral spread within it is difficult, if not impossible, to determine. The situation within the TEM is more straightforward in this respect. As the sample is a thin section (of measurable thickness) the excitation volume can be fairly accurately estimated from the section thickness and the probe diameter (because of the thinness of the sample, the lateral spread of the beam within it is assumed to be negligible). This does, however, assume that the beam penetrates (and X-rays are generated) throughout the whole specimen thickness. This may not necessarily be the case in tissues with high mineral content where many incident electrons may fail to penetrate

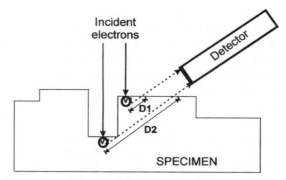

Figure 7.7. *Diagram showing how path length of an X-ray photon within the specimen can vary depending upon whether the excitation volume is situated on the exposed surface (D1) or in a 'trough' (D2).*

the specimen. The problem with bulk samples is increased by the internal absorption of emitted X-rays within the specimen itself. Unfortunately, not only does this absorption decrease the overall signal intensity but also distorts the spectrum. Low energy photons from light elements will tend towards greater absorption. In addition the relative surface topography of the specimen may also affect internal absorption. The path length for an individual photon within the specimen may vary depending upon whether it originates from a 'trough' or 'peak' on the surface of the specimen (figure 7.7).

A typical EDS spectrum derived from a bone sample is shown in figure 7.8. The prominent peaks are the calcium K_α and phosphorus K_α. At the low energy end of the spectrum are the carbon and oxygen K peaks. There are

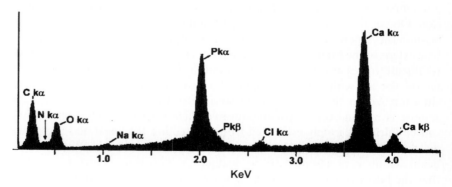

Figure 7.8. *Typical EDS spectrum of bone acquired via a detector fitted with an 'ultrathin' hydrocarbon window.*

also minor peaks for nitrogen, sodium, magnesium and chlorine. Several factors complicate the interpretation, however.

1.   Elements, particularly the heavier ones, generally do not generate a single peak but a series of peaks. This arises because there are numerous possible excitation states of the shell electrons produced by interaction with the incident electrons. Thus a K shell electron, for example, may be energized by the incident electrons, resulting in a vacancy in the K shell and an energetically unstable atom. This vacancy may be filled by an electron dropping from either the L or M shells resulting in X-ray photons of two different, discrete energies. As they both arise from a K shell vacancy they are termed K_α and K_β X-rays. An example of this is seen most clearly for calcium in figure 7.8 where a small K_β peak is seen at an energy slightly above the main K_α peak. A K_β peak for phosphorus is also present but it cannot be clearly distinguished from the K_α peak.

2.   The characteristic intensity of an element within the sample (i.e. peak integral) is proportional to the number of atoms in the irradiated volume. However, determination of the atomic (or weight) ratios of elements within the specimen cannot be determined by simple measurement of the peak integrals alone. This is for two reasons. (a) Lighter elements generally have a lower fluorescence yield than heavier elements and their characteristic X-rays are more likely to be absorbed within the specimen itself. This is a problem more for bulk specimen analysis than thin sections. Some correction for this can be made mathematically, i.e. the so-called ZAF correction. However, in tissues such as bone, dentine and particularly enamel where the mineral content is extremely high, this approach may be difficult. (b) Detectors become increasingly inefficient at X-ray detection with photon energies below approximately 1 keV. The situation is improved considerably with windowless or ultrathin window detectors but these may still be only in the order of 50% efficient at 0.75 keV (taking the value for the calcium K_α peak to be 100%). Thus for the spectrum in figure 7.8 the raw peak integrals for oxygen, nitrogen and carbon in particular are likely to severely underestimate the true atom % for these elements within the irradiated volume. In this respect it is possible to calibrate microscope/detector combinations with the aid of standards of known stoichiometry. For example, it is possible to make 'pills' of Analar (high purity) grade compounds such as $CaCo_3$, $Ca(NO_3)_2$ and $Ca_3(PO_4)_2$ to act as standards, although possible differences in sample and standard topography may lead to some inaccuracy which may be difficult to quantify.

3.   Underlying the peaks is a background or continuum. The intensity of this background radiation is directly related to the total mass of all

elements in the irradiated volume. Thus this offers the potential for quantitative analysis, i.e. determination of relative and absolute mass fraction of an element (element mass per unit mass of specimen) within the analysed volume—the so-called 'continuum method' [58, 59]. Again, the mineralized tissues may prove to be difficult subjects in this type of analysis, where the elements of potentially the greatest interest (Ca and P) constitute a significant proportion of the elemental composition of the sample. Furthermore, in the mineralized tissues (particularly enamel), porosity may be a significant feature. Changes in porosity will affect the true concentration of mineral (i.e. mass per volume) but will not affect the mass fraction. Thus lack of change in the mass fraction may mask an underlying significant change in the mineral density.

While the above technique is used primarily as a research tool, there are clinical applications: for example, in determining the accumulation of fluoride in fluorosis or of aluminium in renal dialysis patients. For an excellent overview of X-ray microanalysis of biological samples see references [60, 61].

7.7. MINERAL MORPHOLOGY

7.7.1. *Scanning electron microscopy*

This technique provides a three-dimensional image that complements two-dimensional histology and has been used at low power for many years for the examination of hard tissues, primarily to demonstrate the three-dimensional structure of cancellous bone in terms of plates and rods (e.g. Whitehouse *et al* [62]). As a tool for bone pathology it has been reviewed by Sela [63] and Boyde *et al* [64]. The imaging of the surface topography presented has enabled identification of the sites of cellular activity, especially osteoclastic resorption cavities (Howship's lacunae), and has provided structural information about woven and lamellar bone, microcallus associated with healing *in vitro* trabecular fracture, tendinous insertions, characteristics of osteocyte lacunae and mineral morphology, including clusters of calcospherites at the calcification front of bone and cartilage [65–67]. It has also less commonly been applied to cultured cells of hard tisues. For example, osteosarcoma cells were grown on glass coverslips, fixed (2.5% glutaraldehyde), dehydrated in acetone, critical point dried (an essential step for cellular preservation), 'sputter'-coated with carbon (or gold which is coarser and gives more density but less resolution; both also conduct electrons away and reduce heating) and examined in the SEM for matrix vesicles released from cell processes [37]. Another application was to examine dehydrated calcified microspheres isolated from bone and directly applied by means of double-sided tape to the surface of an

Table 7.9. *Preparation procedure for scanning electron microscopy. (After Jayasinghe et al [67].)*

Preservation
Store fixed (1% glutaraldehyde in phosphate buffer, pH 7.2; 10% buffered formalin) or unfixed material in 70% ethanol until required.

Sectioning
Cut into slices, 3–4 mm thick (e.g. mid-sagittal for a vertebral body) using a low speed diamond saw (Buehler Isomet-11-1180; Struers Accruton 5). Small developing animal long bones may be simply cut in half longitudinally with a sharp blade.

Deplasticization
For material previously embedded in plastic (e.g. methylmethacrylate), the resin may be removed *after* slicing, by immersion in xylene/chloroform (1 : 1 v/v) for about 5 days in an agitated bath.

Preparation
Remove marrow tissue with a fine water jet or with xylene/chloroform. Superficial organic material may be digested by 5–7% sodium hypochlorite solution (1–2 h) or by 2% hydrogen peroxide (24 h at 37 °C), when osteoid tissue will also be removed. Wash in distilled water and defat for a few hours at room temperature in chloroform/methanol (1 : 1 v/v) before air drying.

Examination
Attach the prepared slice to an SEM aluminium stub with epoxy resin and coat for 1–2 min with a conductive metal, e.g. gold using a sputter coater (e.g. Polaron SEM Sputter Coater). View in the SEM at 10 kV (e.g. Cambridge Stereoscan S4–10; Jeol 840A). With an additional detector this may be operated to provide secondary electron (SE) imaging and backscattered electron (BSE) imaging. For stereopair micrographs a tilt angle difference of 10° is recommended.

[In the absence of an SEM, the above bone slices may be examined under a low power optical microscope (e.g. the Zeiss Tessovar) when the three-dimensional trabecular architecture is demonstrated.]

aluminium stub [65]. Technical details of specimen preparation for SEM are summarized in table 7.9.

7.7.2. High velocity impact ('slam') freezing

The morphology of the inorganic phase of bone can only be resolved in detail in the electron microscope, for which it is commonly prepared by prolonged and rigorous chemical procedures originally developed for soft tissues. This overlooks a long history of cautionary reports that the nature, content and distribution of the bone salt is altered by such methods; see Aaron [69] for a review. In consequence attention has centred around the sheets of small, uniform crystals that permeate the collagen and apparently stiffen the bone in a predictable way. A number of authors in the past have attempted to reduce mineral elution and the possibility that the crystal-like forms are artefacts of chemical fixation and staining by using cryo-techniques (for example, Schraer and Gay [70]). More recently high velocity impact ('slam') freezing has become an established procedure and when combined with freeze-substitution and low temperature embedding its effectiveness at conserving the tissue chemistry has been widely acknowledged [71–73] making it the method of choice for a wide range of materials, including certain hard tissues. For example, the technique has been applied to pieces of rat incisor dentine which were subsequently freeze-dried at −80 °C and vacuum-embedded in Araldite resin at room temperature before ultramicrotomy [74]. Slam freezing involves the rapid compression of the specimen on to a copper block chilled by liquid nitrogen to −190 °C, and is reported to produce a cooling rate within the tissue of between 25 000 and 50 000 °C per second to a depth of 10–20 μm. The process was used to prepare neonatal mouse calvarial bone by Carter *et al* [75]; this tissue is so thin that prior slicing is unnecessary, reducing manipulation and delay and thereby minimizing both mechanically- as well as chemically-induced changes (table 7.10). Ultrathin resin-embedded sections (i.e. <0.25 μm thick and about 1 mm² in area) are cut from the specimen using, for example, a Reichert Ultracut E ultramicrotome. By most TEM standards, sections 0.25 μm are too thick for cell ultrastructural detail. However, they have the advantage of improved section stability in the electron beam and more reliable X-ray microanalysis and mineral morphology due to better mineral retention, since a proportion of bone mineral is in a nascent state and is readily lost during processing. For this reason sections may be cut dry, rather than the more usual flotation on a water bath, to minimize mineral dissolution. The small sections are manipulated using a fine bristle on to copper, aluminium, nylon or gold grids and any flattening of the sections needed may be achieved by chloroform vapour. Areas of interest are examined by means of, for example, a Phillips CM10 TEM with, as an attachment, a PV9800 EDAX microanalysis system and using an accelerating voltage of 100 kV, a spot size of 200 nm and tilting the specimen holder 12° to optimize the collection of X-rays. Background spectra are obtained from areas of peripheral resin devoid of material and there is a device for point counting and mapping elements of interest in relation to tissue topography [75].

To those outside the field such attention to preparative detail in relation to a tissue as ostensibly rock-like as bone mineral must seem pedantic. However, while the chemically fixed and stained bone matrix is permeated by needle-shaped crystals of uniform density, this is not confirmed by its cryopreserved counterpart, where a number of structural dissimilarities combine to suggest an entirely different conclusion. Fresh frozen mineral morphology is characterized by clusters of sinuous and segmented calcified filaments about 5 nm wide (figure 7.9), displaying a variable density along their length and often linked by regularly repeated bridges into ladder-like

Table 7.10. *The procedures for cryopreservation and more traditional chemical preservation for the examination of bone mineral substructure in the TEM.*

Slam freezing/freeze substitution	Chemical processing and embedding
SLAM FIX immediately after removal using an MM80 Freezing Apparatus (Leica UK Ltd, Milton Keynes, UK).	*FIX (primary)* in 3% glutaraldehyde in 0.1 M sodium cacodylate buffer pH 7 for 30 min. Wash three times in fresh buffer. *FIX (secondary)* in saturated osmium tetroxide for 1 h.
FREEZE SUBSTITUTE in methanol or acetone at −80 °C for 10 days using a CS-Auto (Leica UK Ltd, Milton Keynes, UK).	*DEHYDRATE* through graded alcohols, 1 h in each, with three changes of absolute EtOH.
INFILTRATE with Lowicryl K4M at −80 °C using a CS-Auto. Polymerize using UV irradiation [76].	*INFILTRATE* with Spurr's resin (or LR White, methylmethacrylate, or araldite) at room temperature. Allow to polymerize.
SECTION 0.25 μm thick, on an ultramicrotome without aqueous flotation on to collodion-coated grids.	*SECTION* 80 nm thick, on an ultramicrotome with aqueous flotation on to collodion-coated grids.
(Flatten in chloroform vapour.) *STAIN* None.	*STAIN* Uranyl acetate (10% in methanol, to aid penetration into resin, 5–10 min. Rinse in methanol. Dry grid on filter paper). Lead citrate (float grid on lead sulphate drops on a dental wax platform surrounded by damp NaOH pellets in a closed Petri dish, 5–10 min. Rinse in NaOH, then distilled water. Dry grid on filter paper.

binary pairs and more extensive parallel arrays [68, 75, 77]. Moreover, these compartmentalized calcified filaments in bone are entirely in accord with the 'chains of dots' described repeatedly by Hohling and colleagues [74, 78] in developing dentine. However, while these authors perceived their 'chains of dots' as merely transitory features of immaturity dependent upon collagen topography (see also Boyde [79]) our evidence indicates the widespread persistence and permanence of the calcified filaments in the mature skeleton, as illustrated, for example, in the bovine cortex [68]. Furthermore, their developmental independence from collagen fibres is indicated by their occurrence intracellularly as well as extracellularly at bone forming sites [80] and their recurrence lower down the evolutionary chain within those prokaryotes and protozoa that calcify with phosphate [81, 82]. At the same time, as particularly prominent features of calcified keratin (figure 7.9 inset), they have long been known to be present at a location that is demonstrably collagen-free [82, 83], thereby confirming their dissociation from collagen. While the individual filaments are submicroscopic in size, the

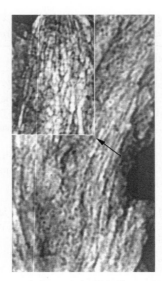

Figure 7.9. *Mineral morphology, calcified filaments and cryopreservation.
Calcified filaments, 5 nm in diameter in mature human bone, showing the
segmentation (arrowed) and the frequent bridging of the sinuous parallel filaments into binary pairs. Mature human cortex, slam-frozen, unstained, TEM,
×200 000. Inset: Calcified filaments in keratin, a tissue that unlike bone
contains no collagen. Baleen whale, sectioned without treatment, unstained,
TEM, ×130 000. (Courtesy of Drs F G E Pautard and D H Carter.)*

filamentous clusters into which they are aggregated are about 1 μm in diameter
and may be resolved under the optical microscope. At this level they are seen to
be distributed as deformable microspheres that colonize the organic matrix
(figure 7.10). Here they are often linked in looped assemblies around the
collagen bundles tying them together to form an interlocked arrangement of
coarse collagen fibres and fine calcified filaments of considerable biomechanical
complexity. Also confirmed optically is their Golgi-directed intracellular origin
(figure 7.11) as the culmination of a sequence of events [75, 84, 85] that leads to
their frequent possession of a less dense centre and 'tail'-like projection characteristic of the future microsphere [80]. Indeed the early method of identifying
the Golgi apparatus in the first place used silver staining reminiscent of the
von Kossa reaction for calcium phosphate/carbonate, whereby adsorption of
silver on specific sites enhances the detail.

There is increasing evidence to suggest that some of the non-collagenous
proteins of bone, including osteocalcin, bone sialoprotein and osteopontin,
are integral to the calcified microsphere populations where they are instrumental in mineral containment within the segmented filaments, preventing
uncontrolled crystal growth into an enameloid form. The capacity for the
substructural metamorphosis of the bone salt may be demonstrated optically

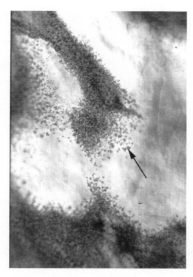

Figure 7.10. *Mineral morphology and calcified microspheres. The filamentous clusters are resolved under the optical microscope as populations of microspheres about 1 μm in diameter (arrowed) that colonize the extracellular matrix. Human ilium, plastic embedded, von Kossa stain, Nomarski optics, ×900.*

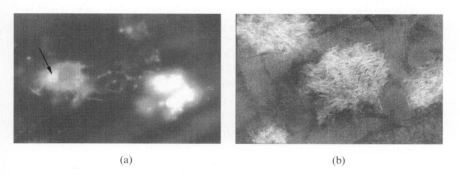

           (a)   (b)

Figure 7.11. *Golgi-directed mineralization in the young mouse calvarium [228]. (a) The prominent juxtanuclear apparatus (Golgi body; arrowed) of a 'young' osteocyte located near the calcification front contains calcified microspheres about 1 μm in diameter which stain with tetracycline; these are exported along the cell processes to the extracellular matrix where they fluoresce at the calcification front in UV light, ×700. (b) In the electron microscope each calcified microsphere is composed of a filamentous cluster of variable density, ×21 500. (Part (b) courtesy of Dr D H Carter.)*

Table 7.11. *Outlining extraction procedure methods for the isolation of calcified microspheres.*

Enzyme digestion

Immature bone
Remove six fresh 5-day-old mice calvaria and using a sterile technique cut in 2 mm squares, freeze dry and weigh.

Incubation medium
(After Ali *et al* [88] for isolating matrix vesicles.) 50 mM TES buffer, pH 7.5, 0.25 M sucrose, 50 ml/dm^3 of 1000 U/ml penicillin and 10 000 U/ml streptomycin. Store in a sterile bottle and use 5 ml/g of calvarial bone.
Add Type I collagenase (or other hydrolytic enzyme, e.g. papain or trypsin) at a concentration of 5000 IU/g of calvarial bone (collagenase enzyme derived from *Clostridium histolyticum* as Clostridiopeptidase A: Sigma Chemicals). Incubate at 37 °C for up to 3 days with brief agitation.

Mature bone
Saw rectangles of fresh bovine femoral cortex about $1 \times 1 \times 2.5$ cm and freeze in isopentane/liquid nitrogen at -178 °C.
Attach to a heavy duty cryomicrotome chuck with 2% carboxymethylcellulose as a supporting medium. Cut cryosections, 20 μm thick, thaw groups in TES buffer at room temperature, weigh wet in sterile containers and incubate as above with hydrolytic enzyme.

Separation
Centrifuge at 3000 rpm for 5 min, remove supernatant, resuspend the dense pellet in TES buffer in Eppendorff tubes and respin. The dense pellet consists of isolated microspheres.

Chemical digestion
Incubate mouse calvarial pieces or bovine cryosections above at 37 °C in 20% sodium hydroxide (or hydrogen peroxide or sodium hypochlorite; Boyde and Sela [89]) for 24–48 h, followed by centrifugation and three washes in distilled water.

Mechanical separation
Cut slice of fresh bovine bone, 1 mm thick, on a band saw, reflux in petroleum ether for 24 h and fragment to a fine powder in a Tema ring mill for 15 min. Place about 20 g of the anhydrous powder in physiological saline or distilled water and mill in an IS stainless steel chilled Attritor for 15 min (Pautard [83]).
Clean the calcified particles of debris for up to 24 h in TES-buffered collagenase to which toluene is added as an antiseptic. Remove the dense fraction from the supernatant and store under acetone.

Separation by incineration
Place the calvaria or segments of mature bone on platinum dishes and incinerate in a muffle furnace at 600 °C for 24 h. Crush lightly into a powder in a pestle and mortar to separate the calcified particles.

Separation by fungi
Certain fungi are capable of releasing the calcified microspheres when incubated with bone in TES buffer as above.

in polarized light by the rapid change from the natural non-birefringent state of fresh hydrated calcified microspheres to an uncharacteristic birefringence when they are dehydrated in alcohol [68], suggesting an unnatural state of uncontrolled physicochemical order [86]. Similarly, removal of the enshrouding proteins by, for example, sodium hydroxide apparently allows the slender filaments to fuse laterally into crystal-like plates reminiscent of enamel [68, 87]. These artificially-induced mineral modulations may provide insight into clinical concerns about the substructural basis of age-related atrophy and osteoporosis and present a facet of bone quality that is as elusive as it is functionally fundamental. To this end the calcified microspheres (or filamentous clusters) can be isolated from mature bone [68] in a number of ways (table 7.11; figure 7.12(a)). There is also preliminary evidence that the chains of calcified nanoparticles of which the 5 nm filaments are composed can in turn be fragmented by freezing and milling procedures and the separating 5–15 nm units isolated (figure 7.12(b); Carter and Aaron, in preparation). Tightly packed particles of identical dimension to this were

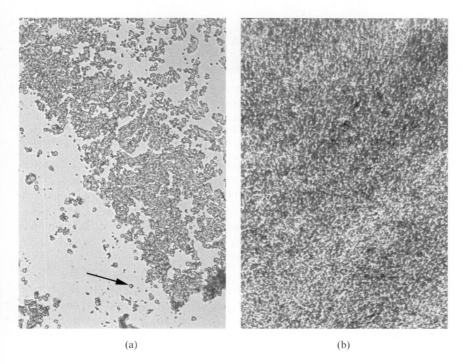

(a)   (b)

Figure 7.12. *Isolated bone particles. (a) Calcified microspheres about 1 µm in diameter (arrowed). Optical microscope, ×1000. (b) Calcified nanoparticles (individually and in bridged hexagonal rings) 5–15 nm in diameter. Electron microscope, ×85 000. (Part (b) courtesy of Dr D H Carter.)*

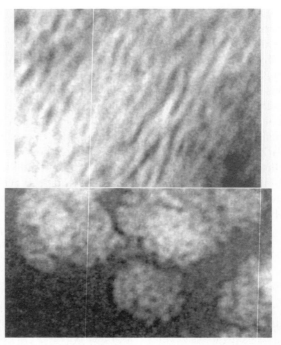

Figure 7.13. *Image-enhanced filaments, 5 nm wide, of keratin calcified with phosphate resembling filamentous mineral in bone and apparently resolving subnanometre detail. Transmission electron microscope. (Courtesy of Dr D H Carter.)*

described as amorphous calcium phosphate (since they could not be interpreted in terms of a crystal model) in adult rat bone by Termine *et al* [90] following deproteination in 95% hydrazine at 55 °C for up to 24 h, confirming earlier observations of Amprino using X-ray scattering. The nanoparticles also recapitulate synthetic preparations prior to crystal formation from supersaturated synthetic fluids after prolonged incubation under physiological conditions [91, 92] which apparently confirms the basic physicochemical process. However, commencement within the Golgi apparatus of the cell will organically enshroud and expose, accelerate and constrain the chemistry, thereby engineering its modulation and metamorphosis. We are left with the conclusion that the state of calcium phosphate is probably a subcrystalline unit, hexagonal, 4 nm × 4 nm and on the threshold of a true crystal with suggestions (figure 7.13) of yet a further substructure within.

Depending on the techniques described above, an open-minded investigator is guided into two general views as to how bone functions and how the associated biochemistry and biophysics are related to the histological patterns. On the one hand is the familiar concept of a *status quo* of apatite crystal epitaxy on collagen fibres [93, 94] or within static submicroscopic deposits of mineral in transitory membrane-bound vesicles in the matrix [95–97]. On the other hand is the evidence for a traffic of large numbers of small independent vehicles composed of numerous filaments (see also [98, 99]) within which physicochemical events are constantly modulated by biological engines transferring energy and products. As we begin to isolate the primary calcified units from the filamentous clusters, we may find that these too have a substructure below TEM resolution and the conventional histology and histomorphometry will need to be replaced by a sub-nanohistology at molecular dimensions where the physics is uncertain at the boundary of the organic and the inorganic. In this dynamic substructural molecular world, apatite may indeed be 'the deceiver' (chapter 6, p. 186) and new techniques will be required to understand how bone and other similarly calcified cells and tissues function.

7.7.3.   Atomic and chemical force microscopy (by specialist Prof. Jennifer Kirkham)

The atomic force microscope (AFM) has found many applications in materials science, physical chemistry, physics and biology as a tool that not only provides images of surfaces in ambient conditions or under fluids with near molecular resolution but also, following modification of the tips, probes intermolecular forces and maps surface properties. Unlike other ultra-high resolution imaging technologies, specimens for AFM need not be fixed or stained and imaging can take place under biologically relevant fluids, simulating *in vivo* conditions. Solutions can be exchanged or modified during imaging, providing unprecedented opportunities for exploring biological processes.

How does the AFM work?   The AFM is a scanning probe microscope which employs a flexible cantilever (usually made of silicone or silicone nitride) on the end of which is a sharp tip. When the tip is brought close to a sample surface, forces between the tip and the sample surface cause the cantilever to bend and this movement is detected optically by the deflection of a laser beam which is reflected off the back of the cantilever to a photodetector. If the tip is scanned over the sample surface, the cantilever deflections can be recorded as a function of position and a three-dimensional image of the sample surface is generated. The spatial resolution of AFM depends upon a number of factors but mainly on the sharpness of the tip, which is routinely of the order of 20 nm but with new nanofabrication technologies and the use of carbon nanotubes may be as little as 1 nm [100]. Atomic resolution may be achieved on robust, flat samples such as gold or graphite but soft biological samples provide a greater challenge because the forces exerted on the sample

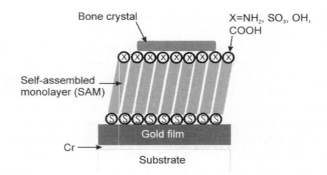

Figure 7.14. *Immobilization of crystals from developing enamel on self-assembled monolayer substrates for imaging in the AFM under fluids.*

by the tip can cause deformation/destruction. These problems have been overcome in part by the introduction of 'tapping' mode AFM imaging, where the tip has only intermittent contact with the sample surface, greatly reducing lateral forces. The AFM is capable of better than 1 nm lateral resolution on ideal samples and 0.01 nm resolution in height measurement.

Imaging of mineralized tissues. Investigation of matrix protein–crystal associations with the relatively small crystals from bone (and dentine) require that imaging be carried out under fluid. In order to anchor the specimens to the substrate during scanning, self-assembled monolayer (SAM) technology is used. Self-assembled monolayers are routinely used to provide 'designer' substrates with specific surface chemistry. These are generated from ω-functionalized alkyl thiols which spontaneously form monolayers on gold surfaces by the formation of a covalent bond between the sulphur and gold atoms [101] (figure 7.14). A range of ω-functionalized alkyl thiols are commercially available and include molecules with methyl, amine, carboxylic acid, hydroxyl and phenyl head groups, offering a wide range of potentially modified surfaces. Similar technology is used in the modification of AFM tips in chemical force microscopy (see below). Amine-terminated SAMs appear to be optimum for the stabilization of crystals from bone and dentine, suggesting an overall negative surface charge for these crystals (figure 7.15). The resulting images reveal the presence of steps or spiral growth sites of the order of size of the unit cell for hydroxyapatite.

Chemical force microscopy and the mapping of surface chemistry. Under normal imaging conditions, the force of interaction between the sample surface and the silicone nitride AFM tip is kept to a minimum to avoid artefact. However, the force of interaction between a chemically functionalized tip and the sample surface can provide information on specific interactions including surface chemistry and the strength of ligand–receptor binding [102]. Chemical modification of AFM tips is most commonly achieved

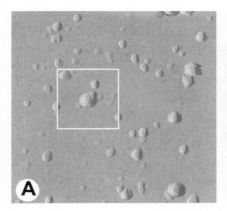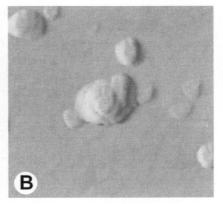

Figure 7.15. *(a) AFM image of crystals isolated from rat bone on SAM surface and viewed under ethanol. Scan area = 2 μm × 2 μm. (b) AFM image of a single crystal isolated from human coronal dentine showing evidence of step or spiral growth. Step height = 1.15 nm. Scan area = 300 nm × 300 nm.*

using SAMs as described previously. This provides a probe with known surface chemistry, which can be used to further characterize sample surfaces. Most methods of surface characterization utilize modified tips to generate force–distance curves at specific points on the sample surface. The tip is approached towards the sample and contact is made with a controllable level of applied force and the tip is then retracted from the surface. During this approach–retract cycle, the cantilever deflection is recorded to provide a plot of force applied (tip deflection is converted to applied force by calculation of the cantilever spring constant) versus distance. The force–distance curve can then be plotted.

If the adhesion force between tip and sample is measured as a function of pH (by changing the fluid in the flow cell between measurements), then a chemical force titration of the groups at the sample surface can be carried out. This can be used to calculate the dissociation constant, pK_a, for dissociating groups at surfaces. Under normal imaging conditions, cantilever deflection is monitored in the vertical direction, providing topographical information. However, the torsional forces (i.e. 'twisting') experienced by the cantilever during scanning can also be monitored by the photodetector, and this provides a frictional image of the sample surface. This is lateral force microscopy. If lateral force microscopy is carried out using functionalized tips, a friction image, which is dependent upon surface charge distribution, will be obtained.

Nanomechanics. Atomic force microscopy has been widely used in the characterization of mechanical properties of materials, including dental materials [103]. Force curve measurements, as described above, can also be used to investigate the elastic and plastic properties of a material, by determining the force required to indent or deform the surface (nano-indentation). Dedicated instrumentation for nano-indentation is commercially available but use of the AFM for indentation combines increased depth and force resolution (~ 0.1 nm and < 50 nN respectively) with imaging of the indentation geometry. Taken together, force–distance curves recorded during indentation and unload cycles and combined topography permit valuable information of visco-elastic properties of the sample to be obtained. A series of such force–distance curves can be compiled to generate maps of surface stiffness with ultra-high resolution. Such mapping is not restricted to synthetic materials. Phase imaging, which results from the phase lag between the driving oscillation and actual cantilever oscillation, can also be used to characterize differences in surface elasticity.

Fact or artefact?   All imaging technologies are susceptible to artefact and the AFM is no different. Tip convolution is the most obvious of these. *Tip geometry* is an important determinant in image quality, producing 'rounding' effects at edges and overestimating x–y distances (though not z) due to *convolution* effects. Deconvolution can be carried out where tip dimensions are known, permitting absolute values to be obtained. Similarly, tip geometry may preclude accurate probing of depth in narrow grooves or valleys, which may be inaccessible if less than the tip diameter. This may be improved by using specially prepared sharpened (and therefore narrow) tips. The advent of the carbon nanotube (tip diameter 0.5–1.5 nm), mounted on conventional AFM tips, will make a significant impact in improving spatial resolution in the future. Any features smaller than the size of the tip cannot, in theory, be accurately imaged (rather, the tip itself is imaged). However, it is well accepted that the presence of *tip asperities* (resulting from the manufacturing process) can permit imaging with atomic resolution on appropriate surfaces. A number of imaging modes, including lateral force imaging as described here, are affected by surface roughness of the sample and it can be difficult to separate the *effects of topography* from the frictional effects when interpreting the data. *Sample deformation* due to applied forces has been significantly improved using tapping mode (and related technologies) but it is still not possible to obtain atomic resolution on biological samples. *Tip contamination*, especially in contact mode, is a real problem. *Image acquisition times* are currently limited to the range of seconds up to minutes, precluding imaging in real time for many biological processes, but it is likely that the development of shorter cantilevers and other technical improvements will permit faster imaging in the millisecond range. *Sample stability* can be a problem when scanning under fluids and the need to *immobilize samples* may interfere with downstream biological events, for example, when imaging polymerase activity on DNA. *Thermal noise* is a major contributor to system *hysteresis*, introducing real (frequently insurmountable) difficulties in returning to the same position on the sample surface after interrupting the imaging process. Finally, the full implications of this technology are not yet fully understood and *interpretation* of data can be difficult due to the relative contributions of capillary, electrostatic, magnetic, hydrophobic and van der Waals forces which vary according to sample, tip and imaging conditions. Existing *mathematical models* are also limited in their application to intermolecular forces such as those being measured here.

Nonetheless, AFM technology represents a valuable research tool which has already brought new insights into a number of previously intransigent problems in mineralized tissue research. Continued technological development, coupled with better understanding of its limitations and applications, will permit interrogation of biological and biomaterials' surfaces with molecular resolution in the future.

7.8. ACKNOWLEDGMENTS

The valuable contribution of a number of undergraduate students in Human Biology at Leeds is gratefully recognized, especially that of Sarah Clancy, Catherine Meer, Jacqueline Allen, Matthew Armstrong, Amy Champaneri, Brian Oliver, Nicholas Clarke, Hayley Turner, Adele Rushworth and David Simpson.

REFERENCES

[1] Aaron J E and Pautard F G E 1973 *Calcified Tissue. Proceedings of the Ninth European Symposium on Calcified Tissues* (Austria: Facta-Publication) p 197
[2] Aaron J E, Rogers J and Kanis J A 1992 *Am. J. Phys. Anthrop.* **89** 325–31
[3] Bordat C, Bouet O and Cournot G 1998 *Histochem. Cell Biol.* **109** 167–74
[4] Stewart R F and Sutton D 1986 *Part. Sci. Technol.* **4** 251–64
[5] Tough R J A, Pusey P N, Lekkerkerker H N W and van den Broeck C 1986 *Mol. Phys.* **59** 595–619
[6] Pautard F G E 1963 *Nature* **199** 531–35
[7] Owen M 1963 *J. Cell Biol.* **19** 19–29
[8] Chen J, Jin H, Ranly D M, Sodek J and Boyan B 1999 *J. Bone Min. Res.* **14** 221–9
[9] Egan M J and Crocker J 1988 *J. Pathol.* **154** 247–53
[10] Evans R A, Dunstan C R and Hills E E 1980 *Bone Histomorphometry* (Paris: Société Nouvelle de Publications Médicales et Dentaires) p 29
[11] Franklin R M and Martin M-T 1980 *Stain Technnol.* **55** 313–21
[12] Vidal O, Kindblom L-G and Ohlsson C 1999 *J. Bone Min. Res.* **14** 923–9
[13] Carter D H, Barnes J M and Aaron J E 1989 *Calcif. Tissue Intl.* **44** 387–92
[14] Aaron J E and Carter D H 1987 *J. Histochem. Cytochem.* **35** 361–9
[15] Kohler G and Milstein C 1975 *Nature* **256** 495–7
[16] Bendayan M 2000 *Biotechnic. Histochem.* **75** 203–42
[17] Carter D H, Sloan P and Aaron J E 1991 *J. Histochem. Cytochem.* **39** 599–606

[18] Carter D H, Scully A J, Davies R M and Aaron J E 1998 *Histochem J.* **30** 1–10
[19] Braidman I, Davenport L K, Carter D H, Selby P, Mawer E B and Freemont A J 1995 *J. Bone Min. Res.* **10** 74–80
[20] Kosternuik P J, Harris J, Halloran B P, Turner R T, Morley-Holton E R and Bikle D D 1999 *J. Bone Min. Res.* **14** 21–31
[21] Newhall K M, Rodnick K J, van der Meulen M C, Carter D R and Marcus R 1991 *J. Bone Min. Res.* **6** 289–96
[22] Hert J, Pribylova E and Liskova M 1972 *Acta Anat.* **82** 218–30
[23] Carter D H, Sloan P and Aaron J E 1992 *Anat. Embryol.* **186** 229–240
[24] Luther F 1998 *Bone Arophy, Trabecular Architecture and Matrix Proteins* (University of Leeds; PhD thesis)
[25] Luther F, Saino H, Carter D H and Aaron J E 2003 *Bone* **32** 652–9
[26] Saino H, Luther F, Carter D H, Natali A J, Shahtaheri M S, Turner D L and Aaron J E 2003 *Bone* **32** 660–8
[27] Pinero G J, Farach-Carson M C, Devoll R E, Aubin J E, Brunn J C and Butler W T 1995 *Arch. Oral Biol.* **40** 145–55
[28] Riminucci I, Silvestrini G, Bonucci E, Fisher L W, Gehron-Robey P and Bianco P 1995 *Calcif. Tissue Intl.* **57** 277–84
[29] Petrusz P 1983 *J. Histochem. Cytochem.* **31** 177–9
[30] Faulk W P and Taylor G M 1971 *Immunohistochemistry* **8** 1081–3
[31] Saito C, Hayashi M, Sakai A, Fujie M, Kuroiwa H and Kuroiwa T 1998 *Biotechnic. Histochem.* **74** 40–8
[32] Kaneko M, Tomita T, Nakase T, Takeuchi E, Iwasaki M, Sugamoto K, Yonenobu K and Ochi T 1998 *Biotechnic. Histochem.* **74** 49–54
[33] Sherwin A F, Carter D H, Poole C A, Hoyland J A and Ayad S 1999 *Histochem. J.* **31** 623–32
[34] Kashiwa H K 1970 *Clin. Orthop.* **70** 200–11
[35] Aaron J E 1973 *Calc. Tissue Res.* **12** 259–79
[36] Wuthier R E 1993 *Symposium: Avian Bone Metabolism: Cell-Mediated Mineralization and Localized Regulatory Factors* (American Institute of Nutrition) p 301
[37] Ringbom-Anderson T, Jantti J and Akerman K E O 1994 *J. Bone Min. Res.* **9** 661–70
[38] Aaron J E 1980 *Metab. Bone Dis. Rel. Res.* **2** S109–16
[39] Birkenhager-Frenkel D H, Groen J J, Bedier de Prairie J A and Offerijns F G J 1961 *Voeding* **22** 634–9
[40] Dequeker J, Remans J, Franssen R and Waes J 1971 *Calc. Tissue Res.* **7** 23–30
[41] Mosekilde L I, Mosekilde L and Danielsen C C 1987 *Bone* **8** 79–85; 1986 *Bone* **7** 207–12
[42] Richelle L J 1964 *Clin. Orthop.* **33** 211–19
[43] Fincham A G 1969 *Calc. Tissue Res.* **3** 327–39
[44] Nusgens B, Chantraine A and Lapiere C M 1972 *Clin. Orthop.* **88** 252–74
[45] Grynpas M D, Patterson-Allen P and Simmons D J 1986 *Calcif. Tissue Intl.* **39** 57–62
[46] Raymaekers G, Aerssens J, Van den Eynde R, Peeters J, Guesens P, Devos P and Dequeker J 1992 *Calcif. Tissue Intl.* **51** 269–75
[47] Vogt J H 1949 *Acta Med. Scand.* **135** 221–30
[48] Dequeker J 1972 *Bone Loss in Normal and Pathological Conditions* (Leuven University Press)
[49] Jowsey J 1977 *The Bone Biopsy* (New York, London: Plenum Medical Book Company)

[50] Li J, Mori S, Kaji Y, Mashiba T, Kawanishi J and Norimatsu H 1999 *J. Bone Min. Res.* **14** 969–79

[51] Jowsey J, Kelly P J, Riggs B L, Bianco A J, Scholz D A and Gershon-Cohen J 1965 *J. Bone Joint Surg.* **47A** 785–806

[52] Bohatirchuk F and Jeletzky T 1971 *Invest. Radiol.* **6** 122–32

[53] Bohatirchuk F 1965 *Amer. J. Anat.* **117** 287–310

[54] Boyde A and Jones S J 1983 *Metab. Bone Dis. Rel. Res.* **5** 145–150

[55] Reid S A and Boyde A 1987 *J. Bone Min. Res.* **2** 13–22

[56] Boyde A, Howell P G T, Bromage T G, Elliot J C, Riggs C M, Bell L S, Kneissel M, Reid S A, Jayasinghe J A P and Jones S J 1992 *Chemistry and Biology of Mineralized Tissues* (Amsterdam: Elsevier) p 47

[57] Jones S J, Glorieux F H and Boyde A 1999 *Calcif. Tissue Intl.* **64** 8–17

[58] Hall TA and Gupta BL 1979. In: *Introduction to Analytical Electron Microscopy* eds J J Hren, J I Goldstein and D C Joy (London: Plenum Press) pp 169–97.

[59] Hall TA and Gupta BL 1982 *J. Microsc.* **126** 333–45.

[60] Morgan AJ 1985 *X-ray Microanalysis in Electron Microscopy for Biologists* (Microscopy Handbooks No. 5) (Oxford: Oxford University Press)

[61] Budd PM and Goodhew J 1988 *Light Element Analysis in the Transmission Electron Microscope: WEDX and EELS* (Microscopy Handbooks No. 16) (Oxford: Oxford University Press) .

[62] Whitehouse W J 1977 *Calc. Tissue Res.* **23** 67–76

[63] Sela J 1977 *Calc. Tissue Res.* **23** 229–34

[64] Boyde A, Maconnache E, Reid S A, Delling G and Mundy GR 1986 *Scanning Electron Microscopy* **4** 1537–54

[65] Lester KS and Ash MM 1980 *J. Ultrastruct. Res.* **72** 141–50

[66] Aitkin I and Ornoy A 1981 *Metab. Bone Dis. Rel. Res.* **3** 199–207

[67] Jayasinghe J A P, Jones S J and Boyde A 1993 *Virchows Archiv. A. Pathol. Anat.* **422** 25–34

[68] Aaron J E, Oliver B, Clarke N and Carter D H 1999 *Histochem. J.* **31** 455–70

[69] Aaron J E 1978 *Vitamin D* (London, New York: Academic Press) p 201

[70] Schraer H and Gay C V 1977 *Calc. Tissue Res.* **23** 185–8

[71] Ross G D, Morrison G H, Sacher R F and Staples R C 1983 *J. Microsc.* **129** 221–8

[72] Sitte H, Neumann K and Edelman L 1984 *Science of Biological Specimen Preparation* (Chicago: A M F O'Hare SEM Inc.) p103

[73] Hayat M A 1989 *Principles and Techniques of Electron Microscopy. Biological Applications* (London: Macmillan) p 395

[74] Hohling H J, Arnold S, Hohling J M, Plate U and Weisman H P 2001 *Materials Science and Engineering Technology (MATWER)* **32** 149–53

[75] Carter D H, Hatton P V and Aaron J E 1997 *Histochem. J.* **29** 783–93

[76] Harvey D M R 1982 *J. Microscopy* **127** 209–21

[77] Carter D H, Scully A J, Hatton P V and Aaron J E 2000 *Histochem. J.* **32** 253–61

[78] Hohling H J 1989 *Teeth* (Berlin: Springer) p 475

[79] Boyde A 1974 *Cell Tissue Res.* **152** 543–50

[80] Aaron J E and Pautard F G E 1972 *Israel J. Med. Sci.* **8** 625–9

[81] Ennever J and Creamer H 1967 *Calc. Tissue Res.* **1** 87–93

[82] Pautard F G E 1975 *Colloques Internationaux CNRS* **230** 93–100

[83] Pautard F G E 1978 *New Trends in Bio-inorganic Chemistry* (London: Academic Press) p 253

[84] Aaron J E 1973 *Calc. Tissue Res.* **12** 259–79
[85] Aaron J E and Pautard F G E 1973 *The Cell Cycle in Development and Differentiation* (Cambridge University Press) p 325
[86] Termine J D and Posner A S 1967 *Calc. Tissue Res.* **1** 8–23
[87] Carter D H, Scully A J, Heaton D A, Young M P J and Aaron J E 2002 *Bone* **31** 389–95
[88] Ali S Y, Sajdera S W and Anderson H C 1970 *Proc. Natl. Acad. Sci.* **67** 1513–20
[89] Boyde A and Sela J 1978 *Calc. Tissue Res.* **26** 47–9
[90] Termine J D, Eanes E D, Greenfield D J and Nylen M U 1973 *Calc. Tissue Res.* **12** 73–90
[91] Nylen M U, Eanes E D and Termine J D 1972 *Calc. Tissue Res.* **9** 95–108
[92] Termine J D and Eanes E D 1974 *Calc. Tissue Res.* **15** 81–4
[93] Glimcher M J 1989 *Anat. Rec.* **224** 139–53
[94] Kuhn L T, Wu Y, Rey C, Gerstenfeld L C, Grynpas M D, Ackerman J L, Kim H-M and Glimcher M J 2000 *J. Bone Min. Res.* **15** 1301–9
[95] Bonucci E 1971 *Clin. Orthop.* **78** 108–39
[96] Anderson H C 1969 *J. Cell Biol.* **41** 59–72
[97] Aaron J E 1980 *Metab. Bone Dis. Rel. Res.* **2** S151–7
[98] Appleton J 1970 *Calc. Tissue Res.* **5** 270–6
[99] Bonucci E 2002 *J. Bone Miner. Metab.* **20** 249–65
[100] Ajayan P M 1999 *Chem. Rev.* **99** 1787–99
[101] Ulman A 1996 *Chem. Rev.* **96** 1533–54
[102] Noy A, Vezenov D V and Lieber C M 1997 *Ann. Rev. Mat. Sci.* **27** 381–421.
[103] Jandt K D 1998 *Mater. Sci. Eng.* **R21** 221–95

SECTION 3

IONIZING RADIATION
TECHNIQUES

Chapter 8

Absorptiometric measurement

Christopher F Njeh and John A Shepherd

8.1. INTRODUCTION

The aim of this chapter is to introduce the reader to absorptiometric methods of the measurement of bone. For completeness, we include a very brief review of some of the basic physical characteristics of the interaction of X-rays with materials. This basic introduction is useful for the whole section including chapters 8–12. The reader should have been previously introduced to these physical principles.

SECTION A: FUNDAMENTAL PRINCIPLES OF RADIATION PHYSICS

The intention of the introduction section is to make the reader aware of the basic physics of bone densitometry. The reader is referred to the references such as Johns and Cunnigham [1] for a detailed description. We will provide a brief review of electromagnetic radiation to set the groundwork for the theory of single and dual energy absorptiometry.

8.2. FUNDAMENTALS OF RADIATION PHYSICS

Measurement of bone is usually carried out with either γ-rays or X-rays. Both are penetrating electromagnetic radiation. Radiation is the propagation of energy. If an object or instrument emits energy in any form then it is said to be emitting radiation. An example of radiant energy that is very familiar is that of matter waves in the form of acoustic radiation or sound (see chapter 14). Sound can only propagate in materials or media that will sustain a

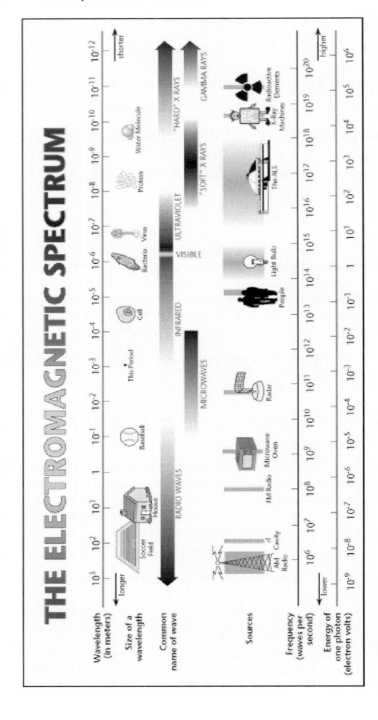

Figure 8.1. *Electromagnetic spectrum.*

pressure wave, such as air, water and many solids. Electromagnetic (EM) radiation differs from sound in that EM radiation is propagated as travelling and coupled electric and magnetic fields, needing no medium through which to travel. The EM spectrum is extremely broad and includes radio waves, infrared (heat) radiation, visible light, ultraviolet radiation, X-ray and γ-rays. Their only difference is their wavelength (figure 8.1). EM can be considered as a transverse wave or as particles propagating at the speed of light, c, where $c \approx 3.00 \times 10^8$ m/s. Considering EM radiation as waves, the speed of propagation, that is c the speed of light, related to the radiation frequency, ν, and wavelength, λ, as

$$c = \lambda\nu \tag{8.1}$$

where ν is measured in units of cycles/s, and λ in units of metres (m). Since c is constant, EM wavelength is inversely proportional to frequency. Furthermore, there are instances where it is more appropriate to consider EM as discrete energy packages called photons. This particle/wave duality of EM will not be discussed in any detail in here but can be studied further using introductory physics textbooks such as Haliday and Resinik [2]. For our purposes, it suffices to say that the two concepts can be related using

$$E = h\nu \quad \text{eV}. \tag{8.2}$$

That is, the energy of a photon is related to its frequency by h, a universal constant named Planck's constant. Plank's constant is in units of eV s and is equal to $h = 4.14 \times 10^{-15}$ eV s. We can relate photon energy to wavelength by substituting in equation (8.1) to equation (8.2) such that

$$E = hc/\lambda \quad \text{eV}. \tag{8.3}$$

From equation (8.3), as photon wavelength gets shorter, the photon energy increases. Electromagnetic radiation can be formed from a variety of mechanisms. We will focus on the two EM categories that are principally used to measure bone, γ-rays and X-rays, and discuss how they are generated.

8.2.1. γ-rays

γ-rays are the result of changes of energy of the nuclei of radioactive atoms. Gamma radiation is always emitted as characteristic radiation for a particular isotope.* Unstable isotopes give off radiation spontaneously, or are radioactive, and are commonly referred to as radioisotopes. This energy may be radiated from the nucleus as three types of radiation, α-, β- or γ-rays. α- and β-rays are not EM, do not penetrate the body sufficiently to be useful for imaging, and because of their high kinetic energy, can

---

* Isotopes are atoms with the same number of protons (Z) but different number of neutrons (N). All isotopes of a particular atom species have similar chemical properties but differ in mass. Isotopes that give off energy (such as γ-radiation) spontaneously are known as unstable isotopes.

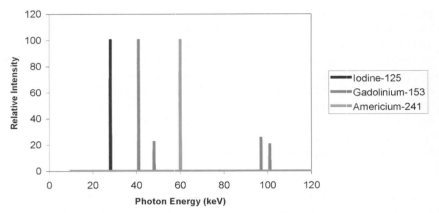

Figure 8.2. *Spectral emission from* ^{153}Gd, ^{125}I *and* ^{241}Am.

cause serious tissue damage. On the other hand, γ-rays are EM and have been useful for medical imaging. With respect to bone measurement, three sources have been widely used in the past for bone densitometry because of their unique properties, iodine-125 (^{125}I), americium-241 (^{241}Am) and gadolinium-153 (^{153}Gd). ^{125}I emits monochromatic (single energy) gamma radiation with energy at 28 keV while ^{241}Am has a characteristic peak at 60 keV. They were used as single photon sources for a technique called single (energy) photon absorptiometry (SPA) discussed later in this chapter. ^{153}Gd emits γ- and X-ray radiation centred around two specific energies, 44 and 100 keV. The spectra of ^{125}I, ^{241}Am and ^{153}Gd are compared in figure 8.2. Gadolinium was used as a source for dual (energy) photon absorptiometry (DPA). Both of these techniques will be discussed below.

Using γ-rays as a source has both advantages and disadvantages. One advantage is that since they are monochromatic (or bi-chromatic), they greatly simplify the equations used on bone densitometry. However, radio-isotopes are constantly decaying and thus can only be used with correction factors for the change in intensity over time. In addition, their fluence for practical sources for imaging is small in comparison with X-ray tubes.

8.2.2. X-rays

X-rays and γ-rays are both EM photons with similar energies, and hence indistinguishable to a detector. For example, an 80 keV X-ray is identical to an 80 keV γ-ray. The distinction between these two has mostly to do with their origin. X-rays are generated external to the nucleus by several different processes. Generally, when X-rays are produced as a radiation source for imaging or bone densitometry, the X-rays are the result of free electrons being accelerated into a target material such as tungsten in an

X-ray tube. The free electrons interact with the target atoms in one of several ways to produce radiation. First, the free electrons may collide with a shell electron with enough energy to free the shell electron. The energy needed to free a shell electron is called the electron's binding energy. The binding energy is unique for each shell electron in each atom. When an incident electron interacts with a shell electron, it will free that electron from its orbit, and ionize the atom, resulting in two free electrons: the original (or primary) free electron and the 'secondary' electron. An outer shell electron will now de-excite the atom by filling the inner shell vacancy and giving off the difference in energy between the two orbital states. These X-rays are called characteristic X-rays since their X-ray energies are unique for each atom. Characteristic X-rays are discrete energies related to energy transition levels between electron shells. Free electrons can also be inelastically scattered by the nucleus in which the path of the electron is deflected by the positively charged nucleus with a loss in kinetic energy. The energy loss is emitted as EM and is necessary and predicted by the conservation of energy. Bremsstrahlung, a German word for 'braking radiation', is broadband radiation that ranges in energy from the electron's peak energy to 0. The probability of Bremsstrahlung radiation is proportional to Z^2 of the target. The ratio of electron energy loss by Bremsstrahlung radiation to that lost by excitation and ionization can be calculated from

$$\frac{\text{Bremsstrahlung radiation}}{\text{excitation and ionization}} = \frac{E_k Z}{820} \qquad (8.4)$$

where E_k is the kinetic energy of the incident electron in MeV and Z is the atomic number of the absorber. Thus, for a tungsten target ($Z = 74$) and accelerating the incident electrons to 100 keV, one gets approximately 1% of the deposited energy radiated out of the tube as Bremsstrahlung radiation. Although only a small fraction of the deposited energy, Bremsstrahlung radiation makes up the majority of the X-rays produced by X-ray tubes.

8.2.3. *Inverse square law*

In free space, all EM radiations obey the inverse square law. This states that the photon flux rate (photons $\text{cm}^{-2}\,\text{s}^{-1}$) measured from a point source decreases by the inverse square of the distance from the source. That is,

$$I = \frac{N}{4\pi r^2} \qquad (8.5)$$

where N is the number of photons emitted from the source per second and r is the distance from the source from which we are taking our measurement. This is essentially just a statement of conservation of energy. Another way of thinking about this is that the total number of photons per unit time passing through any sphere surrounding a point source is always the same.

To get the number of photons per second which our detector of area A will register, we just multiply equation (8.5) by the detector area. One practical way this relationship is used is when predicting the flux rate at a distance r_2 from the source when you have measured it at a distance r_1. In this case, the expected fluence is

$$I_{r_2} = I_{r_1} \left(\frac{r_1}{r_2}\right)^2. \tag{8.6}$$

For example, the X-ray flux rate will be four times smaller at 2 m from the source than at 1 m from the source. Equation (8.6) is valid for practical X-ray sources, such as X-ray tubes.

8.3.   INTERACTION OF X-RAYS AND γ-RAYS WITH MATTER

8.3.1.   Introduction

X-rays and γ-rays are indirectly ionizing radiation: they do not produce large numbers of ions themselves, but they interact with matter to produce charged particles that in turn ionize the material. When a beam of X-rays or γ-rays passes through matter, its intensity is reduced by an amount that is determined by the physical properties, notably thickness, density and atomic number, of the material through which the beam passes. This is a random interaction governed by the laws of probability, and the loss of photons from the beam interacting with matter follows an exponential law. So, if a fairly well collimated beam passes through a particular thickness of an absorbing material, the change in the photon flux, dI, is equal to the change in thickness, dx, times the starting flux, I_0, and a constant, μ,

$$dI = -\mu I_0 \, dx. \tag{8.7}$$

This can be rewritten to solve for the exit flux by integrating to result in the equation

$$I = I_0 \, e^{-\mu x} \tag{8.8}$$

where I is the intensity after passing through a thickness x and μ is a constant for the absorbing material known as the linear attenuation coefficient. If x is measured in mm, μ has units of mm^{-1}. It is evident from equation (8.8) that X-rays and γ-rays do not have finite ranges, whatever their energy. Important points to note are the following.

1.   In the diagnostic range, μ decreases with increasing energy, i.e. the radiation becomes more penetrating.
2. μ increases with increasing density, i.e. the radiation is less penetrating because there are more molecules per unit volume in the material with which to collide.

Table 8.1. *Mass attenuation coefficients for typical materials useful for bone measurement. The density of air (dry) and water is given for 25 °C at 1 atm.*

Photon energy (keV)	Material (ρ in g/cm^3)				
	Air (1.205×10^{-3})	Water (1.000)	Muscle (1.040)	Cortical bone (1.650)	Fat (0.916)
10	4.91	5.066	5.154	19.79	3.081
15	1.522	1.568	1.604	6.193	1.009
20	0.7334	0.7613	0.7777	2.753	0.5332
30	0.3398	0.3612	0.3651	0.9534	0.2959
40	0.2429	0.2629	0.2635	0.5089	0.2353
50	0.2053	0.2245	0.224	0.3471	0.2102
60	0.1861	0.2046	0.2036	0.2727	0.1961
80	0.1658	0.1833	0.1819	0.2082	0.1794
100	0.154	0.1706	0.1692	0.1803	0.1684
150	0.1356	0.1505	0.1492	0.1493	0.1497
200	0.1234	0.137	0.1358	0.1334	0.1366

3. Variation of μ with atomic number is complex, although it clearly increases quite sharply with atomic number at very low energies.

It is sometimes more convenient to separate the effect of density (ρ) from other factors. This is achieved by using a mass attenuation coefficient, (μ/ρ), in units of g cm^{-2}. (μ/ρ) depends solely on the energy of the incident photons and the atomic composition of the attenuating medium such that it is valid for any actual density of the material (i.e. solid, liquid or gas!). The mass attenuation is multiplied by the mass per unit area, commonly referred to as the areal density, σ, where $\sigma = \rho x$. Then equation (8.8) can be written as

$$I = I_0 \, e^{-(\mu/\rho)\sigma}. \tag{8.9}$$

Tables of the mass attenuation coefficient are readily available in X-ray physics textbooks as well as on internet sites. Table 8.1 gives the mass attenuation coefficients most useful for bone densitometry. Other (μ/ρ) values can be derived using the NIST database on the internet (see http://physics.nist.gov/PhysRefData/Xcom/Text/XCOM.html). Note that there are many processes that can attenuate the X-ray beam, such as coherent and incoherent scatter, photoelectric absorptions, pair production, etc. When referring to mass attenuation coefficient tables like those available from the NIST website, there are separate attenuation coefficients given for each of these attenuation processes. When taking measurements with actual detectors through attenuators, your measure should be consistent for the total attenuation *with* coherent scattering coefficients. This is because you cannot separate out the non-attenuated beam from the forward scatter component in your measure.

8.3.2. *Interaction mechanism*

The nature of photon interactions with matter explains the dependence of attenuation on atomic number and photon energy. When a beam of collimated photons interacts with matter, some of the photons may be scattered and some absorbed. Both scatter and absorption result in beam attenuation, i.e. reduction in the intensity of the collimated beam. The beam attenuation is due to processes such as the Compton effect, the photoelectric effect, elastic scattering and pair production. Bone densitometry is carried out at the diagnostic energies where only Compton scattering and the photoelectric effect are important. Only these two processes will be described here.

Useful concepts in understanding photon interaction with matter are those of 'bound and free' electrons. Strictly speaking, there are normally no 'free' electrons in matter. Each electron is 'bound' in the atom by the electrostatic attraction between itself and the positive charge on the nucleus and it can only become 'free' if it receives enough energy to overcome this binding force. For the outer electrons of any atom, this binding energy is only a few electron-volts, which is small compared with the energy of X-ray photons. This leads to the concept that an electron may be considered to be 'free' when its binding energy is small compared with the energy of a photon with which it interacts. For practical purposes, all electrons in the elements that make up the greater part of the soft tissue of the body may be regarded as 'free' for all X-ray energies encountered in radiology [3].

8.3.2.1. *Elastic scattering (coherent scattering)*

When X-rays pass close to an atom, they may cause the 'bound' electrons to vibrate (resonate) at a frequency corresponding to that of the X-ray photon. The electron re-radiates this energy in all directions and at exactly the same frequency as the incoming photons (figure 8.3(a)). Thus, energy is taken from the beam and is scattered in all directions. This is a process of scatter and attenuation without absorption. This process occurs when the binding energy of the electron is high (i.e. high atomic number of the scattering material) and when the quantum energy of the bombarding photons is relatively low. Although a certain amount of elastic scattering occurs at all X-ray energies, it never accounts for more than 10% of the total interaction processes in diagnostic radiology.

8.3.2.2. *Compton (inelastic) scattering*

Unlike elastic scattering, in Compton scattering the photon loses some of its energy to the electron and itself continues in a new direction (i.e. it is

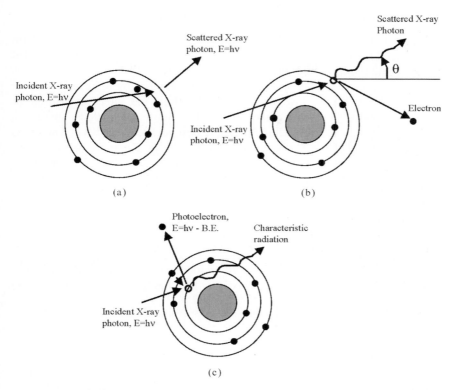

Figure 8.3. *Scatter and absorption processes in an atom. (a) Coherent scattering of an X-ray photon, (b) Compton scattering of an X-ray photon and (c) photoelectric absorption of an X-ray by an atom.*

scattered) but with reduced wavelength and hence with increased energy by the equation

$$E' = \frac{E}{1 + \left(\dfrac{E}{m_e c^2}\right)(1 - \cos\theta)} \qquad (8.10)$$

where E is the energy of the incident photon, E' is the energy of the scattered X-ray photon, θ is the scatter angle of the scattered photon, and m_e is the rest mass of the electron equivalent to 511 keV of energy (figure 8.3(b)). The scattered wavelength can be found by substituting in equation (8.3). Compton scatter creates two major problems in X-ray imaging. First, it reduces the contrasts in the image unless it is collimated out before the detector. Second, it presents a major radiation risk to the personnel using the equipment.

8.3.2.3.   Photoelectric effect

When a photon interacts with the atom it is totally absorbed and an electron is dislodged from its orbit around a nucleus. Part of the photon energy is used to overcome the binding energy of the electron, the remainder is given to the electron as kinetic energy and is dissipated locally (figure 8.3(c)). Whenever the photon energy is just slightly greater than the energy required to remove an electron from a particular shell around the nucleus, there is a sharp increase in the photoelectric absorption coefficient. This is known as an absorption edge. There are two reasons for the sudden increase in absorption. First, the number of electrons available for release from the atom increases. Second, a resonance phenomenon occurs whenever the photon energy just exceeds the binding energy of a given shell.

8.3.3.   Attenuation in tissue

The total X-ray attenuation in tissue is a result of all three processes. Such a total attenuation coefficient can be defined as the sum of the attenuations from all the attenuation processes. For medical X-ray imaging, the

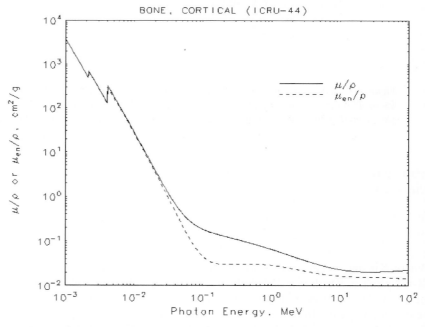

Figure 8.4. *Plot of the mass attenuation coefficient and the mass energy-absorption coefficient as a function of photon energy for cortical bone. Note the strong k-edge absorption at 40 keV corresponding to the calcium K_α edge. The plotted values were adapted from ICRU Report 44 [66].*

predominant attenuation results from photoelectric absorption and Compton scattering. Thus the total mass attenuation, i.e. the mass attenuation coefficient, can be written as

$$\left(\frac{\mu}{\rho}\right)_{\text{total}} = \left(\frac{\mu}{\rho}\right)_{\text{P.E.}} + \left(\frac{\mu}{\rho}\right)_{\text{Compton}} + \left(\frac{\mu}{\rho}\right)_{\text{other}} \qquad (8.11)$$

and substituted into equation (8.5). Figure 8.4 shows the total attenuation for bone over a broad energy range. (Note that figure 8.4 also shows the mass *absorption* coefficient, (μ_{en}/ρ), which is related to the total kinetic energy deposited in the material by the scattered electrons. It is used for dose. (See Hobbie [4], p. 393 for more details.) Since soft tissue of the body has a low effective atomic number ($\bar{Z} = 7.5$), photons of energies greater than 30 keV interact predominantly by Compton scattering. Bone has a higher effective atomic number ($\bar{Z} = 13$) than soft tissue, which means that at low energies, where photoelectric effect predominates, the absorption in bone can be many times higher than in soft tissue.

SECTION B: INSTRUMENTATION AND PRINCIPLES

8.4. GENERATION OF X-RAY

8.4.1. Introduction

X-rays for diagnostic uses are produced when a stream of electrons accelerates through a potential difference ranging from 20 to 120 kV and strikes a target (figure 8.5). The typical X-ray tube consists of a cathode, where the electrons are generated, and an anode, where the electrons are

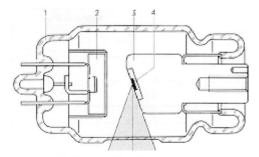

Figure 8.5. *Internal diagram of an X-ray tube. An evacuated glass envelope (1) surrounds the cathode (2) and anode (3). The target (4) is the part of the anode that is actually bombarded by electrons and is the source location of X-rays.*

stopped and X-rays generated. The X-rays originate principally from rapid deceleration of the electrons when they strike the material. The X-rays generated by the deceleration of electrons (versus photoelectric absorption) are known as Bremsstrahlung. The cathode electrons are produced from a tungsten filament by thermionic emission. They are accelerated in the vacuum of the tube by the electric field force between the electrodes, then strike the tungsten target. Most medical X-ray tube anodes are made from tungsten (symbol W). Tungsten is typically used for X-ray accelerating voltages above about 35 kVp because of its high stopping power and good heat dissipation. Special purpose X-ray tubes exist for lower energy X-ray production. For example, molybdenum is commonly used as an anode for mammography tubes because of its strong k-edge emission lines at 17 and 23 keV. The X-rays are emitted in all directions from the anode. Most X-rays are stopped by internal shielding in the X-ray tube. Only a small fraction of the X-rays are emitted through a collimator, escape from the tube head, and are available to the user.

8.4.2.   X-ray spectrum

X-rays are produced by the interaction of the electrons from the filament with the material of the target. One may think that the X-rays would emerge from the X-ray tube at a single energy (monochromatic) similar to the energy of the electrons bombarding the target. However, X-ray output is a complex spectrum (polychromatic) such as that shown in figure 8.6. The spectrum is actually made up of two distinct parts resulting from two

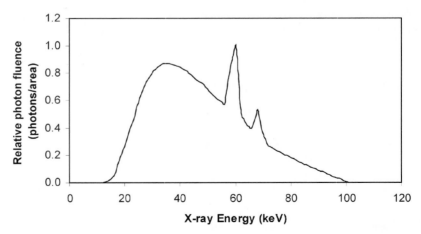

Figure 8.6. *X-ray spectra from a tungsten X-ray tube. The plot was generated from a computer model in Boone et al [67] for a tube voltage of 100 kVp, no filtration, and direct current (no ripple).*

different mechanisms. There is the continuous spectrum of energies resulting from the Bremsstralung radiation, with the maximum energy dependent on the maximum kilovoltage. Superimposed on this continuous spectrum are the characteristic radiation peaks that are specific to the target material. The spectrum shown in figure 8.6 is for tungsten.

8.4.2.1. Continuous spectrum

The continuous spectrum can be understood in terms of the interactions of electrons with the target. When the fast moving electrons strike the anode, they interact with either the orbital electrons or the nucleus. Interactions with orbital electrons will result in the transfer to the target of a small amount of energy that will appear eventually as heat. At diagnostic energies at least 99% of the electron, energy is converted into heat. However, energy lost in the interaction with the nucleus is emitted as Bremsstralung X-rays. The amount of energy lost by the electron in such a collision is very variable and hence the energy given to the X-ray can take a wide range of values. However, for any given accelerating voltage across the X-ray tube, there is a well-defined maximum X-ray energy equal to the energy of a single electron. It is worth noting that before the X-ray emerges from the tube its intensity will be modified in two ways. First, X-rays produced deep in the anode will be attenuated in reaching the surface of the anode. Second, the X-rays will be attenuated in penetrating the window of the X-ray tube.

8.4.2.2. Line spectra

Line spectra result from the photoelectric absorption of an incoming electron interacting with the bound orbital electron in the target. This is directly analogous to the photoelectric absorption of a photon. If the incoming electron has sufficient energy to overcome the binding energy, it can remove a bound electron, thus creating a vacancy in the shell. The vacant energy level is then filled by an electron from a higher energy level falling into it and the excess energy is emitted as an X-ray.

8.4.3. Factors affecting the X-ray spectrum

There are two aspects of X-rays that can be altered: quantity and quality. There is a change in quantity if the spectrum changes in such a manner that its shape remains unaltered, i.e. the intensity at every photon energy changes by the same factor. If, on the other hand, the shape of the spectrum changes, there has been a change in the radiation quality—the penetrating power of the X-ray beam. The quality of the X-ray spectrum is primarily affected by applied voltage, anode material and beam filtering or 'hardening'. The quantity is primarily affected by tube current, time of exposure and applied anode voltage.

8.4.3.1. Tube current

There are two circuits in an X-ray tube that produce tube current: the cathode (filament) current and the cathode-to-anode current. The cathode current is used to heat the filament to the point at which electrons have the thermal energy to escape the conduction band and act as free electrons. The current between the anode and the cathode determines the number of electrons striking the anode. Electrical current is measured in units of amperes as the amount of charge (coulombs) passing though a conductor per second where

$$\text{ampere} = \text{coulombs/s}. \tag{8.12}$$

Both of these currents must exist to create X-rays. The X-ray exposure is proportional to the tube current and time of exposure, but only the quantity of X-ray is affected. Thus, one typically speaks of the tube current in milliamperes (mA) and the time of exposure in seconds as one value called 'M-A-S' and written as mAs for milliamp-seconds. In commercial bone densitometers, different tube currents and exposure times are recommended for different body sizes. For example, paediatric scan protocols use lower mAs than are used for adult scans because of the thinner tissue. This reduces the radiation exposure without reducing image quality.

8.4.3.2. Applied voltage

If other tube operating conditions are kept constant, the flux of X-rays produced (or exposure), increases approximately as the square of the applied anode voltage. That is,

$$I = k(V_{\text{p}})^2 \tag{8.13}$$

where I is the relative intensity of X-rays at some measurement point in front of the X-ray tube, V_{p} is the applied anode voltage and k is a constant. This is due to the increased electron energy and increased efficiency of conversion of electrons into X-rays. For example, increasing an X-ray tube's applied anode voltage from 60 to 80 kVp increases the intensity by 1.78 times. Increasing the kV also alters the radiation quality since the high-energy cut-off has now increased. The profile of the applied voltage also affects both quantity and quality. In commercial bone densitometers, the operator cannot alter the tube voltage.

8.4.3.3. Filtration

Filtration of the X-ray spectrum occurs when any material is placed in the beam between the X-ray anode and the object being imaged. Filtration always exists from the X-ray tube output window (i.e. glass, beryllium) and can be intentionally added to 'harden' the X-ray beam. That is, since all materials absorb low-energy X-rays in preference to high-energy X-rays, the average energy of the X-ray spectrum is higher after passing through a filter. Therefore, to reduce the dose, beam hardening is often used to remove the low-energy X-rays that would not make it through the patient.

8.5. PHYSICAL PRINCIPLES OF ABSORPTIOMETRY

The main principle of absorption is based on the interaction of photons (γ-ray or X-ray) with tissue. Currently the type of photons used are X-ray—however, γ-rays will be introduced here from a historical perspective. As previously mentioned, the behaviour of γ-rays and X-rays in matter is the same.

8.5.1. Single energy (γ-ray or X-ray) absorptiometry

8.5.1.1. Theory

The attenuation of monoenergetic and narrow photon beam can be represented by equation (8.8). This equation is simplistic and, in practice, the substance the beam is traversing is not homogeneous, but composed of layers or mixtures of different materials each with its own mass attenuation coefficient, density and thickness. If the beam passes through N different materials, equation (8.8) can be written as

$$I = I_0 \exp\left(-\sum_{i=1}^{N} \mu_i \rho_i x_i\right). \tag{8.14}$$

In the case of single energy absorptiometry, it is assumed that there are only two materials contributing to the attenuation, namely bone and soft tissue [5]. Then equation (8.14) becomes

$$I(x) = I_0 \exp[-(\mu_s \rho_s x_s(x) + \mu_b \rho_b x_b(x))] \tag{8.15}$$

where the s and b subscripts represent soft tissue and bone respectively measured at pixel x. If the total thickness, W, through soft tissue and bone is kept constant as in figure 8.7, then x_s can be eliminated from equation (8.15). Rearranging and taking the log of both sides, we get

$$\ln\left|\frac{I_0}{I(x)}\right| = (\mu_s \rho_s (W - x_b(x)) + \mu_b \rho_b x_b(x)). \tag{8.16}$$

The path length $w_b(x)$ can be solved to get

$$w_b(x) = \frac{\ln\left(\frac{I_0}{I(x)}\right) - \mu_s \rho_s W}{(\mu_b \rho_b - \mu_s \rho_s)}. \tag{8.17}$$

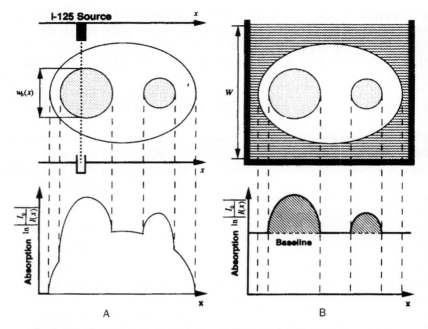

Figure 8.7. *Principles of operation of an SPA/SXA device (from Blake et al [5]).*

Where W can be directly related to the attenuation in soft tissue only,

$$\mu_s \rho_s W = \ln \left| \frac{I_s}{I_0} \right|. \tag{8.18}$$

Substituting equation (8.18) into (8.17), we get the SXA relationship,

$$w(x)_b = \frac{\ln \left(\dfrac{I_s}{I(x)} \right)}{(\mu_b \rho_b - \mu_s \rho_s)}. \tag{8.19}$$

If we integrate $w(x)_b$ over all pixels and multiply by the bone volumetric density, ρ_b, we get the total bone areal density, BMD_{SXA},

$$\text{BMD}_{\text{SXA}} = \rho_b \int w_b(x)\,dx = \frac{\rho_b}{(\mu_b \rho_b - \mu_s \rho_s)} \int \ln \left(\frac{I_s}{I(x)} \right) dx \tag{8.20}$$

measured in g/cm^2.

8.5.1.2. *Practical application: single photon absorptionmetry*

Single photon absorptiometry (SPA) was first introduced by Cameron and Sorenson in 1963 [6]. A highly collimated photon beam from a single-energy radionuclide source, such as ^{125}I (photon energy 27.3 keV) or ^{241}Am (60 keV) coupled to a radiation detector (usually a sodium iodide crystal mounted on a photomultiplier tube), is used to measure radiation attenuation at the measurement site. To correct for overlying soft tissue, the anatomical site at which the bone mineral density (BMD) is being measured has to be immersed in a water bath or surrounded by water bags or water equivalent mouldable materials [7]. On the other hand, SPA techniques are limited in that the photon source is a radionuclide. This source decays and requires regular replacement. It also has a low photon fluence, which causes scanning times to be long and spatial resolution to be poor. The method overcame the problems for RA caused by polychromatic X-rays (temporal and spatial non-uniformity and beam hardening) and non-uniformity of film sensitivity and development. SPA was a widely used and established bone density technique but it has now been superseded by single X-ray absorptiometry (SXA) or peripheral dual X-ray absorptiometry (DXA).

The shortcomings of SPA have been overcome by the introduction of a low dose X-ray source. The physical principles of SXA are the same as SPA except that the radionuclide source has been replaced by an X-ray tube. This has imparted better precision and improved spatial resolution to these systems and has reduced examination time [8]. SXA makes possible a quantitative assessment of bone mineral content at peripheral sites of the skeleton (e.g. distal or ultradistal radius and calcaneus) (figure 8.7). SXA has proven to be a valuable method in the diagnosis of osteoporosis, providing reasonable precision and exceptionally low radiation exposure. However, recently peripheral DXA systems have been supplementing SXA systems, eliminating the need for a water bath or water bolus.

8.5.2. Dual energy absorptiometry

8.5.2.1. Theory

The assumption of constant thickness of the photon pathway restricted the use of SXA/SPA to peripheral sites. Even then, constant thickness conditions were not always realized. The assumption could not be applied to important fracture sites such as the spine, proximal hip and whole body because of widely varying soft tissue thickness and soft tissue composition. Dual photon absorptiometry (DPA) techniques were introduced to overcome the restriction of constant overall thickness of measurement site. This approach uses a radionuclide source, typically ^{153}Gd with photon energies of 44 and 100 keV. The simultaneous measurement of gamma radiation at two different energies allows for the correction of soft tissue and fat of the torso without the need for a water bath (figure 8.8). However, two fundamental assumptions are also used to determine bone density using two energies:

• Transmission through the body of the X-rays within the two energy windows can be accurately described by a monoexponential attenuation process (equation (8.14)).

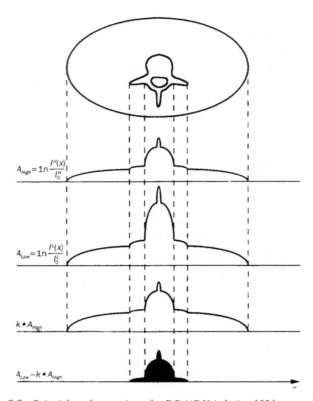

Figure 8.8. *Principles of operation of a DPA/DXA device [52].*

- The human body is a two-component system, i.e. soft tissue and bone mineral.

Two X-ray beams with different energies (high and low) result in the two equations shown below:

$$I^L(x) = I_0 \exp\left[-\left(\left(\frac{\mu}{\rho}\right)^L_s \sigma_s(x) + \left(\frac{\mu}{\rho}\right)^L_b \sigma_b(x) \right) \right] \qquad (8.21)$$

$$I^H(x) = I_0 \exp\left[-\left(\left(\frac{\mu}{\rho}\right)^H_s \sigma_s(x) + \left(\frac{\mu}{\rho}\right)^H_b \sigma_b(x) \right) \right] \qquad (8.22)$$

where the H and L superscripts represent the high and low energy X-ray beams respectively, and σ is the areal density in units of g/cm^2. Equations (8.21) and (8.13) are analogous to equation (8.16), where $\sigma(x)$ has been substituted for $w(x)\rho$ and the linear attenuation coefficient has been replaced by the mass attenuation coefficient. We can solve equations (8.21) and (8.22) simultaneously for the bone areal density:

$$\sigma_b(x) = \frac{\left[\left(\frac{\mu}{\rho}\right)_s^L \Big/ \left(\frac{\mu}{\rho}\right)_s^H\right] \ln\left(\frac{I^H(x)}{I_0^H}\right) - \ln\left(\frac{I^L(x)}{I_0^L}\right)}{\left(\frac{\mu}{\rho}\right)_b^L - \left(\frac{\mu}{\rho}\right)_b^H \left[\left(\frac{\mu}{\rho}\right)_s^L \Big/ \left(\frac{\mu}{\rho}\right)_s^H\right]}. \tag{8.23}$$

Equation (8.23) is known as the DPA equation and is not dependent on the sample being a constant thickness. We can define the '*R*-factor' R_s as

$$R_s = \frac{\left(\frac{\mu}{\rho}\right)_s^L}{\left(\frac{\mu}{\rho}\right)_s^H} \tag{8.24}$$

and then we can rewrite equation (8.23) as

$$\sigma_b(x) = \frac{R_s \ln\left(\frac{I^H(x)}{I_0^H}\right) - \ln\left(\frac{I^L(x)}{I_0^L}\right)}{\left(\frac{\mu}{\rho}\right)_b^L - \left(\frac{\mu}{\rho}\right)_b^H R_s}. \tag{8.25}$$

Equation (8.25) states that all the soft tissue measure is reduced to the R_s term. All the other terms in equation (8.25) are either directly measured or are defined by the known mass attenuation coefficient of bone. To finish our solution, we must make one assumption. We assume that the soft tissue overlaying the bone has the same R_s value as the tissue near the bone and not overlaying it, since the composition of the soft tissue is made up of an unknown percentage of fat and lean tissue. In this region of no bone, we can measure R_s using equation (8.25), since then $\sigma_b = 0$, $I(x) = I_s(x)$ and, in this region of no bone,

$$R_S = \frac{\ln\left(\frac{I_S^L}{I_0^L}\right)}{\ln\left(\frac{I_S^H}{I_0^H}\right)}. \tag{8.26}$$

Thus, R_S is a measure of the percentage of fat of the soft tissue.

Dual X-ray absorptiometry (DXA) is based on the method of X-ray spectrophotometry developed in the 1970s. It was introduced commercially as the direct successor to DPA in 1987 [9]. Similarly to DPA, the fundamental physical principle behind DXA is the measurement of the transmission of X-rays with high (and low) photon energies. The main advantages of an X-ray system over a DPA radionuclide system are shortened examination time due to an increased photon fluence of the X-ray tube and greater accuracy and precision resulting from higher resolution and removal of errors due to source decay [10]. The preferred anatomical sites for DXA measurement of bone mineral include the lumbar spine, the proximal femur and the whole body, but peripheral sites can also be scanned.

8.5.3. Implementation of DXA

8.5.3.1. K-edge absorption

Two methods of generating a dual energy X-ray spectrum have been implemented using either K-edge filters or kVp switching. K-edge absorption filters, made of a rare earth material such as cerium (Ce) and samarium (Sm), are used to split the polyenergetic X-ray beam into high and low energy components that mimic the emissions from ^{153}Gd. The two components have inherently narrow spectral distribution and hence the problems associated with beam hardening are minimized. Lunar DPX (Lunar Corp, Madison) systems have a cerium filter and use pulse height analysis at the detector to discriminate between high and low energy photons. Norland XR (Atkinson, Wisconsin) systems use a samarium filter and separate detectors for high and low energy X-rays [11].

8.5.3.2. Voltage switching

The second way of producing a dual energy X-ray beam is to switch the high voltage generator between high and low kVp during alternate half cycles of the mains supply. Hologic (Waltham, USA) uses this method in its QDR series of DXA systems [5]. The spectral distribution is wider than with the K-edge filter method and the consequent effect of beam hardening is corrected by a rotating calibration wheel containing bone and soft tissue equivalent filters that measures the attenuation coefficient in equation (8.23) and calibrates the scanned image pixel by pixel [12]. Pulse height analysis is not required, giving the instrument an inherently wide dynamic range.

8.5.3.3. Image formation

First generation DXA systems, such as the Hologic QDR 1000, Norland XR-36 (figure 8.9) and Lunar DPX, scan patients rectilinearly using a pinhole collimator producing a pencil beam coupled to a single detector in the scanning arm. When a DXA scan is analysed, the basic data analysis process creates a pixel-by-pixel map of BMD over the entire scanned field calculated from equation (8.23). An edge detection algorithm is first used to find the bone edges. The total projected area of bone is then derived by summing the pixels within the bone edges and a reported value of BMD calculated as the mean BMD over all the pixels identified as bones. Finally, bone mineral content (BMC) is derived by multiplying BMD by projected area.

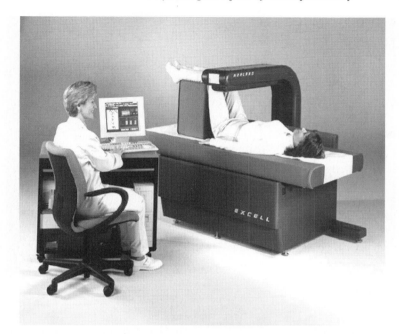

Figure 8.9. *A pencil-beam DXA scanner (Norland EXCEL) illustrating AP spine patient positioning (courtesy of Norland Medical Systems).*

8.5.3.4. Fan beam DXA

A recent significant development in DXA technology has been the introduction of a new generation of systems, such as the Hologic QDR 4500 and the Lunar Expert-XL (figure 8.10), which use a slit collimator to generate a fan beam coupled to a linear array of detectors. Fan beam studies are acquired by the scanning arm performing a single sweep across the patient instead of the two-dimensional raster scan required by pencil beam geometry. As a result, scan time has been shortened from about 5–10 min for the early pencil beam scanners to 10–30 s for the latest fan beam system. Shorter scan time and higher patient throughput, around two to three times that for first generation scanners, are the major advantages of the fan beam DXA system. Another advantage of the fan beam system is the higher image resolution. This allows easier identification of vertebral structure and artefacts resulting from degenerative disease.

8.5.3.5. Cone beam DXA

There is further development using cone beam geometry. Instead of using a fan beam, a cone beam that encompasses the region of interest is used. No more scanning is required. This technology has been implemented in the

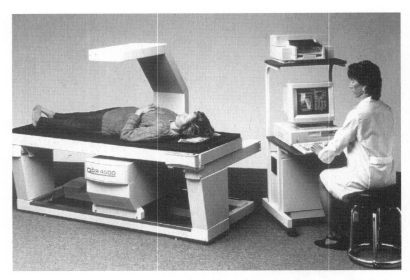

Figure 8.10. *A fan-beam scanner, the Hologic QDR-4500 (courtesy of Hologic Inc).*

Lexxos Bone Densitometer (DMS, Montpelier, France). This densitometer uses a two-dimensional digital flat panel radiographic detector made of a Gd_2O_2S scintillator coupled to a grid of amorphous silicon photodiodes. The detector is $20\,cm^2$ and provides 512×512 pixels; each pixel is $0.4\,mm$

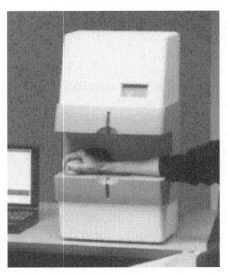

Figure 8.11. *Peripheral DXA scanner (Lunar PIXI) illustrating forearm measurement position (courtesy of Lunar Corp).*

on a side. Pencil beam and fan beam densitometers use a collimation in order to limit the contribution of scatter to the measurement. In cone beam, scatter due to the patient cannot be eliminated by collimators. In order to reduce the scatter, an antiscatter grid is used. The possible advantage of the cone beam systems is further reduction in scanning time.

8.5.3.6. pDXA

DXA is also employed for measurements of the appendicular skeleton. Most standard DXA densitometers allow for highly precise measurement of the radius or calcaneus using regions of interest like those derived from SPA and SXA measurements as well as user-defined sub-regions [13, 14]. Recently peripheral DXA (pDXA) densitometers specially designed for the forearm or calcaneus have been introduced and may provide these measurements at a lower cost (figure 8.11). They have the advantage compared with SXA systems that they dispense with the need for a water bath. Furthermore, these devices are compact and could fit in some practices.

SECTION C: CLINICAL APPLICATIONS

8.6.   SITES MEASURED

The choice of the optimum skeletal site for the prediction of osteoporosis is a subject of great debate. However, the consensus is that sites with higher contents of cancellous (high turnover) bone are more sensitive to osteoporotic changes. Longitudinal studies suggest that most of the skeletal sites currently measured (spine, femur, radius and calcaneus) are useful for predicting the risk of fracture at any site. The best assessment of risk at a specific site, however, is to measure the BMD at that site [15].

8.6.1.   Lumbar spine

The lumbar spine is probably the most frequently studied skeletal site for DXA measurement because of its sensitivity to the effects of ageing, the menopause and secondary causes of bone loss. The region of interest (ROI) studied is usually the lumbar vertebral (L1–L4). In children and adults, projectional BMD increases from L1 to L4 (figure 8.12). With advancing age, however, the distribution pattern of bone mineral in the spine becomes less uniform, due to compression or degenerative changes. When significant non-uniformity of bone mineral is present, there is less confidence that the BMD measurements are representative of true skeletal status.

A detailed guide on the correct measurement procedure is found in the manufacturer's manual. However, this section provides an overview of the

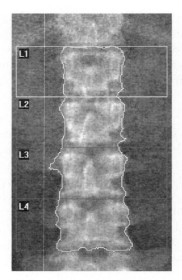

Figure 8.12. *A spine DXA scan showing the ROI analysed (lumbar spine L1–L4).*

proper procedure in acquiring the data. Prior to performing a DXA study, a short interview should establish that there are no contra-indications to the scan (table 8.2).

Before positioning the patient, all metal items should be removed. Acquisition and scan analysis should be performed according to the manufacturer's protocol. In general, posterior–anterior (PA) spine scans are acquired with the patient lying supine on the imaging table. **The patient should be aligned in the middle of the table with the spine straight and parallel**

Table 8.2. *Contra-indications for spinal BMD measurement (adapted from Blake et al [5]).*

1. Pregnancy (abdominal thickness, radiation).
2. Recent oral contrast media (2–6 days). Intravenous contrast media are rapidly excreted and rarely interfere after a few hours.
3. Recent nuclear medicine test (depends on isotope used):
   $^{99}Tc^m$ MDP bone scan 48 h
   $^{99}Tc^m$ MAA lung scan 24 h
   $^{99}Tc^m$ Sc liver scan 48 h
   ^{131}I more than 4 MBq 72 h or longer if there are bone metastases in the lumbar spine
   $^{99}Tc^m$ DTPA or Mag 3 renal scan 24 h.
4. Inability to remain supine on the imaging table for 5 min without movement.
5. Spinal deformity or disease, orthopaedic hardware in the lumbar spine as diagnosed by radiograph, since results will not be useful for a standard osteoporosis work-up.

to the longitudinal table axis (figure 8.9). A soft block is used to raise the patient's legs by supporting them under the knees. This reduces the physiological spinal lordosis and aligns the disc spaces with the X-ray beam to improve separation of individual vertebrae on the scan image. Important anatomical markers, which should be clearly visible on the scan image, are the upper border of the pelvis and the 12th rib. L5 can often be identified by the characteristic 'M' or butterfly shape.

If a compression fracture or other spinal abnormality exists on the lumbar spine radiographs, this vertebra should be excluded regardless of the appearance of the DXA image or the BMD results of the individual vertebra [16]. **DXA scans do not replace a lumbar spine radiograph and should not be used to rule out or confirm a compression fracture.** If there are artefacts overlying bone, this part of the bone must be excluded from the ROI.

In DXA of the spine, the BMD is the areal density of the integral bone, which includes the vertebral body. Extraneous calcification such as in the walls of the aorta, particularly with degenerative disc and apophyseal joint disease with consequent hyperostosis, will cause inaccuracies and over-estimation of BMD [17]. Falsely elevated spinal BMD on PA DXA may be caused by other etiologies such as vertebral wedge (crush fracture), Paget's disease of bone, sclerotic metastases and vertebral haemoglobin.

8.6.2.   *Lateral spine*

A lateral examination of the lumbar spine makes possible an evaluation of the vertebral body, also ensuring almost exclusive measurement of trabecular bone. The development of lateral DXA scanning of the lumbar spine was aimed at reducing the errors intrinsic in the PA or anterior–posterior (AP) examination of the lumbar spine. This is especially true in cases where degenerative and hypertrophic changes in the spine such as osteoarthritis of the articular facets, hypertrophy of the spinous processes and degenerative sclerosis of the end-plate, all lead to increases in BMD as measured by the PA projection [18–21]. In this instance PA BMD are not reliable. Lateral scans allow ROI placement that excludes the posterior processes from the vertebral body, and hence reduce artefacts of posterior degenerative disease and measure a higher percentage of trabecular bone. Vertebral bodies are the actual sites where fractures occur. Lateral spine measurements can be performed with standard instruments with the patient in the lateral decubitus position or in the supine position with a fan-beam scanner. The ROI commonly used for lateral scanning is L2–L3 or L2–L4 (figure 8.13). Overlap of ribs and pelvis might impact L2 and L4 by falsely increasing the BMD measurement and affect the interpretation of the scan [22, 23].

Lateral spine scanning has not replaced PA spine scanning as routine for a number of reasons. First, the measurement site is more demanding in

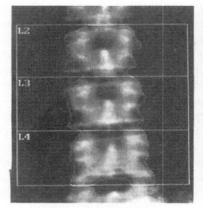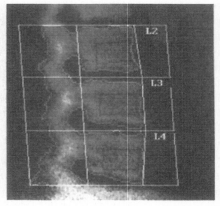

Region	Est.Area (cm²)	Est.BMC (grams)	BMD (gms/cm²	Region	Est.Area (cm²)	Est.BMC (grams)	BMD (gms/cm²)
L2	14.89	16.82	1.129	L2	10.95	8.47	0.774
L3	16.91	18.49	1.094	L3	11.65	9.54	0.819
L4	17.49	19.31	1.104	L4	11.71	9.61	0.821
TOTAL	49.29	54.62	1.109	TOTAL	34.31	27.62	0.805

Figure 8.13. *PA (left) and lateral (right) scans of lumbar spine using Hologic QDR 2000, fan-beam mode.*

acquisition and processing. Second, accurate repositioning of subjects is difficult (improve for fan beam DXA with a C arm) and, coupled with lack of an anatomical landmark for identifying vertebrae, results in inferior measurement precision [22, 24]. Furthermore, the lateral sampling site is smaller and the interpopulation variance is larger. The effect of the variable composition of the soft tissue reference baseline on the accuracy of the BMD measurement is still a significant problem in lateral DXA. Nevertheless, in cross-sectional studies, age-related bone loss from lateral scans is twice that in AP projection, approaching that of QCT, and the difference between normal and osteoporotic group is larger [25, 26]. The correlation between lateral DXA and quantitative computed tomography (QCT), both measures of the vertebral body, has been found to be stronger than that between PA DXA and QCT [27]. However, overlap of the iliac crest at level L4 and ribs at level L2 may substantially increase the measured bone density. Nevertheless, the inclusion of L2–L4 usually yields the best precision and diagnostic sensitivity [23, 28]. This suggests that lateral scans may be more sensitive for detection of early bone loss.

Lateral spine BMD measurements can be performed in either the supine or the decubitus position, depending on the device available. Instruments with C arms allow the patient to be scanned supine. In this case, patient positioning is the same as for PA/AP spine scan. The patient's arm should

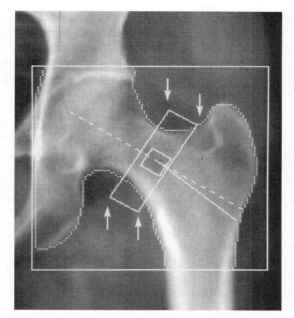

Figure 8.14. *A proximal femur DXA scan, showing the ROI analysed femoral neck (oblong box), Ward's area (box) and trochanter.*

be raised above the shoulders so that they do not interfere with the view of the spine on the lateral image. The decubitus scan is now rarely used, but more information can be found in Blake *et al* [5].

8.6.3. Proximal femur

The proximal femur is also a frequent site of osteoporotic fracture. The structure of the femur is very complex, with two major trabecular systems arranged along the lines of compressive and tensile stresses produced during weight bearing [29]. Bone loss in the proximal femur also follows a pattern that reflects a hierarchy of trabecular groups in meeting the demands of weight bearing [30]. The proximal femur is usually measured with the BMD assessed at the femoral neck, trochanter, Ward's triangle and total hip (figure 8.14). These femoral region of interest were chosen to capture these different loss patterns. These are also the sites in the hip where most fractures occur (figure 8.14). Ward's triangle measures the earliest site of post-menopausal bone loss in the hip and should in theory give the best measure of trabecular bone in the proximal femur. However, poor precision has limited the utility of this site. Femoral neck BMD has been the hip parameter most frequently used for making the diagnosis of osteoporosis. Other ROIs are intertrochanter and total hip: the latter an area weighted mean of the trochanter, intertrochanter and femoral neck. The international committee for standards in bone measurements recommended that all manufacturers should standardize BMD measurements on a common scale using the total hip ROI [31]. Apart from the total hip ROI, the determination of femur ROI is not yet standardized and significant differences exist between different makes of equipment.

As with the spine measurements the patient lies supine on the imaging table. The leg is slightly abducted and internally rotated using a positioning device in order to bring the femoral neck parallel to the scan table. This avoids foreshortening of the femoral neck, which would cause BMD to increase. It is important to stress that the entire leg should be rotated, and not just the foot or lower leg. Positioning of the femoral neck is therefore critical to maintaining good precision. The effect of varying the angles of rotation and abduction of the leg have been investigated [32, 33]. Errors of up to 6.7 and 3.5% were introduced by careless angles of rotation and abduction respectively. The area scanned should include the entire femoral head, the greater trochanter and the proximal end of the femoral shaft at least 2.5 cm below the lesser trochanter. The ROI are automatically generated by the software and thus are highly dependent on proper scanning to include all the landmarks.

8.6.4. Peripheral sites

Peripheral sites measured include distal forearm, calcaneus and phalanges. These sites are accessible for SXA and DXA. Details of how to perform a DXA/pDXA bone density scan of the distal forearm vary with the make of the instrument. For axial systems designed for spine and femur studies, forearm scanning is performed with the patient sitting on a chair next to the scanner table with the forearm resting on the table top (imaging table), the hand pronated and, on some scanners, secured on the positioning board with a restraining strap (figure 8.11). The scan is then acquired in air and BMD measurements are made in the ultradistal, distal (mid-radius) and shaft (1/3-radius) (figure 8.15).

In dedicated peripheral DXA systems a hand-grip is usually provided. Studies of excised bones show that the percentage of compact and trabecular bone show marked changes at the end of long bones [34]. So scan analysis includes two or more ROI, usually comprising a site with a high percentage of trabecular bone—usually called the ultra-distal (UD)—and a predominantly cortical site—usually the one-third radius (figure 8.14) [35]. The UD region is a rectangular strip 1.5 cm wide adjacent to the endplates of the radius and ulna and the one-third region is a 2 cm wide strip centred over cortical bone at a point one-third the distance between the ulna styloid and the olecranon.

The calcaneus is also a site of interest, because it is load-bearing and has a high content of cancellous bone [36]. There are new pDXA machines such

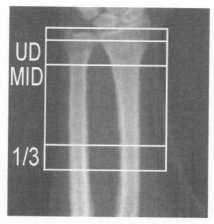

Figure 8.15. *A forearm scan showing the ROI analysed. UD = ultradistal, MID = midshaft and 1/3 = proxomal shaft at one-third the length of the forearm from the distal end.*

as the PIXI (Lunar) dedicated to measuring the calcaneus and forearm. The phalanges, on the other hand, are measured mostly by radiographic absorptiometry techniques. However, special software on DXA devices allows measurements of hand BMD. This is of special interest for the study of rheumatoid arthritis where the hand is a main site of involvement [37].

8.6.5. Total body and body composition

Total body DXA is of interest because it offers a comprehensive view of changes across the whole skeleton and in certain respects supplements calcium balance studies [38] (figure 8.16). For precise results, it is imperative that the patient be placed on the scanning table in a supine position, with all parts of the body including the arms included in the scan field. Total body scans measure BMC and average BMD in the total skeleton together with subregions that include the skull, arms, ribs, thoracic and lumbar spine, pelvis and legs [39]. Total body DXA has interesting application in body composition [40]. In those areas of a whole body scan where the X-ray beam does not intersect bone, it is possible to use the R-value (ratio of the attenuation at the two energies) to estimate separately the masses of fat and lean tissue [41]. Over bone, however, only BMD and total (fat and lean) soft tissue mass can be measured. Extrapolation of measurements of percentage of body fat in soft tissue over adjacent bone means that a whole body DXA scan can provide estimates of total body fat and lean mass as well as BMC [11].

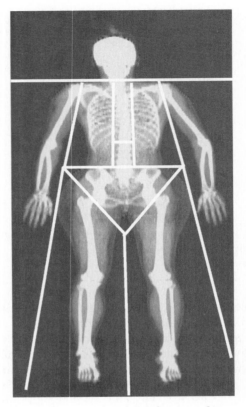

Figure 8.16. *Whole body scan and analysis of regions of interests.*

8.6.6.   *Vertebral morphometry*

Because of the relatively high resolution of fan-beam DXA scanners, anatomical details of the examined region are depicted clearly. Using DXA to obtain lateral images of the lumbar spine allows the scanning beam—in contrast to conventional cone beam radiography—to be generally parallel to the vertebral endplates. This allows a better definition of vertebral dimensions for a morphometric analysis, even though resolution and signal-to-noise ratio are both worse than X-ray. In reference to the DXA approach, this method has been called morphometric X-ray absorptiometry, or MXA [42]. Overlying structures such as ribs or the iliac crest may have an adverse effect on the morphometric analysis. To enhance the accuracy of MXA, technical modifications of the X-ray tube and the detector system may provide images with higher resolution, thereby enhancing the analysis of vertebral deformities. These techniques are still in the developmental and early clinical evaluation phase.

Table 8.3. *Effective doses (typical) for various DXA scan modes (adapted from Blake et al [5]). The effective dose from naturally occurring radiation sources (cosmic, natural radioisotopes, etc.) is included as a comparison.*

Technique	Measurement site	Effective dose (µSv)
pDXA	Forearm	0.05
DXA	Spine	10
	Lateral spine	10
	Femur	2
	Total body	3
QCT	Spine	100
pQCT	Forearm	1
Naturally occurring radiation	Whole body	2400 per year at sea level

8.7.   RADIATION DOSE TO THE PATIENT

Studies of the radiation dose to patients from absorptiometry scans have confirmed that patient exposure is small compared with many other sources of exposure including most radiological investigations involving ionizing radiation (table 8.3) [43–48]. A detailed discussion of radiation doses and radiation safety was made in chapter 3. Studies of radiation dose to patient from DXA confirm that patient dose is small (0.08–4.6 µSv) compared with that given by many other investigations involving ionizing radiation. Fan-beam technology with increased resolution has resulted in increased patient radiation dose (6.7–31 µSv) but this is still relatively small. Carrying vertebral morphometry using DXA also incurs less radiation dose (<60 µSv) than standard lateral radiographs. QCT has a radiation dose (25–360 µSv) comparable with simple radiological examination such as chest X-ray but lower than imaging CT. Radiation doses from other techniques such as RA and SXA are in the same order of magnitude as pencil-beam DXA. For pencil-beam DXA and SXA systems the time average dose to staff from scatter is very low even with the operator sitting as close as 1 m from the patient during measurement. However, the scatter dose from fan-beam DXA systems is considerable higher and approaches limits set by regulator bodies for occupational exposure.

8.8.   SOURCES OF *IN VIVO* MEASUREMENT ERROR

There have been numerous reports on error sources in bone densitometry. Error as used here means either a difference in expected accuracy or precision in the measurement. Detailed discussion of the concepts of errors in physical measurements is presented in chapter 4.

8.8.1. Accuracy

Poor accuracy means a systematic measure of a different value from truth and is often reported as a percentage difference from truth. Absolute truth in bone densitometry is defined as a comparison of the DXA results to a more accurate measure such as ashed bone weight from cadaver bones. Since validation of absolute truth involves destructive testing of cadaver bones, this validation is rarely done. It is sometimes more important to show that a clinical densitometer is accurate, or gives the same values as the device used to create *in vivo* reference data. This would be thought of as clinical accuracy. The accuracy of devices of exactly the same make and model can usually be compared using anthropomorphic phantoms. Blake *et al* [49] showed that Hologic devices in the UK measured the same value on a spine phantom with an accuracy within approximately ±2%. Similar results have been found for Lunar devices (figure 8.17). The accuracy of devices of different makes and models needs to be verified by an *in vivo* study. Comparisons between makes and models are common in the literature. A few examples are given in references [24, 35, 50, 51]. In general, measurement values across manufacturers are neither directly comparable nor interchangeable. Thus, measurements of the same patient on two different devices should not be used to infer a change in bone density. Several attempts have been made to eliminate inter-manufacturer accuracy differences. Standardized DXA bone mineral density values have been reported between Hologic, Lunar and Norland densitometers for the lumbar spine [52], proximal femur [53] and forearm [50]. These standardized values are typically reported as 'standardized BMD' or sBMD in units of mg/cm^2 versus g/cm^2. However, total body BMD, lateral BMD and peripheral sites (fingers, heel, etc.) have not been standardized at present. Tothill [54, 55] found significant differences in bone area, BMC, BMD, %FAT and Total Lean Mass for whole body DXA. At this time, there is no published information comparing the different finger densitometers. Genant *et al* [56] reviewed the accuracy errors for common DXA measurement sites shown in table 8.4.

8.8.2. Precision

Precision errors are differences in the repeat serial measurements on the same individual or object. Precision is typically expressed as either a standard deviation or coefficient of variation in percent. Precision is best quantified in the scan mode of interest using *in vivo* scans of subjects with similar characteristics to the clinical or study population. That is, a paediatric

(a)

(b)

Figure 8.17. *(a) The aluminium spine phantom used by Lunar/GE DXA instruments. (b) Hologic anthropomorphic spine phantom used for daily quality control.*

clinic should quantify precision using *in vivo* paediatric population. This is because the principal limitation to precision is not the intrinsic X-ray noise that is related to the number of dose. The precision is typically limited by the user's ability to reposition the patient in exactly the same position on repeat visits. Guidelines for quantifying precision for DXA devices is given by Glüer [57]. Precision measuring by taking repeat measurements on the same individual with repositioning in between each measure (to simulate repeat visits) is referred to as short-term precision. It has been quantified for virtually all makes and models of densitometers using small populations. For example, Maggio *et al* [58] reported on the short-term reproducibility of femur BMD in the elderly. Long-term precision is defined as the expected reproducibility for measurements taken over extended periods (months to

Table 8.4. *Accuracy and precision for common DXA measurement sites (from Genant et al [56]).*

Measurement	Precision (coefficient of variations in %)	Accuracy (% difference from accurate reference)
PA lumbar spine	1–2	4–10
Lateral lumbar spine	2–3	5–15
Femur	1.5–3	6
Forearm	1	5
Total body	1	3

years.) It differs from short-term precision in that the subject may have had non-bone changes that could affect the accuracy of the bone density measure such as changes in the composition of the surrounding soft tissue [59, 60]. Although long-term precision is the most appropriate to use conceptually when estimating the precision of periodic patient visits, the estimation of long-term precision is difficult since there are always some actual bone changes occurring over time (or at least they cannot be ruled out). However, Patel *et al* [60] found that long-term precision was only slightly worse than is typically found for short-term precision: 1.12% for the total spine, 2.21% for the femoral neck BMD and 1.32% for the total hip BMD. Table 8.4 gives the approximate precision for different scan modes.

8.8.3. Other error sources

In general, errors occur when the assumptions and/or simplifications made by the manufacturer break down in the extremes of the measurement range. Each measure modality (SXA, DXA, QCT, pQCT, RA, etc.) has its own specific technology or technique error sources. Furthermore, errors can occur due to artefacts that are part of the patient and each region of interest has its own common artefacts. We will survey the more common errors for both specific modalities and for specific regions of interests (table 8.5).

1. *Patient positioning*: By far the most common limiting factor to the precision of DXA scans is patient positioning. It is common practice that the repeatability of phantom scans is approximately 0.5% for AP spine while the short-term *in vivo* precision is typically 1–1.5% because of the errors associated with projecting the patient's bones slightly differently.
2. *Regions of interest*: Additional error can be added to the analysis by the regions of interest being placed slightly differently between baseline and follow-up scans.

Table 8.5. *Possible sources of errors on DXA scans.*

Hardware	Malfunction, e.g. 'jagged contours'
	Different or wrong scan mode
Software	Software malfunction
	Change in software
Acquisition	Poor patient positioning
	Movement in the scan field
	Insufficient size of the scan field
	Not enough soft tissue in the scan field
	Artefact in the scan field: metal objects such as belt buckles, coins, truss, corset, underwired bra, buttons or zipper
	Recent ingestion of calcium-containing tablets, which can still be found undissolved in the GI tract
	Air in the scan field
	Scan not properly centred
	Implants (non-removable artefacts such as breast implants, pacemakers etc.)
	Patient changed weight during clinical trial or follow-up visit
	Patient very obese, BMI > 30
Analysis	Wrong placement of ROI
	Correction of bone edges (bone map, editing)
	Wrong labelling (L1 instead of L2)
	Wrong angle of midline (femur scans)
Others	Presence of vertebral crush fracture and generative diseases
	Wrong patient scanned
	Scoliosis

3. *Patient movement*: If there is any patient movement during the scan, the bone projection will change and cause the results to be invalid. Movement usually looks like discontinuities in the bone when you know from visual observation that there are no physical anomalies with the bone (i.e. missing arms, misalignment of long bones, etc.)

4. *Contrasts*: Accuracy errors can also occur in bone density measures based on artefacts in the scan field such as radiographic contrast [61]. In general, contrast will invalidate bone density results and it is advisable to wait approximately 72 h after a contrast injection before measuring bone density.

5. *Artefacts*: Jacobson *et al* [62] reviewed the topic and presented many examples of common artefact found in DXA images. In short, artefacts, especially metal artefacts, can decrease accuracy since the DXA algorithm typically assumes metal to be bone. The major DXA manufacturers all have algorithms to specifically help the user remove metal artefacts from the scan and thus restore accuracy.

6. *Weight change*: Weight change has been shown to affect the accuracy of DXA measurements [54, 63]. Body composition measurement will be significantly affected by weight loss.

8.9.   QUALITY ASSURANCE AND QUALITY CONTROL

8.9.1.   *Quality assurance*

The concepts of quality assurance and control have been discussed extensively in chapter 4. However, there are device-specific procedures to be followed to maintain the device at optimal performance. Each DXA device comes with specific protocols for routine QA and requires the measurement of a phantom. There are currently available both manufacturer-specific and the generic phantoms that can be used to monitor machine stability on a regular basis. These include the Hologic, Norland and Lunar spine phantoms and European spine phantom (ESP) (figure 8.16). Phantoms have also been developed to test body compositions [68]. An ideal phantom should simulate as closely as possible the conditions encountered during routine use. So, for DXA, it should resemble the human body or body parts and is composed of material that approximate the density and attenuation properties of both mineralized and soft tissue. Two of the most widely used phantoms for daily QA will be described.

The Lunar spine phantom is made of aluminum, simulating four vertebrae, increasing in size and density from L1 to L4. The spine phantom is scanned in 15 cm of water to mimic the soft tissue thickness. The Lunar spine phantom has some limitations. First, it is of limited anthropomorphic design. Second, the phantom is formed from a flat piece of aluminium and each vertebra has a uniform density with sharp edges. Thus, it does not provide a good test of the edge-detection algorithm. However, the density of the vertebrae does increase from L1 to L4 to allow for a test of system linearity. Lunar has since developed an equivalent phantom made of hydroxyapatite and encased in a solid block of tissue-equivalent plastic.

The Hologic phantom is composed of four hydroxyapatite vertebrae embedded in a tissue that mimics an epoxy-resin block 17.5 cm thick. The vertebrae are of homogeneous density but closely resemble true vertebrae in size and shape. This phantom also has its limitations. The vertebrae show none of the heterogeneity of the real spine. As a consequence, the edge-detection algorithms are still not adequately tested. Also the vertebrae are all of similar density; thus the phantom does not allow testing of system linearity.

8.9.2.   *Cross calibration*

It is important that a clinician requesting DXA studies is aware that results for the same patient measured on different manufacturers' equipment give different BMD results. These differences for DXA arise from differences in scanner design, edge-detection algorithms and calibration. In many situations, the differences among different DXA devices are not important. For example, in a clinic with only one machine on which all patients are measured, the relative differences between manufacturers are not important. As long as the scanner used is stable, the analysis technique is consistent, and patients are not scanned on any other systems, any observed changes in BMD over time can be assumed to be real and not due to machine artefact. However, investigators often do not have the luxury of using only a single machine for the duration of a study: many older DXA devices will be replaced in the coming years, and a hospital or clinic may choose to switch to another manufacturer's system. In either case, the continued monitoring of bone status in patients would be complicated by this change in equipment. In clinical drug trials, multiple centres are typically used in order to enrol the number of patients needed to establish the efficacy of a particular pharmaceutical. More often than not, the centres enrolled will be equipped with a variety of different scanners. Direct comparisons of centres with different equipment cannot be made without performing cross calibration measurements.

To resolve the problems caused by these disagreements, the International DXA Standardization Committee (IDSC) was established. A study was carried out on 100 healthy women evenly distributed over the age range 20–80 years on whom PA spine and hip scans were measured [52]. Standardized PA spine (L2–L4) BMD results (denoted sBMD and expressed in units of mg/cm^2) were derived from the manufacturers' existing BMD figures (expressed in units of g/cm^2) using the following conversion equations [52, 64, 65].

Hologic: $sBMD = 1000(BMD_{hologic} \times 1.0755)$

Lunar: $sBMD = 1000(BMD_{lunar} \times 0.9522)$

Norland: $sBMD = 1000(BMD_{norland} \times 1.0761)$.

In general, lumbar spine sBMD values obtained by scanning a patient on any one of the three manufacturers' systems are expected to agree to within 2–5% [65].

The ICSBM has recently issued guidelines for the standardization of femur BMD [31]. However, the femur equations are different from the spine due to different effects of edge detection and the adoption of a more sophisticated statistical approach to comparing values from different systems [53]. Femur standardization is based on the total hip ROI rather than the femoral neck ROI. The equations are as follows:

Hologic: $sBMD = 1000(1.008 \times BMD_{hologic} + 0.006)$

Lunar: $sBMD = 1000(0.979 \times BMD_{lunar} - 0.031)$

Norland: $sBMD = 1000(1.012 \times BMD_{norland} + 0.026)$.

The ICSBM report on the standardization of the femur BMD includes a recommendation for the use of standardized reference data, thereby making T and Z score figures derived from different manufacturers' equipment compatible.

A similar study evaluating the correlation among the various peripheral devices has recently been reported [50].

REFERENCES

[1] Johns H E and Cunningham J R 1983 *The Physics of Radiology* 4th ed (Springfield, IL: Charles C Thomas)
[2] Haliday D and Resnick R 1981 *Fundamentals of Physics* 2nd edn (New York: Wiley)
[3] Meredith W J and Massey J B 1972 *Fundamental Physics of Radiology* 2nd edn (Baltimore: Williams and Wilkins)
[4] Hobbie R K 1997 *Intermediate Physics for Medicine and Biology* 3rd edn (New York: Springer)
[5] Blake G M, Wahner H W and Fogelman I 1999 *The Evaluation of Osteoporosis: Dual Energy X-ray Absorptiometry and Ultrasound in Clinical Practice* 2nd ed (London: Martin Dunitz)
[6] Cameron J R and Sorenson J A 1963 Measurement of bone mineral in vivo: an improved method *Science* **142** 230–2
[7] Adams J E 1997 Single and dual energy X-ray absorptiometry *Eur. Radiol.* **7** Suppl 2 S20–31
[8] Kelly T L, Crane G and Baran D T 1994 Single X-ray absorptiometry of the forearm: precision, correlation, and reference data *Calcif. Tissue Intl.* **54**(3) 212–8
[9] Cullum I D, Ell P J and Ryder J P 1989 X-ray dual photon absorptiometry: a new method for measurement of bone density *Br. J. Radiol.* **62** 587–92
[10] Kelly T, Slovick D, Schoenfield D and Neer R 1988 Quantitative digital radiography versus dual photon absorptiometry of the lumbar spine *J. Clin. Endocr. Metab.* **67** 839–44
[11] Blake G M and Fogelman I 1997 Technical principles of dual energy X-ray absorptiometry *Semin. Nucl. Med.* **27**(3) 210–28
[12] Blake G M, McKeeney D B, Chhaya S C, Ryan P J and Fogelman I 1992 Dual energy X-ray absorptiometry: the effects of beam hardening on bone density measurements *Med. Phys.* **19**(2) 459–65
[13] Yamada M, Ito M, Hayashi K, Ohki M and Nakamura T 1994 Dual energy X-ray absorptiometry of the calcaneus: Comparison with other techniques to assess bone density and value in predicting risk of spine fracture *Am. J. Roentgenol.* **163**(12) 1435–40
[14] Faulkner K G, McClung M R, Schmeer M S, Roberts L A and Gaither K W 1994 Densitometry of the radius using single and dual energy absorptiometry *Calcif. Tissue Intl.* **54**(3) 208–11
[15] Miller P D, Bonnick S L, Rosen C J, Altman R D, Avioli L V, Dequeker J *et al* 1996 Clinical utility of bone mass measurements in adults: consensus of an international panel *The Society for Clinical Densitometry Semin. Arthritis Rheum.* **25**(6) 361–72
[16] Ryan P J, Evans P, Blake G M and Fogeman I 1992 The effect of vertebral collapse on spinal bone mineral density measurements in osteoporosis *Bone Miner.* **18**(3) 267–72
[17] Franck H, Munz M and Scherrer M 1995 Evaluation of dual-energy X-ray absorptiometry bone mineral measurement–comparison of a single-beam and fan-beam design: the effect of osteophytic calcification on spine bone mineral density *Calcif. Tissue Intl.* **56**(3) 192–5

[18] Reid IR, Evans M C, Ames R and Wattie D J 1991 The influence of osteophytes and aortic calcification on spinal mineral density in postmenopausal women *J. Clin. Endocrin. Metab.* **72**(6) 1372–4

[19] Drinka P J, DeSmet A A, Bauwens S F and Rogot A 1992 The effect of overlying calcification on lumbar bone densitometry *Calcif. Tissue Intl.* **50**(6) 507–10

[20] Ito M, Hayashi K, Yamada M, Uetani M and Nakamura T 1993 Relationship of osteophytes to bone mineral density and spinal fracture in men *Radiology* **189**(2) 497–502

[21] Yu W, Gluer C C, Fuerst T, Grampp S, Li J, Lu Y *et al* 1995 Influence of degenerative joint disease on spinal bone mineral measurements in postmenopausal women *Calcif. Tissue Intl.* **57**(3) 169–74

[22] Larnach T A, Boyd S J, Smart R C, Butler S P, Rohl P G and Diamond T H 1992 Reproducibility of lateral spine scans using dual energy X-ray absorptiometry *Calcif. Tissue Intl.* **51**(4) 255–8

[23] Jergas M, Breitenseher M, Gluer C C, Black D, Lang P, Grampp S *et al* 1995 Which vertebrae should be assessed using lateral dual-energy X-ray absorptiometry of the lumbar spine *Osteoporos. Intl.* **5**(3) 196–204

[24] Blake G M, Herd R J and Fogelman I 1996 A longitudinal study of supine lateral DXA of the lumbar spine: a comparison with posteroanterior spine, hip and total-body DXA *Osteoporos. Intl.* **6**(6) 462–70

[25] Uebelhart D, Duboeuf F, Meunier P J and Delmas P D 1990 Lateral dual-photon absorptiometry: a new technique to measure the bone mineral density at the lumbar spine *J. Bone Min. Res.* **5**(5) 525–31

[26] Slosman D O, Rizzoli R, Donath A and Bonjour J P 1990 Vertebral bone mineral density measured laterally by dual-energy X-ray absorptiometry *Osteoporos. Intl.* **1**(1) 23–9

[27] Yu W, Gluer C C, Grampp S, Jergas M, Fuerst T, Wu C Y *et al* 1995 Spinal bone mineral assessment in postmenopausal women: a comparison between dual X-ray absorptiometry and quantitative computed tomography *Osteoporos. Intl.* **5**(6) 433–9

[28] Rupich R C, Griffin M G, Pacifici R, Avioli L V and Susman N 1992 Lateral dual-energy radiography: artifact error from rib and pelvic bone *J. Bone Min. Res.* **7**(1) 97–101

[29] Singh M, Riggs B L, Beabout J W and Jowsey J 1973 Femoral trabecular pattern index for evaluation of spinal osteoporosis A detailed methodologic description *Mayo Clin. Proc.* **48**(3) 184–9

[30] Wahner H W 1996 Use of densitometry in management of osteoporosis. In *Osteoporosis* eds R Marcus, D Feldman and J Kelsey (New York: Academic Press) pp 1055–74

[31] Hanson J 1997 Standardization of femur BMD *J. Bone Min. Res.* **12**(8) 1316–17

[32] Goh J C, Low S L and Bose K 1995 Effect of femoral rotation on bone mineral density measurements with dual energy X-ray absorptiometry *Calcif. Tissue Intl.* **57**(5) 340–3

[33] Wilson C R, Fogelman I, Blake G M and Rodin A 1991 The effect of positioning on dual energy X-ray bone densitometry of the proximal femur *Bone Miner.* **13**(1) 69–76

[34] Schlenker R A and VonSeggen W W 1976 The distribution of cortical and trabecular bone mass along the lengths of the radius and ulna and the implications for *in vivo* bone mass measurements *Calcif. Tissue Res.* **20**(1) 41–52

[35] Heilmann P, Wuster C, Prolingheuer C, Gotz M and Ziegler R 1998 Measurement of forearm bone mineral density: comparison of precision of five different instruments *Calcif. Tissue Intl.* **62**(5) 383–7

[36] Vogel J M, Wasnich R D and Ross P D 1988 The clinical relevance of calcaneus bone mineral measurements: a review *Bone Miner.* **5**(1) 35–58

[37] Peel N F, Spittlehouse A J, Bax D E and Eastell R 1994 Bone mineral density of the hand in rheumatoid arthritis *Arthritis Rheum.* **37**(7) 983–91

[38] Kelly T L, Berger N and Richardson T L 1998 DXA body composition: theory and practice *Appl. Radiat. Isot.* **49**(5–6) 511–3

[39] Herd R J, Blake G M, Parker J C, Ryan P J and Fogelman I 1993 Total body studies in normal British women using dual energy X-ray absorptiometry *Br. J. Radiol.* **66**(784) 303–8

[40] Laskey M A 1996 Dual-energy X-ray absorptiometry and body composition *Nutrition* **12**(1) 45–51

[41] Pietrobelli A, Formica C, Wang Z and Heymsfield S B 1996 Dual-energy X-ray absorptiometry body composition model: review of physical concepts *Am. J. Physiol.* **271**(6 Pt 1) E941–51

[42] Steiger P, Cummings S R, Genant H K and Weiss H 1994 Morphometric X-ray absorptiometry of the spine: correlation *in vivo* with morphometric radiography. Study of Osteoporotic Fractures Research Group [see comments] *Osteoporos. Intl.* **4**(5) 238–44

[43] Kalender W A 1992 Effective dose values in bone mineral measurements by photon absorptiometry and computed tomography *Osteoporos. Intl.* **2**(2) 82–7

[44] Lewis M K and Blake G M 1995 Patient dose in morphometric X-ray absorptiometry [letter comment] *Osteoporos. Intl.* **5**(4) 281–2

[45] Lewis M K, Blake G M and Fogelman I 1994 Patient dose in dual x-ray absorptiometry *Osteoporos. Intl.* **4**(1) 11–5

[46] Njeh C F, Apple K, Temperton D H and Boivin C M 1996 Radiological assessment of a new bone densitometer—the Lunar EXPERT *Br. J. Radiol.* **69**(820) 335–40

[47] Njeh C F, Fuerst T, Hans D, Blake G M and Genant H K 1999 Radiation exposure in bone mineral density assessment *Appl. Radiat. Isotopes* **50**(1) 215–36

[48] Huda W and Morin R L 1996 Patient doses in bone mineral densitometry *Br. J. Radiol.* **69**(821) 422–5

[49] Blake G M 1996 Replacing DXA scanners: cross-calibration with phantoms may be misleading *Calcif. Tissue Intl.* **59**(1) 1–5

[50] Shepherd J A, Cheng X G, Lu Y, Njeh C, Toschke J, Engelke K *et al* 2002 Universal standardization of forearm bone densitometry *J. Bone Min. Res.* **17**(4) 734–45

[51] Grampp S, Genant H K, Mathur A, Lang P, Jergas M, Takada M *et al* 1997 Comparisons of noninvasive bone mineral measurements in assessing age-related loss, fracture discrimination, and diagnostic classification *J. Bone Min. Res.* **12**(5) 697–711

[52] Genant H K, Grampp S, Gluer C C, Faulkner K G, Jergas M, Engelke K *et al* 1994 Universal standardization for dual x-ray absorptiometry: patient and phantom cross-calibration results [see comments] *J. Bone Min. Res.* **9**(10) 1503–14

[53] Hui S L, Gao S, Zhou X H, Johnston C C Jr, Lu Y, Gluer C C *et al* 1997 Universal standardization of bone density measurements: a method with optimal properties for calibration among several instruments *J. Bone Min. Res.* **12**(9) 1463–70

[54] Tothill P, Hannan W J, Cowen S and Freeman C P 1997 Anomalies in the measurement of changes in total-body bone mineral by dual-energy X-ray absorptiometry during weight change *J. Bone Min. Res.* **12**(11) 1908–21

[55] Tothill P, Avenell A and Reid D M 1994 Precision and accuracy of measurements of whole-body bone mineral: comparisons between Hologic, Lunar and Norland dual-energy X-ray absorptiometers *Br. J. Radiol.* **67**(804) 1210–7

[56] Genant H K, Engelke K, Fuerst T, Glüer C C, Grampp S, Harris S T *et al* 1996 Noninvasive assessment of bone mineral and structure: state of the art *J. Bone Min. Res.* **11**(6) 707–30

[57] Glüer C C, Blake G, Lu Y, Blunt B A, Jergas M and Genant H K 1995 Accurate assessment of precision errors: how to measure the reproducibility of bone densitometry techniques *Osteoporos. Intl.* **5**(4) 262–70

[58] Maggio D, McCloskey E V, Camilli L, Cenci S, Cherubini A, Kanis J A *et al* 1998 Short-term reproducibility of proximal femur bone mineral density in the elderly *Calcif. Tissue Intl.* **63**(4) 296–9

[59] Blake G M, Herd R J, Patel R and Fogelman I 2000 The effect of weight change on total body dual-energy X-ray absorptiometry: results from a clinical trial *Osteoporos. Intl.* **11**(10) 832–9

[60] Patel R, Blake G M, Herd R J and Fogelman I 1997 The effect of weight change on DXA scans in a 2-year trial of etidronate therapy *Calcif. Tissue Intl.* **61**(5) 393–9

[61] Andrich M P, Cawley M and Chen C C 1996 Artifacts caused by nonionic contrast media and a portacath on a dual-energy x-ray absorptiometry whole-body composition study *Clin. Nucl. Med.* **21**(5) 407–8

[62] Jacobson J A, Jamadar D A and Hayes C W 2000 Dual X-ray absorptiometry: recognizing image artifacts and pathology *Am. J. Roentgenol.* **174**(6) 1699–705

[63] Tothill P and Avenell A 1998 Anomalies in the measurement of changes in bone mineral density of the spine by dual-energy X-ray absorptiometry *Calcif. Tissue Intl.* **63**(2) 126–33

[64] Steiger P 1995 Standardization of measurements for assessing BMD by DXA [letter] *Calcif. Tissue Intl.* **57**(6) 469

[65] Genant H K 1995 Universal standardization for dual X-ray absorptiometry: patient and phantom cross-calibration results [letter] *J. Bone Min. Res.* **10**(6) 997–8

[66] International Commission on Radiation Units and Measurements 1989 *Tissue Substitutes in Radiation Dosimetry and Measurements* (Bethesda, MD: ICRU) Report No 44

[67] Boone J M and Seibert J A 1997 An accurate method for computer-generating tungsten anode X-ray spectra from 30 to 140 kV *Med. Phys.* **24**(11) 1661–70

[68] Diessel E, Fuerst T, Njeh C F, Tylavsky F, Cauley J, Dockrell M and Genant H K 2000 Evaluation of a new body composition phantom for quality control and cross-calibration of DXA devices *J. Appl. Physiol.* **89**(2) 599–605

Chapter 9

Quantitative computed tomography

Thomas F Lang

9.1. INTRODUCTION

Computed tomography (CT) scanners image the distribution of linear attenuation coefficents in a thin axial slice of tissue. As shown in figure 9.1, the patient cross-section is contained within a fan of X-rays defined between the edges of the detector array and an X-ray point source. The log-attenuation

$$\log A = \ln\left(\frac{I_0}{I}\right)$$

where I_0 and I are the radiation intensities incident on the patient and detector respectively, is measured along each line, or 'ray-path', defined between the X-ray point source and each element of the detector array. The beam is variably collimated so that the thickness of the fan along the length of the patient ranges from 10 mm down to 1 mm depending on the type of acquisition. An axial cross-sectional image is acquired as the X-ray tube and detector rotate around the patient. The log attenuation values for each ray path are continuously measured by a computer, and a mathematical process called 'back-projection' is employed to reconstruct the CT cross-section on to a rectilinear array of picture elements. Because the fan-beam has a finite thickness along the length of the patient, the picture elements are really volume elements, or 'voxels'. Typical image arrays are square matrices of 512×512 elements of 0.2–0.9 mm dimensions in-plane, with the thickness of the voxels ranging from 10 mm down to 1 mm. CT images of large volumes of tissue may be acquired by imaging a cross-section and moving the patient table through the CT scanner between images, generating a 'stack' of cross-sectional images. In older models of CT scanners, the patient table stepped in discrete increments, and a 360° rotation of the source/detector was performed at each position. In

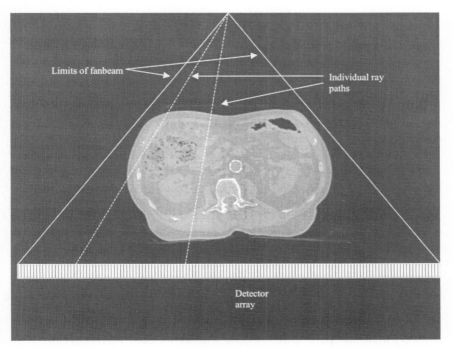

Figure 9.1. *CT scanner geometry superimposed on axial image through torso.*

the newer model of helical CT scanners, the table and source/detector system move continuously, resulting in significant reductions in image acquisition time [1, 2].

Because the absolute values of the linear attenuation coefficients measured by the CT scanner are functions of the effective X-ray energy (which varies between CT scanner models and different kVp settings of the same scanner), a simple scale, known as the Hounsfield scale, is used to standardize them. The grey-scale value of each voxel is represented as a Hounsfield unit (HU), given by

$$\mathrm{HU_T} \equiv \left(\frac{\mu_\mathrm{T} - \mu_\mathrm{w}}{\mu_\mathrm{w}}\right) \times 1000$$

where $\mathrm{HU_T}$ is the HU of a volume element of tissue and μ_T and μ_w are the linear attenuation coefficients of the tissue and of water, respectively. The HU scale is a linear scale in which air has a value of -1000, water 0, muscle 30, with bone typically ranging from 300 to 3000 units.

The value of the HU for a given tissue type depends on the composition of the voxel volume. If the sizes of the structures in the tissue are smaller than the dimensions of the voxel, the HU value is subject to partial volume averaging, in which the HU value is the average HU of the constituent tissues of the voxel, weighted by their volume fractions. For example, a 0.78 mm × 0.78 mm × 10 mm voxel of trabecular bone is a mixture of bone, collagen, cellular marrow and fatty marrow, and the HU is the volume-weighted average of these four constituents. Also, as a result of the poly-chromatic X-ray energy spectrum, the values of the HU are affected by beam hardening. In a CT image, the result of this is that, for the same tissue, the HU can vary as a function of location within the body, and the presence and shape of bony structures surrounding that tissue. Although manufacturers of CT equipment have implemented beam-hardening correc-tions, the efficacy of these corrections varies between manufacturers and between technical settings on different machines.

CT is a highly useful modality for bone imaging because its high attenuation coefficient allows for excellent contrast from the surrounding soft tissue. Thus it is possible to segment the bone from the surrounding soft tissue for highly accurate and reliable measures of bone cross-sectional area and, using high resolution techniques, to extract measures of the micro-architecture of the trabecular bone. Moreover, because of the highly linear relationship between the attenuation coefficient and the density of bone mineral, HU values, when properly calibrated, can be used to estimate the apparent volumetric density of bone mineral. In this chapter, we will discuss use of CT imaging to measure volumetric bone mineral density and bone geometry in the axial and peripheral skeleton. We will also discuss the use of high-resolution CT imaging to assess the micro-architecture of spinal trabecular bone.

9.2. SINGLE-SLICE SPINAL BONE MINERAL DENSITY MEASUREMENT

Quantitative computed tomography (QCT) is a long-standing method for measuring BMD in the metabolically-active trabecular bone in the vertebral bodies [3–6]. A lateral projection scan is first utilized to localize the lumbar vertebral bodies and then scans of 8–10 mm thickness are acquired through 2–4 contiguous lumbar vertebral levels (figure 9.2). A bone mineral calibra-tion standard is placed under the lower back of the patient and is used to convert the native HU scale of the CT image to units of bone mineral density. The CT image is then processed using a software program which analyses the calibration phantom to convert HU to BMD and then places a region of interest in the trabecular bone of the vertebral body [6–8]. The program then calculates the mean BMD of the vertebral region of interest. This is either presented as an average of two or four vertebral levels. The radiation dose for this procedure has been reported as 80 μSv, a value which includes the dose of the lateral localizer scan [1].

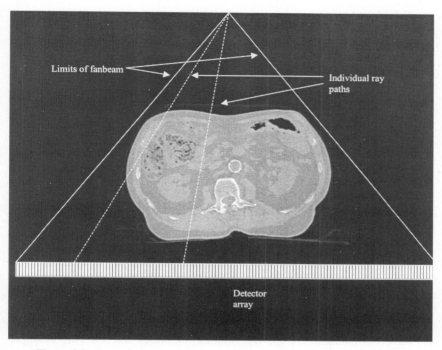

Figure 9.2. *Lateral scout view used to delineate lumbar spine anatomy for vertebral slice selection. Mid-planes of T12–L3 are superimposed in yellow on the image.*

9.3. PHYSICAL SIGNIFICANCE OF QCT MEASUREMENTS

CT BMD assessment is based on quantitative analysis of the HU in volumes of bone tissue. Typically, the BMD is quantified using a bone mineral reference phantom which is scanned simultaneously with the patient. In order to minimize the impact of beam hardening, the calibration phantom is placed as close as possible to the vertebrae, and is normally located under the lumbar spine of the patient. The calibration standard originally developed by Cann and Genant [4] at UCSF, and which is currently marketed by Mindways (South San Francisco, CA, USA), consists of an acrylic wedge containing cylinders of solutions with varying concentrations (200, 100, 50 and 0 mg/cm^3) of dipotassium hydrogen phosphate in water. An additional cylinder contains alcohol as a reference material for fat. A solid calcium hydroxyapatite-based calibration standard was later developed by Image Analysis (Columbia, KY, USA) and by Siemens Medical Systems (Erlangen, Germany). The Image Analysis standard consists of rods with varying concentrations (200, 100 and 50 mg/cm^3) of calcium hydroxyapatite

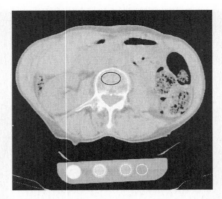

Figure 9.3. *Axial QCT image showing regions of interest placed in calibration phantom and anterior mid vertebral body.*

mixed in a water-equivalent solid resin matrix [9]. During the analysis of the QCT image, regions of interest are placed (figure 9.3) in each of the calibration objects, and linear regression analysis is used to determine a relationship between the mean HU measured in each region and the known concentrations of bone-equivalent material. This calibration relationship is then used to convert the mean HU in the patient region of interest (for example, vertebra or proximal femur) into a concentration (reported in mg/cm^3, i.e the mass of bone per unit tissue volume) of bone equivalent material in the region of interest. Unlike areal bone mineral density, the QCT density measurement is independent of bone size, and thus is a more robust measure for comparisons of bone density between populations and potentially for growing children as well.

The major source of error in the QCT bone measurement is the phenomenon of partial volume averaging. Because the voxel dimensions in QCT measurements (0.8–1.0 mm in the imaging plane, 3–10 mm slice thicknesses) are larger than the dimensions and spacing of trabeculae, a QCT voxel includes both bone and marrow constituents. Thus a QCT measurement is the mass of bone in a volume containing bone, red marrow and marrow fat. A single-energy QCT measurement is capable of determining the mass of bone in a volume consisting of two components (e.g. bone and red marrow), but not in a three-component system [10]. Resolving the mass fractions of bone, red marrow and marrow fat in the QCT voxel requires a dual-energy QCT measurement [11, 12]. Because fat has a HU value of −200, compared with 30 HU for red marrow and 300–3000 HU for bone, the presence of fat in the QCT volume depresses the HU measurement. Thus, the presence of marrow fat causes single-energy QCT to underestimate the mass of bone per unit tissue volume, an error which can be corrected using dual-energy acquisitions. The effect of marrow fat on QCT measurements is larger at the spine than at the hip or peripheral skeletal sites. Whereas the conversion from red to fatty marrow tends to finish by the mid-20s in the hip and peripheral skeleton, the vertebrae show a gradual age-related increase in the proportion of fat in the bone marrow which starts in youth and continues through old age [13]. The inclusion of fatty marrow in the vertebral BMD measurement results in accuracy errors ranging from 5 to 15% depending on the age group. However, because the increase in marrow fat is age-related, single-energy CT data can be corrected using age-related reference databases, and the residual error is not considered to be clinically relevant. Provided that the QCT scan is acquired at low effective energies (i.e. 80–90 kVp), the population SD in marrow fat accounts for roughly 5 mg/cm^3 of the 25–30 mg/cm^3 population SD in spinal trabecular BMD. This residual error is not considered large enough to merit clinical use of dual-energy techniques, which are more accurate, but which have larger radiation doses and precision errors [14].

Precision errors ranging from 1 to 2% have been reported for spinal QCT [7, 8]. This precision error is attributable to several sources including the reproducibility of slice and region of interest positioning as well as scanner instabilities [15]. Simultaneous calibration corrects to some extent for scanner instabilities, as well as for variable beam hardening depending on patient size and shape. Using a simultaneous calibration technique, the long-term CV of a well-maintained CT scanner should be close to 1%. The effect of variable patient positioning can be minimized by careful review of lateral localizer scans to ensure consistent slice placement and gantry angulations. Computer programs that place the region of interest automatically or semi-automatically may also be used to reduce precision errors.

9.4. MEASUREMENT OF BMD USING VOLUMETRIC CT IMAGES OF THE SPINE AND HIP

Unlike relatively simple skeletal structures such as the spine and distal radius, the hip is a difficult site for single-slice measurements because minor variations in slice positioning can result in large variations in BMD due to increased operator dependence. A volumetric acquisition encompassing the entire hip can address this issue in that the volume to be quantified can be determined in software. Three-dimensional acquisitions and analyses have become feasible with the advent of fast helical CT scanners and inexpensive computer workstations which allow for the rapid processing of large data sets. Procedures have been developed for three-dimensional scanning of the spine and hip. In typical spine and hip protocols, the L1/L2 vertebrae and proximal femora (from superior aspect of femoral head to inferior aspect of lesser trochanter) are scanned with contiguous 3 mm slices. Using a radiation dose calculation developed by Kalender *et al* [2], the radiation

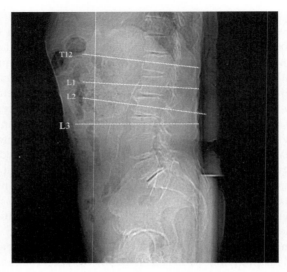

Figure 9.4. *Total femur trabecular region of interest for hip QCT superimposed in white on CT volume reformatted along femoral neck axis. Vertical red lines on left-most and middle images indicates direction of femoral neck axis.*

doses for these procedures have been estimated at 350 and 1200 μSv for spine and hip protocols respectively.

The primary advantage of helical-CT densitometry is for quantification of the proximal femur [16, 17]. An algorithm developed by Lang *et al* [16] processes volumetric CT images of the proximal femur to measure bone mineral density in the femoral neck, the total femur, and the trochanteric regions (figure 9.4). Bone mineral density, bone mineral content and volume are computed for trabecular, cortical and integral bone within each of these regions. In addition, the algorithm quantifies geometric properties of the proximal femur such as femoral neck, cross-sectional area and axis length. For trabecular BMD measurements, the precision of this method *in vivo* was found to range from 0.6 to 1.1% depending on the volume of interest assessed. The ability to measure trabecular and cortical bone may have interest both in terms of understanding the aetiology of hip fractures as well as the effect of therapeutic interventions. Several studies have reported that trabecular bone may be important in the aetiology of trochanteric as opposed to femoral neck fractures [18–20]. In a study *in vitro* examining the effect of bone mineral density and geometry on proximal femoral fracture load, Lang *et al* [16] found that volumetric trochanteric trabecular BMD was highly correlated with fracture load when the femora were loaded simulating a fall to the side. On the other hand, volumetric bone mineral density was only moderately correlated ($r^2 = 0.45$) with failure load in femora loaded in a single-legged stance condition. However, femoral neck BMD, cross-sectional area and axis length together explained 93% of the variation in failure load [16]. The ability to examine both trabecular and cortical bone is an advantage for QCT of the hip, particularly in the case of therapies such as parathyroid hormone (PTH). In a QCT study of patients undergoing PTH therapy, Cann *et al* [21] found that cortical bone maintained its density but significantly increased its mass and volume. Another interesting application of hip QCT is patient-specific finite element modelling (FEM), in which the QCT image provides a mechanical model of the proximal femur based on the bone geometry and spatial distribution of material properties, which are calculated by converting the CT image from bone density units to elastic moduli (and other material properties) using relationships measured *in vitro* [22–24]. When subjected to specific loading conditions simulating a normal stance or a fall to the side, the FEM can be used to estimate the distribution of stress and strain in the femur, the ultimate strength and the region of predicted failure, among other variables of interest. In a cross-sectional study of patients with hip fracture and normal controls, Cody *et al* [25] found that this technique provided better discrimination of hip fracture cases than bone density measured by either DXA or QCT.

The primary advantage of volumetric QCT of the spine is improved reliability in measurement of trabecular BMD. Because the three-dimensional image can be corrected for rotation of the vertebra in the coronal and sagittal planes, it is possible to measure with greater reliability in patients with scoliosis or lordotic curvature. The precision is also better because the volumetric analysis can also correct for operator error in slice placement and angulation. There is no evidence that use of volumetric spine data for BMD quantification improves discriminatory capability over single-slice QCT. An example of a three-dimensional QCT approach is that of Lang *et al* [8], in which an image of the entire vertebral body is acquired and anatomic landmarks such as the vertebral endplates and the spinous process are used to fix the three-dimensional orientation of the vertebral body, allowing for definition of new trabecular and integral regions which contain most of the bone in the vertebral centrum, as shown in figure 9.5. Although measuring a larger volume of tissue may enhance precision, these new regions are highly correlated with the mid-vertebral sub-regions assessed with standard QCT techniques and may not contain significant new information about vertebral strength. Consequently, volumetric studies of regional BMD, which examine specific sub-regions of the centrum that may vary in their contribution to vertebral strength [26–28], and studies of the cortical shell [29], the condition of which may be important for vertebral strength in osteoporotic individuals, are of interest for future investigation.

The two principal QCT manufacturers in the United States are Image Analysis (Columbia, KY) and Mindways (San Francisco, CA). In the United States and Europe, Siemens provides bone mineral measurement as an upgrade to their CT scanners. Mindways offers a single-slice spinal

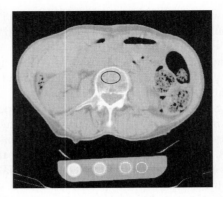

Figure 9.5. *Trabecular and integral vertebral body regions of interest super-imposed in white on vertebral body scan which has been reformatted so that slicing is perpendicular to endplates. The vertical red lines in the top and middle images (sagittal and coronal views respectively) indicate the slicing axis.*

QCT product for the spine as well as volumetric CT products for both the hip and spine. The Mindways volumetric QCT product, known as QCTPRO^{TM}, reconstructs the mid 10 mm slice through the vertebra based on operator delineation of the position and coronal and sagittal angle of the vertebra. Using the corrected slice, the user may then place a region of interest in the trabecular bone of the anterior vertebral body. The volumetric hip package offered by Mindways, known as CTXA^{TM}, basically simulates an AP DXA measurement by projecting the CT volume into the coronal plane. The calibration technology used by Mindways is the original UCSF liquid calibration phantom design described previously. The QCT manufacturer Image Analysis offers a single-slice QCT module in which the calibration and region of interest placement is performed automatically. Image Analysis is planning to offer a volumetric product in which a slice corrected for sagittal and coronal rotation is used as input to the automated single slice analysis. This product is in the final stages of beta testing. For calibration technology, Image Analysis offers bone mineral reference standards consisting of varying concentrations of calcium hydroxyapatite embedded in a water equivalent resin.

REFERENCES

[1] Kalender W A 1992 Effective dose values in bone mineral measurements by photon absorptiometry and computed tomography *Osteoporos. Intl.* **2** 82–7
[2] Kalender W A, Schmidt B, Zankl M and Schmidt M 1999 A PC program for estimating organ dose and effective dose values in computed tomography *Eur. Radiol.* **9**(3) 555–62

[3] Cann C E 1981 Low-dose CT scanning for quantitative spinal mineral analysis *Radiology* **140** 813–5

[4] Cann C E and Genant H K 1980 Precise measurement of vertebral mineral content using computed tomography *J. Comput. Assist. Tomogr.* **4** 493–500

[5] Genant H K, Cann C E, Pozzi-Mucelli R S and Kanter A S 1983 Vertebral mineral determination by quantitative computed tomography: clinical feasibility and normative data *J. Comput. Assist. Tomogr.* **7** 554

[6] Kalender W A, Klotz E and Süss C 1987 Vertebral bone mineral analysis: an integrated approach *Radiology* **164** 419–23

[7] Steiger P, Block J E, Steiger S, Heuck A, Friedlander A, Ettinger B, Harris S T, Glüer C C and Genant H K 1990 Spinal bone mineral density by quantitative computed tomography: effect of region of interest, vertebral level, and technique *Radiology* **175** 537–43

[8] Lang T F, Li J, Harris S T and Genant H K 1999 Assessment of vertebral bone mineral density using volumetric quantitative CT *J. Comput. Assist. Tomogr.* **23**(1) 130–7

[9] Faulkner K G, Glüer C C, Grampp S and Genant H K 1993 Cross calibration of liquid and solid QCT calibration standards: corrections to the UCSF normative data *Osteoporos. Intl.* **3** 36–42

[10] Goodsitt M, Hoover P, MS V and Hsueh S 1994 The composition of bone marrow for a dual-energy quantitative computed tomography technique: a cadaver and computer simulation study *Invest. Radiology* **29** 695–704

[11] Goodsitt M M, Rosenthal D I, Reinus W R and Coumas J 1987 Two postprocessing CT techniques for determining the composition of trabecular bone *Invest. Radiol.* **22** 209–15

[12] Glüer C C, Reiser U J, Davis C A, Rutt B K and Genant H K 1988 Vertebral mineral determination by quantitative computed tomography (QCT): Accuracy of single and dual energy measurements *J. Comput. Assist. Tomogr.* **12**(2) 242–58

[13] Dunnill M, Anderson J and Whitehead R 1967 Quantitative histological studies on age changes in bone *J. Pathol. Bacteriol.* **94** 275–91

[14] Glüer C C and Genant H K 1989 Impact of marrow fat on accuracy of quantitative CT *J. Comput. Assist. Tomogr.* **13**(6) 1023–35

[15] Laval-Jeantet A M, Genant H K, Wu C , Glüer C C, Faulkner K and Steiger P 1993 Factors influencing long-term *in vivo* reproducibility of QCT (vertebral densitometry) *JCAT* **17**(6) 915–21

[16] Lang T F, Keyak J H, Heitz M W, Augat P, Lu Y, Mathur A and Genant H K 1997 Volumetric quantitative computed tomography of the proximal femur: precision and relation to bone strength *Bone* **21**(1) 101–8

[17] Heitz 1995 Bestimmung von Dichte und Stabilitaet des Menschlichen Oberschenkelknochens auf der Basis von CT-Daten Physics (Tuebingen: University of Tuebingen) p 130

[18] Lang T F, Augat P, Lane N E and Genant H K 1998 Trochanteric hip fracture: strong association with spinal trabecular bone mineral density measured with quantitative CT *Radiology* **209**(2) 525–30

[19] Mautalen C, Vega E and Einhorn T 1996 Are the etiologies of cervical and trochanteric hip fractures different? *Bone* **18**(3) 133S–7S

[20] Hans D, Dargent P, Schott A, Breart G and Meunier P 1996 Ultrasound parameters are better predictors of trochanteric than cervical hip fracture: the EPIDOS

prospective study In *World Congress on Osteoporosis* vol 1 eds S Papapolous, P Lips, H Pols, C Johnston and P Delmas (Amsterdam: Elsevier) pp 161–5

[21] Cann C, Roe E, Sanchez S and Arnaud C 1999 PTH effects in the femur: envelope-specific responses by 3DQCT in postmenopausal women *J. Bone Min. Res.* **14** s137

[22] Cody D D, Gross G J, Hou F J, Spencer H J, Goldstein S A and Fyhrie D P 1999 Femoral strength is better predicted by finite element models than QCT and DXA *J. Biomechanics* **32**(10) 1013–20

[23] Keyak J 1996 Prediction of femoral neck strength using automated finite-element modelling Bioengineering Graduate Group (San Francisco: University of California, San Francisco)

[24] Keyak J 1999 Precise prediction of femoral fracture load using non-linear finite element models *Trans. Orthop. Res. Soc.*

[25] Cody D D, Divine G W, Nahigian K and Kleerekoper M 2000 *Bone* density distribution and gender dominate femoral neck fracture risk predictors *Skeletal Radiol.* **29**(3) 151–61

[26] Sandor T, Felsenberg D, Kalender W and Brown E 1991 Global and regional variations in the spinal trabecular bone: single and dual energy examinations *J. Clin. Endocrinol. Metab.* **72** 1157–68

[27] Sandor T, Felsenberg D, Kalender W and Brown E 1992 *Heterogeneity of the Loss of BMD from Spinal Cortical Bone* 9th International Workshop of Bone Density, Traverse City

[28] Sandor T, Felsenberg D, Kalender W A, Clain A and Broen E 1992 Compact and trabecular components of the spine using quantitative computed tomography *Calcif. Tissue Intl.* **50** 502–6

[29] Faulkner K G, Cann C E and Hasegawa B H 1991 The effect of bone distribution on vertebral strength: assessment with patient-specific nonlinear finite element analysis *Radiology* **179** 669–74

Chapter 10

Peripheral quantitative computed tomography and micro-computed tomography

Christopher C Gordon

10.1. INTRODUCTION

Bone mineral density (BMD) of the central and peripheral skeleton can be evaluated with a high degree of accuracy and precision using quantitative computed tomography (QCT). As it provides cross-sectional images, QCT is unique among the methods for measuring bone mineral density because it provides separate estimates of cortical and trabecular bone. While BMD of the peripheral and axial skeleton can be reliable estimates of bone strength and reliably estimate the propensity to fracture, numerous studies indicate that bone strength is only partially explained by BMD. Quantitative assessment of bone geometry and microstructure characteristics of the trabecular network, such as trabecular bone volume, trabecular spacing and connectivity, may improve estimates of bone strength. As such, in the past decade, considerable progress has been made in the development of cross-sectional imaging methods for assessing bone density, bone geometry and the micro-architecture of trabecular bone.

This chapter reviews a special group of QCT systems, peripheral quantitative computed tomography (pQCT), developed for volumetric bone mineral assessment at the peripheral skeleton in both humans and various animals models. In addition the details of the ultra-high resolution micro-computed tomography (μCT) systems developed to image trabecular micro-architecture *in vitro* are also examined.

10.2. DEVELOPMENT OF pQCT

QCT is an established technique for determination of bone mineral density in the axial spine and appendicular skeleton such as the forearm and tibia. Unlike

Table 10.1.

Machine and characteristics	Sample diameter (mm)	Voxel size (μm)	Slice thickness (mm)
XCT Research SA, variable slice thickness option, useful for mice to rabbits.	90	70–500	1.0 (0.1–0.8)
XCT Research M, softer X-ray beam, useful for higher precision *ex vivo* studies.	50	70–500	0.5
XCT Microscope, higher mechanical precision, useful for thin cortical shell studies, and histomorphometric studies.	50	30–300	0.1
XCT 3000 Research, useful for larger animals such as monkeys.	300	200–800	1.0
XCT 2000, clinical unit used to measure distal and mid-point of radius and distal and mid-point along the tibia.	140	200–800	2.2
XCT 3000 clinical scanner used to measure the tibia and distal to mid-femur and limited access to the femoral neck. Also has positioners for dental applications.	300	200–800	2.2

DXA, QCT utilizes a transaxial image to allow separate measurement of the true volumetric density (mg/cm^3) and cross-sectional area of trabecular and cortical bone without superposition of other tissues and provides exact three-dimensional localization of the target volume.

To some extent, the high capital cost and limited access to conventional all-purpose CT scanners prompted the development of dedicated peripheral QCT (pQCT) instrumentation specifically for measurements of purely trabecular and cortical BMC and BMD at peripheral sites. Such an instrument also has the advantage of delivering a lower dose to the patient than standard spinal QCT because only the appendicular skeleton is irradiated.

Since the mid 1970s pQCT has been used on a research basis [1–4]. Currently Stratec Medizintechnik (Pforzheim, Germany) produces the widest range of pQCT systems commercially. These pQCT systems were developed from the original XCT-960 model (the only pQCT device with FDA approval by 1993), which allowed a research option. Derived from the original 960 model the current systems were adapted for research and clinical applications [5]. Table 10.1 lists the various pQCT systems commercially available. The distinctive characteristics of each scanner are described. For example, the XCT SA (SA refers to small animals) is designed to study the bones of animals ranging in size from sheep or primates to rats *in vivo* or *in vitro*. The high-resolution XCT Research M (M refers to mouse) is used for *in vivo* studies of specimens that range in size from mice to ferrets. The XCT Microscope was developed for quasi-histomorphometric analysis of small excised bones at a resolution sufficient to resolve individual trabeculae.

10.3.   pQCT MACHINE DESCRIPTION

A basic pQCT machine operates on second generation CT scanning principles. The pQCT machine consists of two major components, the scanner hardware and a computer system for hardware control and image data analysis. The scanner hardware contains a collimated X-ray source that emits a very narrow X-ray beam, a fixed set of solid state detectors and a mechanical system allowing transverse, radial and axial displacements of the source–detector assembly.

The computer system controls the complete scanning procedure. The pQCT image is derived from a series of X-ray transmissions recorded as the X-ray tube and detectors rotate in discrete steps around the object being scanned. The X-ray transmissions represent the fraction of X-rays that pass through various lengths of the object. Taken together, these transmitted fractions acquired at various angles around the object are referred to as projection data. As with standard computed tomography, the technique of back-projection with filtration is applied to the projection data to generate the final tomographic image. The pQCT image is composed of a discrete number of picture elements (pixels). However, since the tomographic bone slice has a predetermined thickness, the pixels that comprise the reconstructed image actually represent individual volume units referred to as 'voxels' (see table 10.1).

Like standard body CT, each voxel in the pQCT image represents the linear attenuation coefficient (μ) of the tissue being studied. The attenuation coefficient is dependent on the energy of the X-ray beam as well as the amount of the absorbing material. The final density of the object imaged is estimated from the linear attenuation coefficients. This is done by comparing the linear attenuation coefficient of each voxel in the cross-sectional image of the bone with a calibration equation that relates the attenuation value to the volumetric density of a calcium hydroxyapatite phantom. On summing up the individual bone mineral content values assigned to each voxel, the pQCT analysis software can determine the mineral mass (BMC in mg) and the mineral density (BMD in mg/cm^3) of the whole bone slice or of selected cortical and trabecular regions of interest.

Figure 10.1 shows a current XCT SA system used for pre-clinical scanning of various animal models. This unit has a tabletop design and can be used to scan animals ranging in size from mice to rabbits. As well,

Figure 10.1. *XCT SA pQCT scanner used for pre-clinical testing (courtesy of Stratec Medizintechnik).*

various excised bone specimens can be scanned. The same second generation CT scanning principles and hardware are built into the largest of the XCT scanners, the XCT 3000 shown in figure 10.2. As illustrated in this picture the XCT 3000 unit can be easily integrated in an office-based setting that requires no additional shielding from scattered radiation. The XCT 3000 has an extended scan length in the z direction of 300 mm, which when fitted with various positioning devices enables it to perform measurements along the tibia and distal end of the femur [6]. Although it has a smaller gantry opening, this flexibility in scanning is also available on the XCT 2000, which can be used to scan the distal tibia as well as the midshaft and distal part of the forearm.

pQCT machines manufactured by Stratec Medizintechnik are calibrated so that a zero density value is assigned to every voxel in the image that corresponds to a measurement of pure fat. With this in mind it is important to interpret correctly what kind of bone density is assessed. Solid bone matter has a specific density of approximately $1.9 \, \text{g/cm}^3$. This value can be estimated based on the fractional composition of bone tissue. Practically, it is assumed to be 42% mineral with a density of $3.2 \, \text{g/cm}^3$ and 58% matrix with a density of $1.0 \, \text{g/cm}^3$. Therefore the specific density of solid bone matter can be determined by combining the mineral density with the matrix density in the following way:

$$\text{density of solid bone matter} = (\text{mineral} + \text{matrix}) \text{ density}$$
$$\text{mineral density} = 0.42 \times 3.2 \, \text{g/cm}^3 = 1.34 \, \text{g/cm}^3$$
$$\text{matrix density} = 0.58 \times 1.0 \, \text{g/cm}^3 = 0.58 \, \text{g/cm}^3$$
$$\text{density of solid bone matter} = 1.92 \, \text{g/cm}^3.$$

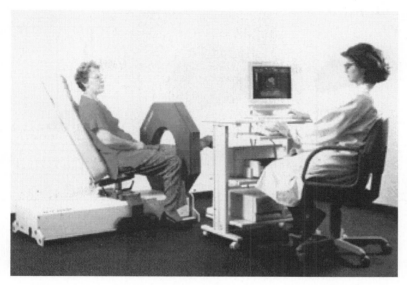

Figure 10.2. *XCT 3000 pQCT scanner operated in an office setting (courtesy of Stratec Medizintechnik).*

In practice pQCT machines measure the mineral density of bone. As such, cortical bone is determined with a nominal mineral density of $1200 \, \text{mg/cm}^3$. In individuals with normal bone mineral status, trabecular bone ranges in density from 200 to $300 \, \text{mg/cm}^3$. This represents the average density of the trabecular network including its marrow content. The total or integral density of the cortical and trabecular compartments is determined as approximately $400–500 \, \text{mg/cm}^3$ in individuals without bone disease [7].

10.4.   BONE PROPERTIES AND VARIABLES MEASURED BY pQCT

The most distinct advantage of pQCT is the capability to separate cortical and trabecular bone compartments and thus to be able to monitor very small changes in a short period of time [8, 9]. This is important because trabecular bone has an 8–10 times larger surface area than cortical bone and therefore reacts much faster to metabolic changes brought on by disease or a pharmacological agent. pQCT analysis provides separate volumetric measures of trabecular and cortical BMD and BMC. As shown in figure 10.3, once the cross-sectional image is acquired, special analysis procedures involving thresholding and edge detection are applied to the image to separate bone from soft tissue and, most importantly, trabecular bone from cortical bone.

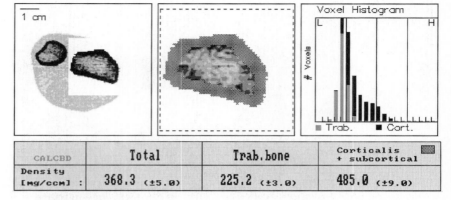

Figure 10.3. *Forearm scan taken through the distal radius with the XCT 2000. The analysis software separates the imaged bone into total, trabecular and cortical compartments.*

Geometric parameters related to the cortical shell, such as axial and polar moments of inertia and the mean thickness of the cortical shell, are also assessed with pQCT [10–13]. The extended scan length available on the units enables them to perform measurements at several sites along the bone. In figure 10.4 an example of a combination of two measurement sites is shown for the radius and tibia. With this multiple slice technique, geometry changes of the bone can be detected and monitored at different sites along the length axis.

Variables combining the density and geometry of the cortical shell are also derived from the pQCT image. The moment of inertia weighted by the cortical density of each voxel in the cortical shell has been shown to capture the strength of the bone. This so-called strength strain index (SSI) can be measured non-invasively and correlates strongly with bone strength [14–16]. Its exact formulation is given in figure 10.5. As shown, for each voxel in the cortical shell, the distance from the centre of mass of the cortical shell is calculated, combined with the area of each voxel and weighted by the mineral density determined from the linear attenuation coefficient of the voxel. The SSI directly correlates with three-point bending strength [16]. In a similar way, the pQCT derived bone strength index (BSI), an index similar to the SSI, is able to predict the fracture load of bones in three-point bending tests. BSI and fracture load compared in more than 200 rat femurs resulted in a correlation exceeding 0.9 [17].

There is sufficient contrast between muscle and fat in the pQCT image to allow for the measurement of muscle cross-sectional area. Using the XCT 2000, scans taken at a point 66% of the tibia length proximal to the ankle provide a cross-sectional image through the calf muscle. Using

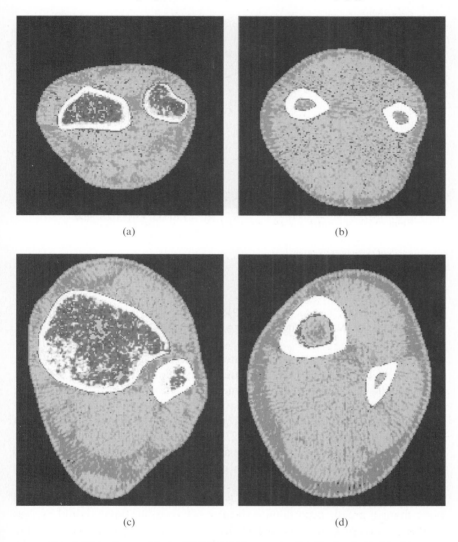

Figure 10.4. *Scans with the XCT 2000 taken at the distal and proximal radius (top) and distal and proximal tibia (bottom).*

a special analysis procedure based on thresholding and edge detection, pQCT is able to determine the cross-sectional area of muscle, fat and bone [18]. Measurements taken through the calf with pQCT have been used to establish a strong correlation between bone area and muscle cross-sectional area [19, 20]. These measurements have also been applied to the forearm [21].

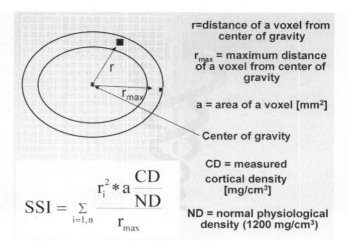

Figure 10.5. *Formulation of the SSI parameter.*

10.5. pQCT ACCURACY AND PRECISION FOR BONE MINERAL AND BONE GEOMETRY ASSESSMENTS

A number of studies have demonstrated the accuracy of pQCT. *In vitro*, the accuracy of the method was calculated to be about 2% [10, 22, 23]. In a cadaver study in which radii were measured with pQCT and then ashed, Takada *et al* [22] found high correlations between total pQCT BMC and ash weight ($r = 0.90$) and between pQCT total BMD and ash weight ($r = 0.82$). Another cadaver study [23] compared the pQCT measurement with neutron activation and flame atomic absorption spectroscopy and revealed that pQCT measured both cortical and total BMC with a high degree of accuracy ($r \approx 0.86$–0.96).

For forearm measurements, the proportion of trabecular bone to total bone changes greatly according to the distance from the ulnar styloid process [24]. Therefore, precision with pQCT depends on careful positioning to ensure that the same target volume is scanned. The short-term *in vivo* precision of the clinical pQCT has been measured using groups of healthy young volunteers. Butz *et al* [7] found relative precision errors (CV) of 1.7% for trabecular, 0.8% for total and 0.9% for cortical BMD measurements. Lehmann *et al* [25] and Schneider *et al* [26] calculated absolute precision errors for trabecular regions of interest between 2.6 and 3.1 mg/cm^3, which resulted in CVs of under 1%. In a study by Grampp *et al* [27] of pre- and post-menopausal women, the average absolute precision errors for the trabecular and total region were of the same order as in the previous studies (1.8–3.4 and 3.8–8.5 mg/cm^3 respectively), but the resulting CVs of the post-menopausal population were higher

(0.9–2.1% and 1.1–2.6%, respectively), because of lower average BMD in their groups.

The precision errors from pQCT can be lowered if multiple-slice scans are used to ensure the evaluation of a larger volume [28, 29]. For example, lower precision errors for trabecular BMD have been reported for normal (0.3%), osteoporotic (0.6%) and severely osteoporotic (0.9%) women scanned using multi-slice techniques.

10.6.   CLINICAL UTILITY OF pQCT

Several reviews have concluded that the low radiation dose and high degree of precision achievable with clinical pQCT scanners make them effective tools for detecting cortical and trabecular bone changes in various paediatric and adult populations [30–32]. For example, the effects of ageing and the differences between the sexes can be examined with pQCT. Also, pQCT measurements of bone mineral density and geometry have been shown to be useful for distinguishing osteoporotic and non-osteoporotic patients and for fracture discrimination.

Although DXA has established itself as the most widely used method for measuring BMD, it is not reliable for children [33]. This unreliability is due to the following inherent limitation. With photon absorptiometry techniques such as DXA or peripheral dual X-ray absorptiometry (pDXA), the bone mineral mass per scanned segment is normalized to the projected area of the scanned segment to produce a measure of areal BMD in units of $g\,cm^{-2}$. Normalization to the projected area means that DXA cannot account for bone depth and may be affected by dramatic changes in body size within the same person and by extreme body types across individuals. For paediatric applications specifically, DXA underestimates the true BMD in subjects during early childhood, puberty and adolescence due to the dramatic growth-related changes in bone size that occur during these periods. CT, however, adequately adjusts for bone size effects by using cross-sectional imaging to provide equitable comparisons on a per unit volume basis regardless of age, gender or changes in skeletal size.

The effective dose received from a spinal QCT is approximately $60\,\mu Sv$. Although this amount is far less that that received from background radiation ($2400\,\mu Sv$), it is far greater than that of DXA ($\sim 1\,\mu Sv$). This significant radiation exposure precludes the routine use of QCT for paediatric applications. The effective dose for a typical pQCT exam is also comparable with spinal DXA ($< 2\,\mu Sv$) but far less than spinal QCT because only the appendicular skeleton is irradiated. This low dose plus its ease of use makes pQCT ideal for paediatric measurements. In addition, the time required to perform these pQCT scans fits well within the limits that a child can tolerate. As a result of these points, pQCT is quickly gaining recognition as an exclusive, non-invasive modality for children receiving treatment for diseases such as juvenile diabetes, asthma and growth hormone deficiency [34, 35]. The availability of well defined paediatric reference data sets will further advance its use in children [36, 37].

Using pQCT, age-related bone loss can be studied separately for total, cortical and trabecular bone at the same site. For example, in women studied with pQCT annual changes of 0.5–0.9% were found in the trabecular BMD and 0.9–1.1% in the total BMD [7]. In a cross-sectional study Grampp *et al* [38] found that the annual loss of bone mass in healthy women was twice as much for cortical bone (0.5%/year) as the loss of mass in trabecular bone (0.22%/year). Interestingly, in a study of elderly females, Boonen found the opposite effect, with there being greater trabecular bone loss [39].

pQCT measurements of BMC and BMD at the radius are successful in distinguishing between osteoporotic and non-osteoporotic patients [7, 39], in monitoring subjects during treatment [40, 41] and in predicting fracture loads *in vitro*. In a biomechanical study Sparado *et al* [12] found that the cortical shell contributes substantially to the mechanical strength of the distal radius. In a study of 20 cadaver specimens Augat *et al* [11] examined the combination of the geometric properties and BMD values determined from pQCT at the distal radius and found that the strength of the radius could be best predicted by considering the second moment of inertia and cortical BMD.

The importance of measurement of cortical bone by pQCT has also been found *in vivo*. Studies have suggested that the thinning of the cortical rim at the radius was a potential mechanism contributing to osteoporotic changes [30, 45], and have identified this compartment as a promising location for BMC and thickness measurements. In a study that examined the ability of BMD, BMC and cross-sectional area to detect osteoporotic changes, Grampp *et al* [38] found that only the cortical area and BMC significantly distinguished between women with non-traumatic vertebral fractures and healthy post-menopausal women. Overall, these *in vivo* and *in vitro* data suggest that pQCT measurement of cortical rather than trabecular bone at the radius may have greater diagnostic sensitivity in terms of appendicular measurements.

10.7. USE OF pQCT IN PRE-CLINICAL TESTING

Pre-clinical tests in drug research need a reliable tool to measure changes in bone density and geometry. Ideally, the resulting changes in density or geometry should reach statistically significant levels in short time periods and with small animal groups. Several studies have demonstrated that both of these aims can be readily achieved with the various pQCT units available for pre-clinical research development. As shown in table 10.1, the

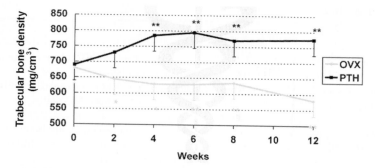

Figure 10.6. *Effect of ovariectomy and PTH treatment on trabecular bone density in the proximal tibia of rats in vivo (see reference [42]).*

XCT units designed for animal work have a higher spatial resolution and smaller slice thickness than the clinical units because much smaller bones are being measured [5].

The ovariectomized rat or mouse is a commonly used model for post-menopausal osteoporosis. In combination with the pQCT technology, this mouse and rat model has been used to study the effects of different drugs such as bisphosphonates, oestrogen, Raloxifene and parathyroid hormone (PTH) [42–44]. For example, Gasser investigated the effect of PTH in ovari-ectomized (OVX) rats. In groups of 12 animals each in an OVX and PTH treated group, significant differences in bone mineral density could be detected within two weeks using pQCT. Changes in cortical geometry, such as moments of inertia, showed significant changes after eight weeks of treatment. The results for trabecular density are shown in figure 10.6. By contrast DXA was unable to detect statistically significant changes within 24 weeks [42].

The pQCT device is also used in models for larger animals such as Cynomolgus monkeys. For example, the XCT 3000 Research system can be used to assess the bone mineral density of the femoral neck [45, 46]. These devices have been used to demonstrate the fracture repair mechanism in sheep [47].

10.8.   INTRODUCTION TO µCT

To fully understand the influence of bone structure and bone geometry on the mechanical integrity of bone, the three-dimensional characteristics of its structure must be assessed. µCT was developed for this purpose and allows for the examination of the response of trabecular bone to various pharmaco-logical agents or to various disease states. µCT gives insight into trabecular

Table 10.2.

Company	Scanner	Comments
Imtek Inc	MicroCAT	35–50 mm field of view, 50 μm resolution, small animal imaging
Enhanced Vision Systems	MS-8 and RS-9 SmallSpecimen Animal Scanner	Up to 50 mm field of view, studies of small animals such as mice and rats
Skyscan	Model 1074	Biological and geological applications
Scanco Medical	μCT 20, μCT 80	10–20 μm resolution, studies of trabecular bone structure *in vitro*

structure with a spatial resolution better than 50 μm. Currently there are four main manufacturers of commercial μCT units (table 10.2). The information provided by each of these units compares strongly with histomorphometric parameters with the advantage that these parameters are measured three-dimensionally and non-destructively [51].

A μCT scanner, typically, consists of a micro-focused X-ray source, a detector array, an object stage, computing resources for reconstruction and visualization, as well as an optical bench to isolate environmental vibrations. The source sends X-rays through the object mounted on the stage. The detector array records attenuated intensities of the X-ray beam from various orientations. The source, detector and objects are moved under precise control in such a way that the data needed to reconstruct the content of

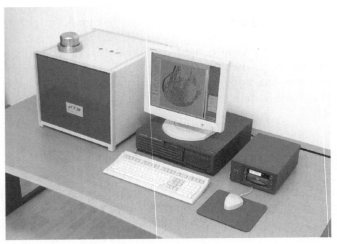

Figure 10.7. *In vitro μCT 20 tabletop scanner (courtesy of Scanco Medical).*

the object are collected. Overall, the imaging system is contained in a lead-shielded box that is interfaced to a computer for image reconstruction and data analysis. Figure 10.7 illustrates the typical footprint and layout of a μCT analysis station. There is one important point to note from this figure: these units can be housed in an office setting that requires no lead shielding.

10.9. WHAT CAN BE MEASURED WITH μCT?

With these μCT units the first objective is to image the sample at a resolution sufficient to resolve individual trabeculae. Once imaged, the second objective is to characterize the imaged structure with established histomorphometric parameters [52–54].

Figure 10.8 shows a typical trabecular specimen rendering achievable with current μCT systems. Figure 10.9 shows that larger specimens can also be imaged. It is clear that the true three-dimensional characteristics and the micro- and macro-architecture of the bone can be captured with μCT. The three-dimensional data sets generated from these μCT systems can be used to reproducibly calculate parameters such as trabecular thickness, trabecular separation, trabecular bone volume, and various indices of connectivity and orientation such as the degree of anisotropy and the Euler number [55]. In addition the three-dimensional μCT data sets have been used in finite element modelling to examine the impact of the various histomorphometric parameters on bone strength determined *in vitro* [56].

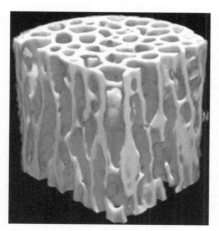

Figure 10.8. *Three-dimensional rendering of the trabecular architecture contained in the vertebral body of a sheep (courtesy of Stratec Medizintechnik).*

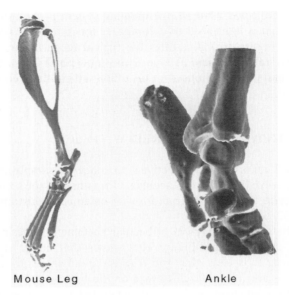

Mouse Leg   Ankle

Figure 10.9. *Mouse ankle and mouse leg rendered in three dimensions (courtesy of Scanco Medical).*

10.10. SUMMARY

In summary, pQCT is a dedicated CT system that acquires a cross-sectional image to measure BMD and BMC at peripheral sites such as the distal radius and tibia. By acquiring a cross-sectional image pQCT has four advantages over a projection base technique such as DXA. First, metabolically more active trabecular bone can be measured separately from cortical bone. Thus, the effects of ageing, disease and treatment on cortical and trabecular bone can be followed separately. Second, the true volumetric bone mineral density (mg/cm^3) is assessed. Third, cross-sectional imaging allows geometric parameters such as moments of inertia and cortical thickness to be calculated. Fourth, cross-sectional capabilities allow peripheral quantitative computed tomography to bypass imaging obstacles related to bone size. The resulting advantage is that pQCT can provide accurate and precise measurements of bone density and geometry in growing children.

In addition, there is growing use of extracting a cross-sectional muscle area from the pQCT image and using this area to infer muscle strength. While use of pQCT for assessing the interrelationship between bone and muscle remains a research tool, in the near future it is likely that physicians will have the option to consider a bone–muscle strength index along with the other risk factors in osteoporotic patients.

In the field of pre-clinical research and drug testing, the ability to image various animal models and bone samples in detail opens new dimensions in understanding bone metabolism and bone disease. The pQCT units developed for scanning various animal models have reduced the time needed for new drug development. They offer the advantage of examining drug effects on modelling and remodelling processes. Combined with the uses of the ultra-high resolution μCT scanners, these pQCT scanners used in animal testing have all but replaced the need for standard histomorphometry.

REFERENCES

[1] Ruegsegger P 1988 Quantitative computed tomography at peripheral measuring sites *Ann. Chir. Gynaecol.* **77** 204–7
[2] Hosie C J and Smith D A 1986 Precision of measurement of bone density with a special purpose computed tomography scanner *Br. J. Radiol.* **59** 345–50
[3] Schneider P and Borner W 1991 Peripheral quantitative computed tomography for bone mineral measurement using a new special QCT-scanner: methodology, normal values, comparison with manifest osteoporosis *Rofo. Fortschr. Geb. Rontgenstr. Neuen Bildgeb. Verfahr.* **154** 292–9
[4] Hangartner T N, Overton T R, Harley C H, Van den Berg L and Crockford P M 1985 Skeletal challenge: an experimental study of pharmacologically induced changes in bone density in the distal radius, using gamma-ray computed tomography *Calcif. Tissue Intl.* **37** 19–24
[5] Tysarczyk-Niemeyer G 1997 New noninvasive pQCT devices to determine bone structure *J. Jpn. Soc. Bone Morphom.* **7** 97–105
[6] Sievanen H, Koskue V, Rauhio A, Kannus P, Heinonen A and Vuori I 1998 Peripheral quantitative computed tomography in human long bones: Evaluation of *in vitro* and *in vivo* precision *J. Bone Min. Res.* **13** 871–82
[7] Butz S, Wüster C, Scheidt-Nave C, Götz M and Ziegler R 1994 Forearm BMD as measured by peripheral quantitative computed tomography (pQCT) in a German reference population *Osteoporos. Intl.* **4**(4) 179–84
[8] Breen S A, Millest A and Loveday B 1996 Regional analysis of bone mineral density in the distal femur and proximal tibia using peripheral quantitative computed tomography in the rat *in vivo Calcif. Tissue Intl.* **58** 449
[9] Ferretti J L, Gaffuri O and Capozza R 1995 Dexamethasone effects on structural, geometric and material properties of rat femur diaphyses as described by peripheral quantitative computed tomography (pQCT) and bending tests *Bone* **16** 119
[10] Louis O, Willnecker J, Soykens S, Van Den Winkel P and Osteaux M 1995 (1995) Cortical thickness assessed by peripheral quantitative computed tomography: Accuracy evaluated on radius specimens *Osteoporos. Intl.* **5** 446–9
[11] Augat P, Reeb H and Claes L E 1996 Prediction of fracture load at different skeletal sites by geometric properties of the cortical shell *J. Bone Min. Res.* **11** 1356–63
[12] Spadaro J A, Werner F W, Brenner R A *et al* 1994 Cortical and trabecular bone contribute strength to the osteoporotic distal radius *J. Orthop. Res.* **12** 211–8

[13] Bouxseim M L, Myburg K H, Vandermeulen M C H, Lindenberg E and Marcus L R 1994 Age-related differences in cross-sectional geometry of the forearm in healthy women *Calcif. Tissue Intl.* **54** 113–8

[14] Schiessl H, Ferretti J L, Tysarczyk-Niemeyer G and Willnecker J 1996 Noninvasive bone strength index as analyzed by peripheral quantitative computed tomography. In *Pediatric Osteology. New Developments in Diagnostic and Therapy* ed E Schoenau (Amsterdam: Elsevier) p 141

[15] Schiessl H, Ferretti J L, Tysarczyk-Niemeyer G *et al* 1998 The role of the muscles to the mechanical adaptation of bone. In *Advances in Osteoporosis* vol 1 ed G P Lyritis (Athens) p 53

[16] Wilhelm G, Felsenberg D, Bogusch G, Willnecker J, Thaten J and Gummert P 1998 Biomechanical examinations for validation of the bone strength index SSI, calculated by peripheral quantitative computed tomography. In *Musculoskeletal Interactions* vol 2 ed G P Lyritis (Athens) pp 105–10

[17] Feretti J L, Capozza R F and Zanchetta J R 1996 Mechanical validation of a tomographic (pQCT) index for noninvasive estimation of rat femur bending strength *Bone* **18** 97–102

[18] Schiessl H and Willnecker J 1998 Muscle cross sectional area and bone cross sectional area in the human lower leg measured with peripheral computed tomography. In *Musculoskeletal Interactions* vol 2 ed G P Lyritis (Athens) pp 47–52

[19] Schiessl H, Frost H M and Jee W S 1998 Estrogen and bone-muscle strength and mass relationship *Bone* **21**(1) 1–6

[20] Rittweger J, Beller G, Ehrig J, Jung C, Koch U, Ramalla J, Schmidt F, Newitt D, Majumdar S, Schiessl H and Felsenberg D 2000 Bone-muscle strength indices for the human lower leg *Bone* **27** 319–26

[21] Schoenau E, Neu C M, Mokov E, Wassmer G and Manz F 2000 Influence of puberty on muscle area and cortical bone area of the forearm in boys and girls *J. Clin. Endocrinol. Metab.* **85** 1095–8

[22] Takada M, Engelke K, Hagiwara S, Grampp S and Genant H K 1996 Accuracy and precision study *in vitro* for peripheral quantitative computed tomography *Osteoporos. Intl.* **6** 207–12

[23] Louis O, Soykens S, Willnecker J, van den Winkel P and Osteaux M 1996 Cortical and total bone mineral content of the radius: accuracy of peripheral computed tomography *Bone* **18** 467–72

[24] Schlenker R A and Von Seggen W W 1976 The distribution of cortical and trabecular bone mass along the lengths of the radius and ulna and the implications for *in vivo* bone mass measurements *Calcif. Tissue Res.* **20** 41–52

[25] Lehman R, Wapniarz M, Kvansnicka H M *et al* 1992 Reproducibility of bone density measurements of the distal radius using a high-resolution special scanner of peripheral quantitative computed tomography (single energy pQCT) *Radiology* **32** 177–81

[26] Schneider P and Borner W 1991 Peripheral quantitative computed tomography for bone mineral measurement using a new special QCT-scanner: methodology, normal values, comparison with manifest osteoporosis *Rofo. Fortschr. Geb. Rontgenstr. Neuen Bildgeb. Verfahr.* **154** 292–9

[27] Grampp S, Jergas M, Lang P *et al* 1996 Quantitative CT assessment of the lumbar spine and radius in patients with osteoporosis *Am. J. Roentenol.* **167** 133–40

[28] Ruesegger P, Durand E and Darnbacher M A 1991 Localization of regional forearm bone loss from high resolution computed tomography images *Osteoporos. Intl.* **1** 76–80

[29] Ruesegger P 1994 The use of peripheral QCT in the evaluation of bone remodeling *Endocrinologist* **4** 167–76

[30] Ito M, Tsurusaki K and Hayashi K 1997 Peripheral QCT for the diagnosis of osteoporosis *Osteoporos. Intl.* **7**(S3) 120–7

[31] Augat P, Fuerst T and Genant H K 1998 Quantitative bone mineral assessment at the forearm: a review *Osteoporos. Intl.* **8** 299–310

[32] Genant H K *et al* 1996 Noninvasive assessment of bone mineral and structure: state of the art *J. Bone Min. Res.* **11**(6) 707–30

[33] Schoenau E 1998 Problems of bone analysis in childhood and adolescence *Pediatric Nephrol.* **12** 420–9

[34] Lettgen B, Hauffa B, Mohlmann C, Jeken C and Reiners C 1995 Bone mineral density in children and adolescents with juvenile diabetes. Selective measurements of bone mineral density of trabecular and cortical bone using peripheral quantitative computed tomography *Hormone Res.* **43** 173–5

[35] Reilly S M, Hambleton G, Adams J E and Mughal M Z 2001 Bone density in asthmatic children treated with inhaled corticosteroids *Arch. Disease Childhood* **84** 183–4

[36] Neu C, Manz F, Rauch F, Merkel A and Schonau E 2001 Bone density and bone size at the distal radius in healthy children and adolescents: A study using peripheral quantitative computed tomography *Bone* **28**(2), 227–32

[37] Binkley T L and Specker B L 2000 PQCT measurement of bone parameters in young children *J. Clin. Densitom.* **3** 9–14

[38] Grampp S, Lang T, Jergas M *et al* 1995 Assessment of skeletal status by peripheral quantitative computed tomography of the forearm: short-term precision *in vivo* and comparison to dual X-ray absorptiometry *J. Bone Min. Res.* **10** 1566–76

[39] Takagi Y, Fujii Y, Miyauchi A, Goto B, Takahashi K and Fujita T 1995 Trans-menopausal change of trabecular bone density and structural pattern assessed by peripheral QCT in Japanese women *J. Bone Min. Res.* **10** 1830–4

[40] Schneider P, Fischer M, Allolio B, Felsenberg D, Schroder U, Semmler J and Itter J 1999 Alendronate increases bone density and bone strength at the distal radius in postmenopausal women *J. Bone Min. Res.* **14** 1387–93

[41] Ruesegger P, Keller A and Dambacher M A 1995 Comparison of the treatment effect of ossein–hydroxyapatite compound and calcium carbonate in osteoporotic females *Osteoporos. Intl.* **5** 30–4

[42] Gasser J A 1995 Assessing bone quality by pQCT *Bone* **17** S145–54

[43] Sato M 1995 Comparative X-ray densitometry of bones from ovariectomized rats *Bone* **17** S157–62

[44] Sato M, Bryant H U, Iverson P, Helterbrand J, Smietana F, Bemis K, Higgs R, Turner C H, Owan I, Takano Y and Burr D B 1996 Advantages of raloxifene over alendronate or estrogen on nonreproductive and reproductive tissues in the long-term dosing of ovariectomized rats *J. Phamacol. Exp. Ther.* **279** 298–305

[45] Jerome C P, Johnson C S and Lees C J 1995 Effect of treatment for 3 months with human parathyroid hormone I-34 peptide in ovariectomized cynomolgus monkeys (*Macaca fascicularis*) *Bone* **17** S415–20

[46] Hotchkiss CE 1999 Use of quantitative computed tomography for densitometry of the femoral neck and spine in cynomolgus monkeys *Bone* **24** 101–7

[47] Augat P, Merk J, Genant H K and Claes L 1997 Quantitative assessment of experimental fracture repair by peripheral computed tomography *Calcif. Tissue Intl.* **60** 194–9

[48] Feldkamp L A, Goldstein S A, Parfitt A M, Jesion G and Kleerekoper M 1989 The direct examination of three-dimensional bone architecture *in vitro* by computed tomography *J. Bone Min. Res.* **4** 3–11

[49] Kuhn J L, Goldstein SA, Feldkamp L A, Goulet R W and Jesion G 1990 Evaluation of a microcomputed tomography system to study trabecular bone structure *J. Orthop. Res.* **8** 833–42

[50] Ruesegger P, Koller B and Muller R 1996 A microtomographic system for non-destructive evaluation of bone architecture *Calcif. Tissue Intl.* **58** 24–9

[51] Muller R, Hahn M, Vogel M, Delling G and Ruesegger P 1996 Morphometric analysis of noninvasively assessed bone biopsies: comparison of high resolution computed tomography and histologic sections *Bone* **18** 215–20

[52] Odgaard A and Gundersen H G 1993 Quantification of connectivity in cancellous bone, with special emphasis on 3D reconstructions *Bone* **14** 173–82

[53] Parfitt A M, Matthews C and Villanueva A 1983 Relationships between surface, volume and thickness of iliac trabecular bone in aging and in *osteoporosis J. Clin. Invest.* **72** 1396–409

[54] Parfitt A M 1983 The stereologic basis of bone histomorphometry, theory of quantitative microscopy and reconstruction of the third dimension In *Bone Histomorphometry: Techniques and Interpretations* ed R A Recker (Boca Raton, FL: CRC) pp 53–87

[55] Muller R, Hidebrand T, Hauselmann H J and Ruesegger P 1996 *In vivo* reproducibility of three-dimensional structural properties of noninvasive biopsies using 3D-pQCT *J. Bone Min. Res.* **11** 1745

[56] Muller R and Ruesegger P 1996 Analysis of mechanical properties of cancellous bone under conditions of simulated atrophy *J. Biomech.* **29** 1053–60

Chapter 11

Radiogrammetry

Jonathan A Thorpe and Christian M Langton

11.1. OVERVIEW

Radiogrammetry is a comparatively simple bone assessment technique involving the comparative measurement of cortical bone thickness and medullary cavity width. Radiogrammetry has the advantage over other techniques of being largely insensitive to variations in image quality or to the presence of other disease, so making it highly suitable for application to routine medical images. This chapter gives an overview of the approach and technical application of the technique.

11.2. INTRODUCTION

With the decoupling of bone deposition and resorption that occurs through age or metabolic bone disease, the resulting increase in periosteal bone formation and endosteal resorption causes an increase in the overall circumference of the long bones, whilst also decreasing the thickness of the cortical shell [1–3]. This effect can be observed in medical images (figure 11.1), and measurements taken from such images can aid in the investigation of the underlying age or disease process. Such measurements form the basis for radiogrammetry.

    In the strictest sense, radiogrammetry is the measurement of cortical bone width or geometry from a radiographic image, and is one of the oldest methods of bone measurement. Although probably performed *ad hoc* since the earliest days of radiography, the technique was not formally described until Barnett and Nordin did so in 1960 [4]. Unfortunately, although simple and cheap to perform, the original techniques were plagued by the poor precision resulting from the unavoidable operator variability

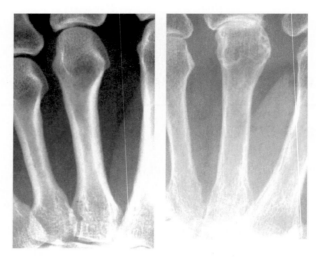

Figure 11.1. *Change in cortical thickness with onset of osteoporosis demonstrating the potential of the radiogrammetric technique. Plane radiographs of the middle metacarpals of two female subjects are shown, one a 42 year old without osteoporosis (left), and one a 90 year old with (right). For the osteoporotic subject, the thickness of the cortical bone along the metacarpal shaft is noticeably thinner, whilst the overall circumference of the cortical bone relative to the bone length is somewhat increased. Significantly, these changes are clearly visible despite the differences in radiographic exposure times, and the considerable joint disease of the older subject, both of which might typically confound photodensitometric or bone mineral density measurement techniques.*

associated with physically taking single measurements from radiographs. As a result, the advent, beginning in the 1970s, of the dedicated and more precise absorptiometric methods of bone assessment such as SPA and DPA and later SXA and DXA saw radiogrammetry slide into decline. In recent years, however, the reductions in operator variability achieved through automated radiographic bone analysis software have reversed this trend towards obsolescence [5], and thus radiogrammetry is currently enjoying something of a revival.

Radiogrammetry has traditionally been performed by taking direct measurements from plane radiographs, hence the name, but the technique is more versatile than this. Certainly, despite what the name would suggest, there is no reason why radiogrammetry should be limited to purely radiographic modalities. Indeed, radiogrammetry can theoretically be performed with any technique capable of reproducibly measuring the cortical and medullary dimensions of a representative bone. To encompass this, we will therefore stretch the definition beyond that of radiography alone to say that, in modern usage, radiogrammetry is the relational measurement of key cortical bone and/or/to medullary cavity dimensions.

This then raises the question of when and how radiogrammetry can be correctly employed, and it is these issues that will be discussed in this chapter. This chapter is not intended to provide a literature review of radiogrammetry; rather, it is intended to provide the reader with an overview on how the technique might be approached and applied, and which technical considerations have to be made to achieve that application.

SECTION A: FUNDAMENTAL PRINCIPLES OF RADIOGRAMMETRY

In the most common applications, radiogrammetry measurements are taken from reproducibly-acquired plane radiographic images of long bones [1, 6–8]. These measurements are then compared with some form of reference scale to produce a comparative result. The original technique requires no specialist equipment, but more modern approaches obtain improvements in accuracy and precision through the use of additional equipment, techniques and automation software.

11.3. BASIC ONE-DIMENSIONAL RADIOGRAMMETRIC MEASUREMENTS FROM TWO-DIMENSIONAL PLANAR IMAGES

At the simplest level, there are three steps to performing even basic radiogrammetry.

1. Obtain an image of a target long bone.
2. Perform one-dimensional measurements of cortical and possibly medullary cavity width.
3. Compare measurements with a reference scale.

The problems that occur in obtaining an image, and the need for a comparative reference scale, are discussed in the later sections. For now, we will look at simple one-dimensional measurements taken from our two previous images. Both are of simple bones: two-dimensional plane radiographs of the middle metacarpal.

Dimensional measurements are taken across the narrowest portion of the shafts of the bones (figure 11.2). Three measurements can be taken directly, namely:

1. W, the total width of the bone
2. M, the width of the medullary cavity
3. C, the total width of cortical bone

and in addition

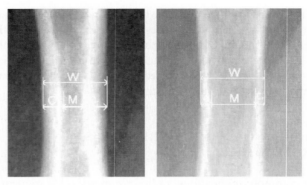

Figure 11.2. *Basic radiogrammetric measurements of the rotated and enlarged (×2) middle metacarpals of the two subjects shown in figure 11.1, with normal on the left and osteoporotic on the right. The measurements are W (total width), M (medullary width), C_l (width of left side cortical bone) and C_r (width of right side cortical bone).*

4. C_l, the thickness of the left-side cortical bone (in these images)
5. C_r, the thickness of the right-side cortical bone (in these images).

Of course, it is possible to derive some of these measurements from others:

$$C = W - M \tag{11.1}$$

or

$$C = C_l + C_r. \tag{11.2}$$

In practice, taking the total bone width and subtracting the medullary cavity width (equation (11.1)) is typically more accurate than measuring and summing the two cortical widths (equation (11.2)).

11.4.   THE CORTICAL INDEX

As can be seen from figure 11.2, and as shall be expanded upon in Section C, there is much potential for error when relying upon absolute measures of cortical bone width. Variations in image geometry, difficulties in obtaining a line of measurement perpendicular to the cortical bone, and irregularities on the internal surface of the cortical shell itself all act to confound accurate measurement. Errors resulting from image geometry and the measurement line can, for the most part, be compensated for by comparing the cortical bone width with the medullary width to give what is known as the cortical index (CI). Unfortunately, errors resulting from surface irregularity cannot be reduced in this way.

The CI is derived from comparison of the cortical and medullary widths:

$$\text{cortical index} = \text{CI} = C/W. \tag{11.3}$$

For metacarpal studies, the cortical index is often referred to as the metacarpal cortical index (MCI). As comparative ratios, the CI and MCI have the advantage of fully compensating for magnification errors in measurement and as such are typically used as the basis for comparison between the results from different individuals or slightly differing beam geometry.

11.5. PRECISION OF BASIC ONE-DIMENSIONAL RADIOGRAMMETRY MEASUREMENT

As we have described, the potential inaccuracies resulting from the imprecision of single measurements are likely to be high, and coefficients of variation (CV) of up to 11% have been reported [9]. Repeat measurements on either the same or similar bones can improve precision to a limited degree (around 8.4% [9]), but such repeat measurements are both time consuming and heavily operator dependent.

Improvements in software have allowed the gradual automation of the analysis process, increasing the reliability of single measurement as well as allowing large numbers of measurements to be taken across several bones. This has improved precision considerably, with a CV of 0.6–0.65% achievable from analysis of three regions of interest on the three middle metacarpals. This forms the basis of the Pronosco X-posure technique (Pronosco A/S Vedbaek, Denmark) [5, 10, 11].

11.6. EXTENDING RADIOGRAMMETRY FROM ONE-DIMENSIONAL TO TWO-DIMENSIONAL MEASUREMENT

The Pronosco X-posure system provides a good example of how a single measure can be improved by extending it to multiple regions of interest (ROIs). The X-posure technique calculates cortical and medullary width from automatically placed ROIs, thus effectively extending the standard radiogrammetry measure from a one-dimensional measurement to a two-dimensional measurement. This technique is reliant on the reliable placement of the three ROIs on the mid-shaft of the three middle metacarpals.

Jorgensen *et al* [11] report that reliable placement is achieved through an active shape model algorithm [12], adapted specifically to the metacarpals. The model identifies the diaphysis of each of the three metacarpals and then places three ROIs of fixed heights, 2.0, 1.8 and 1.6 cm, to cover them.

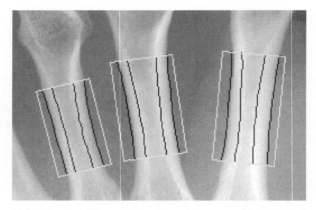

Figure 11.3. *Three middle metacarpals, with regions of interest and cortical bone automatically identified by the Pronosco X-posure software.*

The bone width for each line of each ROI is calculated and summed for the whole ROI, and the ROIs are then successively moved along the metacarpals, with the sum bone width recalculated at each step. The final position of each ROI is then set at the point where the sum bone width was at a minimum—typically the middle of the metacarpal shaft (figure 11.3). Once positioned, the cortical thickness, bone width and hence the medullary cavity width can be calculated for each line, and then averaged for each ROI. Finally, assuming the expected cross-sectional profile of the bone is known, a calculation can be made of the total bone volume within the projected area.

11.7. CONVERSION OF TWO-DIMENSIONAL RADIOGRAMMETRIC MEASUREMENTS TO BONE VOLUME PER AREA

As mentioned, once radiogrammetric measurements have been obtained, it becomes possible to estimate the actual volume or area of cortical bone, provided that certain assumptions are made about the bone cross-sectional shape. This technique is again fundamental to the Pronosco X-posure system and is well described by Jorgensen *et al* [11], Rosholm *et al* [5] and Bouxsein *et al* [13].

In the X-posure system, the key measure of bone volume per area (VPA) for the region of interest is derived from the cortical thickness (C), bone width (W) and the region of interest length, with VPA defined as

$$\text{VPA} = \frac{\text{bone volume}}{\text{area}}. \tag{11.4}$$

Bone volume can be calculated from the equations of a hollow cylinder volume whilst the area can be calculated from the width and length of the ROI. In practice, the ROI length simplifies out of the equation, and so, assuming the bone has a circular cross-section, the equation to derive VPA becomes

$$\text{VPA} = \pi * C * \left(\frac{1 - C}{W} \right). \tag{11.5}$$

As most long bones have an elliptical rather than circular cross section, however, it is useful to expand the equation to include ellipses, thus:

$$\text{VPA} = \pi * e * C * \left(\frac{1 - C}{W} \right) \tag{11.6}$$

where e is the eccentricity ratio of the ellipse at the projected angle.

Rosholm *et al* [5] acknowledge that this technique does make some assumptions about the shape of the bone. In particular, it is assumed only the cortical thickness and bone width change along the length of the bone, not the overall cross-sectional shape of the bone itself—that is, if the bone were to change from an elliptical profile at one end to a circular profile at the other, the results would become increasingly less accurate as the cross-sectional shape changed from that expected. Good modelling information would therefore be required to apply the technique to more complicated bones.

11.8. CONVERSION OF CALCULATED BONE VOLUME TO BONE MINERAL DENSITY (BMD)

For the time being at least, diagnosis of osteoporosis typically requires confirmation by dual energy X-ray absorptiometry (DXA), and thus there is a strong urge for clinicians or researchers to convert or compare the results of radiogrammetry with DXA-derived BMD. In so doing, the conversion procedure has to make certain assumptions about bone structure which may not hold true in reality. The advantages to be gained from a close, if not exact, analogy with bone mineral density are considerable, however, provided such comparisons are not given exaggerated importance.

A rough estimate of BMD may be drawn for bones which have only limited variation in shape and size between individuals, by a simple conversion equation that accounts for the volume of bone (VPA), the expected mineral density per unit volume for the bone in question (p), and a 'conversion factor' dependent on both the radiogrammetry and the DXA imaging techniques (c). Thus:

$$\text{BMD}_{\text{derived}} = \text{VPA} * c * p. \tag{11.7}$$

A more accurate estimate is somewhat more problematic, as changes in the cortical bone porosity that occur with age and disease exhibit a subtle effect on DXA BMD which is not observed in radiogrammetry [5, 14]. Conversely, changes in the cortical bone surface which may become visible on radiographic film as 'striations' exert a potential influence on radiogrammetry but not DXA results [14]. These differences can be corrected for by careful comparative study [5], but require correction factors tailored to the site, disease, subject age and applied image technique and as such are not for the faint hearted.

11.9. EXTENDING RADIOGRAMMETRY TO TWO-DIMENSIONAL AREAS AND THREE-DIMENSIONAL VOLUMES FROM TWO-DIMENSIONAL CROSS-SECTIONAL SLICES

Although performing a CT examination for the sake of radiogrammetry alone can probably not be justified in terms of either cost or radiation dose, it may well be possible to perform radiogrammetry on retrospective CT data. CT certainly has the advantage of providing an exact cross-section of the bone, thus making calculations of cortical and medullary area and hence volume fairly straightforward, provided that the cross-sectional slice is perpendicular to the bone.

If we assume the cross-sectional profile of the bone to be elliptical, with the widest dimension aligned to the image x axis and narrowest to the y axis, and where W_x and W_y are the total widths of the bone in the respective x and y axes, M_x and M_y are the widths of the medullary cavity in the respective x and y axes, and M_{area} and C_{area} are the respective medullary cavity and cortical bone area, we can calculate M_{area} as

$$M_{area} = \pi * \left(\frac{M_x * M_y}{4} \right) \tag{11.8}$$

and so derive the cortical area as

$$C_{area} = \left[\pi * \left(\frac{C_x * C_y}{4} \right) \right] - M_{area}. \tag{11.9}$$

Calculating cortical or medullary volume then becomes straightforward; we simply multiply C_{area} and M_{area} by the slice thickness. The cortical and medullary volume of entire bones could then be calculated by simply summing the results of successive slices together, provided the slices covered the entire bone, without gaps or overlaps.

11.10. EXTENDING RADIOGRAMMETRY FROM TWO-DIMENSIONAL SLICE MEASUREMENT TO TRUE THREE-DIMENSIONAL

Although it would appear that the 'ultimate' in radiogrammetric measurement would be to calculate cortical and medullary volumes from a complete three-dimensional reconstruction of the bone, actually, at least for simple tubular bones, there would be little information to gain from three-dimensional reconstruction that would not be achieved from simply summing the results of successive CT two-dimensional cross-sectional slices. However, for more complicated bones, or if faced with CT data that do not follow the axis of the bone, three-dimensional reconstruction would provide the only means to make any realistic measure of bone volume. Such techniques often require a fair amount of interpolation and, with the risk of error this introduces, realistically should only be attempted when absolutely necessary.

SECTION B: LIMITING FACTORS IN RADIOGRAMMETRY

Ultimately, as with all measurements taken from diagnostic images, there are three main limiting factors in radiogrammetry—precision, accuracy and validity. Precision is the reproducibility of the measure, whilst accuracy is how close to the 'true' value our measurements fall. Validity is whether our measurements, however accurate and precise, are of any actual clinical value.

Simply put, the degree of measurement precision attainable is determined by the sharpness of the image and the reliability of our measuring technique. The accuracy attainable for any single test iteration is in turn determined by the image geometry (including any post-acquisition processing) and the measurement precision. Thus the degree of both precision and accuracy attainable are ultimately determined by the interrelated factors of image sharpness and geometry, as well as the method used to take measurements from the image. Validity remains a clinical issue, determined by the choice of target disease, bone and image modality.

11.11. IMAGE SHARPNESS AND IMAGE GEOMETRY

Radiogrammetry relies on the accurate and precise measurement of the distances between two edges. Image sharpness, the capability of an image modality to define such edges, is therefore of fundamental importance to good radiogrammetric measurement. The factors which influence image sharpness are, of course, dependent on the modality employed but,

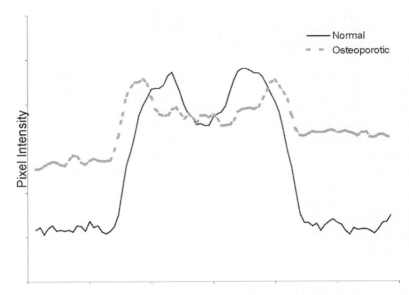

Figure 11.4. *Plot of pixel intensities taken across a sample width of the osteo-porotic and non-osteoporotic metacarpals shown in figure 11.1. Two points of note: (i) The underexposure of the osteoporotic image gives a higher average but lower dynamic range of pixel intensities than the non-osteoporotic image. The rapid rate of change (steep gradient) and larger dynamic range would make edges appear sharper and thus easier to identify. (ii) Although both bones can be seen to have the same overall width, for the osteoporotic bone the peaks of maximum cortical density are farther apart due to the compensa-tory broadening that occurs with long term bone loss.*

regardless of modality, the blurring of the visible edge between the cortical bone and the external soft tissue, or the cortical bone edge and the internal medullary cavity, reduces the image sharpness and hence the maximum achievable precision and accuracy. Radiogrammetry, however, is typically performed as a comparative measure between cortical bone thickness and medullary cavity width—the cortical index. Thus the key question becomes not 'how sharp is the image of the bone edge?', but rather 'how sharp is the image of the bone edge as compared with the width of the bone image?', and that is how we shall discuss it here. Certainly, regardless of other factors, the edge of a low-density bone is likely to be harder to reliably identify than that for a high-density bone (figure 11.4), and the desired level of sharpness should always be considered with respect to the subject group.

At a fundamental level, sharpness is limited by the resolution of the system at the detector plane, as shown in figure 11.5. Poorer resolution increases the proportional width of the partial edge effect pixels around the outside and inside borders of the bone and thus reduces the precision of

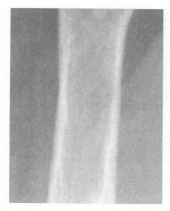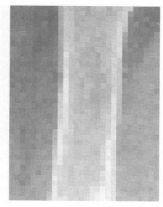

Figure 11.5. *Sharpness is limited by system resolution at the image plane. The image on the left is the osteoporotic bone image from figure 11.1, at the original radiographic resolution. The image on the right is the same image, but with resolution degraded by a factor of 10, to a level roughly equal to a high-resolution DXA system.*

any radiogrammetric measure. For any cone-beam or fan-beam point source transmission modalities, such as plane radiography or fan-beam DXA, image resolution can be improved by changing the geometry—magnifying the image by moving the beam point source closer to the target bone.

This magnification comes at a price, however. First, for ionizing radiation modalities, moving the source closer also significantly increases the radiation dose to the subject. Second, with degree of magnification inversely related to distance from source to object, moving the object close to the source exaggerates the effect of slight differences in the source–object distance, causing different parts of the bone to be magnified by very different amounts. Similarly, geometric distortion can also be seen if an object is not 'flat' relative to the image plane, or if the object is too close to the edge of the field of view (figure 11.6). There is also an absolute limit to which the image resolution—and thus sharpness—can be improved by moving the object; the improvement achievable is determined by the width of the source focal point: the wider the focus, the more penumbra (partial shadows) and the lower the sharpness.

Similarly, moving the object too close to the detector or film can cause problems for modalities which 'project' an image. A certain amount of random scatter is inevitable with photons interacting with bone and soft tissue and, assuming no collimation, any scatter effects will be present in the final image, unless scattered beyond the field of the imaging plane. If the scatter angle is wide, and the image plane far beyond the object plane, then the resulting scatter will appear as no more than diffuse 'fog' in the

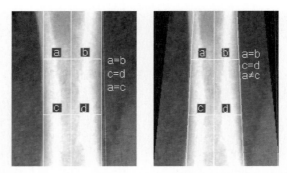

Figure 11.6. *Effect of image distortion on measurement validity. The left image shows the non-osteoporotic metacarpal from figure 11.1, with an overlayed grid and four measurements a, b, c, d. The right image shows the same image under simulated geometric distortion around a focal point (not shown) located somewhere above the image. For the undistorted image, valid comparison between a, b, c or d could be made. For the distorted image, only measurements taken at the same distance from the focal point would be valid, thus a can be compared with b or c with d but neither a nor b with either c or d, unless the distortion can be compensated for.*

image, reducing the overall image quality somewhat, but little else. But for low scatter angles, or image planes close to the object planes, the scatter effect will be more serious, softening the image of the edge of the bone, making it appear wider than it actually is and making it harder to measure the width. This fogging is known as 'exposure fog' rather than the strict radiographic definition of fogging, which results from bad film or bad processing. Anti-scatter grids or other techniques can reduce scatter somewhat, but at a cost of reducing photon count, so requiring longer exposure times and again increasing radiation dose. It is normally better to reduce the effect of scatter by restricting the field size, or else maintain a reasonable distance between object and detector.

The above may create the impression that any attempt to improve image sharpness can often distort the image geometry to such an extent as to degrade, rather than improve, radiogrammetric measurement. This is not necessarily the case, as a knowledge of the shape of the target bone coupled with a good understanding of the types of distortion to which a particular image modality is prone can produce superior image protocols. However, in practice, these protocols are often similar to the established imaging protocols for the bone of interest and, when in doubt, it is normally best to use these. It is fair to say though, that radiogrammetry always requires a good understanding of the image geometry if any useful comparisons are to be made between subjects.

SECTION C: THE CLINICAL APPLICATION OF
RADIOGRAMMETRY

11.12. IMPLEMENTING A NEW RADIOGRAMMETRY TECHNIQUE IN A CLINICAL SETTING

Although the technique at its simplest requires little or no dedicated equipment, there are still several steps to be followed when deciding whether and how to attempt radiogrammetry measurement:

1. Choosing an appropriate target condition.
2. Choosing the target bone.
3. Choosing the modality.
4. Establishing the image geometry.
5. Choosing the means of measurement.
6. The need for comparative reference.

Although this is the ideal order for implementing radiogrammetry, in reality many of the steps are predetermined. For example, if only plane radiography is available, this will automatically determine the image modality and heavily influence the choice of target bone and conditions which can be investigated. One of the attractions of radiogrammetry, however, is that it can often be retrospectively applied to images which have been acquired for a different clinical purpose—to 'routine' diagnostic radiographs, for example.

11.13. CHOOSING AN APPROPRIATE TARGET CONDITION

At least within the clinical setting, the decision to implement radiogrammetry should always stem from the underlying pathology. Clinically, we are not interested in measuring bone simply for the sake of measurement, and must therefore consider the changes that the expected pathology would produce. Do we know if bone thickness will change between pathology and non-pathology? If so, by how much will it change?

Radiogrammetry in any form can only measure the gross dimensions of cortical bone and the medullary cavity and so is not appropriate for studying all bone diseases or conditions. To be suitable for study, a condition or disease must therefore cause a measurable degree of change to the cortical or medullary thickness. Some potential examples include osteoporosis (either post-menopausal, age-related or any other form), Paget's disease, primary hyperparathyroidism, or simply the response of the bone to injury, disuse or exercise.

Most of the diseases concerned are of bone metabolism, and radiogrammetry is effective at identifying bone deterioration associated with advanced metabolic bone diseases, although it is not necessarily appropriate for the diagnosis of a specific disease type. Some diseases may be suitable for some forms of radiogrammetry, but not for others. For example, primary hyperparathyroidism causes irregular cortical resorption, thus radiogrammetry performed from a single measurement would most likely be unsuitable due to the associated high error in repeatability; but radiogrammetry drawn from a region of interest would smooth out some of this variability, perhaps enough to show significant differences between subjects and controls.

11.14. CHOOSING THE TARGET BONE

Once the condition of interest has been decided, together with the likely degree to which the condition will change cortical bone thickness, it then becomes possible to specify the bone or bones to be studied. All other things being equal, this would be the bone which would be expected to show the greatest changes with disease onset. The choice would, however, likely be heavily influenced by the difficulty involved in imaging the desired bone. Even if the stapes were the most suitable bone, there are few techniques which could image it with any real accuracy.

In general, bones or bone regions with a good balance of cortical bone and medullary cavity area are most suitable, typically any tubular (long) bone or any other bone with a cortical proportion of between 25 and 75%. Bones with too high a proportion of cortical bone are generally unsuitable as they do not tend to show much variation under a disease process. Bones with the thinnest cortical shells are similarly unsuitable as they are far more subject to repeatability errors, with small measurement inaccuracies producing a proportionally higher effect, although higher precision techniques could potentially be performed on some of these bones.

As the dimensions of cortical bone and medullary cavity change with age and disease, the age and health status of the potential subject group should also be accounted for when choosing the target bone—it should not be assumed that just because the bone of a fit healthy adult is ideal for study, that that of an aged or diseased individual would be likewise. Furthermore, the presence of the target disease process does not preclude the presence of other disease processes. The possible presence of a confounding disease process must also therefore be taken into account when choosing the target bone.

11.15. CHOOSING THE MODALITY

In an ideal world, the imaging modality would be chosen once the target bone and disease condition had been decided upon. In reality, modality would typically be predetermined by availability. In any case, the choice would depend upon the degree of accuracy and precision required for the desired bone, traded off against any cost and health implications (such as radiation dose) associated with the selected image modality. Knowing in advance the reference range and the difference expected for symptomatic individuals, it should then be possible to deduce which imaging techniques will produce an image of appropriate definition. At least within the research arena, however, the reference range and expected difference are unlikely to be known in advance and so modality selection can become a process of trial-and-error, dictated by availability.

Although plane radiography is the obvious choice, it is far from the only one. Indeed, any technique can be considered for radiogrammetry, if capable, by whatever means, of measuring cortical thickness with an acceptable degree of precision, with the exact precision dependent on the condition under investigation and the bone selected. To meet the precision criteria, the modality would be expected to clearly distinguish between cortical bone, the inner medullary cavity and the outer surrounding tissue. Further to this, the methodology chosen for the technique should be reproducible to such a degree that the combined inaccuracies of the modality and the methodology do not degrade the precision below an acceptable level. Certainly, deducing which imaging modalities are appropriate is considerably easier if the difference between symptomatic individuals and the reference range is known already.

Potential modalities obviously include plane radiography, but also X-ray CT, μCT, higher resolution DXA systems, and MRI. For any modality, a good knowledge of any modality-specific image artefacts and distortions would be required. In addition, any 'proprietary' modifications for the specific system would also need to be accounted for—for example, it is quite common with modern digital radiography systems for proprietary software algorithms to 'enhance' the edges of the bone to make them more visible—but distorting the visible dimensions somewhat in the process. Ideally, radiogrammetry should always be performed on the 'raw' image and, at least for radiography, DICOM images will normally suffice. Radiogrammetry can still be performed on processed images, but doing so would require the accumulation of a reference data set at least specific to the imaging machine, and potentially to the imaging software version. This would typically be impractical for any long-term or multi-centre study, unless machine and software could be standardized throughout.

11.16. ESTABLISHING THE IMAGE GEOMETRY

In practice, for prospective studies, the established clinical imaging protocols for a particular site of interest would be expected to offer the best balance of sharpness and geometry, provided the target bone is roughly central in the image. Thus it is normally better to stick with the established protocols unless there are clear benefits to be gained from doing otherwise. For retrospective radiogrammetry studies performed on previously acquired images, however, the target bone may not be in the centre of the image, and in such cases the potential influence of image sharpness and geometry should be taken into account. In general, though, even geometric distortions can be compensated for, provided the target bone appears in roughly the same part of the image for successive subjects.

11.17.   CHOOSING THE MEANS OF MEASUREMENT

All radiogrammetry relies on the measurement of bone dimensions; however, which measurements to be taken and how they are to be taken varies from technique to technique. As we have discussed, both cortical bone and medullary cavity can be measured in terms of width (one dimension), area (two dimensions) or volume (three dimensions). Typically, measurement in fewer axes is simpler, but more reliant on good positioning, whilst more axes require more complex equipment, but offer greater flexibility in terms of approach.

The available choices of measurement are heavily influenced by the choice of modality and the number of dimensions that modality employs. One-dimensional measurements have the advantage of simplicity, but are highly subject to positioning and repeatability errors. Two- or three-dimensional measures are more robust, but typically require dedicated software. Thus a one-dimensional measure might be a good choice for a pilot study, due to the ease of use, but larger research studies or true clinical diagnosis are likely to require some degree of process automation.

Indeed, in general, the measurement of a single value from each individual is never enough, as radiogrammetry is itself the measured relation between cortical bone and medullary or total bone thickness, thus requiring at least two measurements from each individual. Furthermore, the measurement precision and validity of most investigations might well be improved by examining more than one bone from the same subject, and the most successful techniques have indeed employed automated multiple measurements taken from several bones.

If any meaningful clinical information is to be found, it is also vital to choose a type of measure suitable for drawing comparisons between individuals or between an individual and a reference range. As individuals obviously vary in bone size, age, height and weight, direct comparison of bone dimensions are largely meaningless unless attempting to calculate actual bone ash weight or fracture load. Most meaningful comparisons between individual and population as a whole would probably employ the cortical index, or a similar comparative ratio.

11.18. THE NEED FOR COMPARATIVE REFERENCE

Radiogrammetry is the comparative measurement of cortical bone thickness as a means of identifying pathology or providing quantitative information on bone structure. As such, most radiogrammetric techniques rely on comparison of the individual with a known reference range. This implies that a reference has to be available, and valid for comparison with subjects of sex and age. If not available, a reference range must be established. As we have already mentioned, radiogrammetric measurements can vary considerably, dependent on such factors as modality, beam geometry, target bone and the subject group under investigation. As a result, most research studies would require an age- and sex-matched control group to provide normative data, if any meaningful comparison with the general population were to be drawn.

11.19. MEASUREMENT VALIDITY

It would be remiss to discuss the limitations of any clinical imaging technique without at least touching on the issue of measurement validity. Broadly speaking, for radiogrammetry, as with any bone assessment technique, there are four comparisons that can be made, namely the identification of disease, the comparison with fracture incidence/prediction of fracture risk, the prediction of fracture load, or the prediction of bone mineral density/ash weight. For a technique to be valid, it must be able to predict at least one of these factors to a reasonable degree. Conversely, when implementing a new radiogrammetry technique, we must accept the risk that even with all limiting factors accounted for and all errors reduced to a minimum, the investigation might still prove invalid simply because an inappropriate target condition or bone was chosen, or because the modality was inappropriate even when optimally applied. In short, it may simply be that there was nothing there to see or that we are unable to see it.

11.20. FURTHER RESEARCH OPPORTUNITIES IN RADIOGRAMMETRY

Radiogrammetry research is scattered throughout many fields and across more than five decades. Although the technique is enjoying somewhat of a revival, current research is largely focused on the prediction of post-menopausal osteoporosis from automated radiogrammetry of the metacarpals. Some work has been conducted on other sites, but little work has been conducted on three-dimensional data, non-radiographic images or on diseases other than post-menopausal osteoporosis. All of these areas may warrant further investigation, making radiogrammetry a promising area for future research.

REFERENCES

[1] Einhorn T A 1992 Bone strength: the bottom line *Calcif. Tissue Intl.* **51** 331–9
[2] Smith R W and Walker R 1980 Femoral expansion in aging women: implications for osteoporosis and fractures *Henry Ford Hosp. Med. J.* **28** 168–70
[3] Ruff C B and Hayes W C 1982 Superiosteal expansion and cortical remodelling of the human femur and tibia with ageing *Science* **217** 945–8
[4] Barnett E and Nordin B 1960 The radiological diagnosis of osteoporosis: a new approach *Clin. Radiol.* **11** 166–74
[5] Rosholm A, Hyldstrup L, Baeksgaard L, Grunkin M and Thodberg H H 2001 Estimation of bone mineral density by digital X-ray radiogrammetry: theoretical background and clinical testing *Osteoporos. Intl.* **12** 961–9
[6] Extonn Smith A N, Millard P H, Payne P R and Wheeler E F 1969 Method for measuring quantity of bone *Lancet* **2** 1153–4
[7] Dequeker J 1982 Precision of the radiogrammetric evaluation of bone mass at the metacarpal bones. In *Non-Invasive Bone Measurements: Methodological Problems* eds J Dequeker and C C Johnston (Oxford: IRL) pp 27–32
[8] Meema H E and Meindock H 1992 Advantages of peripheral radiogrammetry over dual-photon absorptiometry of the spine in the assessment of prevalence of osteoporotic vertebral fractures in women *J. Bone. Min. Res.* **7** 897–903
[9] Bloom R A 1980 A comparative estimation of the combined cortical thickness of various bone sites *Skeletal Radiol.* **5** 167–70
[10] Johnell O, Onnby K and Redlund-Johnell I 2000 Superior short-term *in-vivo* precision with digital X-ray radiogrammetry compared to DXA *J. Bone Min. Res.* **15**(S1) 5282
[11] Jorgensen J T, Andersen P B, Rosholm A and Bjarnason N H 2000 Digital X-ray radiogrammetry: a new appendicular bone densitometric method with high precision *Clin. Physiol.* **20** 330–5
[12] Cootes T, Taylor C, Cooper D and Graham J 1995 Active shape models: their training and application *Computer Vision and Image Understanding* **61** 38–59
[13] Bouxsein M L, Palermo L, Yeung C and Black DM 2002 Digital X-ray radiogrammetry predicts hip, wrist and vertebral fracture risk in elderly women: a prospective analysis from the study of osteoporotic fractures *Osteoporos. Intl.* **13** 358–65
[14] Black D M, Palermo L, Sorensen T, Jorgensen J T, Lewis C, Tylavsky F, Wallace R, Harris E and Cummings S R 2001 A normative reference database study for Pronosco X-posure system *J. Clin. Densitom.* **4** 5–12

Chapter 12

In vivo neutron activation analysis and photon scattering

Ian M Stronach and Alun H Beddoe

12.1. INTRODUCTION

In this chapter the use of two 'fringe' physical techniques in the measurement of bone are briefly described. *In vivo* neutron activation analysis (IVNAA) was the principal technique used in the 1970s [1] but it has been superseded by dual X-ray absorptiometry (DXA) or dual photon absorptiometry (DPA) in the past two decades (see earlier chapters), largely because of the radiation effective doses required for IVNAA and because DXA/DPA provides imaging (although not used for diagnosis) as well as bone density information. IVNAA in this context actually measures total body or partial body calcium, not mineral density, and so does not actually differentiate osteomalacia from osteoporosis. On the other hand, photon scattering techniques enable volumetric bone density to be measured in small defined bone volumes (via Compton scattering; refer to [2]) and also the volumetric bone mineral density (via the coherent: Compton ratio; refer to [3]). The scattering techniques have not been followed up in the clinical setting, presumably because of the difficulties in maintaining reproducibility in the clinical situation due to the need to precisely define the bone volume of interest over repeated investigations. IVNNA and the scattering techniques are not generally used nowadays but they are discussed in this book because of the historical part they have played in the development of bone measurement techniques and because they provide information which is not available via DXA/DPA (both of which effectively measure areal density). These techniques also provide a tool for research into body composition.

The remainder of this chapter is split into two parts, the first dealing with IVNAA and the second with the scattering techniques.

12.2. *IN VIVO* NEUTRON ACTIVATION ANALYSIS (IVNAA)

The IVNAA technique was first applied in human subjects to measure the total body contents of sodium, chlorine and calcium by Anderson and coworkers in 1964 [4]. Four years later a Birmingham University group reported the clinical usefulness of IVNAA in the study of bone disease [5]. These seminal papers introduced a whole new field of medical research into the composition of the human body, in particular it became possible to measure a whole range of important elements including calcium, hydrogen, carbon, oxygen, phosphorus, sodium, iodine, nitrogen and chlorine. IVNAA can be separated into whole body and partial body techniques (where the region of interest might be part of the appendicular skeleton) and into so-called delayed gamma and prompt gamma techniques (see below).

IVNAA is an analytical tool based on nuclear rather than chemical or biochemical reactions. Several mechanisms exist by which neutrons from an external source can interact with the nuclei of constituent elements of the body. These include:

- *Elastic* collisions (n, n') where the incident neutron collides with the nucleus and scatters such that overall kinetic energy is conserved. The kinetic energy lost by the neutron is transferred to the nucleus, which recoils. No radiation is emitted as the energy transfer is purely as kinetic energy.
- *Inelastic* collisions $(n, n'\gamma)$ where energy is absorbed internally by the nucleus, and kinetic energy is not conserved. The nucleus is left in an excited state, and γ-rays are emitted as it de-excites.
- *Neutron capture* $((n, 2n), (n, p), (n, \alpha)$ or $(n, \gamma))$, where the neutron is captured, or absorbed, by the nucleus, leading to a nuclear reaction. Neutron capture changes the atomic mass and/or the atomic number of the nucleus. Neutron capture may or may not be accompanied by subsequent γ-ray emission. The most important for IVNAA is radiative capture (n, γ) where a new isotope of the element is produced, which subsequently decays emitting a γ-ray.

The probability of a given interaction occurring (known as the 'cross-section') depends on the energy of the neutron and on the target nucleus, with some interactions being characterized by a threshold neutron energy below which they cannot occur. By undergoing repeated elastic collisions with nuclei (predominantly hydrogen nuclei) fast neutrons (typically with a mean energy of 2–10 MeV depending on the neutron source) can be slowed down rapidly in a few microseconds to 'thermal' energies (having a mean energy of 0.025 eV, defined as the mean kinetic energy of neutrons in thermal equilibrium with their surroundings at room temperature). Thermal neutrons can diffuse through body tissues for a few hundred microseconds and are eventually captured by the nuclei they encounter. The process of radiative neutron capture increases the atomic weight of an element by one nucleon, which may lead to a stable isotope after prompt de-excitation (e.g. $^{14}N(n, \gamma)^{15}N$ which produces the stable nitrogen isotope ^{15}N) or to a radio-isotope which subsequently decays by β^- or γ emission (e.g. $^{23}Na(n, \gamma)^{24}Na$ which produces the radioactive isotope ^{24}Na, which has a half-life of 15 h and decays by β^- and γ emission).

All neutron activation techniques depend ultimately on the emission of a γ-ray characteristic of the product nucleus. This emission can occur almost immediately (within 10^{-15} s) in which case the technique is usually referred to as 'prompt γ analysis' or over a longer time period (of the order of minutes) where the technique is generally referred to as 'delayed γ analysis'. By counting the γ-ray photon emission under controlled conditions the atomic composition of the subject under analysis can be accurately determined.

The prospective success of any IVNAA technique is a function of the product of the abundance of the element of interest in the body and its neutron cross-section. In this case the cross-section can be considered to be the probability of a neutron of the appropriate energy interacting with a nucleus of the element of interest in a reaction which will produce a detectable γ-ray photon. The degree of success also depends on achieving a sufficiently high and uniform neutron fluence rate over the region of interest. Fluence rate can be defined as the number of neutrons per unit area per second crossing a plane perpendicular to the direction of travel of the neutrons. The neutron energy spectrum (i.e. whether the neutrons are predominantly thermal or fast) is selected or modified depending on the reaction being monitored. In addition an acceptable detection efficiency must be achieved—that is, sufficient γ-ray photons emitted from the element of interest must be detected in comparison with those γ-ray photons from other activated elements and from natural background radiation. Finally, the limitation in any clinical work involving ionizing radiation is the effective dose which can reasonably be given to patients and/or the normal populations used as controls (see chapter 13).

12.2.1. Delayed gamma techniques

The delayed gamma analysis of calcium in the human body is generally carried out by utilizing one of two activation reactions. The first, which was by far the most frequently used, involves the $^{48}Ca(n, \gamma)^{49}Ca$ reaction, where ^{49}Ca decays with a half life of 8.72 min and emits a γ-ray at 3.084 MeV. The activation occurs primarily at thermal neutron energies [5–7]. The second method utilizing the $^{40}Ca(n, \alpha)^{37}Ar$ reaction was developed at the University of Washington [8]. The radioactive isotope ^{37}Ar has a half-life of 35.02 days and decays emitting characteristic chlorine X-rays of energy 2.82 keV. Air exhaled by the subject containing the ^{37}Ar is collected and

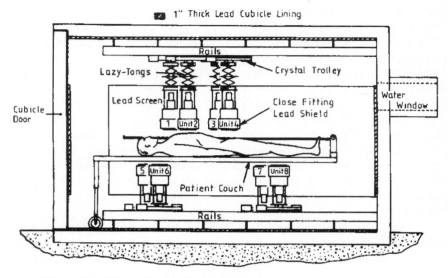

Figure 12.1. *The Leeds whole body counter.*

measured *in vitro*. Though the effective dose is very much less than for the first technique (^{40}Ca is much more abundant than ^{48}Ca) there are many unresolved problems with the technique and it never became popular, so that the discussion which follows is limited to the first method.

The induced activity, $I(t)$, is proportional to the mass of calcium, M, present in the skeleton (or part of the skeleton) such that

$$I(t) = \frac{MN}{A} \phi \sigma_{\text{th}} f [1 - \exp(-\lambda t)] \qquad (12.1)$$

where N is Avogadro's number and A is the atomic weight of ^{48}Ca, ϕ and σ_{th} are the thermal neutron fluence and the thermal neutron capture cross-section of ^{48}Ca, f is the isotopic abundance of ^{48}Ca (0.187%) and λ is the decay constant of ^{49}Ca ($1.32 \times 10^{-3}\,\text{s}^{-1}$ given the 8.72 min half-life of ^{49}Ca). The short half-life of ^{49}Ca necessitates the patient being transferred from the neutron exposure facility to the whole body or shadow shield counter (figures 12.1 and 12.2) as quickly as possible so that the counts yield is maximized. For a system with overall efficiency, E, the yield, C, over the 3.084 MeV peak is given by

$$C = E \int_{t_d}^{T} I(t)\, dt \qquad (12.2)$$

where t_d is the delay time between end of irradiation and commencement of counting and T is the total counting time. In practice it is difficult to derive the efficiency accurately, so calibration is achieved by counting

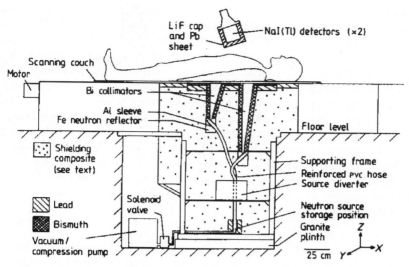

Figure 12.2. *The prompt gamma IVNAA system at Swansea which utilizes a* $4\,GBq\,^{252}CF$ *neutron source [9].*

anthropomorphic phantoms containing known amounts of calcium under identical irradiation and counting conditions and deriving body habitus corrections. The total (or partial) body calcium is therefore given by

$$M = \frac{C - C_{\mathrm{BG}}}{C_{\mathrm{std}} - C_{\mathrm{BG}}} H M_{\mathrm{std}} \qquad (12.3)$$

where C_{std} is the counts obtained with the anthropomorphic phantom, C_{BG} is the counts obtained over the same peak in the absence of calcium (including any interfering reaction), H is a body habitus correction usually derived from phantom studies and M_{std} is the mass of calcium present in the anthropomorphic phantom. The precision of the technique is generally limited by the effective neutron dose one accepts for the subject, so that the overall precision is not affected by the calibration phantom such that

$$\sigma_{\mathrm{M}}\% = [C + 2C_{\mathrm{BG}}]^{1/2} \frac{100}{C}\%. \qquad (12.4)$$

Precisions of the order of 3% are achievable with an effective dose of 10 mSv.

Neutron sources

Although the (n, γ) reaction involves thermal neutrons it is necessary to use sources of fast neutrons and use the body as a thermalizing medium in order to get a sufficiently uniform thermal neutron fluence. For delayed gamma techniques, ideally neutron output needs to be of the order of 10^{10} neutrons per second, and this is achievable with nuclear reactors, heavy charged particle accelerators and so-called (d,T) neutron generators (where deuterons are accelerated on to a tritium target) [5, 10, 11]. Isotopic sources have also been used [12] typically plutonium–beryllium (^{238}Pu–Be), americium–beryllium (^{241}Am–Be) or californium (^{252}Cf). The ^{238}Pu–Be and ^{241}Am–Be neutron sources contain a mixture of an α-particle emitting radioisotope (^{238}Pu or ^{241}Am) and beryllium. Neutrons are produced by the ^{9}Be$(\alpha, n)^{12}$C nuclear reaction, in which an α-particle is absorbed by a beryllium nucleus, and a neutron is emitted, leaving behind a carbon nucleus. In the case of ^{252}Cf, the neutrons are produced by the spontaneous nuclear fission of the californium nucleus. These sources produce fast neutron spectra of mean energy around 4.5 MeV for ^{238}Pu–Be and ^{241}Am–Be, and around 2.3 MeV for ^{252}Cf. The main advantage of such systems is that they can be readily set up in a hospital or clinic and require virtually no technical support. Isotopic sources also enable a lower patient dose to be achieved for an equivalent activation.

Sources of error
The major source of imprecision and inaccuracy of the delayed gamma technique is the variability of the fast and thermal neutron flux within the patient and this is particularly true of total body techniques. Since calcium is concentrated in the skeleton and the latter has varying amounts of soft tissue overlying it, it is not possible to achieve uniform thermal flux throughout the skeleton, as the incident neutrons are thermalized and absorbed by different amounts in the different thicknesses of soft tissue. The use of neutron moderators such as polyethylene, aids thermal neutron flux uniformity, as does using a higher initial neutron energy source (but at the expense of patient exposure). Bilateral irradiation also improves uniformity [13]. Even if uniform thermal neutron flux through the skeleton is achieved, the obvious non-uniformity of calcium distribution throughout the subject under study is a further source of error, arising from both geometrical and self-absorption effects. While these sources of error affect the absolute determination of calcium, they become insignificant when monitoring serial changes.

Interference from other neutron activation reactions can be a problem, at least in principle, although for the element calcium this only has to be considered when using a (d,T) 14 MeV neutron generator. The fast neutron interaction with body chlorine, ^{38}Cl$(n, \gamma)^{39}$S, produces a delayed 3.103 MeV gamma which, using NaI detectors, is not resolvable from the 3.084 MeV gamma from ^{49}Ca.

Detectors
Delayed gammas emitted from ^{48}Ca$(n, \gamma)^{49}$Ca reactions are usually counted in a whole body counter to minimize background; such a counter situated at University of Leeds is shown in figure 12.1 [14]. While the designs of whole body counters vary, the Leeds system consists of two arrays of four 15 cm NaI(Tl) detectors (to maximize collection efficiency) positioned

above and below the patient in a steel- or lead-lined cubicle which effectively shields out background radiation from terrestrial and extra-terrestrial sources (to around 2% of normal background). NaI(Tl) detectors are used despite the limited energy resolution because of their much higher collection efficiency than, say, hyperpure germanium detectors, though these are used in some situations where good resolution is the most important parameter [9]. For further discussion of NaI(Tl) and other scintillation detectors the reader is referred to the general radiation detection literature, for example Knoll [15].

An alternative approach [16] is to place the patient on a moving couch drawn at constant speed past one or more stationary detectors (typically two detectors around 30 cm diameter placed above and below the patient and heavily shielded).

With either system it is necessary to have highly stable counting electronics ideally coupled to a multi-channel analyser (single-channel analysers are sometimes used), counting over 10–30 min. The detector arrays must be arranged to maximize coverage for the typical subject. The counting efficiency of the detector system depends directly on the number and volume of detectors. Curiously whole body and partial body counters vary quite considerably between the various centres but all seem to achieve broadly comparable performances.

Total body techniques
Total body techniques have been used primarily to derive calcium indices for populations of patients with metabolic bone disorders such as osteoporosis, osteomalacia and renal osteodystrophy or in groups of patients with endocrine disorders [17]. Total body calcium indices relate calcium in given groups of patients to those in comparable normal patients matched for age and sex and perhaps other nutritional or body habitus parameters (cf. section 12.2.3).

Absolute measurement of calcium can be achieved by applying equation (12.3) both to patients and to various anthropomorphic phantoms of appropriate sizes and containing known amounts of calcium. Many calibration techniques have been described but the most sophisticated was that in use at the Brookhaven National Laboratory [12]. An Alderson anthropomorphic phantom was used; measured amounts of Ca were irradiated in a standardized thermal neutron beam from a nuclear reactor, then the irradiated calcium was distributed homogeneously in the phantom and the total counts due to ^{49}Ca were measured in the whole body counter. After each patient was irradiated and counted, subject-specific attenuation and geometry corrections were applied to the net counts to derive total calcium. Because calcium is virtually limited to the skeleton it is not possible to introduce internal standardization (i.e. elemental count ratios) as is possible with the prompt gamma measurement of nitrogen [18].

Partial body techniques

Partial body techniques can be more useful in that they provide calcium data for specific body regions, and an additional advantage is that the effective dose to the patient can be very much less than for total body techniques, particularly when the region of interest is in the appendicular skeleton. Regions studied have included the hand, forearm and spine, and most centres used several isotopic neutron sources to achieve uniform thermal neutron flux across the region of interest [17]. Measurement of the induced activity is achieved by using shielded NaI detectors or in the case of spinal measurements by using shadow shield counters. Precisions of the order of 2.5–3% were obtainable with entrance skin equivalent doses ranging from 75 to 100 mSv for hand or arm and 5 to 30 mSv for spine [13]; in these equivalent doses the assumed radiation weighting factor was 10, so that using the current ICRP60 recommended factor of 20 would double the equivalent dose [19]. It should be noted that effective doses were generally not measured or derived in the 1960s and 1970s, at least not in body composition research centres.

12.2.2. Prompt gamma techniques

The methods discussed above to measure calcium are all delayed gamma techniques. In 1977 Zamenhof *et al* [20] showed theoretically that it would be possible to measure calcium by a prompt gamma technique. This was later confirmed experimentally by the Swansea group [21] in a feasibility study which demonstrated that total body calcium could be measured with a precision of $\sim$5% with a skin dose averaged over the body of 1–1.3 mSv. The Swansea team continued the theme to develop a multi-element IVNAA system for both prompt and delayed gamma analysis and which included the prompt gamma analysis of calcium [9]. A schematic diagram of this instrument is shown in figure 12.2.

For prompt measurement Ryde *et al*'s system had two detection systems depending on the element one wished to measure. For calcium two n-type hyperpure germanium (HP Ge) detectors were positioned above the scanning couch depicted in figure 12.2 but out of the direct fast neutron beam (to minimize radiation damage to the detectors). The detectors were shielded by caps of ^{6}Li-enriched LiF and further shielded by lead, wax and bismuth. Using this set-up it was possible to achieve a precision of 2.6% for an entrance equivalent dose of 6.4 mSv.

12.2.3. Clinical applications and conclusion

The clinical applications of the measurement of calcium by IVNAA were largely carried out in the 1970s and early 1980s and have been documented in some detail by Cohn and Parr [17], Cohn [1, 13] and McNeill and Harrison [22]. Both partial and total body calcium measurements have been reported.

Table 12.1. *Syndromes which have been studied using IVNAA.*

Osteoporosis	Post-menopausal
	Senile osteoporosis
	Osteogenisis imperfecta
	Drug-induced osteoporosis
Osteomalacia	Vitamin D deficiency states
	Anticonvulsant drug induced
	Hypophosphatemic rickets
Renal osteodystrophy	
Paget's disease	
Endocrine disorders	Cushing's syndrome
	Acromegaly
	Parathyroid disorders
	Thyroid disorders
	Hypogonadism
Other disorders	Myotonic dystrophy
	Thalassaemia
	Alchoholic cirrhosis
	Cadmium-induced bone loss

Baseline data on normal calcium levels have also been obtained because these are essential for assessment of calcium deficiency in the various syndromes where bone loss or calcium loss are featured. These syndromes are summarized in table 12.1, which has been adapted from a table in Cohn and Parr [17]. In any such assessments factors such as age, sex, ethnicity, not to mention body habitus and weight, must all be considered if the measurements are to be of clinical value.

The IVNAA technique has, however, been superseded by DXA and DPA, and was never routinely available as a population screening technique in the way the new techniques are currently used. Though IVNAA is in a sense a more direct technique than DXA/DPA, the radiation effective dose required, especially for whole body studies, as well as its cost, poor spatial resolution and the lack of internal standardization (as with, say, prompt gamma analysis of nitrogen [18]) have all encouraged researchers towards using DXA/DPA in the clinical setting. For all these reasons the measurement of calcium by IVNAA has had to join the many other superseded techniques which litter the history of medicine.

12.3. PHOTON SCATTERING METHODOLOGIES IN MEASUREMENT OF BONE DENSITY

Simple transmission based approaches to measurement of bone density such as dual photon absorptiometry (DXA, used here to include both radioisotope source and X-ray tube based measurement systems) provide an integral measure of the amount of bone in the path of the incident beam of radiation, in terms of the areal density of the bone ($g\,cm^{-2}$) (see chapter 6). This is potentially affected by variations in bone size and shape which could result in the same total amount of bone tissue lying in the path of the incident radiation for different actual bone density and dimensions. Furthermore, while the density of trabecular bone is considered to provide the best determinant of the severity of osteoporosis [23], DXA provides a single measurement combining the cortical and trabecular components of the bone.

In response to these issues, techniques which can measure the density of a volume within the bone have been developed. Two such techniques are quantitative computed tomography (QCT) (see chapter 7) and photon scattering methodologies; the latter form the subject of this section.

The scattering approach
In photon scattering methodologies, a γ-ray or X-ray source is used to direct a beam of photons into the bone. A radiation detector (such as a hyperpure germanium semiconductor detector) measures the amount of this radiation which is scattered through a particular angle, which gives a measure of the density of the bone. The volume interrogated is the intersection of cones projected from the collimation of the source and detector, and is affected in addition by the angle through which the detected radiation has been scattered. In contrast to transmission based measurements, where the attenuation of a radiation beam is the quantity used to assess bone density, attenuation of the radiation is an interference for which corrections must be made in the measurement technique. In addition, detection of radiation which has scattered multiple times in regions outside the volume of interest can also provide an interfering signal, particularly where polychromatic radiation sources such as X-ray tubes are used.

By careful design, the scattering methodology can provide a measurement of bone density in a well defined volume inside a bone, without contribution from cortical bone, and which is relatively unaffected by bone size and shape or the thickness of overlying tissue. These techniques have, however, remained a minority interest in the face of the very widespread use of DXA in the clinical context.

Techniques
A range of techniques have been developed to assess bone density using scattered γ-rays or X-rays, each of which applies different approaches to solving the problem of attenuation of the incident and scattered radiation (see for example [2, 3, 24–26]). An additional approach, involving quantitative assessment of the shape of the Compton scattered spectrum, has also been applied [27, 28].

These techniques are restricted to use on peripheral bone sites such as the calcaneus or mandibles, where the bone can be relatively easily localized, as the location of the scattering volume well within a volume of trabecular bone is essential to provide comparable results. Incorrect positioning resulting in inclusion of cortical bone within the scattering volume, or some of the scattering volume lying within the surrounding soft tissue, would result in significant errors in the measurement.

12.3.1 Theory

Compton scattering
In Compton scattering a photon interacts with an individual atomic electron, with some of the energy of the photon being taken up by the recoil of the electron. The photon is scattered with reduced energy, dependent on the angle through which scattering takes place such that the energy of photon after Compton scattering is given by

$$E' = \frac{E}{1 + k(1 - \cos\theta)} \tag{12.5}$$

where E is the energy of the unscattered photons (γ-rays or X-rays), and E' is the energy of the Compton scattered radiation after scattering through an angle θ. The constant k is the initial energy of the photons normalized to the rest mass energy of the electron (511 keV), that is:

$$k = \frac{E}{m_e c^2} \tag{12.6}$$

where m_e is the electron rest mass, and c is the speed of light.
  The differential cross-section for this process (the probability of interaction per atom per unit solid angle) is given by the Klein–Nishina differential cross-section modified by the incoherent scattering function, which allows for the interaction between the electron and other electrons and the nucleus of the atom. Thus the Compton scattering differential cross-section $(\mathrm{cm}^2\,\mathrm{atom}^{-1}\,\mathrm{sr}^{-1})$ is

$$\frac{\mathrm{d}\sigma_{\mathrm{Comp}}}{\mathrm{d}\Omega} = \frac{\mathrm{d}\sigma_{\mathrm{KN}}}{\mathrm{d}\Omega} S(x, z) \tag{12.7}$$

where $S(x, z)$ is the incoherent scattering function as tabulated by Hubbell *et al* [29, 30], and $\mathrm{d}\sigma_{\mathrm{KN}}/\mathrm{d}\Omega$ is the Klein–Nishina differential cross-section for scattering from an isolated electron, given by

$$\frac{\mathrm{d}\sigma_{\mathrm{KN}}}{\mathrm{d}\Omega} = \frac{r_e^2}{2} [1 + k(1 - \cos\theta)]^{-2} \left[1 + \cos^2\theta + \frac{k^2(1 - \cos\theta)^2}{1 + k(1 - \cos\theta)} \right] \tag{12.8}$$

where r_e is the classical electron radius (2.817×10^{-15} m).

Coherent scattering

In coherent scattering, a photon interacts with a bound atomic electron, with the interaction involving the whole atom acting as a coherent unit. The scattered photon has the same energy as the incident photon, and the differential cross-section is given by the Thompson differential cross-section modified by the atomic form factor, $F(x, z)$, which takes account of electron binding effects, giving the coherent, or Rayleigh, scattering differential cross-section (cm^2 atom^{-1} sr^{-1}) as

$$\frac{\mathrm{d}\sigma_{coh}}{\mathrm{d}\Omega} = \frac{r_e^2}{2}(1 + \cos^2\theta)[F(x, z)]^2 \qquad (12.9)$$

for scattering through an angle θ, from an atom of atomic number z, where r_e is the classical electron radius (2.817×10^{-15} m). The momentum transfer, x, of the scattering interaction is given by

$$x = \left(\frac{1}{\lambda}\right)\sin\left(\frac{\theta}{2}\right) \qquad (12.10)$$

where λ is the wavelength of the radiation in angstroms.

Tabulated values for $F(x, z)$ are given by Hubbell *et al* [29, 30] and Hubbell and Overbo [31].

12.3.2.   Techniques

Scattering methodology

In general the methodology for the photon-scattering measurement of bone density depends on the fact that the scattering is proportional to the cross-section for scattering to an angle θ, and the number of electrons per unit volume (the electron density) in the region produced by the intersection of the projected source and detector collimation cones. The electron density is dependent on the atomic number of the material and the physical density. Thus a measurement of the amount of radiation scattered from a volume inside an object can be used to give the density of that volume of the object, given an appropriate calibration factor. Unfortunately various factors complicate the process, one of which being the attenuation of the radiation in the body tissues as it passes to and from the region of interest. If we consider the simplified case shown in figure 12.3, the attenuation of incident radiation along the path of length a, and of scattered radiation along the path of length b, must be taken into account. Clearly the neglect of this effect will result in changes in the measured density if the amount of tissue surrounding the bone changes, or indeed if the density of the bone changes. The attenuation effect increases as the density of the bone increases, so this effect would act to offset the expected increase in scattered radiation as the density increases.

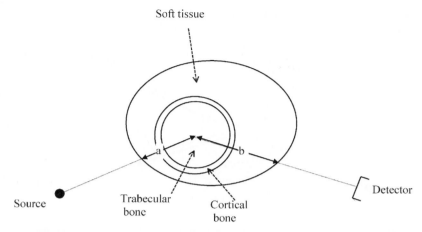

Figure 12.3. *Simplified diagram of scattering geometry, showing source and detector, and the paths in tissue of incident and scattered radiation.*

Various methodologies have been applied to the solution of this problem, and have involved the use of multiple sources and detectors, rotation of the source/detector assembly or the volume being measured, or both, and use of detectors with good energy resolution to measure separately different types of scattering. Two techniques which are still being developed or used will be described in some detail, with a summary of other approaches. The first technique involves Compton scattering, with attenuation correction made using various transmission measurements, while the second involves measurements of both Compton and coherent scattering, with correction for attenuation made by taking the ratio of these two.

It should also be noted that when the radiation dose is quoted for a scattering methodology, this is the absorbed dose to the very small volume of tissue irradiated, usually of the order of a few cubic centimetres.

Measurement using Compton scattering
In this technique, described by Clarke and Van Dyk in 1973 [2] for *in vitro* measurements, two radiation sources and two detectors were used, with measurements of both the scattered and transmitted radiation through the body being made to allow correction for attenuation. This early work used sources of relatively high γ-ray energy (200–2000 keV), with one source selected to provide radiation of energy similar to that of scattered radiation from the first source. The source-detector assembly or the body, was rotated through 180° part way through the measurement to allow the correct measurement paths to be produced. A simplified methodology using a single source and detector, but requiring small angle rotations of the both body and the source-detector assembly, was described by Olkkonen and

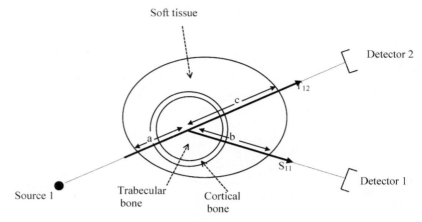

Figure 12.4. *Bone density measurement by Compton scattering, stage one, measurements for source 1.*

Karjalainen [32]. Hazan *et al* [23] made measurements of 50 subjects using a ^{137}Cs γ-ray source (660 keV) and detecting the Compton scattered radiation at 90°, with a measurement dose of 20 mGy.

Further development of the technique used lower energies to reduce the measurement dose. An *in vivo* measurement system was described by Webber and Kennett [26] and Kennett and Webber [33] for measurement of the calcaneus. This used the original high energy methodology as described by Clarke and Van Dyk [2], and required rotation of the foot through 180° after the initial scattering and transmission measurements. A 90° scattering geometry was used with ^{153}Sm (103 keV) as the primary source, and ^{170}Tm (85 keV) as the secondary source. A measurement precision of 1.5% was achieved for repeated phantom measurements over a six month period, with a radiation dose of 1.6 mGy.

A simplified low γ-ray energy measurement was described in principle for *in vitro* measurements by Clarke and Van Dyk [2]. This technique again uses two transmission and two scattering measurements, as shown in figures 12.4 and 12.5. However, the sample is not rotated between measurements and only a single γ-ray energy is used. The assumption is made that the energy of the Compton scattered γ-rays is sufficiently close to that of the transmitted γ-rays that the attenuation effects will be the same for each. Initially, as shown in figure 12.4, measurements are made of the transmission (T_{12}) and scattering (S_{11}) from a γ-ray or X-ray source placed at the position marked 'Source 1'. This source is then removed and replaced at position 'Source 2', where it is again used to produce transmission (T_{21}) and scattering (S_{22}) measurements as shown in figure 12.5. An

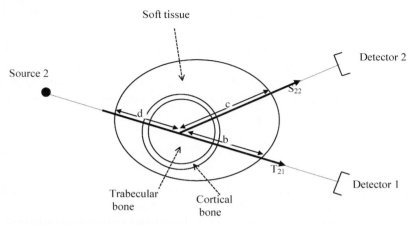

Figure 12.5. *Bone density measurement by Compton scattering, stage two, measurements for source 2.*

alternative design would use two sources, at the positions marked, with each one shielded while the measurements were made with the other source.

In simple terms the amount of transmitted and scattered radiation measured at detectors 1 and 2 from the sources are given by (ignoring detector efficiency factors)

$$S_{11} = e^{-\mu(E)a} \, \Sigma_{\text{Comp}} \, e^{-\mu(E')b} I_{01}$$

$$T_{12} = e^{-\mu(E)a} \, e^{-\mu(E)c} I_{01}$$

$$S_{22} = e^{-\mu(E)d} \, \Sigma_{\text{Comp}} \, e^{-\mu(E')c} I_{02}$$

$$T_{21} = e^{-\mu(E)d} \, e^{-\mu(E)b} I_{02}$$

$$(12.11)$$

where S_{11}, S_{22} are the scattered radiation counts measured in detector 1 originating from source 1 and measured in detector 2 originating from source 2 respectively; T_{12}, T_{21} are the transmitted radiation counts measured in detector 2 originating from source 1 and measured in detector 1 originating from source 2 respectively; E and E' are the energies and $\mu(E)$ and $\mu(E')$ are the linear attenuation coefficients for the unscattered and Compton scattered γ-rays or X-rays respectively; I_{01} and I_{02} are the incident flux of radiation from sources 1 and 2 respectively; a, b, c, d are path lengths defined in figures 12.4 and 12.5; and Σ_{Comp} is the probability of Compton scattering through angle θ.

If $E' \approx E$, then $\mu(E') \approx \mu(E)$ and so

$$\frac{S_{11}S_{22}}{T_{12}T_{21}} = \Sigma_{\text{Comp}}^2 .$$

$$(12.12)$$

The probability of Compton scattering through angle θ is given from

$$\Sigma_{\text{Comp}} = \frac{N_A}{A} \rho \left[\frac{d\sigma_{\text{Comp}}}{d\Omega} \right] \tag{12.13}$$

where ρ is density, N_A is Avogadro's number and A is the atomic mass of the scattering material. Thus, combining these equations, the density can be given by

$$\rho = h \sqrt{\frac{S_{11} S_{22}}{T_{12} T_{21}}} \tag{12.14}$$

where h is a calibration factor, which is dependent on the cross-section for Compton scattering for the material being measured.

Hence from the measurements of scattered and transmitted radiation as shown in the diagrams, and a suitable calibration measurement, the density of an object can be calculated. It should be noted that the cross-section for Compton scattering is dependent on the effective atomic number of the scatterer as well as the momentum transfer, but this effect can be largely removed by use of a suitable calibration [34]. Duke and Hanson [34] discuss in more detail the theoretical basis for both monochromatic (γ-ray) and polychromatic (X-ray tube) sources, and identify several simplifying assumptions in the calculation. In particular they identify a significant potential source of error in the detection of photons following multiple scattering events, as the theory described above assumes detection of X-rays only after single scattering.

The use of X-ray tubes as the radiation sources was investigated as these can provide greater intensity of incident radiation than can radioisotope γ-ray sources, and thus more rapid or more precise measurements. However, when the polychromatic spectrum from an X-ray tube is used, this increases the need to correct for multiple scattering of X-rays. Work on this technique has continued, with Speller and Horrocks [25] examining correction for multiple scattering using Monte Carlo simulations for phantom measurements. For clinical measurements, correction based on a supplementary detector placed such that only the photons having undergone multiple scatters can be detected has also been studied [35].

Measurement using the coherent/Compton scattering ratio
An alternative approach has been developed which involves measurement of two elements of the scattered radiation spectrum, using a high resolution detector, such as a hyperpure germanium semiconductor detector. In this technique, first reported by Puumalainen *et al* [3], coherent and Compton scattered γ-rays are measured separately. This is made possible with a detector with good energy resolution, because of the shift in energy of the Compton scattered γ-rays. However, the assumption is made that

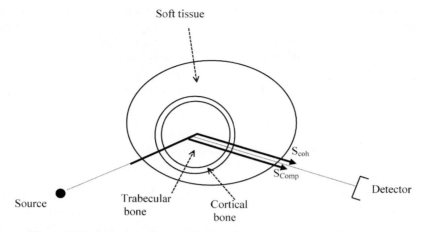

Figure 12.6. *Schematic diagram showing the general process of bone density measurement using the ratio of coherent and Compton scattered radiation.*

the attenuation of coherent and Compton scattered γ-rays will not be significantly different.

The general measurement technique is shown in figure 12.6. As the paths through tissue are the same length for coherent and Compton scattered γ-rays, the attenuation effects cancel out if the ratio of coherent to Compton scattering is taken, and the energies of the coherent and Compton scattered photons are sufficiently close that the attenuation coefficients are insignificantly different. With 60 keV incident γ-rays and a scattering angle of 90°, the energy difference between coherent and Compton scattering is 6.1 keV, which leads to a negligible source of error due to attenuation in the coherent to Compton ratio [3].

The ratio of coherent to Compton scattering, R, is found by making simultaneous measurements in a high resolution detector of the coherent (S_{coh}) and Compton (S_{Comp}) scattered γ-rays as shown in figure 12.6:

$$R = \frac{S_{coh}}{S_{Comp}}. \tag{12.15}$$

The theoretical value of R is found from the ratio of the probabilities of coherent and Compton scattering, and reduces to

$$R = h(\theta, E) \frac{[F(x, z)]^2}{S(x, z)} \tag{12.16}$$

where $h(\theta, E)$ is a function of scattering angle and incident photon energy only [36]. This can be further reduced to $R \propto z^b$ where b depends on the momentum transfer, x (and hence on scattering angle and photon energy). Values of b ranging between 1.7 and 4.0 were found by Leichter *et al* [37] for a range of photon energies and scattering through 47°. Thus the ratio of coherent to Compton scattering will depend strongly on the effective atomic number of the material in which scattering takes place. As the effective atomic number of bone mineral is significantly higher than that of soft tissue (cortical bone has an effective z of about 13.8 compared with 6.4 for yellow marrow [38]), this provides a more sensitive measure of bone mineral status than a method based on Compton scattering alone. A technique using coherent scattering alone would be still more sensitive, but the difficulties of correction for attenuation would recur. Ndlovu *et al* [39] describe such a technique, with correction for attenuation based on a transmission measurement, but an assumption of symmetry of the scattering object has to be made in this case.

For coherent to Compton scattering ratio measurements various scattering angles have been used, ranging from 22.5° [37] through to 150° [27, 28]. Large angle scattering is expected to provide greater sensitivity to changes in bone mineral density, but results in reduced amounts of coherent scattered radiation detected, because the coherent scatter cross-section decreases rapidly as the scattering angle. Poorer measurement precision might therefore be expected for systems based on large angle scattering. Work has mainly involved use of radioisotope sources, in particular [241]Am [3, 40, 41], which emits γ-rays of 59.5 keV. In attempts to improve measurement precision, work using X-ray sources has also been carried out [42, 43], with specially designed filtration used to provide a quasi-monoenergetic spectrum. The difficulty with X-ray sources is distinguishing between coherent and Compton scattering of a polychromatic incident X-ray spectrum and this work does not seem to have progressed beyond initial investigations.

The effect of the marrow content of the bone on the measured ratio has been studied, with Ling *et al* [41] finding that the coherent to Compton ratio decreased by 2.5% for a 10% increase in the yellow marrow content of a bone. Later work, in particular that of Guttmann and Goodsitt [38], shows that the errors that can arise in the ratio due to variations in the fat content of the yellow marrow in trabecular bone are reduced when large angle scattering is used.

The majority of work has been carried out on measurement systems for the calcaneus. In the original work by Puumalainen *et al* [3], using [241]Am and a 90° scattering angle, the measurement was made for a locally absorbed radiation dose of less than 1 mSv. Shukla *et al* [44] quote an accuracy of 5% and precision of 3% for a system using [241]Am and a 71° scattering angle, with a soft tissue radiation dose of 3 mGy [45]. Current work by Morgan *et al* [27, 28] using an [241]Am source, with a backscatter geometry stated as 150° scattering angle, has found that mandibular bone density can be measured with a precision of approximately 1%, for a 0.4 mGy skin surface dose.

12.3.3. Conclusions

The scattering methodologies describe provide some advantages over DXA in terms of the quantity measured. Measurement based on Compton scattering alone gives volumetric bone density while the coherent/Compton scattering ratio gives volumetric bone mineral density. In contrast, DXA measures areal density (in $g\,cm^{-2}$), and is therefore affected by variations in size and shape of bone, and in extra-osseal calcium. Of these techniques the coherent/Compton scattering ratio might therefore be expected to be most sensitive to early changes in bone mineral status. In addition, the measurement of bone density using γ-ray or X-ray scattering provides the potential to make a measurement of trabecular bone independent of the surrounding cortical sheath. However, thus far very few clinical measurements have been made using measurement systems based on these techniques and, while research continues, there are relatively few researchers working in this field.

Current work on the Compton scattering methodology focuses on the use of polychromatic sources, with correction for multiple scattering being a key factor [35]. Development work continues examining the coherent/Compton ratio method, in the backscatter geometry [27, 28]. In addition Morgan *et al* [27, 28] are developing measurements based on the shape of the Compton peak to provide tissue composition information.

REFERENCES

[1] Cohn S H 1981 *In vivo* neutron activation analysis; state of the art and future prospects *Med. Phys.* **8** 145–54

[2] Clarke R L and Van Dyk G 1973 A new method for measurement of bone mineral content using both transmitted and scattered beams of gamma-rays *Phys. Med. Biol.* **18** 532–9

[3] Puumalainen P, Uimarihuhta A, Alhava E and Olkkonen H 1976 A new photon scattering method for bone mineral density measurements *Radiology* **120** 723–4

[4] Anderson J, Osborn S B, Tomlinson R W S, Neutron D, Rundo J, Salmon L and Smith J W 1964 A neutron activation analysis *in vivo*: A new technique in medical investigation *Lancet* **2** 1201–5

[5] Chamberlain M J, Fremlin J H, Peters D K and Phillip H 1968 Total body calcium by whole body neutron activation *Br. Med. J.* **2** 581–5

[6] Palmer H E, Nelp W B, Murano R and Rich C R 1968 The feasibility of *in vivo* activation analysis of total body calcium and other elements of body composition *Phys. Med. Biol.* **13** 269–79

[7] Comar D, Riviere R, Raynaud C and Kellershohn C 1968 Recherches préliminaires sur la composition et le métabolisme do l'os étudiés par radioactivation *in vivo* chez l'homme *Radioaktive Isotope in Klinik und Forschung* (Munich: Urban and Schwarzenberg) **8** 186

[8] Palmer H E 1973 Feasibility of measuring total body calcium in animals and humans by measuring Argon-37 in expired air after neutron activation *J. Nucl. Med.* **18** 522–7

[9] Ryde S J S, Morgan W D, Evans C J, Sivyer A and Dutton J 1989 Calibration and evaluations of a ^{252}CF based neutron activation analysis instrument for the determination of nitrogen *in vivo*. *Phys. Med. Biol.* **34** 1429–41

[10] Boddy K, Holloway I and Elliott A 1973 A simple facility for total body *in vivo* activation analysis *Int. J. Radiat. Res* **24** 428–30

[11] Cohn S H, Dombrowski C S and Fairchild R G 1970 *In vivo* activation analysis of calcium in man *Int. J. Appl. Radiat. Isotop.* **21** 127–36

[12] Cohn S H, Shukla K K, Dombrowski C S and Fairchild R G 1972 Design and calibration of a broad beam ^{238}Pu,Be neutron source for total body neutron activation analysis *J. Nucl. Med.* **13** 487–92

[13] Cohn S H 1980 The present state of *in vivo* neutron activation analysis *Atomic Energy Rev.* **18** 599–660

[14] Burkinshaw L 1978 *In vivo* analysis and the whole body radiation counter. In *Liquid Scintillation Counting* vol 5, eds A Crook and P Johnston (London: Heyden) pp 111–22

[15] Knoll G F 1989 *Radiation Detection and Measurement* 2nd edn (New York: Wiley)

[16] Boddy K 1967 The development and performance of a prototype shadow shield whole body monitor *Phys. Med. Biol.* **12** 43–50

[17] Cohn S H and Parr R M 1985 Nuclear based techniques for the *in vivo* study of human body composition *Clin. Phys. Physiol. Meas.* **6** 275–301

[18] Vartsky D, Ellis K J and Cohn SH 1979 *In vivo* measurement of body nitrogen by analysis of prompt gammas from neutron capture *J. Nucl. Med.* **20** 1158–65

[19] ICRP *1990 Recommendations of the International Commission on Radiological Protection* Publication ICRP60 (London: Pergamon)

[20] Zamenhof R G, Deutsch O L and Murray E W 1977 Feasibility study of prompt gamma capture *in vivo* neutron activation analysis *Med. Phys.* **6** 179–96

[21] Morgan W D, Ryde S J S, Dutton J, Evans C J and Sivyer A 1984 *In vivo* measurement of calcium by prompt-gamma neutron activation analysis. *Proc. 5th Int. Conf. on Nuclear Methods in Environmental and Energy Research* ed J R Vogt (US DOE Conf 840408) 751–8

[22] McNeill K G and Harrison J E 1981 Partial body neutron activation. In *Non-invasive Measurements of Bone Mass and their Clinical Application* ed S H Cohn (Boca Raton: CRC) 165–90

[23] Leichter I, Bivas A, Giveon A, Margulies J Y and Weinreb A 1987 The relative significance of trabecular and cortical bone density as a diagnostic index for osteoporosis *Phys. Med. Biol.* **32** 1167–74

[24] Hazan G, Leichter I, Loewinger E, Weinreb A and Robin GC 1977 The early detection of osteoporosis by Compton gamma ray spectroscopy *Phys. Med. Biol.* **22** 1073–84

[25] Speller R D and Horrocks J A 1988 A Monte Carlo study of multiple scatter effects in Compton scatter densitometry *Med. Phys.* **15** 707–12

[26] Webber C E and Kennett T J 1976 Bone density measured by photon scattering. I. A system for clinical use *Phys. Med. Biol.* **21** 760–9

[27] Morgan H M, Shakeshaft J T and Lillicrap S C 1998 Gamma-ray backscatter for body composition measurement *Appl. Radiat. Isot.* **49** 555–7

[28] Morgan H M, Shakeshaft J T and Lillicrap S C 1999 Gamma-ray scattering for mandibular bone density measurement *Br. J. Radiology* **72** 1069–72

[29] Hubbell J H, Veigele W J, Briggs E A, Brown R T, Cromer D T and Howerton R J 1975 Atomic form factors, incoherent scattering functions and photon scattering cross sections *J. Phys. Chem. Ref. Data* **4** 471–538

[30] Hubbell J H, Veigele W J, Briggs E A, Brown R T, Cromer D T and Howerton R J 1977 Erratum: Atomic form factors, incoherent scattering functions and photon scattering cross sections *J. Phys. Chem. Ref. Data* **6** 615–6

[31] Hubbell J H and Overbo I 1979 Relativistic atomic form factors and photon coherent scattering cross sections *J. Phys. Chem. Ref. Data* **8** 69–105

[32] Olkonnen H and Karjalainen P 1975 A ^{170}Tm gamma scattering technique for the determination of absolute bone density *Br. J. Radiology* **48** 594–7

[33] Kennett T J and Webber C E 1976 Bone density measured by photon scattering. II. Inherent sources of error *Phys. Med. Biol.* **21** 770–80

[34] Duke P R and Hanson J A 1984 Compton scatter densitometry with polychromatic sources *Med. Phys.* **11** 624–32

[35] Mooney M J, Naga S S D, Speller R D and Koligliatis T 1996 Monitoring and correction of multiple scatter during clinical Compton scatter densitometry measurements *Phys. Med. Biol.* **41** 2399–410

[36] Kerr S A, Kouris K, Webber C E and Kennett T J 1980 Coherent scattering and the assessment of mineral concentration in trabecular bone *Phys. Med. Biol.* **25** 1037–47

[37] Leichter I, Karellas A, Craven J D and Greenfield M A 1984 The effect of momentum transfer on the sensitivity of a photon scattering method for the characterization of tissues *Med. Phys.* **11** 31–6

[38] Guttmann G D and Goodsitt M M 1995 The effect of fat on the coherent-to-Compton scattering ratio in the calcaneus: A computational study *Med. Phys.* **22** 1229–34

[39] Ndlovu A M, Farrell T J and Webber C E 1991 Coherent scattering and bone mineral measurement: The dependence of sensitivity on angle and energy *Med. Phys.* **18** 985–9

[40] Leichter I, Karellas A, Shukla S S, Looper J L, Craven J D and Greenfield M A 1985 Quantitative assessment of bone mineral by photon scattering: calibration considerations *Med. Phys.* **12** 466–8

[41] Ling S-S, Rustgi S, Karellas A, Craven J D, Whiting J S, Greenfield M A and Stern R 1982 The measurement of trabecular bone mineral density using coherent and Compton scattered photons *in vitro. Med. Phys.* **9** 208–15

[42] Puumalainen P, Uimarihuhta A, Olkkonen H and Alhava E M 1982 A coherent/Compton scattering method employing an X-ray tube for measurement of trabecular bone mineral content *Phys. Med. Biol.* **27** 425–9

[43] Webster D J and Lillicrap S C 1985 Coherent-Compton scattering for the assessment of bone mineral content using heavily filtered X-ray beams *Phys. Med. Biol.* **30** 531–9

[44] Shukla S S, Karellas A, Leichter I, Craven J D and Greenfield M A 1985 Quantitative assessment of bone mineral by photon scattering: Accuracy and precision considerations *Med. Phys.* **12** 447–8

[45] Shukla S S, Leichter I, Karellas A, Craven J D and Greenfield M A 1986 Trabecular bone mineral density measurement *in vivo*: Use of the ratio of coherent to Compton scattered photons in the calcaneus *Radiology* **158** 695–7

SECTION 4

NON-IONIZING TECHNIQUES

Chapter 13

Magnetic resonance imaging

Laurent Pothuaud and Sharmila Majumdar

13.1. INTRODUCTION

Bone quality, which depends on several factors pertaining to both cortical and trabecular bone, may potentially be important in the study of osteoporosis, both in the prediction of fracture risk and for the assessment of therapy. Although the skeleton is composed of about 80% cortical bone, trabecular bone is highly responsive to metabolic stimuli and has a turnover rate approximately eight times that of cortical bone [1]. Hence, trabecular bone is a prime site for detecting early bone loss and monitoring response to therapeutic interventions. The evaluation of trabecular bone structure (TBS) could contribute to defining bone quality, and thus complement bone density in characterizing skeletal status.

The primary imaging modalities used to measure bone properties, especially in the clinical arena, are X-ray or ultrasound based. In X-ray based imaging techniques the image intensity reflects tissue density, while in ultrasound images the image intensity is a depiction of the attenuation and speed alteration of sound waves. In this chapter, the potential of magnetic resonance imaging (MRI) will be explored. Typically, in clinical MRI images, the signal detected originates from the hydrogen atom which is found in tissue water (70% of the human body), body fat, cholesterol etc. Techniques of MRI will not be reviewed here in detail, as excellent reviews may be found elsewhere [2, 3]. However, a summary is in order to create the context for the research review that follows. The nucleus of the hydrogen atom contributes to the MR signal. Nuclei with odd numbers of protons and neutrons have an associated net magnetic movement and hence behave like magnets when placed in a strong magnetic field giving rise to a net magnetization. This magnetization is the physical entity in MRI that is manipulated using radiofrequency (RF) pulses and gradients

379

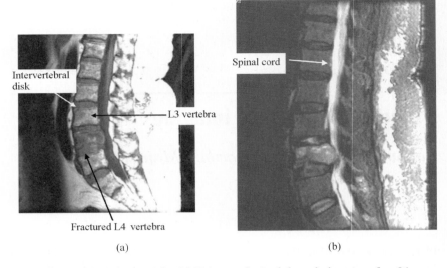

Figure 13.1. *(a) T_1 weighted MR image obtained through the spine of an 84 year old female subject with an atraumatic fracture of L4 vertebra. (b) T_2 weighted MR image obtained through the spine of an 84 year old female subject with an atraumatic fracture of L4 vertebra (courtesy Cynthia Chin, UCSF, Department of Radiology).*

in the magnetic field. A sequence of magnetic field gradients is switched on in order to encode the three-dimensional evolution of the net magnetization. The evolution of the net magnetization may be detected by using coils and then, via a series of mathematical reconstructions, an image reflecting the spatial distribution of the magnetization is generated. The signal at any location depends on the precise sequence of RF pulses, gradients and timing factors, but also, significantly, reflects the characteristics of the surrounding tissue. These characteristics reflect several complex intrinsic factors, such as the density of the protons, spin–lattice relaxation time (T_1), spin–spin relaxation time (T_2), magnetic field inhomogeneity induced relaxation time (T_2^*), chemical shift and diffusion effects.

The three-dimensional non-invasive imaging capabilities of MRI have been widely used clinically to assess and diagnose osteoporotic and vertebral fractures. Figure 13.1 shows an example of such clinical images, depicting the morphology and signal differences that are seen in MR images of vertebral fractures. In recent years, MRI has also been developed to assess the characteristics of trabecular bone. It permits not only the depiction but also the quantification of TBS, and hence reveals its biomechanical properties. MR can be used to assess the properties of trabecular bone in two fundamental ways. The first is an indirect measure, often termed relaxometry or quantitative magnetic resonance (QMR). This method takes advantage of the fact that trabecular bone alters the adjoining marrow relaxation properties in proportion to its density and structure, and thereby provides information regarding trabecular bone network. The second is the direct visualization of the dark, trabecular bone which, because of its low water content and short MR relaxation times, appears in stark contrast to the bright marrow fat and water, in high resolution MR images.

In the subsequent sections, we will review the field of MRI, as it pertains to TBS, biomechanical properties and bone quality assessment. The chronological development of research in this area follows a path of identifying and then overcoming various barriers to utilizing MRI in the study of osteoporosis and fractures in normal ageing bone, and in the assessment of therapeutic efficacy.

13.2.   QUANTITATIVE MAGNETIC RESONANCE (QMR)

Cortical bone and trabecular bone have very short intrinsic T_2 (relaxation time) values, low water content, and thus have relatively low MR detectable magnetization. However, the presence of the trabecular bone network and its impact on marrow relaxation times is what is often utilized in assessing trabecular bone properties using QMR. Thus, understanding that the presence of trabecular bone affects marrow relaxation times in MR is essential, although the intricate mechanisms of tissue relaxation processes are beyond the scope of this discussion. Imagine bone marrow as a liquid in the presence of solid trabecular interfaces. The interactions between the 'liquid' marrow and the solid interfaces result in an alteration of marrow relaxation times. The liquid (marrow) on the surface of each solid interface has particular qualities, due to a number of effects such as hydrogen bonding and dipole–dipole interactions related to the molecular behaviour of the liquid that adheres to or is close to the solid interfaces. This change in molecular motion characteristically manifests itself in MR as a modification of T_1 relaxation times, which have particular relevance at high field strengths. The magnitude of T_1 modification depends on the surface area to volume ratio of this solid–liquid interface, increases at higher magnetic field strengths and increases when the number of bone and marrow interfaces increase, that is, as bone density increases.

T_2 relaxation times may also be governed by similar mechanisms to T_1 relaxation, but are additionally affected by processes such as diffusion. Magnetic susceptibility—the property that defines the ability of materials to become magnetized—of trabecular bone is substantially different from that of bone marrow. If a sample consisting of a fine network of bone with marrow present in the cavities of the network is placed in a magnetic field (or a MR scanner), the magnetic lines of force are distorted through the sample, particularly at the interfaces of the bone and marrow. This gives rise to localized inhomogeneities in the magnetic field which depends on the number of trabecular bone–marrow interfaces, the size of the individual trabeculae and the field strength. The diffusion of water (protons) in these magnetic field inhomogeneities results in an irreversible loss of magnetization and thus shortens the marrow relaxation time T_2. This effect also depends on magnetic field strength and is greater at higher magnetic fields.

In addition to these effects on marrow relaxation there is an additional effect that may occur in the presence of trabeculae, that is, the modification of the marrow relaxation time T_2^*. In specific types of MRI sequences, in addition to diffusion mediated loss of magnetization, the magnetization is further lost irreversibly as a result of the field inhomogeneities. This results in a characteristic relaxation (decay) time T_2^* which includes the additional contribution due to field inhomogeneities as well as the T_2 relaxation properties. This effect forms the basis of QMR. In table 13.1, a partial summary of the studies relating T_2^* to bone properties is presented.

The impact of bone on the MR properties of marrow was first investigated in an *in vitro* experiment in 1986 by Davis *et al* [16], at a field strength of 5.8 T. Bone from autopsy specimens was ground up and sifted into a series of powders with graded densities ranging from 0.3 to 0.8 g/cm^3 and then immersed in normalized saline or cottonseed oil to simulate bone marrow. Spectroscopic techniques showed that as the density of suspended bone increased, T_1 relaxation time decreased as a result of an increase in surface area-to-volume ratio (increased surface interactions increased since higher bone density was composed of smaller particles, thus more surface area). As the suspended bone density increased, there were concomitant increases in the magnetic field inhomogeneities, and thus also decreases in T_2^*, an effect more profound than on T_1.

At a field strength of 0.6 T, Rosenthal *et al* [17] showed that in excised cadaveric specimens the relaxation time T_2^* of saline present in the marrow spaces was shorter than that of pure saline. Clinically, Sebag *et al* [18] showed qualitatively that bone marrow in the presence of trabecular bone showed increased signal loss in gradient echo images, where T_2^* effects predominate. Subsequently, quantitative estimates of T_2^* in regions of varying bone density such as in the epiphysis, metaphysis and diaphysis were measured by Ford *et al* [19] using a technique known as interferometry, and localized proton spectroscopy. They demonstrated that in the distal femur of a normal volunteer, T_2^* in the epiphysis was 10.5 ms, compared with 14.3, 72.2 and 79.9 ms respectively, in the metaphysis, diaphysis and subcutaneous fat. These findings correspond to the fact that increased trabeculation in the epiphysis (compared with the metaphysis and diaphysis) resulted in a shortening of T_2^* [19]. In a small sample size, it has also been shown that T_2^* values may potentially distinguish osteoporotics from normals [20].

Calibration of T_2^* with measures of BMD have been undertaken both *in vitro* [4–6, 8, 14, 17, 21] and *in vivo* [7, 9, 19, 20, 22]. Dried excised specimens

Table 13.1. *A partial summary of the studies relating T_2^* to trabecular bone properties.*

Study	Skeletal site	Bone density	Bone structure	Bone strength	Ref.
In vitro human	Vertebra	BMD $r = 0.92^{\#\#}$			[4]
In vitro sheep	Vertebra	BMD $r = 0.87^{\#}$	TB area $r = 0.80^{**}$ TbSp $r = 0.62^{*}$		[5]
In vitro rat	Proximal tibia		TB fraction $r = 0.85^{\#}$		[6]
In vivo human	Proximal tibia	BMD $r = 0.96^{**}$			[7]
In vitro human	Lumbar vertebra	BMD $r = 0.70^{*}$			[8]
In vivo human	Lumbar vertebra	Calcium density $r = 0.59^{*}$			[9]
In vitro human	Tibia	(Ash density)$^{-1}$ $r = 0.88^{*}$		$(YM)^{-1}$ $r > 0.87^{*}$	[10]
In vitro human	Lumbar vertebra		TB area $r = 0.78^{(-)}$ TP[1] density $r = 0.74^{\#\#}$	YM $r = 0.91^{\#\#}$	[11]
In vivo human	Distal radius		TB area $r = 0.90^{*}$		[12]
In vivo human	Calcaneus	BMA[2] – hip $r = 0.66^{**}$			[13]
In vitro porcine	Lumbar vertebra			Stress[3] $r = 0.83^{\#}$	[14]
In vitro human	Lumbar vertebra	BMD $r = 0.70^{**}$		Stress[3] $r = 0.91^{\#\#}$	[15]

$^{\#\#}\,p < 0.0001$; $^{\#}\,p < 0.001$; $^{**}\,p < 0.01$; $^{*}\,p < 0.05$.

[1] Trabecular plate.

[2] Bone mineral per area.

[3] Ultimate compressive stress.

of vertebral bodies have been immersed in saline, an emulsion of peanut oil and water to replicate marrow [4]. In order to measure susceptibility mediated effects on marrow relaxation times T_1, T_2 and T_2^* experiments have been conducted at 1.5 T. Standard single energy quantitative CT (QCT) measures of BMD were then correlated with measures of T_1, T_2 and T_2^*. While at 1.5 T there was no variation of the estimated T_1 or T_2 with trabecular density, at a lower field of 0.1 T, Remy *et al* [23] have demonstrated a dependence of T_2 on bone mineral density. Investigators have also found that $1/T_2^*$ increased with bone density at a rate of $0.20 \pm 0.02\,\text{s}^{-1}/\text{mg}/\text{cm}^3$ (correlation coefficient was 0.92, $p < 0.0001$) [4]. Several investigators have further correlated T_2^* in the both animal and human vertebrae [5, 8, 14], ovariectomized rat models [6], and found correlations with measures of BMD, trabecular separation, and also a dependence on the orientation of the trabeculae in the magnetic field. T_2^* variations with bone density are also dependent on the spatial resolution at which the images are obtained [22], as well as on the three-dimensional distribution of the trabecular bone, or structure, as shown in computer studies [21, 24] and phantom experiments [17, 25–27]. Furthermore, the choice of echo time and the precise model selected to obtain the T_2^* relaxation time measure, also affect the rate of change of T_2^* with bone density [27, 28].

In the area of osteoporosis, the biomechanical properties of trabecular bone are of ultimate importance. Using specimens from the human tibia [10] and vertebrae [11], it has been shown that $1/T_2^*$ increases linearly as the elastic modulus increases. However, the elastic modulus increases at a rate of $19.9\,\text{MPas}^{-1}$ [11] in the vertebral specimens, but only at a rate of $8.7\,\text{MPas}^{-1}$ in the tibial specimens [10]. However, this is not surprising since correlation between biomechanical strength and bone density in vertebral specimen and tibial specimen have shown differences, possibly due to the regional variations and heterogeneity of tibial bone, and differences in trabecular orientation and structure [29]. Correlations between ultimate compressive strength and T_2^* have been studied in porcine bone [14] and in human vertebral samples [15]. These studies demonstrate strong correlations between ultimate strength and measures of relaxation time. All of these studies are indicative of the potential use of T_2^* as an indirect measure of bone strength, a parameter of importance in osteoporosis.

In vivo calibration of T_2^* with trabecular bone density has been obtained from coincident measurements in the forearm, distal femur and proximal tibia using MR and QCT [30]. The relaxation rate, $1/T_2^*$, increased as the bone density increased, at a rate $0.20 \pm 0.01\,\text{s}^{-1}/\text{mg}/\text{cm}^3$ (correlation coefficient was 0.88, $p < 0.0001$). In a study involving six normal and nine post-oophorectomy patients, Sugimoto *et al* [9] showed a change in $1/T_2^*$ with bone density at a rate of $0.114\,\text{s}^{-1}/\text{mg}/\text{cm}^3$. Fransson *et al* [7] correlated T_2^* in the tibia with measures of BMD in the proximal femur and calcaneal ultrasound. The investigators found good correlations between T_2^* with BMD, but relatively lower correlations with ultrasound measures. This could be due to the significant heterogeneity of bone structure in the calcaneus, as well as in the tibia, and the fact that the ultrasound measure is a single point measure and could be measuring a small and variable region between subjects. The heterogeneity in the bone density and its impact on T_2^* in the calcaneus was quantified by Guglielmi *et al in vivo* [31], who showed that the shortest relaxation time occurs in the superior talar region

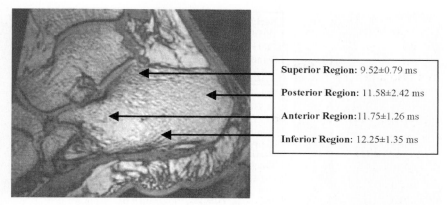

Figure 13.2. *Sagittal image through the calcaneus showing the heterogeneity of T_2^* values in different regions. (Adapted from [31].)*

(figure 13.2), corresponding to the highest bone mineral density. They also demonstrated a linear correlation between MR and DXA measurements ($r = 0.66$ for $1/T_2^*$ versus BMD). Song *et al* [32] corroborated the regional differences of T_2^* in the calcaneus, and also showed the impact of the trabecular bone orientation on marrow relaxation time. In a case control study, T_2^* measures of the proximal femur distinguished between subjects with hip fractures and normal subjects [33], as shown in figure 13.3. In addition, a combination of relaxation time measures and BMD improved the ability to discriminate subjects with vertebral fractures from those without [34]. The use of relaxation time measures as an indirect means of characterizing trabecular bone, and to some extent trabecular bone orientation, has evolved and grown in the past several years, and does have potential use in the study of osteoporosis.

Other methods of characterizing marrow relaxation properties and their impact on discerning trabecular bone properties have also been studied. These methods rely on further analysing the MR signal and, rather than calculating a single relaxation parameter, quantifying either a distribution of relaxation times or analysing in detail the phase of the MR signal. Using the difference in phases in the MR images, Allein *et al* [12, 35–37] have shown another potential method for measuring the impact of trabecular bone density and structure on the marrow properties. Fantazzini *et al* [38] have calculated a distribution of relaxation times, in an *in vitro* setting, and ascribed differences to relaxation time distributions to differences in trabecular bone structure measures. Further investigation of these properties and unravelling the role of TBS on the detailed behaviour of the MR signal and relaxation is clearly warranted, and likely to have applications in *in vivo* studies.

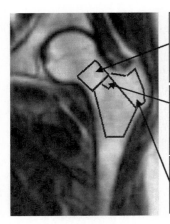

Neck: 14.7±1.0 ms (Pre)
18.5±3.0 ms (Post)*
21.0±3.3 ms (Post, Hip Fx)**
Significant Difference between * Pre and Post and ** Post and Post, Hip fx.

Wards: 15.4 ±1.6 ms (Pre)
18.7±2.4 ms (Post)*
22.7±3.4 ms (Post, Hip Fx)**
Significant Difference between * Pre and Post and ** Post and Post, Hip fx.

Trochanter: 14.7±1.0 ms (Pre)
18.6±2.5 ms (Post)*
19.8 ±3.7 ms (Post, Hip Fx)
Significant Difference between * Pre and Post

Figure 13.3. *Coronal image through the proximal femur showing the T_2^* values in different regions, in different groups of subjects: pre-menopausal, post-menopausal normal, post-menopausal fracture subjects, showing significant differences between pre- and post-menopausal women, and between post-menopausal normal and hip fracture patients. (Adapted from [33].)*

13.3.   IMAGING OF TRABECULAR BONE STRUCTURE

As described above, cortical and trabecular bone have a low water content and short T_2 and are not detectable using routine proton imaging methods. However, the marrow surrounding the trabecular bone network, if imaged at high resolution, reveals the trabecular network as seen in the representative image of a human femoral specimen in figure 13.4, obtained using a small bore scanner (resolution of 78 μm isotropic). In the figure, the dark network represents the trabecular bone network, while the higher intensity background represents marrow-equivalent material in the trabecular spaces. Using such images, multiple different image processing and image analysis algorithms have been developed, the goal of all of these being to quantify the trabecular bone structure in two or three dimensions. The measures that have been derived so far are many. Some of them are synonymous with the histomorphometric measures such trabecular bone volume fraction (BV/TV), trabecular thickness (TbTh), trabecular spacing (TbSp) and trabecular number (TbN); others include connectivity or Euler number, fractal dimension, tubularity, maximal entropy, etc. Some of the results using these descriptors and their utility thus far will be summarized below. Another aspect of computerized analysis also deals with visualization and depiction of the three-dimensional structure imaged using MR. There are numerous ways of surface and volume rendering such images, and figure 13.5 shows an example of three-dimensional renderings of MR images obtained at 1.5 T using a clinical scanner. The specimens were obtained from a young and

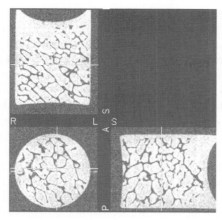

Figure 13.4. *Axial, coronal and sagittal images at a resolution of 78 μm isotropic obtained at 2.35 T. The trabeculae in the human femoral sample appear dark, as a result of the short T_2, with the doped saline appearing bright (courtesy Patrick Porion, Centre de Recherche sur la Matière Divisée, Orléans, France).*

an elderly subject, as labelled in the figure. Larger trabecular spaces and loss of trabecular bone in the elderly subject are clearly visible.

A number of calibration and validation studies (*in vitro* and *in vivo*) have been undertaken in which MR-derived measures of structure are compared with measures derived from other modalities, such as histology, μCT,

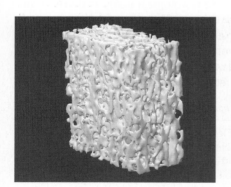 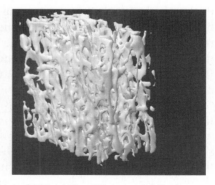

Figure 13.5. *Images obtained at resolutions of 117 × 117 × 300 μm at 1.5 T of two vertebral body samples from a 36 year old (left) and a 69 year old (right) male were surface rendered. The loss of trabecular bone volume in the elderly subject (trabecular bone volume = 14%) compared with the younger subject (trabecular bone volume fraction 28%) is apparent by the large spaces seen in the sample on the right (courtesy Olivier Beuf, Andres Laib, Magnetic Resonance Science Center, UCSF).*

BMD and with biomechanics. These include cross-sectional studies in established animal models, studies in human subjects (both cross-sectional and longitudinal) and methods assessing different image processing techniques that form an integral core of the derivation of TBS from MR images.

One of the primary issues in MR-derived visualization and quantification of TBS arises from the fact that the spatial resolution of the MR images is often comparable with the thickness of the trabecular bone itself, which gives rise to partial volume effects in the image. Thus, the image may not depict very thin trabeculae or may represent an average or a projection of a few trabeculae. Recognizing that MR-derived measures are not identical to histological dimensions, a major focus in the field has been using established measures to investigate the resolution-dependence of TBS measures and then calibrating MR-derived measures of TBS.

In the subsequent sections, *in vitro* studies, studies in animal models and *in vivo* human studies will be reviewed.

13.3.1.   *In vitro studies*

Small-bore high field magnets (typically vertical bore spectrometers with micro-imaging inserts) have been widely used for imaging specimens of trabecular bone but, since the potential of MR lies in *in vivo* human imaging, a large number of *in vitro* studies have also been performed using clinical scanners with enhanced hardware and software. A partial summary of studies undertaken using small-bore animal scanners (spectrometers) is shown in table 13.2.

Hipp *et al* [40] have compared the morphological analysis of 16 bovine trabecular bone parallelopipeds using both three-dimensional MR reconstruction ($92 \times 92 \times 92\,\mu m^3$) and two-dimensional optical images ($23 \times 23\,\mu m^2$) of the six faces of the samples [40]. The volume fraction measured by MRI was linearly related to and not different from the average area fraction measured from optical images ($R^2 = 0.81$, $p = 0.96$).

Table 13.2. *Micro-MRI of trabecular bone with small-bore scanners.*

Samples	Tesla	Skeletal site	Field of view (cm^3)	Resolution (μm^3)	Ref.
Human	9.4	Verteb/calca	$1.0 \times 1.0 \times 1.0$	$78 \times 78 \times 78$	[39]
Bovine	9.4	Tibia	$1.0 \times 1.0 \times 1.0$	$78 \times 78 \times 78$	[39]
Bovine	8.6	Femur/tibia	$1.2 \times 1.2 \times 1.2$	$92 \times 92 \times 92$	[40]
Bovine	9.4	Tibia	$0.8 \times 0.8 \times 0.8$	$61 \times 61 \times 61$	[41]
Human	9.4	Distal radius	$1.0 \times 1.0 \times 1.0$	$78 \times 78 \times 78$	[42, 43]
Human	2.35	Femur	$1.0 \times 1.0 \times 1.0$	$78 \times 78 \times 78$	[44, 45]
Minipigs	4.7	Vertebra	$1.1 \times 1.1 \times 2.2$	$85 \times 85 \times 85$	[46]
Small rats	9.4	Tibia	$0.6 \times 0.6 \times 0.6$	$46 \times 46 \times 46$	[47]

Resolution-dependent effects in comparing TBS measurements of human distal radius specimens with MRI of $156 \times 156 \times 300\,\mu m^3$ and high-resolution X-ray tomographic microscopy (XTM) at resolutions of $18 \times 18 \times 18\,\mu m^3$ [48] have been undertaken. In these studies, high-resolution XTM images were used to simulate resolution degradation using low-pass filtering. It was observed that a decrease of the spatial resolution induced an over-estimation of BV/TV and TbTh, and an under-estimation of TbSp and TbN. Furthermore, an increase in slice thickness had a greater impact on TbTh, TbSp and TbN. For a factor of 9 in spatial resolution between the two sets of images, the morphological measures ranged from a factor 1.6 (for TbSp) to 5 (for TbN), with a factor 3 for BV/TV and TbTh. The two sets of parameters were not statistically different, except for TbTh ($p = 0.02$). Using a large set of trabecular bone samples from several skeletal sites and different densities and structures, TBS measures of 18 trabecular bone specimens from MRI ($117 \times 117 \times 300\,\mu m^3$) and optical imaging ($20 \times 20\,\mu m^2$) were compared [49]. It was shown that there was good correlation between the MR derived and optical imaging derived measures such as TbSp ($r = 0.89$, $p < 0.01$) and TbN ($r = 0.78$, $p < 0.01$), moderate correlation for BV/TV ($r = 0.69$, $p < 0.01$) and very poor correlation for TbTh ($r = 0.06$, $p = 0.84$). These correlations indicate that MR images can depict trabecular bone structure, although the absolute measures differ from the measures obtained at higher resolution, except in the measures of trabecular thickness. This is explained by the fact that the image resolution is comparable with the dimensions of the trabeculae being measured, and a small sampling error, or a partial volume effect, in the estimation leads to a large percentage or fractional error.

The effect of slice thickness on standard morphological measurements has been investigated by Kothari *et al* [50], who artificially degraded from 100 to 1000 μm (with a spatial resolution of $100 \times 100\,\mu m^2$), using three-dimensional optical reconstruction of trabecular bone specimens with an initial isotropic resolution of $40 \times 40 \times 40\,\mu m^3$. Measurements such as BV/TV, TbSp and TbN had weak resolution-dependency, while TbTh was strongly affected by the resolution in rising rapidly with increasing slice thickness, requiring very high resolution for precise evaluation.

Vieth *et al* [51] have compared standard morphological TBS measurements (BV/TV, TbTh, TbSp, TbN) of 30 calcaneus specimens using MRI ($195 \times 195\,\mu m^2$ in-plane resolution and $300/900\,\mu m$ slice thickness) and contact radiographs (digitized with $50 \times 50\,\mu m^2$ spatial resolution) of sections obtained from the same specimens. The results of this study show that MR-based measurements were significantly correlated with those obtained from digitized contact radiographs. However, partial volume effects due to slice thickness as well as image post-processing (thresholding) had a substantial impact on these correlations: the thicker the slice, the poorer the correlation.

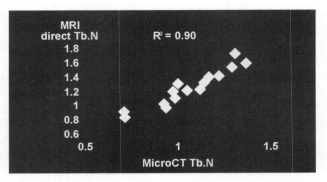

Figure 13.6. *A plot of trabecular number calculated using the distance trans-*
form technique using MR images and also higher-resolution μCT images.
The correlation between the measures from images obtained from two modal-
ities is high; however, the slope is not equal to 1, due to partial volume effects.
(Adapted from [54].)

Lin *et al* [52] have confirmed correlation between TBS parameters derived
from MR images and serial grinding images (three specimens). They also
established that the heterogeneity of calcaneal bone structure (as determined
from MR images) is real and correlated the magnitude of the spatial heterogeneity using higher resolution microscopic images. However, using a single
cadaveric sample, and somewhat lower resolution images, Engelke *et al* [53]
have shown that absolute values of structure measures differ considerably,
when derived from MRI ($195 \times 195 \times 1000\,\mu m^3$) and stained grindings
(digitized at $13 \times 13\,\mu m^2$ resolution). The set of stained grindings images
was used to artificially degrade the resolution and check the effect of the
slice thickness. It was shown that even with $13 \times 13 \times 500\,\mu m^3$, accurate results
could not be obtained using standard morphological evaluation; however
given the single specimen, no correlative assessment was possible.

The accuracy of a new model-independent morphological measure,
based on the distance transformation (DT) three-dimensional technique
applied to high-resolution MR images of human radius specimens with
in vivo resolution of $156 \times 156\,\mu m^2$ in plane and 300 or 500 μm in slice thickness, have been investigated by Laib *et al* [54]. These measures were
compared with high-resolution μCT images ($34 \times 34 \times 34\,\mu m^3$), and good
correlations were found between the two sets of measurements, the best
being $R^2 = 0.90$ for TbN (see figure 13.6). This study has shown that new
model-independent morphological measures applied to high-resolution
MR images give better accuracy than standard morphological measurements. From the same set of radius specimens, Pothuaud *et al* [55] have
shown the ability of the three-dimensional-line skeleton graph analysis
(LSGA) to characterize high-resolution MR images obtained with a

$156 \times 156 \times 300\,\mu m^3$ resolution. They obtained some topological measures that are highly correlated to high-resolution µCT measurements at the isotropic resolution of $34 \times 34 \times 34\,\mu m^3$ ($R^2 = 0.88$ to 0.91).

Using a small-bore MR scanner, Gomberg et al [42] have derived topological TBS parameters from µMR images ($78 \times 78 \times 78\,\mu m$) of the human wrist, and used these measures in the prediction of Young's modulus (YM) evaluated from uniaxial compression tests. Several parameters were used; the strongest single linear predictor of YM gave a correlation coefficient $R^2 = 0.67$ ($p < 0.0001$), while the most predictive of all two-terms models gave $R^2 = 0.79$ ($p < 0.0001$).

Using a clinical scanner, Majumdar et al [48] derived TBS measurements from MR reconstructions ($156 \times 156 \times 300\,\mu m^3$) of human distal radius cubes, and evaluated the explanatory power of these measurements in the prediction of the elastic modulus evaluated from non-destructive testing of the same specimens. Although TBS characteristics were not accurately restored (compared with $18 \times 18 \times 18\,\mu m^3$ XTM-based measures), significant correlations were established with the mean elastic modulus (evaluated in the three directions of the specimens): $r = 0.96$ ($p = 0.002$) for BV/TV, $r = 0.82$ ($p = 0.05$) for TbTh, $r = 0.95$ ($p = 0.003$) for TbSp and $r = 0.78$ ($p = 0.06$) for TbTh.

Hwang et al [43] used probability-based structural parameters to characterize high-resolution MR images ($156 \times 156 \times 391\,\mu m^3$) of human distal radius specimens. Young's moduli of the specimens were measured by non-destructive testing, and the structural parameters were found to account for 91% of the variation of YM. A probability-based evaluation technique is directly applied to the grey level MR images, without the thresholding stage.

The feasibility of using MRI at the resolution of $117 \times 117 \times 300\,\mu m^3$ to quantify TBS, as well as better predict mechanical properties, was established by Majumdar et al [49], using a set of 94 specimens of several skeletal sites, with a wide range of bone densities and structures. Figure 13.7 shows examples of trabecular bone specimens from two different skeletal sites, having equivalent bone densities, but significantly different directional elastic moduli, primarily as a result of differences in trabecular bone orientation. Among several results reported in this study, it was shown that MR-based structural measures, used in conjunction with BMD (evaluated from QCT measures), enhanced the prediction of bone strength. Using a stepwise regression model, including structural parameters in addition to BMD, resulted in an improvement of the prediction of the mean elastic modulus (evaluated from non-destructive testing). The adjusted correlation coefficient increased from 0.66 to 0.76 for all the specimens, 0.71 to 0.82 for vertebral specimens, and 0.64 to 0.76 for femoral specimens.

MR images ($117 \times 117 \times 300\,\mu m^3$) of vertebral mid-sagittal sections of lumbar vertebrae, and standard morphological parameters were calculated

DISTAL FEMUR: BONE MINERAL DENSITY : 226 mg/cm³
Mean Elastic Modulus: 42.9 Mpa; Cranio-caudal (SI) Elastic Modulus: 67.62

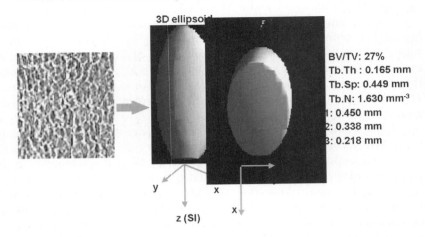

VERTEBRAL BODY: BONE MINERAL DENSITY : 225 mg/cm³
Mean Elastic Modulus: 67.98 Mpa; Cranio-caudal Elastic Modulus: 208.47

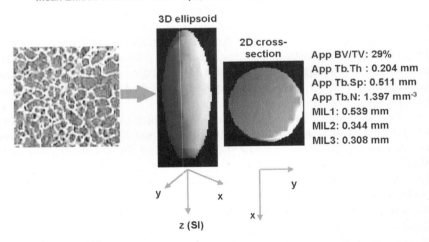

Figure 13.7. *Three-dimensional images from the distal femur (top) and a vertebral body (bottom) were obtained. The two specimens have almost equal bone density, but distinctly different elastic moduli in the direction of primary orientation (SI), and mean elastic modulus. The three-dimensional ellipsoid shows that the primary orientation is in the SI direction in both cases, but the specimen from the distal femur depicts considerable anisotropy in the orthogonal directions as demonstrated by the ellipsoidal shape of the two-dimensional cross-section of the ellipsoid. This anisotropy in trabecular structure partially explains the difference in mechanical properties of the two specimens.*

by Beuf *et al* [15]. Ultimate stress was estimated in two perpendicular directions (horizontal/vertical) using compression testing applied to two cylindrical samples drilled in each vertebra close to the mid-sagittal section. All the morphological parameters were correlated to both horizontal and vertical ultimate stresses ($r > 0.8$, $p < 0.001$). Using the same set of MR images, Pothuaud *et al* [56] compared topological parameters evaluated from a newly developed technique, three-dimensional line skeleton graph analysis (LSGA) [57], with elastic moduli evaluated from finite element analysis. High correlations were found between topological parameters and elastic moduli ($R^2 = 0.94$–0.95), as well as between topological parameters and ultimate stress ($R^2 = 0.85$ for the horizontal direction, $R^2 = 0.76$ for the vertical direction).

Link *et al* [58] employed texture parameters measured on high-resolution MR images ($156 \times 156 \times 300 \, \mu m^3$) of proximal femur and spine specimens. Whereas the correlation between YM and BMD was $R^2 = 0.66$ for the spine specimens and $R^2 = 0.61$ for the femur specimens, a multivariate-regression model combining both BMD and texture parameters increased the correlation to $R^2 = 0.83$ for spine and to $R^2 = 0.72$ for femur. While this study shows that grey-level texture evaluation could be useful for high-resolution MR applications, the relationships between the grey-level characteristics and the true TBS needs to be explored [45, 59].

13.3.2. *Animal models*

Several investigators have conducted studies in *in vitro* animal models of osteoporosis and *in vivo* animal studies. Figure 13.8 shows a coronal *in vitro* image of a rabbit tibia (resolution $156 \times 156 \times 128 \, \mu m$) and an *in vivo* three-dimensional image of a rat tail obtained using a 2 T magnet equipped with accustar gradients at a spatial resolution of 78 µm, both demonstrating the trabecular bone, against the bright marrow background. White *et al* [60] have shown that MRI (on a 2.0 T scanner) could assess TBS in rats both *in vitro* and *in vivo*. TBS was evaluated from $78 \times 78 \times 78 \, \mu m^3$ resolution MR images of rat tibia (*in vitro*) and tail vertebra (*in vivo* under pentobarbital anaesthesia) in both ovariectomized and control matched groups. There was a significant decrease in BV/TV with concomitant decrease in TbTh and increase in TbSp in the tibia and tail vertebra of ovariectomized rats compared with the control group.

Recent studies have shown the potential of MRI to depict TBS changes in the rat tibia following ovariectomy osteoporosis model coupled with oestrogen treatment. These MR experiments were performed on a 9.4 T spectrometer, with isotropic resolution of $39 \times 39 \times 39 \, \mu m^3$ [61, 62].

Luo *et al* [63] used *in vivo* MRI to monitor time-dependent changes of TBS in rat tibia induced by glucocorticoid therapy. MRI was performed

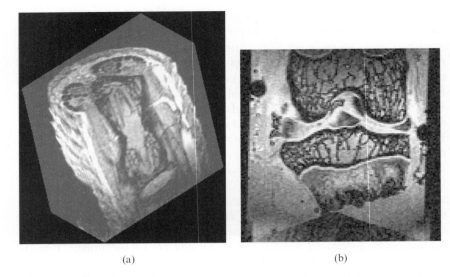

(a)   (b)

Figure 13.8. *(a) Rat tail in vivo, with Gd-DTPA infusion, obtained at 2 T CSI using a three-dimensional SPGR sequence at a TR/TE/flip angle of 110/3.5/4 at an isotropic resolution of 78 μm. Total acquisition time 2 h (courtesy David White, Department of Radiology, UCSF). (b) Single slice from a three-dimensional stack of images from a rabbit knee dissected, with soft tissue removed, in saline obtained at 2 T using a three-dimensional SPGR sequence at a TR/TE/flip angle of 110/5/4. Resolution is 156 μm isotropic, acquisition time is 2 h (courtesy David White, Department of Radiology, UCSF).*

on a 4.7 T scanner with a spatial resolution of 51×51 μm^2 in an axial plane, 100 μm along the tibia. Significant decrease in BV/TV and TbN and increase in TbSp were observed after 15 days in the treated group compared with the control group. These changes reversed between 15 and 30 days, which was consistent with previous histomorphometry studies.

Takahashi *et al* [47] have used μMRI at 9.4 T to monitor the effects of preventive agents in an ovariectomized rat. The results were in accordance with literature data obtained in similar models by histomorphometry, proving that μMRI allows the characterization of trabecular bone structure in small animals with sufficient accuracy.

Borah *et al* [46], using a small-bore scanner, obtained MR images at an isotropic resolution of 85 μm to analyse the three-dimensional trabecular bone structure and mechanical properties of vertebral specimens of young and mature mini-pigs. Accurate characterization of the three-dimensional TBS based on newly-developed non-morphological parameters as well as finite element analysis (FEA) showed significant differences between the two groups of mini-pigs. Furthermore, apparent Young's modulus, evaluated from experimental test and FEA modelling, were similar and TBS parameters contributed to the prediction of compressive strength evaluated experimentally.

These studies have established the feasibility of using MRI in the study of animal models of osteoporosis. In addition to monitoring and depicting trabecular bone structure changes, MR also offers the ability to characterize marrow signal changes [64–68], which may be relevant in the study of osteoporosis and therapeutic interventions.

13.3.4. In vivo human studies

The strength of the magnetic field in tesla and the quality of the radio-frequency (RF) coils govern the signal-to-noise ratio in MR images. With recent advances in phased array coils and higher strength magnets, the potential of MRI of bone structure is ever-increasing. Integral components for MR image acquisition are the detector or receiver coils that are dedicated to each skeletal site. Typical radiofrequency coils are shown in figures 13.9(a) and (b). The image contrast can be manipulated in MR images based on the specific pulse sequence used, and the appearance of trabecular bone can be varied based on whether a spin–echo or gradient–echo sequence is used [3]. The high susceptibility difference between bone and marrow induces suscept-ibility artefacts at their boundary, which in the case of *in vivo* imaging could have a high impact on the bone structure quantification [3]. Although spin–echo images may be preferable to reduce this effect, gradient–echo images acquire an equivalent volume in considerably less time and can be exploited *in vivo* at several skeletal sites and by the optimization of the pulse sequence timing (short echo time) in gradient–echo imaging one can attempt to mini-mize the susceptibility artefact [3].

In vivo MRI of TBS is constrained by perpetual compromises between acquisition time, signal-to-noise ratio (which affects the quality or contrast of the image) and field of view. These constraints limit imaging of TBS to sites that are easily accessible, such as the extremities. External parameters such as blood flow and body motion limit imaging in the central skeleton. At the present time, the skeletal sites most commonly imaged are the radius [42, 69–74] and calcaneus [52, 75, 76]. The distal radius is a site with a large quantity of trabecular bone, and is a common site for osteoporotic fractures. It is easily accessible with localized detection coils and subjects are able to comfortably tolerate immobilization for the period required for high-resolution imaging. The calcaneus, although not a typical site for osteo-porotic fractures, has been used with success to predict fracture at other sites [76, 77] and this skeletal site is well adapted to high-resolution MRI [52, 75, 76]. The phalanges have recently been of increased interest as a site for bone density measurement [78, 79] and can also be imaged by high-resolution MRI. Table 13.3 provides a partial summary of studies performed using a

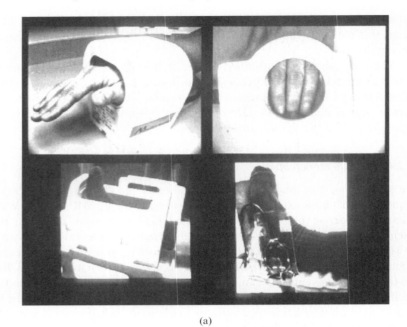

(a)

Figure 13.9. *(a) Top left: quadrature coil used for imaging the wrist. Top right: 75 mm surface coil used for imaging the phalanges. Bottom row: A linear extremity coil (left) and a phased array coil used for high resolution imaging of the calcaneus.*

1.5 T clinical scanner and dedicated RF coils, in both the ultra-distal radius and the calcaneus.

Axial images of the distal radius are generally acquired with the subject placed in a supine position with hands at the side, using a dedicated RF coil. Although the acquisition time is less than 20 min, wrist and fingers are immobilized with the coil and the arm is also immobilized to reduce motion during the scan. A set of fast coronal and/or sagittal localizer images are obtained from which the axial slices can be prescribed. Figure 13.10 shows representative axial scans through the distal radius of two subjects. As seen in the images, and in the graphs alongside, there is considerable variation in the trabecular bone structure in the distal radius. There is greater trabeculation in the distal part, and trabecular bone content decreases towards the shaft. The magnitude of heterogeneity also differs between subjects, as seen in the figure.

For calcaneus imaging, the subject is placed in a supine position while the foot is constrained inside a dedicated RF coil, for example, such as those shown in figure 13.9. Axial and/or coronal scout view locators are

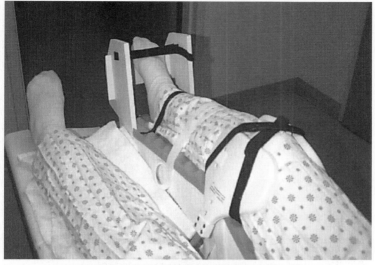

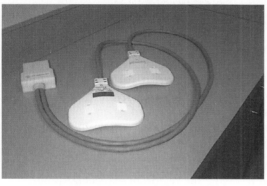

(b)

Figure 13.9. *(b) Four-coil phased array coil used for high-resolution imaging of the knee joint, the tibia and femur trabecular bone structure in particular.*

used to prescribe sagittal or oblique slices in the central calcaneal bone (see figure 13.11).

For imaging the phalanges, Jara *et al* [81] have used a modified spin–echo sequence and shown images that clearly depict the trabecular bone structure, and Stampa *et al* [82] have quantified trabecular bone structure from such MR images, as shown in figure 13.12.

The distal femur and distal and proximal tibia also contain a large amount of trabecular bone. Changes in gait, the onset of arthritis and systemic changes in bone density and quality may also be depicted in these regions. In figure 13.13, a scan in the tibia is shown.

Table 13.3. *High-resolution in vivo imaging of trabecular bone with clinical scanners.*

Site	Pulse seq.	Image size (voxels^3)	Resolution (μm^3)	Time (min)	TE (ms)	TR (ms)	Flip (°)	Ref.
DR	SE	512 × 256 × 32	137 × 137 × 500	10.9	7.8	80.0	140	[72]
DR	SE	512 × 256 × 32	137 × 137 × 350	–	9.7	80.0	140	[42]
DR	GE	512 × 384 × 60	156 × 156 × 500	–	6.2	28.6	30	[71]
Cal	GE	512 × 384 × 60	195 × 195 × 500	15.2	6.2	28.6	40	[76]
DR	GE	512 × 512 × 60	156 × 156 × 700	16.0	7.8	31.0	30	[69]
Cal	GE	512 × 512 × 28	195 × 195 × 1000	–	11.0	35.0	30	[75]
Cal	GE	–	195 × 195 × 1000	25.0	7.4	32.6	30	[52]
DR	GE	256 × 256 × 12	195 × 195 × 800	20.0	17.1	60.0	60	[70]
DR	GE	512 × 384 × 60	156 × 156 × 500	12.0	6.7	29.5	30	[73,74]
DR	GE	512 × 288 × 124	156 × 156 × 500	17.5	6.1	28.5	30	[73]
DR	GE	512 × 512 × 64	156 × 156 × 500	12.0	5.6	29.0	30	[80]

DR: distal radius; Cal: calcaneus; SE: spin–echo; GE: gradient–echo; TE: echo time; TR: repetition time; Flip: flip angle.

Although there may be predictable improvements in the field strength of MR scanners, pulse sequence design and RF coil optimization, the resolution achievable *in vivo* on clinical scanners is limited by the inherent signal-to-noise ratio of the images. The quantitative evaluation of TBS from these images also constitutes a major area of research. The processing of high-resolution MR images generally consists of several stages [73, 83] and may include some or all of the following:

1. pre-processing
2. segmentation to define the region of trabecular bone
3. texture analysis or direct grey-level analysis
4. binarization into bone and marrow phases
5. derivation of TBS parameters.

Other stages, such as a manual region adjustment for reproducibility studies or image registration for longitudinal studies, could also be necessary. In the case of high-resolution MR images of TBS, Newitt *et al* [73] have shown that each stage needs to be standardized and normalized in order to ensure a high degree of reproducibility. In particular, these authors describe a standardized analysis system with considerable reduction of human interaction. The efficiency of this system was evaluated in terms of reproducibility (2–4%) and has been successfully applied in several cross-sectional [69, 71, 76, 80, 84] and longitudinal [74, 85–87] studies.

Some noise-reduction based pre-processing techniques have been applied before the binarization stage, such as low-pass filtering [69] or

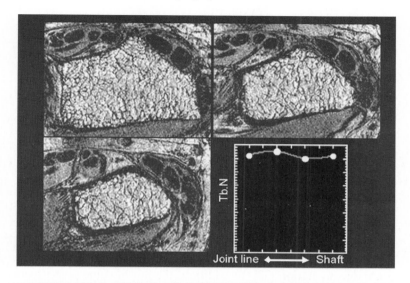

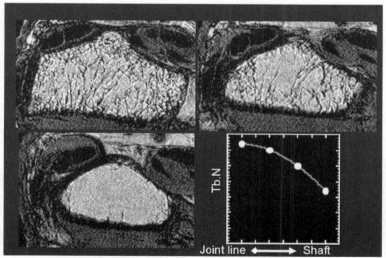

Figure 13.10. *High-resolution MR images obtained in the axial plane through the distal radius, at 1.5 T using the coil shown in figure 13.9(a) top left. The images are obtained for two subjects and show sections from the distal radius starting closer to the joint line and into the shaft. Alongside is a plot showing the variation of TbN from the joint line into the shaft. The differences in bone structure, such as reduced trabecular bone volume and number in the image at the bottom, underlines the considerable heterogeneity in different regions of the radius and also between subjects. The spatial resolution of the images is $156 \times 156 \times 500\,\mu m$ (courtesy David Newitt, Department of Radiology, UCSF).*

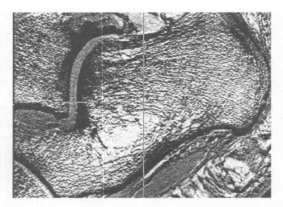

Figure 13.11. *High-resolution MR image obtained in the sagittal plane of the calcaneus, at 1.5 T using the coil shown in figure 13.9(a) bottom right. The spatial resolution of the images is 195 × 195 × 500 μm.*

histogram deconvolution [88]. In addition, the use of some post-processing schemes after the binarization, based on either morphological criterion relative to the shape and morphology of the trabeculae [89], or based on topological criterion relative to the numbers of bone and marrow

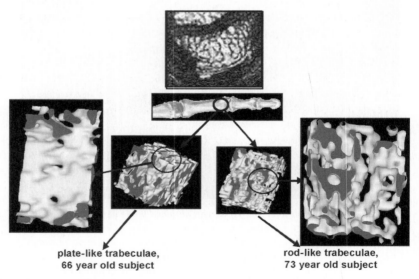

plate-like trabeculae,
66 year old subject

rod-like trabeculae,
73 year old subject

Figure 13.12. *High-resolution MR image obtained in the axial plane in the phalanges, at 1.5 T. Three-dimensional renderings from such images for two different subjects (a 66 year old and a 73 year old) showing differences in structure are shown as well (courtesy Claus Glüer, Bernd Stampa, University of Kiel, Germany).*

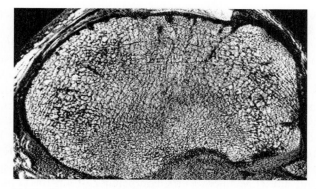

Figure 13.13. *Cross-sectional image in the axial plane of the proximal tibia, obtained at 1.5 T using the coil in figure 13.9(b).*

components [44, 90] have been applied. Wu *et al* [91] proposed a sophisticated histogram model taking into account the partial volume effect characterizing MR images, using a probabilistic approach. The relative solid and void (or marrow) fractions of each voxel were evaluated, allowing classification of each voxel independently, as being either in the solid phase or in the void (or marrow) phase. Hwang *et al* [43] used spatial correlation analysis, which is based on the probability of finding bone at a specific location. Some morphological parameters were deduced from this analysis, such as inter-trabecular spacing, contiguity and tubularity. In a set of 20 subjects classified as normal or osteoporotic on the basis of vertebral deformity, a combination of some of these parameters was predictive of the vertebral deformity ($R = 0.78$, $p < 0.005$) [92].

Grey-level histogram-based binarization is commonly used with high-resolution MR images of trabecular bone, but because it is a scheme based on a single threshold value applied with a limited resolution (compared with the mean size of the trabeculae) it may lead to a global thickening of trabeculae or a loss of the thinner trabeculae. For example, if the image resolution is 100 μm and the trabecular dimensions are of the order of 100 μm, an error of 1 voxel may potentially be reflected as a 100% error in the estimated trabecular width. Similarly, a trabecula that is 50 μm thick will be detected either as a 100 μm structure or as no trabecula (marrow phase). Recognizing that it is not possible to accurately reconstitute the 'true' trabecular bone structure from high resolution MR images, Majumdar *et al* [69] introduced the notion of 'apparent' trabecular bone network. Of note was the fact that while noting the 'apparent' network is not identical to the 'true' histological structure, it none the less reflects some 'apparent' morphological and topological properties that are highly correlated to the 'true' one [48–50, 54, 55]. Furthermore, such 'apparent' evaluation gives sufficient sensitivity and specificity to depict trabecular bone differences [69, 71, 76, 80, 84] and/or changes [74, 85–87] from *in vivo* data.

More recently, the distance transformation (DT) technique was applied to high-resolution *in vivo* MR images of the distal radius $(156 \times 156 \times 500 \, \mu m^3)$ in post-menopausal women [80]. Morphology-based parameters were evaluated without assumption of any TBS model, and the most significant parameter in distinguishing subjects with vertebral fracture $(n = 88)$ from those without vertebral fracture $(n = 60)$ was the intra-individual distribution of separation (Standard Deviation of the Trabecular bone Separation parameter). Using receiver operating curve (ROC) analysis, the competence of this parameter was comparable with radius or spine BMD measures, but not as pertinent as the competence of hip BMD alone.

Finite element analysis (FEA) involves the calculation of mechanical properties of TBS by using finite element (FE) models, generally obtained in converting each voxel to equally shaped brick eight-node elements [93]. Nevertheless, such voxel-based FE models consist of a very large number of elements and special-purpose FE solvers are required to evaluate mechanical properties with reasonable calculation conditions. Although FEA was initially developed and optimized for *in vitro* applications, recent advances in FEA techniques allow the direct evaluation of mechanical properties from high-resolution *in vivo* MR images of trabecular bone [86, 94].

Newitt *et al* [74] studied two groups of post-menopausal women with normal $(n = 22)$ and osteopenic $(n = 37)$ status (based on standard BMD criterion) using high resolution *in vivo* MR images of distal radius $(156 \times 156 \times 500 \, \mu m^3)$. Standard morphological TBS parameters as well as FEA-based mechanical parameters were evaluated. All of the FEA-based Young's and shear moduli were lower in the osteopenic group compared with the normal group, and the anisotropy of the Young's moduli were higher in the osteopenic group compared with the normal group. Although FEA-based parameters were statistically correlated to bone volume fraction, a stepwise regression analysis showed that the combination of bone volume fraction with some of the morphological parameters could improve the prediction of the mechanical properties. Van Rietbergen *et al* [86] used the FEA approach to characterize the longitudinal changes in TBS.

Saha *et al* [95] used a surface skeleton representation to map the trabecular bone network into plate- and rod-like trabeculae. With this modelling approach, called digital topological analysis (DTA), each voxel of the skeletonized image is classified as an isolate, curve, surface or junction type voxel, following its local topology (the numbers of objects, tunnels, and cavities in the immediate neighbourhood). This approach has been applied to *in vivo* MR images of the radius, showing structural differences in patients with similar bone volume fraction [42]. Furthermore, some of the DTA indices seem to be well suited to distinguish rod-like from plate-like trabecular bone networks [95, 96]. Gomberg *et al* [97] reported a

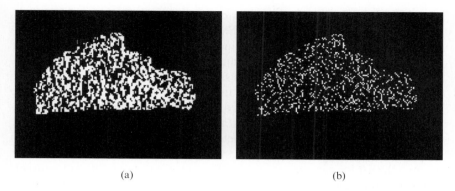

(a)   (b)

Figure 13.14. *High-resolution MR imaging of human ultradistal radius (156 × 156 × 500 μm³): two-dimensional slice of the three-dimensional binarized image (a), and the same slice of the corresponding skeleton graph (b). When observed on a two-dimensional slice, the skeleton graph seems to be highly disconnected compared with the initial structure observed on the same slice. Nevertheless, the connections that appear on the two-dimensional slice (a) go through upper and lower slices when the structure is thinned (b).*

technique using the surface skeleton representation inherent to the DTA approach to evaluate the local orientation of the plate-like trabeculae. This analysis has been validated with MR images of the radius, showing good visual alignment on trabecular bone cross-sectional images. Wehrli *et al* [96] applied the DTA technique to high-resolution *in vivo* MR images ($137 \times 137 \times 350 \, \mu\text{m}^3$) of the radius of 79 women with a wide range of BMD and vertebral deformity status. Although DTA-derived parameters were correlated to bone volume fraction as well as to BMD, they were able to significantly distinguish subjects with deformities from those without deformities ($p < 0.0005$). Due to the nature of the DTA-derived parameters in terms of plate-like or rod-like trabecular models, the changes of TBS observed with post-menopausal osteoporosis were interpreted by the authors as a conversion of trabecular plates to rods.

A topological approach based on deriving some local topological measures from a three-dimensional line skeleton graph analysis (3D-LSGA) has recently been developed [57]. The skeleton graph is obtained using a thinning algorithm with both topological and morphological constraints, an example of such a skeleton graph applied to an *in vivo* MR image of the distal radius is shown in figure 13.14. Other descriptors of the complexity of the network have also been developed such as Entropy (E) or Maximal Entropy (ME) and applied to *in vivo* MR images [98]. These measures have been used in cross-sectional studies to differentiate osteoporotic subjects from non-osteoporotic subjects and in a one-year longitudinal study to assess the impact of a therapeutic drug [84, 87].

In addition to applications in osteoporosis and ageing, MR-based methods have been used to study TBS in patients undergoing transplantation, in subjects with osteoarthritis and may have potential applications in studying patients with spinal cord injury. To characterize the TBS in men with osteoporosis induced by heart transplantation, Link *et al* [85] used high-resolution *in vivo* imaging of the calcaneus ($195 \times 195 \times 1000\,\mu m^3$). Three groups of subjects were studied, 40 men after heart transplantation, 11 men before heart transplantation, and 10 age-matched male volunteers. Standard histomorphometric parameters, as well as BMD of the lumbar spine and vertebral fracture status, were used to compare the different groups. Both TBS parameters and BMD were statistically different between subjects before and after transplantation ($p < 0.05$). Furthermore, differences in TBS ($p < 0.05$) were observed between patients with (42%) and without vertebral fracture in the 40 men studied after heart transplantation. Such differences did not appear with BMD parameter. Finally, moderate correlation was found between time after transplantation and TBS parameters, while no correlation was found with BMD, indicating perhaps changes in trabecular structure precede what is detectable by measures of BMD. Beuf *et al* [99] have studied trabecular bone structure in subjects with mild, moderate and severe osteoarthritis of the knee joint and demonstrated significant differences between TBS in the femur and tibia, which decreased with the degree of osteoarthritis. The authors found that the apparent BV/TV, TbN and TbSp in the femoral condyles could be used to differentiate healthy subjects or subjects with mild osteoarthritis from patients with severe osteoarthritis ($p < 0.05$). Among individuals, the authors demonstrated that the structural variation of the lateral and medial femoral condyle was indicative of the disease extent.

13.4. CONCLUSION

Over the past several years different magnetic resonance methods have emerged as a means of measuring trabecular bone structure, and have been shown to play a role in the assessment of skeletal strength and integrity. While *in vitro* studies and studies in animal models are a mandatory part of technique development, the potential strength and role of MRI lies in its human *in vivo* applications. Not only is the use of MR an attractive alternative for assessing trabecular bone structure, but its potential for quantifying bone marrow composition makes it an attractive modality for the comprehensive characterization of age-related or therapy-related metabolic changes in the skeleton. With the advent of higher field magnets for clinical imaging, and computerized image processing, MR promises to provide an important complement to standard methods of assessing osteoporosis and response to therapy.

13.5. ACKNOWLEDGMENT

We gratefully acknowledge Dr Hillie Cousart for her careful editorial oversight during manuscript preparation.

REFERENCES

[1] Frost H M 1964 Dynamics of bone remodelling. In *Bone Biodynamics* ed H M Frost (Boston: Little Brown) pp 315–34

[2] Brown M A and Semelka R C 1995 MRI: basic principles and applications, ed Wiley-Liss (New York: Wiley)

[3] Majumdar S, Newitt D, Jergas M, Gies A, Chiu E, Osman D, Keltner J, Keyak J and Genant H 1995 Evaluation of technical factors affecting the quantification of trabecular bone structure using magnetic resonance imaging *Bone* 17 417–30

[4] Majumdar S, Thomasson D, Shimakawa A and Genant H K 1991 Quantitation of the susceptibility difference between trabecular bone and bone marrow: experimental studies *Mag. Res. Med.* 22 111–27

[5] Kang C, Paley M, Ordidge R and Speller R 1999 R_2' measured in trabecular bone in vitro: relationship to trabecular separation *Mag. Res. Imaging* 17 989–95

[6] Takahashi M, Wehrli F W, Hwang S N and Wehrli S L 2000 Relationship between cancellous bone induced magnetic field and ultrastructure in a rat ovariectomy model *Mag. Res. Imaging* 18 33–9

[7] Fransson A, Grampp S and Imhof H 1999 Effects of trabecular bone on marrow relaxation in the tibia *Mag. Res. Imaging* 17 69–82

[8] Brismar T B, Karlsson M, Li T and Ringertz H 1999 The correlation between R_2' and bone mineral measurements in human vertebrae: an *in vitro* study *Eur. Radiol.* 9 141–4

[9] Sugimoto H, Kimura T and Ohsawa, T 1993 Susceptibility effects of bone trabeculae. Quantification *in vivo* using an asymmetric spin-echo technique *Invest. Radiol.* 28 208–13

[10] Jergas M D, Majumdar S, Keyak J H, Lee I Y, Newitt D C, Grampp S, Skinner H B and Genant H K 1995 Relationships between Young's modulus of elasticity, ash density, and MRI derived effective transverse relaxation T_2^* in tibial specimens *J. Comput Assist. Tomogr.* 19 472–9

[11] Chung H, Wehrli F W, Williams J L and Kugelmass S D 1993 Relationship between NMR transverse relaxation, trabecular bone architecture, and strength *Proc. Natl. Acad. Sci. USA* 90 10250–4

[12] Allein, S, Majumdar S, De Bisschop E, Newitt D C, Luypaert R and Eisendrath H 1997 *In vivo* comparison of MR phase distribution and $1/T_2^*$ with morphologic parameters in the distal radius *J. Mag. Res. Imaging* 7 389–93

[13] Brismar T B 2000 MR relaxometry of lumbar spine, hip, and calcaneus in healthy premenopausal women: relationship with dual energy X-ray absorptiometry and quantitative ultrasound *Eur. Radiol.* 10 1215–21

[14] Brismar T B, Hindmarsh T and Ringertz H 1997 Experimental correlation between T_2^* and ultimate compressive strength in lumbar porcine vertebrae *Acad. Radiol.* 4 426–30

[15] Beuf O, Newitt D C, Mosekilde L and Majumdar S 2001 Trabecular structure assessment in lumbar vertebrae specimens using quantitative magnetic resonance imaging and relationship with mechanical competence *J. Bone Min. Res.* **16** 1511–9

[16] Davis C A, Genant H K and Dunham J S 1986 The effects of bone on proton NMR relaxation times of surrounding liquids *Invest. Radiol.* **21** 472–7

[17] Rosenthal H, Thulborn K R, Rosenthal D I, Kim S H and Rosen B R 1990 Magnetic susceptibility effects of trabecular bone on magnetic resonance imaging of bone marrow *Invest. Radiol.* **25** 173–8

[18] Sebag G H and Moore S G 1990 Effect of trabecular bone on the appearance of marrow in gradient echo imaging of the appendicular skeleton *Radiology* **174** 855–9

[19] Ford J C and Wehrli F W 1991 *In vivo* quantitative characterization of trabecular bone by NMR interferometry and localized proton spectroscopy *Mag. Res. Med.* **17** 543–51

[20] Wehrli F W, Ford J C, Attie M, Kressel H Y and Kaplan F S 1991 Trabecular structure: preliminary application of MR interferometry *Radiology* **179** 615–21

[21] Ford J C, Wehrli F W and Chung H W 1993 Magnetic field distribution in models of trabecular bone *Mag. Res. Med.* **30** 373–9

[22] Majumdar S and Genant H K 1992 *In vivo* relationship between marrow T_2^* and trabecular bone density determined with a chemical shift-selective asymmetric spin-echo sequence *J. Mag. Res. Imaging* **2** 209–19

[23] Remy F and Guillot G 1998 Trabecular bone characterization with low-field MRI *Mag. Res. Imaging* **16** 639–42

[24] Majumdar S 1991 Quantitative study of the susceptibility difference between trabecular bone and bone marrow: computer simulations *Mag. Res. Med.* **22** 101–10

[25] Engelke K, Majumdar S and Genant H K 1994 Phantom studies simulating the impact of trabecular structure on marrow relaxation time, T_2' *Mag. Res. Med.* **31** 384–7

[26] Selby K, Majumdar S, Newitt D C and Genant H K 1996 Investigation of MR decay rates in microphantom models of trabecular bone *J. Mag. Res. Imaging* **6** 549–59

[27] Yablonskiy D A, Reinus W R, Stark H and Haacke E M 1997 Quantitation of T_2' anisotropic effects on magnetic resonance bone mineral density measurement *Mag. Res. Med.* **37** 214–21

[28] Newitt D C, Majumdar S, Jergas M D and Genant H K 1996 Decay characteristics of bone marrow in the presence of a trabecular bone network: *in vitro* and *in vivo* studies showing a departure from monoexponential behavior *Mag. Res. Med.* **35** 921–7

[29] Ashman R B, Rho J Y and Turner C H 1989 Anatomical variation of orthotropic elastic moduli of the proximal human tibia *J. Biomech.* **22** 895–900

[30] Majumdar S, Thomasson D, Shimakawa A and Genant H K 1991 Quantitation of the susceptibility difference between trabecular bone and bone marrow: experimental studies *Mag. Res. Med.* **22** 111–27

[31] Guglielmi G, Selby K, Blunt B A, Jergas M, Newitt D C, Genant H K and Majumdar S 1996 Magnetic resonance imaging of the calcaneus: preliminary assessment of trabecular bone-dependent regional variations in marrow relaxation time compared with dual X-ray absorptiometry *Acad. Radiol.* **3** 336–43

[32] Song H K, Wehrli F W and Ma J 1997 Field strength and angle dependence of trabecular bone marrow transverse relaxation in the calcaneus *J. Mag. Res. Imaging* **7** 382–8

[33] Link T M, Majumdar S, Augat P, Lin J C, Newitt D, Lane N E and Genant H K 1998 Proximal femur: assessment for osteoporosis with T_2^* decay characteristics at MR imaging *Radiology* **209** 531–6

[34] Wehrli F W, Hopkins J A, Hwang S N, Song H K, Snyder P J and Haddad J G 2000 Cross-sectional study of osteopenia with quantitative MR imaging and bone densitometry *Radiology* **217** 527–38

[35] De Bisschop E, Luypaert R, Allein S and Osteaux M 1996 Quantification of trabecular structure in the distal femur using magnetic resonance phase imaging *Mag. Res. Imaging* **14** 11620

[36] Mihalopoulou E, Allein S, Luypaert R, Eisendrath H, Bezerianos A and Panayiotakis G 1998 Comparison of computer simulated and phantom measured phase variance in the study of trabecular bone *Mag. Res. Imaging* **16** 29–36

[37] Allein S, Mihalopoulou E, Luypaert R, Louis O, Panayiotakis G and Eisendrath H 2000 MR phase imaging to quantify bone volume fraction: computer simulations and *in vivo* measurements *Mag. Res. Imaging* **18** 275–9

[38] Fantazzini P, Viola R, Alnaimi S M and Strange J H 2001 Combined MR-relaxation and MR-cryoporometry in the study of bone microstructure *Mag. Res. Imaging* **19** 481–4

[39] Chung H W, Wehrli F W, Williams J L and Wehrli S L 1995 Three-dimensional nuclear magnetic resonance microimaging of trabecular bone *J. Bone Min. Res.* **10** 1452–61

[40] Hipp J A, Jansujwicz A, Simmons C A and Snyder B 1996 Trabecular bone morphology using micro-magnetic resonance imaging *J. Bone Min. Res.* **11** 286–97

[41] Simmons C A and Hipp J A 1997 Method-based differences in the automated analysis of the three-dimensional morphology of trabecular bone *J. Bone Min. Res.* **12** 942–7

[42] Gomberg B R, Saha P K, Song H K, Hwang S N and Wehrli F W 2000 Topological analysis of trabecular bone MR images *IEEE Trans. Med. Imaging* **19** 166–74

[43] Hwang S N, Wehrli F W and Williams J L 1997 Probability-based structural parameters from three-dimensional nuclear magnetic resonance images as predictors of trabecular bone strength *Med. Phys.* **24** 1255–61

[44] Pothuaud L 2000 Correlation between the 3D microarchitecture and the radiographic projection of the trabecular bone: relation to osteoporosis: University of Orléans

[45] Pothuaud L, Benhamou C L, Porion P, Lespessailles E, Harba R and Levitz P 2000 Fractal dimension of trabecular bone projection texture is related to three-dimensional microarchitecture *J. Bone Min. Res.* **15** 691–9

[46] Borah B, Dufresne T E, Cockman M D, Gross G J, Sod E W, Myers W R, Combs K S, Higgins R E, Pierce S A and Stevens M L 2000 Evaluation of changes in trabecular bone architecture and mechanical properties of minipig vertebrae by three-dimensional magnetic resonance microimaging and finite element modeling *J. Bone Min. Res.* **15** 1786–97

[47] Takahashi M, Wehrli F W, Wehrli S L, Hwang S N, Lundy M W, Hartke J and Borah B 1999 Effect of prostaglandin and bisphosphonate on cancellous bone volume and structure in the ovariectomized rat studied by quantitative three-dimensional nuclear magnetic resonance microscopy *J. Bone Min. Res.* **14** 680–9

[48] Majumdar S, Newitt D, Mathur A, Osman D, Gies A, Chiu E, Lotz J, Kinney J and Genant H K 1996 Magnetic resonance imaging of trabecular bone structure in the distal radius: relationship with X-ray tomographic microscopy and biomechanics *Osteoporos. Intl.* **6** 376–85

[49] Majumdar S, Kothari M, Augat P, Newitt D C, Link T M, Lin J C, Lang T, Lu Y and Genant H K 1998 High-resolution magnetic resonance imaging: three-dimensional trabecular bone architecture and biomechanical properties *Bone* **22** 445–54

[50] Kothari M, Keaveny T M, Lin J C, Newitt D C, Genant H K and Majumdar S 1998 Impact of spatial resolution on the prediction of trabecular architecture parameters *Bone* **22** 437–43

[51] Vieth V, Link T M, Lotter A, Persigehl T, Newitt D, Heindel W and Majumdar S 2001 Does the trabecular bone structure depicted by high-resolution MRI of the calcaneus reflect the true bone structure? *Invest. Radiol.* **36** 210–7

[52] Lin J C, Amling M, Newitt D C, Selby K, Srivastav S K, Delling G, Genant H K and Majumdar S 1998 Heterogeneity of trabecular bone structure in the calcaneus using magnetic resonance imaging *Osteoporos. Intl.* **8** 16–24

[53] Engelke K, Hahn M, Takada M, Vogel M, Ouyang X, Delling G and Genant H K 2001 Structural analysis of high resolution *in vitro* MR images compared to stained grindings *Calcif. Tissue Intl.* **68** 163–71

[54] Laib A, Beuf O, Issever A, Newitt D C and Majumdar S 2003 Direct measures of trabecular bone architecture from MR images

[55] Pothuaud L, Laib A, Levitz P, Benhamou C L and Majumdar S 2003 3D-Line skeleton graph analysis (LSGA) of high-resolution magnetic resonance images: a validation study from 34-mm-resolution micro-computed tomography

[56] Pothuaud L, Van Rietbergen B, Mosekilde L, Levitz P, Benhamou C L, Beuf O and Majumdar S 2003 Relationship between the topological parameters describing trabecular bone network and trabecular bone mechanical properties

[57] Pothuaud L, Porion P, Lespessailles E, Benhamou C L and Levitz P 2000 A new method for three-dimensional skeleton graph analysis of porous media: application to trabecular bone microarchitecture *J. Microsc.* 2000 **199** 149–61

[58] Link T M, Majumdar S, Lin J C, Newitt D, Augat P, Ouyang X, Mathur A and Genant H K 1998 A comparative study of trabecular bone properties in the spine and femur using high resolution MRI and CT *J. Bone Min. Res.* **13** 122–32

[59] Luo G, Kinney J H, Kaufman J J, Haupt D, Chiabrera A and Siffert R S 1999 Relationship between plain radiographic patterns and three- dimensional trabecular architecture in the human calcaneus *Osteoporos. Intl.* **9** 339–45

[60] White D L, Schmidlin O, Jiang Y, Zhao J, Majumdar S, Genant H, Sebastian A and Morris R C 1997 MRI of trabecular bone in an ovariectomized rat model of osteoporosis, 1021

[61] Kapadia R D, Majumdar S, Stroup G B, Hoffman S J, Zhao H, Gowen M and Sarkar S K 1998 Quantitative analysis of MR microscopy images of trabecular bone architecture in a rat model of osteoporosis. 6th ISMRM, Sydney, Australia p 404

[62] Kapadia R D, Stroup G B, Badger A M, Koller B, Levin J M, Coatney R W, Dodds R A, Liang X, Lark M W and Gowen M 1998 Applications of micro-CT and MR microscopy to study pre-clinical models of osteoporosis and osteoarthritis *Technol. Health Care* **6** 361–72

[63] Luo Y, Nuss M, Majumdar S, Wilcox D, Nguyen P, Lane B, Wegner C and Jacobson P 1999 *In vivo* assessment of trabecular bone changes induced by prednisolone in rat tibia using high resolution MRI. 7th ISMRM, Philadelphia, Pennsylvania, USA p 1044

[64] Derby K, Kramer D M and Kaufman L 1993 A technique for assessment of bone marrow composition using magnetic resonance phase interference at low field. *Mag. Res. Med.* **29** 465–9

[65] Ballon D, Jakubowski A, Gabrilove J, Graham M C, Zakowski M, Sheridan C and Koutcher J A 1991 *In vivo* measurements of bone marrow cellularity using volume-localized proton NMR spectroscopy. *Mag. Res. Med.* **19** 85–95

[66] Dooms G C, Fisher M R, Hricak H, Richardson M, Crooks L E and Genant H K 1985 Bone marrow imaging: magnetic resonance studies related to age and sex *Radiology* **155** 429–32

[67] Schick F, Bongers H, Jung W I, Skalej M, Lutz O and Claussen C D 1992 Volume-selective proton MRS in vertebral bodies. *Mag. Res. Med.* **26** 207–17

[68] Schick F, Einsele H, Bongers H, Jung W I, Skalej M, Duda S, Ehninger G and Lutz O 1993 Leukemic red bone marrow changes assessed by magnetic resonance imaging and localized 1H spectroscopy. *Ann. Hematol.* **66** 3–13

[69] Majumdar S, Genant H K, Grampp S, Newitt D C, Truong V H, Lin J C and Mathur A 1997 Correlation of trabecular bone structure with age, bone mineral density, and osteoporotic status: *in vivo* studies in the distal radius using high resolution magnetic resonance imaging *J. Bone Min. Res.* **12** 111–8

[70] Gordon C L, Webber C E, Christoforou N and Nahmias C 1997 *In vivo* assessment of trabecular bone structure at the distal radius from high-resolution magnetic resonance images *Med. Phys.* **24** 585–93

[71] Majumdar S, Link T M, Augat P, Lin J C, Newitt D, Lane N E and Genant H K 1999 Trabecular bone architecture in the distal radius using magnetic resonance imaging in subjects with fractures of the proximal femur. Magnetic Resonance Science Center and Osteoporosis and Arthritis Research Group *Osteoporos. Intl.* **10** 231–9

[72] Song H K and Wehrli F W 1999 *In vivo* micro-imaging using alternating navigator echoes with applications to cancellous bone structural analysis *Mag. Res. Med.* **41** 947–53

[73] Newitt D C, Van Rietbergen B and Majumdar S 2003 Processing and analysis of *in vivo* high resolution MR images of trabecular bone for longitudinal studies: reproducibility of structural measures and micro-finite element analysis derived mechanical properties

[74] Newitt D C, Majumdar S, Van Rietbergen B, Von Ingersleben G, Harris S T, Genant H K, Chesnut C, Garnero P and MacDonald B 2003 *In vivo* assessment of architecture and micro-finite element analysis derived indices of mechanical properties of trabecular bone in the radius

[75] Ouyang X, Selby K, Lang P, Engelke K, Klifa C, Fan B, Zucconi F, Hottya G, Chen M, Majumdar S and Genant H K 1997 High resolution magnetic resonance imaging of the calcaneus: age-related changes in trabecular structure and comparison with dual X-ray absorptiometry measurements *Calcif. Tissue Intl.* **60** 139–47

[76] Link T M, Majumdar S, Augat P, Lin J C, Newitt D, Lu Y, Lane NE and Genant H K 1998 *In vivo* high resolution MRI of the calcaneus: differences in trabecular structure in osteoporosis patients *J. Bone Min. Res.* **13** 1175–82

[77] Pothuaud L, Lespessailles E, Harba R, Jennane R, Royant V, Eynard E and Benhamou C L 1998 Fractal analysis of trabecular bone texture on radiographs: discriminant value in post menopausal osteoporosis *Osteoporos. Intl.* **8** 618–25

[78] Mulder J E, Michaeli D, Flaster E R and Siris E 2000 Comparison of bone mineral density of the phalanges, lumbar spine, hip, and forearm for the assessment of osteoporosis in postmenopausal women *J. Clinic Densitom.* **3** 373–81

[79] Gulam M, Thornton M M, Hodsman A B and Holdsworth D W 2000 Bone mineral measurement of phalanges: comparison of radiographic absorptiometry and area dual X-ray absorptiometry *Radiology* **216** 586–91

[80] Laib A, Newitt D C, Lu Y and Majumdar S 2003 New model-independent measures of trabecular bone structure applied to *in vivo* high-resolution MR images

[81] Jara H, Wehrli F W, Chung H and Ford J C 1993 High-resolution variable flip angle 3D MR imaging of trabecular microstructure *in vivo Mag. Res. Med.* **29** 528–39

[82] Stampa B, Kuhn B, Heller M and Gluer C C 1998 Rods or plates: a new algorithm to characterize bone structure using 3D magnetic resonance imaging. 13th International Bone Densitometry Workshop, Delavan, WI

[83] Newitt D C and Majumdar S 2000 A Semi-automated system for segmenting, registering, thresholding, and analyzing high resolution MRI of trabecular bone. Proceedings of the 8th Annual Meeting of the ISMRM, Denver, CO p 127

[84] Pothuaud L, Newitt D C and Majumdar S 2001 Trabecular bone microarchitecture derived from high-resolution MRI of the ultradistal radius: relationship to osteoporotic status. 23rd Annual Meeting of the ASBMR, Phoenix, Arizona, USA p S461

[85] Link T M, Lotter A, Beyer F, Christiansen S, Newitt D C, Lu Y, Schmid C and Majumdar S 2000 Changes in calcaneal trabecular bone structure after heart transplantation: an MR imaging study *Radiology* **217** 855–62

[86] van Rietbergen B, Majumdar S, Newitt D C and MacDonald B 2001 High-resolution MRI and micro-FE for the evaluation of changes in calcaneal bone mechanical properties in postmenopausal women after one year of idoxifene treatment. 47th Annual Meeting of ORS, San Francisco, California, USA

[87] Pothuaud L, Newitt D C, Chesnut C, Genant H K, MacDonald B and Majumdar S 2001 New descriptors of trabecular bone microarchitecture: an *in vivo* longitudinal study using magnetic resonance imaging. 23rd Annual Meeting of the ASBMR, Phoenix, Arizona, USA p S461

[88] Hwang S N and Wehrli F W 1999 Estimating voxel volume fractions of trabecular bone on the basis of magnetic resonance images acquired *in vivo Int. J. Imaging Syst. Technol.* **10** 186–98

[89] Antoniadis T, Scarpelli J P, Ruaud J P, Gonord P and Guillot G 1999 Bone labelling on micro-magnetic resonance images *Med. Image Anal.* **3** 119–28

[90] Pothuaud L, Porion P, Levitz P and Benhamou C L 2000 3D thresholding of trabecular bone images obtained by magnetic resonance imaging: effect of the threshold value on the 3D microarchitecture. 20th Annual Meeting of ASBM, Toronto, Canada

[91] Wu Z, Chung H W and Wehrli F W 1994 A Bayesian approach to subvoxel tissue classification in NMR microscopic images of trabecular bone *Mag. Res. Med.* **31** 302–8

[92] Wehrli F W, Hwang S N, Ma J, Song H K, Ford J C and Haddad J G 1998 Cancellous bone volume and structure in the forearm: noninvasive assessment with MR microimaging and image processing *Radiology* **206** 347–57 [published erratum appears in 1998 *Radiology* **207**(3) 833]

[93] van Rietbergen B, Weinans H, Huiskes R and Odgaard A 1995 A new method to determine trabecular bone elastic properties and loading using micromechanical finite-element models *J. Biomech.* **28** 69–81

[94] van Rietbergen B, Majumdar S, Pistoia W, Newitt D C, Kothari M, Laib A and Rüegsegger P 1998 Assessment of cancellous bone mechanical properties from micro-FE models based on micro-CT, pQCT and MR images *Technol. Health Care* **6** 413–20

[95] Saha P K, Gomberg B R and Wehrli F W 2000 Three-dimensional digital topological characterization of cancellous bone architecture *Int. J. Imaging Syst. Technol.* **11** 81–90

[96] Wehrli F W, Gomberg B R, Saha P K, Song H K, Hwang S N and Snyder P J 2001 Digital topological analysis of *in vivo* magnetic resonance microimages of trabecular bone reveals structural implications of osteoporosis *J. Bone Min. Res.* **16** 1520–31

[97] Gomberg B R, Saha P K, Song H K, Hwang S N and Wehrli F W 2003. Three-dimensional digital topological analysis of trabecular bone

[98] Pothuaud L, Newitt D C, Levitz P and Majumdar S 2001 Maximal entropy as a predictor of trabecular bone strength from high-resolution MR images. Proceedings of the 9th Annual Meeting of the ISMRM, Glasgow, Scotland

[99] Beuf O, Ghosh S, Newitt D C, Link T M, Steinbach L S, Reis M, Lane N and Majumdar M 2000 Characterization of trabecular bone mico-architecture in the knee in osteoarthrosis using high-resolution MRI. Proceedings of the 8th Annual Meeting of the ISMRM, Denver, CO p 2135

Chapter 14

Quantitative ultrasound

Christian M Langton and Christopher F Njeh

SECTION A: FUNDAMENTALS OF ULTRASOUND PROPAGATION

14.1. TERMINOLOGY

14.1.1. Ultrasound

The term ultrasound describes sound waves above the audible threshold, generally defined as 20 kHz. Sound waves are mechanical pressure waves in origin, which propagate through a material, the velocity and attenuation being strongly dependent upon the mechanical and structural properties of the material.

14.1.2. Frequency

The primary variable that we may alter in ultrasound measurements is the frequency. This is defined as the number of wavelengths (cycles) per unit time. The velocity c (m s^{-1}) is described by the product of frequency f (Hz) and wavelength λ (m). For conventional clinical imaging, the choice of frequency is a compromise between the quality of the image (often referred to as the resolution) and the amount of signal detected (often referred to as signal-to-noise ratio). As the frequency increases, the spatial resolution will increase, but so will the attenuation (see section 14.3.2). For example, in the assessment of the eye the tissues are relatively superficial and hence attenuation is relatively low. This enables us to use frequencies in the region of 10 MHz, providing high-resolution images. For assessment of the kidney, however, there are large overlying tissues that create significant attenuation of the ultrasound signal. This requires a reduction in the frequency of the ultrasound signal, with an associated reduction in spatial resolution. This is not critical since the dimensions of the kidney are significantly larger than those of the eye. Assessment of the kidney will typically be undertaken at frequencies in the range 2.5–3.5 MHz.

In comparison, quantitative ultrasound (QUS) assessment of predominantly cancellous bone sites, such as the calcaneus, is undertaken using frequencies below 1 MHz, typically 0.2–0.6 MHz. For smaller bones that contain less cancellous bone such as the phalanx, the frequency may be increased to 2.5 MHz.

14.2. ULTRASOUND PROPAGATION THROUGH MATERIALS

14.2.1. Spring model propagation

Imagine an array of molecules linked with inter-molecular springs, shown in figure 14.1. As an ultrasound pressure wave propagates through the material, molecules in regions of high pressure will be pushed closer together (compression) whereas molecules in regions of low pressure will be pulled apart (rarefaction), shown in figure 14.2. As the sound wave propagates through the tissue, molecules will exhibit the oscillatory nature of the pressure wave and will therefore vibrate around their equilibrium position.

14.2.2. Modes of wave propagation

There are various ways that an ultrasound wave may propagate through a material, often termed 'modes', illustrated in figure 14.3. The simplest is a *longitudinal* (*compression*) wave where the molecules of the material oscillate in the same direction as the ultrasound wave propagation. There are several other modes of ultrasound propagation through a material. A *transverse* (*shear*) wave describes ultrasound propagation where the tissue molecules vibrate at right-angles to the direction that the wave is travelling. This mode is not supported in soft tissues but may be propagated through bone. Other more complicated modes include *surface* and *plate* waves (particularly relevant to the assessment of the tibial cortex) and bar waves (a complex combination of longitudinal and transverse wave propagation through a sample whose lateral

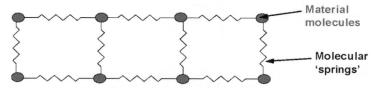

Figure 14.1. *Diagrammatic representation of a tissue where the molecules are connected via 'molecular springs': the stiffer the springs, the faster the ultrasound wave will travel through the tissue.*

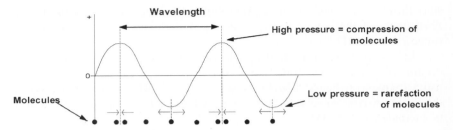

Figure 14.2. *In regions of positive pressure, the molecules are pushed closer together (compression); in regions of negative pressure, the molecules are pulled apart. Once the ultrasound wave has passed through the tissue, the molecules return to their original position.*

dimensions are less than the wavelength of the ultrasound wave). For surface (often termed Rayleigh) waves, the molecules follow a retrograde elliptical motion, the amplitude falling exponentially with depth.

14.2.3. Velocity of ultrasound waves

The velocity of a sound wave through a material is dependent upon the mechanical properties of the tissue simulated by the intermolecular springs:

$$\text{velocity} = \sqrt{\frac{\text{elasticity}}{\text{density}}}. \tag{14.1}$$

The stiffer the 'intermolecular springs', the higher the velocity of ultrasound through it. This explains why soft tissues have a velocity in the region of $1500 \, \text{m s}^{-1}$ whereas cortical bone has a velocity in the region of $3500 \, \text{m s}^{-1}$.

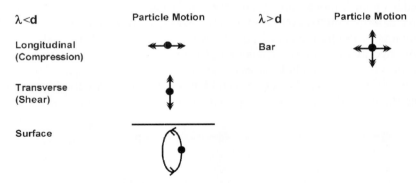

Figure 14.3. *Description of molecular motion for various modes of ultrasound propagation. When the wavelength is less than the lateral dimensions of the sample, longitudinal, transverse and surface waves may be propagated; when the wavelength is greater than the lateral dimensions, a bar wave may be propagated.*

In comparison, a metal such as aluminium has a velocity in the region of $8000 \, \text{m s}^{-1}$.

14.2.4. Propagation velocity dependence

The various transmission modes are related to different forms of equation (14.1), having appropriate elastic moduli. For example, for longitudinal (V_{long}), transverse (V_{trans}) and bar waves (V_{bar}) we have:

$$V_{long} = \sqrt{\frac{B + \frac{4}{3}G}{\rho}} \qquad V_{trans} = \sqrt{\frac{G}{\rho}} \qquad V_{bar} = \sqrt{\frac{E}{\rho}} \qquad (14.2)$$

where ρ is the density of the material, and B, G and E are the bulk, shear and Young's moduli respectively.

14.2.5. Phase and group velocity

Phase velocity refers to the velocity of a continuous wave (single frequency); group velocity refers to the velocity of a pulse (wave packet) which contains a range of frequencies. Group velocity is most often reported since pulse measurements are generally easier to obtain than those for a continuous wave. For certain materials such as water, the phase and group velocities are essentially equal whereas for complex media such as cancellous bone (known as a dispersive material) these two velocities are not the same. As explained in section 14.5.1, it should be noted that a short ultrasound pulse may be described by a series of different frequency waves, with varying amplitudes and phases (Fourier theorem). For a short ultrasound pulse, we may therefore describe either a single *group* velocity or *phase* velocity as a function of the frequency components within the signal.

A comparison of group and phase velocity measurements in bone is provided by Nicholson *et al* [1].

14.3.   AMPLITUDE, INTENSITY AND ATTENUATION

14.3.1. Amplitude and intensity

The amplitude of an ultrasound pulse generally refers to the maximum peak-to-peak voltage of the received electrical signal. Ultrasound intensity has units of W m^{-2} describing the rate of energy flow through a unit cross-sectional area.

14.3.2. Attenuation

As an ultrasound wave travels through a tissue, the intensity will fall due to attenuation processes (similar to the propagation of ionizing waves such as

X-rays through tissue). The intensity of a plane wave propagating in the x direction decreases with distance as

$$I_x = I_0 \, e^{-\mu(f)x} \qquad (14.3)$$

where I_0 and I_x are the intensities incident and at a distance x (cm), and $\mu(f)$ is the frequency dependent intensity attenuation coefficient (dB cm^{-1}).

Attenuation is typically reported in decibels (dB), a logarithmic scale defined as

$$20 \log(A_1/A_2) \text{ for amplitude} \quad \text{(volts)} \qquad (14.4a)$$

or

$$10 \log(I_1/I_2) \text{ for intensity} \quad \text{(W m}^{-2}\text{)}. \qquad (14.4b)$$

There are several attenuation processes involved in the propagation of ultrasound, particularly absorption (the conversion of sound energy to heat mainly due to internal molecular friction) and scattering (the spatial redistribution of ultrasound intensity due to interaction with inhomogeneities). The amount of scattering depends primarily on the ratio of the ultrasound wavelength to the size of the scattering particle and on the acoustic impedance mismatch between the propagation medium and the scattering particle.

Additional factors contributing to the attenuation include beam spreading (diffraction) and mode conversion (e.g. longitudinal wave energy into shear wave energy). The predominant attenuation mechanism in cancellous bone is scattering, whilst absorption predominates in cortical bone. Several theoretical models have been developed to explain the attenuation of ultrasound in cancellous bone (see section C).

14.3.3. Broadband ultrasound attenuation

In the frequency range 0.1–1 MHz, the most useful for bone characterization, the total attenuation is approximately linearly proportional to frequency, as

$$\mu(f) = \alpha f \qquad (14.5)$$

where α is the slope of attenuation against frequency (dB MHz^{-1} cm^{-1}). In clinical practice, this has become known as broadband ultrasound attenuation (BUA).

BUA is measured by recording the amplitude spectrum of an ultrasound pulse through a reference material, chosen to be de-gassed water, and through the sample to be studied, illustrated in figure 14.4. The attenuation (dB) at each frequency (f) is calculated from the amplitude through water and sample respectively, and plotted as a function of frequency between 0.2 and 0.6 MHz. The slope of this plot is defined as the BUA index, with units of dB MHz^{-1}. Dividing this by the sample width provides a volumetric parameter with units of dB MHz^{-1} cm^{-1}.

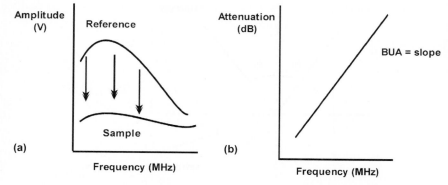

Figure 14.4. *Diagrammatic representation of BUA measurement where (a) describes measurement of frequency spectra through a reference material (usually water) and the test sample. Attenuation is plotted in (b) against frequency, the regression slope being the BUA parameter.*

14.4. INTERFACE BEHAVIOUR

14.4.1. Acoustic impedance

The propagation material offers a resistance to the passage of ultrasound, analogous to the electrical resistance to the flow of electrical charge along a wire. For sound waves the resistance is described by the acoustic impedance (Z) defined as the product of tissue density (ρ) and propagation velocity (c). The acoustic impedance determines the behaviour of an ultrasound wave at a tissue interface, namely transmission, reflection and refraction.

14.4.2. Normal incidence at a tissue interface

For normal incidence the incident wave is split into transmitted and reflected waves which propagate in opposite directions, illustrated in figure 14.5.

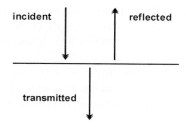

Figure 14.5. *Diagrammatic representation of ultrasound behaviour at an interface where an incident wave is split into reflected and transmitted components.*

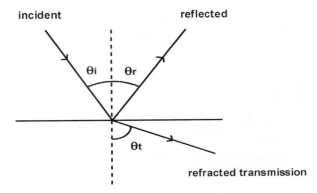

Figure 14.6. *Diagrammatic representation of the behaviour of an ultrasound wave at a non-normal interface. The angle of reflection equals the angle of incidence, the angle of refracted transmitted wave being defined by the ratio of ultrasound velocities either side of the interface.*

The fraction of amplitude ($A\alpha_r$) or intensity ($I\alpha_r$) reflected compared with the incident amplitude or intensity, and the fraction of amplitude ($A\alpha_t$) or intensity ($I\alpha_t$) transmitted are described by equations (14.6). Note that Z is the acoustic impedance.

$$A\alpha_r = \frac{(Z_2 - Z_1)}{(Z_2 + Z_1)} \qquad I\alpha_r = \frac{(Z_2 - Z_1)^2}{(Z_2 + Z_1)^2}$$

$$A\alpha_t = \frac{2Z_2}{(Z_2 + Z_1)} \qquad I\alpha_t = \frac{4Z_2 Z_1}{(Z_2 + Z_1)^2}. \tag{14.6}$$

14.4.3.   Non-normal incidence at a tissue interface

For angled incidence to a tissue boundary, shown in figure 14.6, the behaviour of the transmitted and reflected waves may be described using Snell's law.

$$\frac{\sin\theta_i}{\sin\theta_r} = \frac{c_1}{c_1} \qquad \frac{\sin\theta_i}{\sin\theta_{tl}} = \frac{c_1}{c_2} \qquad \frac{\sin\theta_i}{\sin\theta_{tt}} = \frac{c_1}{c_{2t}} \tag{14.7}$$

where c_1 and c_2 are the longitudinal wave velocities in materials 1 and 2 respectively, c_{2t} is the transverse wave velocity in material 2 and θ_{tl} and θ_{tt} are the transmission refraction angles for longitudinal and transverse waves respectively.

This demonstrates that the angle of reflection equals the angle of incidence (in material 1) at the boundary, the angle of refracted transmission being defined by the ratio of ultrasound velocities either side of the tissue boundary.

If, however, the second material can support a transverse (shear) wave, mode conversion may take place, whereby some of the incident longitudinal wave energy is converted into transmitted transverse wave energy. Since the velocity of a transverse wave is approximately half that for a longitudinal, its angle of refraction (θ_{tt}) is lower (equation (14.7)).

14.4.4 Coupling

The behaviour of an ultrasound wave at an interface explains why coupling is an important requirement for the ultrasonic assessment of tissues. For the assessment of bone, techniques utilizing both direct coupling via a gel and immersion coupling with water have been developed. Since the acoustic impedance is approximately $430\,\mathrm{kg\,m^{-2}\,s^{-1}}$ for air, $1.5 \times 10^6\,\mathrm{kg\,m^{-2}\,s^{-1}}$ for most soft tissues and $7 \times 10^6\,\mathrm{kg\,m^{-2}\,s^{-1}}$ for bone, the presence of air will provide almost total reflection (equation (14.6)) and hence minimal transmission through the skin. Replacing air with water or gel significantly increases the transmission of ultrasound into the tissues. Poor coupling is probably the major cause of measurement artefact in the assessment of bone by ultrasound.

14.5. ULTRASOUND WAVE FORMATS

14.5.1. Continuous, tone-burst and pulsed waves

Waves may be described in terms of their length in time (time-domain), as shown in figure 14.7. A continuous wave (a), by definition, continues for an infinite time. A tone-burst wave (b) is a section of a continuous wave, often regularly repeated, creating a series of bursts. Most ultrasound systems incorporate a pulse consisting of a few ($\sim$3–5) wavelengths (c). The Fourier theorem states that any waveform may be explained by a series of different frequency waves, varying in amplitude and phase (delay). Hence, a continuous wave may be described by a single frequency and a short pulse of ultrasound may be described as a spectrum of individual single frequencies, as shown in figure 14.8.

14.5.2. Bandwidth theorem

Waves may also be expressed in terms of their time frequency content (frequency domain). Two extreme examples are a continuous wave, continuing for infinite time but containing only one frequency; and an infinitely short (Dirac) pulse lasting for an infinitely short period (time ≈ 0) and containing an infinite number of frequencies. The bandwidth theorem states that the product of time content (Δt) and frequency content (Δf) for a wave or pulse is unity. A typical ultrasound pulse will contain a few wavelengths and contain a broad range of frequencies.

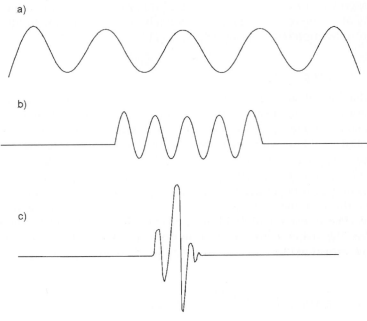

Figure 14.7. *Time-domain representation of (a) continuous, (b) tone-burst and (c) pulse ultrasound waves.*

14.5.3. *Frequency spectrum and Q factor*

By plotting the amplitude of each frequency, we obtain the frequency spectrum, shown in figure 14.9. The frequency content may be quantified in terms of the width of the frequency spectrum ($F_2 - F_1$) and the central (fundamental) frequency (F_0), termed the Q factor, defined as

$$Q = \frac{F_2 - F_1}{F_0}.$$ (14.8)

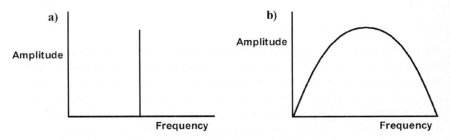

Figure 14.8. *Corresponding frequency-domain representations for (a) continuous and (b) pulse time-domain signals shown in figure 14.7.*

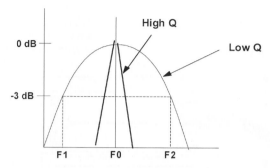

Figure 14.9. *Diagrammatic representation of the frequency spectrum of an ultrasound signal of fundamental frequency F_0. The lower and upper frequencies at half the peak amplitude (signal reduced by 3 dB) are F_1 and F_2 respectively.*

As described above, the shorter the signal in the time domain, the broader the frequency spectrum. Ultrasound transducers utilized for the measurement of BUA at the calcaneus typically have a Q of approximately 2.

SECTION B: INSTRUMENTATION

14.6. THE ULTRASOUND TRANSDUCER AND BEAM PROFILE

14.6.1. Piezoelectric effect and transducer

The term transducer describes a device which interconverts energy forms. The piezoelectric effect describes the interconversion of electrical and mechanical energy. For the generation of ultrasound, an electric signal is applied across a piezoelectric material that induces a mechanical deformation, which in turn produces a pressure wave. Ultrasound piezoelectric transducers for the assessment of bone are generally in the form of a thin circular disc, the thickness of which determines the resonant frequency (F_0); the thickness being half a wavelength. For the detection of ultrasound, the piezoelectric material converts a mechanical deformation, due to the ultrasound pressure wave, into an electrical signal that may then be modified and processed. A naturally occurring piezoelectric material is quartz which is still used for some high-frequency ($\sim$50 MHz region) applications. Piezoelectric ceramic materials are generally utilized, the most popular being lead zirconate titanate (PZT), with barium titanate utilized if particularly short (low Q) pulses are required.

14.6.2. Transducer design

With reference to figure 14.10, the ceramic element is coated on each flat face with electrodes. The front window of the transducer is made of plastic and

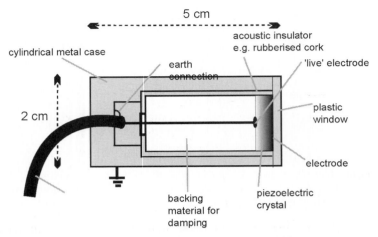

Figure 14.10. *Diagrammatic representation of an ultrasound transducer.*

serves two purposes:

1. to protect the piezoelectric element and
2. to optimize ultrasound propagation, the front window is a quarter wavelength thick.

The insulator dampens any radial mode of vibration of the piezoelectric element. The backing material serves an important role which essentially controls the pulse length (and hence frequency spectrum). The backing is generally made from tungsten-loaded Araldite, chosen to have an acoustic impedance equal to that of the piezoelectric material (minimal reflection) and a high attenuation. The net effect is that the portion of the ultrasound pulse that propagates into the transducer is suppressed, producing a very short transmitted pulse. For comparison, an undamped (air-backed) transducer would 'ring' for a considerable time and have a narrow frequency spectrum.

14.6.3. Beam profile

The generation of an ultrasound wave from a circular transducer may be simulated by a piston source with a complex diffraction pattern emanating from it, described by the near (Fresnel) field and far (Fraunhofer) field diffraction respectively, shown in figure 14.11. The length of the near field (often denoted X_{max}) may be approximated to r^2/λ where r is the transducer radius and λ is the wavelength. The divergence angle in the far field (θ) is described by

$$\theta = \sin^{-1}\left(\frac{0.61\lambda}{r}\right). \tag{14.9}$$

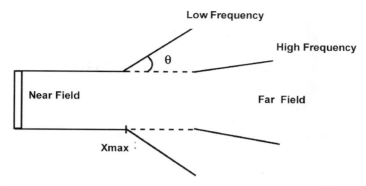

Figure 14.11. *Beam profile of a wave propagating from an ultrasound transducer.*

The combination of near and far fields may be simulated using Huygens' principle, illustrated in figure 14.12, where an ultrasound wave is divided into a number of wavelets. We consider a plane wave where each wavefront is represented as a parallel line. Each wavefront is itself made up of a series of Huygens wavelet sources, each emanating a circular wavefront (similar to dropping a stone on to water). Each wavefront is produced by adding the amplitude of its wavelets. When the plane wave passes through an opening, corresponding to the transducer face, only those wavelets within the opening continue (figure 14.12(a)). Again, by summing the wavelet amplitudes for each wavefront (figure 14.12(b)), both near and far field diffraction patterns are created.

14.6.4. Focusing

Focusing of the ultrasound beam may be implemented to provide a narrower beam, shown in figure 14.13, often utilized in imaging bone analysers (section 14.14.1). Although an acoustic lens may be incorporated, this is generally achieved using a curved transducer element. Focusing can only occur

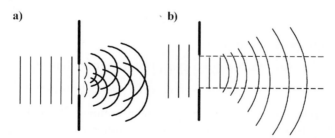

Figure 14.12. *Diagrammatic representation of the Huygens principle, where a wavefront is represented by a number of equal wavelets. By adding the effect of these wavelets, near and far field diffraction patterns are created.*

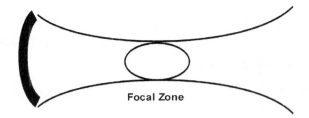

Figure 14.13. *Diagrammatic representation of focusing of an ultrasound wave.*

within the near field and beginning of the far field. If the radius of curvature for the transducer element is R then the width of the focal zone is described by $R(\lambda/r)$. As the width of focusing reduces so does the length of the focal zone. This may have significant implications if focusing is required over a significant length of the ultrasound beam. The dependence of ultrasound intensity upon the degrees of focusing is shown in figure 14.14.

14.7.   INSTRUMENTATION

14.7.1   Pulse–echo technique

The majority of clinical applications of ultrasound are related to the imaging of soft tissue structures using a pulse–echo technique, illustrated in figure 14.15. The ultrasound waves are generated and received by the same transducer, placed upon the surface of the skin with a water-based gel to facilitate coupling of the ultrasound into the tissues. An ultrasound pulse is propagated into the tissues and is reflected at interfaces. For conventional medical ultrasonography, it is the dimensions of the tissues that are generally of interest, for example, the bi-parietal diameter (head width) for prediction of gestation age. The standard equation for velocity is

$$\text{velocity} = \frac{\text{tissue thickness propagated}}{\text{time taken}} \qquad (14.10)$$

which is transposed to provide

$$\text{tissue thickness} = \text{assumed propagation velocity}$$
$$\times \text{ measured propagation time.} \qquad (14.11)$$

For pulse–echo measurements, the propagation time is divided by a factor of 2 since the ultrasound wave will have propagated twice through the tissue of interest (transducer to tissue interface and back to transducer). Clinical ultrasound scanners assume that all tissues have a constant velocity, but this assumption can lead to measurement errors and image artefacts.

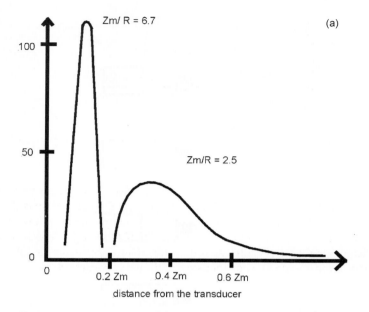

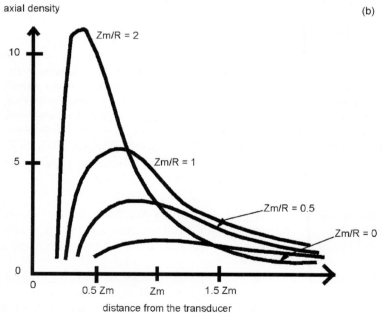

Figure 14.14. *The variation in intensity along the ultrasound beam for various degrees of focusing (expressed as ratios of near field length), where Z_m and R describe the near field length and radius of curvature respectively. Note that a flat disc has $Z_m/R = \infty$.*

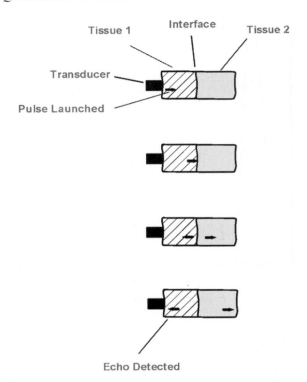

Figure 14.15. *Diagrammatic representation of the pulse–echo technique where a single transducer transmits an ultrasound pulse, and also receives reflected ultrasound pulses from tissue interfaces. The ultrasound signal is launched into the first tissue; when it hits the tissue interface, a portion of the ultrasound signal is transmitted into the second tissue, and a portion is reflected within the first tissue, often termed an echo. Only when the reflected signal is received by the ultrasound transducer is the echo detected.*

14.7.2. Transmission technique

For the assessment of highly attenuating materials such as cancellous bone, transmission rather than pulse–echo techniques are often adopted, whereby two transducers are utilized, one acting as transmitter, the other as receiver.

The simplest form of transmission utilizes separate transducers for transmission and reception of the ultrasound signal, the two transducers being coaxially aligned (figure 14.16(a)). Although more recent instrumentation developments have provided bone velocity measurements along the tibial cortex and other anatomical sites, in principle they are still operating in a transmission mode (figure 14.16(b)) although the operator may handle a single transducer housing.

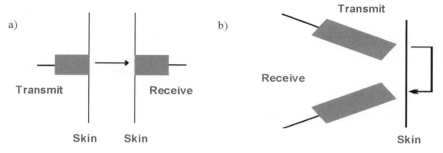

Figure 14.16. *Diagrammatic representation of the transmission technique where separate transducers are utilized to transmit and receive ultrasound pulses. The transducer arrangement may be (a) coaxially or (b) pseudo-reflection aligned. Examples are measurement of the calcaneus and tibial cortex respectively.*

14.7.3. Simple radio-frequency (RF) system

The simplest ultrasound system, illustrated in figure 14.17, consists of a transmitter connected to a transducer with a receive transducer connected directly to an oscilloscope, hence operating in transmission mode (see section 14.7.2).

14.7.3.1. Transmitter

The transmitter may be a signal generator of sufficient output voltage (~ 20 V) providing either continuous or tone-burst signals; or alternatively, a purpose-built electric spike generator of approximately 200 V with a width of approximately 100 ns.

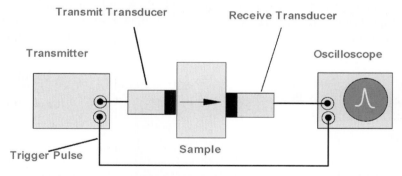

Figure 14.17. *Diagrammatic representation of a simple RF system operating in transmission mode. The transmitter sends a signal to the transmit transducer, along with a trigger pulse to the oscilloscope. After passing through the sample, the ultrasound signal is detected by the receive transducer and fed into the oscilloscope.*

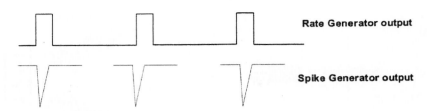

Figure 14.18. *Timing diagram demonstrating that a leading edge of the rate generator triggers an output from the spike generator.*

14.7.3.2. Spike generator

The purpose-built spike generator will incorporate a rate generator which controls the pulse repetition frequency (PRF, the number of ultrasound pulses produced per second), typically in the region of 500–1000 Hz. Ideally, the generator should provide a pulse to trigger the oscilloscope. If the oscilloscope is being operated in conventional time-base mode, the beginning of the trace will always correspond to the generation of an ultrasound pulse. Synchronization of the spike output signal to the rate generator is shown in figure 14.18.

14.7.3.3. Digitization of the received ultrasound signal

As an alternative to analogue display of the received ultrasound signal on a conventional oscilloscope, the received signal may also be digitized, enabling, for example, transfer of the data to a computer. The analogue electrical signal output from the ultrasound receive transducer is digitized by an analogue-to-digital (ADC) convertor, at a defined rate (typically 5–20 MHz) and over a defined number of amplitude levels (typically 8-bit, from 0 to 255), illustrated in figure 14.19. A 'rule of thumb' is that the digitization rate should be five times the maximum frequency of the signal of interest. If successive ADC points are displayed on a computer

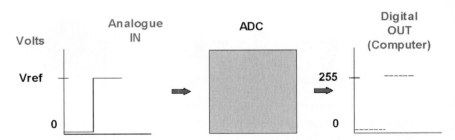

Figure 14.19. *Diagrammatic representation of an 8-bit analogue-to-digital converter. The ratio of an analogue input signal to the reference voltage is converted into a corresponding ratio 8-bit digital output.*

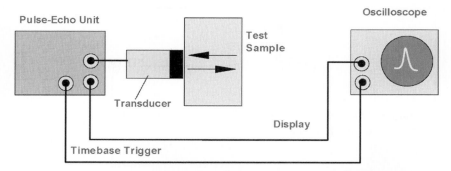

Figure 14.20. *Diagrammatic representation of an integrated pulse–echo A-scan system where a single transducer is utilized to both transmit and receive ultrasound signals. The oscilloscope is triggered such that the beginning of the trace corresponds to the launch of an ultrasound signal.*

screen then the time between each point will be the inverse of the digitization rate, the signal amplitude represented by 0–255 discrete amplitude levels. The ADC has a reference voltage, the 256 levels covering the voltage range from zero to the reference voltage. Since ultrasound signals consist of negative as well as positive voltages, zero volts generally corresponds to half the reference voltage. Amplification is generally applied to the received ultrasound signal prior to the ADC to ensure that the received signal covers the maximum range of levels without saturating (going beyond V_{ref}).

An advantage of digital representation of an ultrasound signal is that further signal processing and analysis may be undertaken, for example, fast Fourier transform (FFT) to obtain the frequency spectrum, particularly suited to the determination of BUA (see section 14.3.3). The FFT transforms digital data in the 'time-domain' into the 'frequency domain', as illustrated in figures 14.7 and 14.8.

14.7.4.   Integrated pulse–echo system

For pulse–echo measurements as utilized, for example, in backscattering analysis (section 14.7.6), a single transducer is used to both generate and detect the ultrasound pulses. The instrumentation is illustrated in figure 14.20.

14.7.4.1.   Transmitter

The transmitter unit is the same spike generator (section 14.7.3.2) as for the simpler RF system, incorporating a rate generator. If the PRF is set too low, the refresh rate of the display may be sufficiently slow to cause flicker; if set

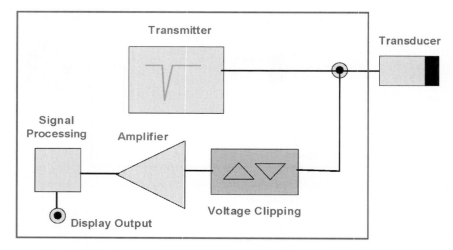

Figure 14.21. *Diagrammatic representation of a pulse–echo receiver where the transmitter spike launches an ultrasound signal, resultant echoes being detected by the same transducer. The received signal is then amplified to a level appropriate for subsequent display. The signal processing may include filtering and conversion of the RF signal into the A-scan representation.*

too high, there is a danger that a new ultrasound pulse could be initiated before all the echoes of interest have been received from the previous initiation, thus potentially overlapping echoes from different transmitted signals.

14.7.4.2.   Pulse–echo receiver

The primary role of the pulse–echo receiver, illustrated in figure 14.21, is to provide 'voltage clipping' (a simple circuit being a pair of crossed diodes) only allowing signals within a small voltage range to pass. This is to ensure that the received echo signals from within the tissues being investigated (typically millivolts) may be identified and not lost within the significantly higher transmission spike (several hundred volts). Effectively, the amplitude of the 'detected' transmission spike is significantly reduced, enabling the received echo signals to be observed. The receiver may also contain signal-processing functions including filtering.

14.7.4.3.   Amplitude modulated (A-scan) display

The basic unmodified ultrasound signal format is often referred to as an RF signal, shown in figure 14.7(c). The simplest method of displaying an ultra-sound pulse is through A-scan representation on an oscilloscope where the vertical axis represents signal amplitude and the horizontal axis represents

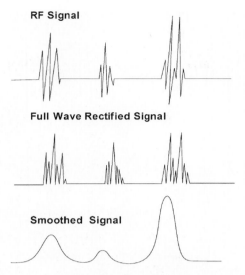

Figure 14.22. *Conversion of RF signal into A-scan representation. The RF signal is initially full-wave rectified and then smoothed.*

depth of material propagation. The RF wave is generally full-wave rectified (all half-wavelengths positive) and smoothed (the simplest electrical circuit being a capacitor) to provide a simpler representation, illustrated in figure 14.22.

14.7.5.   Rectilinear scanning

Rectilinear mechanical scanning has been implemented for ultrasound assessment of the calcaneus, providing images of velocity and BUA data, from which values within a defined region of interest may be reported. This technique is a derivative of the basic transmission technique, except that the co-aligned transducers are scanned in a rectilinear manner over the region of the heel. The basic principle is illustrated in figure 14.23.

14.7.6.   Backscattering analysis

A recent development in the ultrasonic assessment of bone has been back-scatter analysis [2, 3], where a pulse–echo technique is utilized to provide a time-domain representation of ultrasound signals being reflected (back-scattered) from the trabecular structure of the calcaneus. By applying a time-window to the received data, echoes originating from a particular depth from within the calcaneus may be analysed, for example, to calculate BUA.

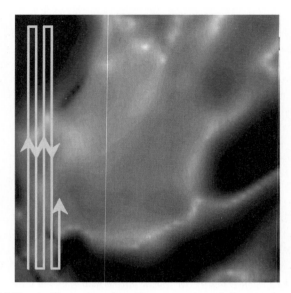

Figure 14.23. *An illustration of rectilinear scanning of ultrasound.*

SECTION C: THEORETICAL MODELLING

In vivo, the total attenuation is measured, that is the combination of absorption and scattering. Although bone mass accounts for part of the attenuation, the architecture of the bone is also responsible for the attenuation through scattering mechanisms [5]. The specific roles played by both absorption and scattering in overall attenuation are still not well known. Wear [6, 7] concluded from the observed frequency dependence of backscatter attenuation that absorption was the main component of attenuation in cancellous bone. However, McKelvie and Palmer [8] established that absorption in calcified bone tissue and bone marrow could not explain the high attenuation observed in cancellous bone. Therefore, it has been hypothesized that energy losses by scattering in the internal structure of cancellous bones or viscous friction at the bone–marrow interfaces are likely to be the principal causes of attenuation [9, 10]. Biot theory, Schoenberg's theory and scattering models have all been proposed to explain ultrasound interactions in cancellous bone.

14.8.   BIOT THEORY

Modelling the frequency dependence of ultrasound attenuation has proved very difficult. Attempts have been made to apply theories such as that developed by Biot [11] and Schoenberg [12] to cancellous bone [13–15]. The application of Biot theory to cancellous bone has recently been reviewed by Haire and Langton [13]. Biot theory is derived by considering the separate motion of the solid elastic frame and the interspersed fluid, induced by the ultrasonic wave [11]. The theory gives rise to three elastic parameters that are dependent on the structural and elastic properties of the porous medium. The Biot theory predicts the existence of two longitudinal waves: the fast wave corresponding to the solid and fluid moving in phase and the slow wave corresponding to the solid and fluid moving out of phase [3]. The Biot theory has been applied to cancellous bone with varying degrees of success [8, 16–18]. Hosokawa [18] found good agreement between the measured and calculated speed of sound but not for attenuation. This could be because Biot theory predicts absorption due to viscous losses at internal interfaces only, but experimental measurements record signal loss through other mechanisms as well. Signal loss may include contributions from reflections at the flat surfaces of the samples [19], scattering, absorption of the scattered field and artefacts of the measurement system, such as diffraction [20, 21] and phase cancellation [22].

14.9. SCHOENBERG'S THEORY

Because of the complexity of the Biot theory, recently Hughes *et al* [14] have suggested using Schoenberg's theory [12]. This theory assumes cancellous bone to be made up of a simple layered structure of alternating plates of bone and marrow. The well-established theory of acoustic propagation in strata could be then applied. In Schoenberg's theory, the fluid is assumed to be inviscid and so viscous absorption is not predicted. Using bovine samples from the tibial and femoral epiphyses, known to contain a well-oriented trabecular structure, Hughes *et al* [14] found a qualitative agreement between measured phase velocity and predictions of Schoenberg's theory. Although a unified theory is needed to explain the signal velocity and signal loss measured *in vivo*, it is clear that they are influenced by the architecture of the material.

14.10 OTHER MODELS

Strelitzki *et al* [9] recently proposed that a scattering model based on velocity fluctuations in a binary mixture (marrow fat and cortical matrix) can be used to estimate the ultrasonic attenuation in cancellous bone. They used data from the literature and predicted attenuation values that were of the same order of magnitude as experimentally determined. Nicholson *et al* [10] also used this scattering model in cancellous bone to look at the relationship between BUA and BMD. The model predicted very similar nonlinear trends to those previously observed experimentally. They also demonstrated that attenuation was dependent on scatterer size in addition to porosity. This further supports the argument that attenuation is influenced by structure. Potential limitations in this approach include the failure to include absorption into the model. Furthermore the model was not used to predict velocity relationships.

SECTION D: *IN VITRO* EXPERIMENTS

Laboratory and clinical studies are presented in sections D and E respectively. *In vitro/in vitro* studies tend to address the more fundamental questions underlying ultrasonic assessment of bone status. *In vivo* studies, on the other hand, are studies carried out on human subjects.

14.11.   BONE SAMPLES

14.11.1.   Source

Bone samples used for *in vitro* studies can be obtained from both animal and human origins. The specific location depends on whether the bone of interest is cortical or cancellous. Animal samples have been obtained from the bovine femur and equine third metacarpus. Human samples from the vertebra, calcaneus, femur, tibia and phalanges have also been studied.

14.11.2.   Sample size and shape

Most *in vitro* studies are carried out on regular shaped bone specimens. The shape and size of the samples are determined by several factors:

1.   *Experimental objectives.* For example, if the purpose is to study ultrasound variables in one direction, then cylindrical samples may be appropriate. On the other hand if we are interested in evaluating the anisotropy, then it is useful to have cubic samples.
2.   *Mechanical testing.* Most samples used for QUS measurements are also used for mechanical testing. There are sample preparation requirements for mechanical testing (refer to chapter 5). Cubic and cylindrical specimens are commonly used in compressive testing. Generally, the specimen must be large enough that a representative section of pores and struts can be analysed, yet small enough to resolve the heterogeneity of the tissue. It has been shown that a 5 mm cubic test sample is the minimum size for which the continuum characterization is realistic [23].

3. *Diameter of the transducer.* For longitudinal measurements it is preferable to have samples whose width or diameter are equal to or greater than the transducer's diameter. This is a requirement if attenuation is to be measured. For an immersion experiment where the sample is smaller than the diameter of the transducer, wave-blocking holders must be used to inhibit the unattenuated signals from passing alongside the specimen. For bar wave experiments, it is necessary to have the lateral dimensions of the sample significantly smaller than the ultrasound wavelength (refer to section 14.2.2)

14.11.3. *Sample preparation*

Preparation of bone samples is also governed by the study objectives, specimen size and shape. Most sample preparation for QUS investigation has three phases: cutting, defatting and degassing.

1. *Cutting.* The samples should be cut under constant water irrigation. This prevents heating and possible damage to the samples. Both soft tissue and the cortical endplates should be removed to leave purely cancellous bone.
2. *Degassing.* Air pockets in bone samples are detrimental to QUS measurement because they are highly reflective. All air bubbles in a sample must be removed to improve the reliability of the measurement. This can be achieved by degassing the samples submerged in water using a vacuum desiccator over a period of time.
3. *Defatting.* Cancellous bone has pores filled with bone marrow, which might make degassing take longer. Most studies are carried out on samples with marrow removed (defatted) to speed up the degassing process. Njeh *et al* [48] defatted bone samples using the following procedure: the sample is subjected to a high speed jet of water, and then compressed air. This process is repeated until no fat is visible. The sample is then tumbled overnight in excess of 2:1 chloroform: methanol mixture. The sample is finally dried in air and stored in a vacuum desiccator.

14.12. MEASUREMENT: METHODOLOGY AND ANALYSIS

14.12.1. *Coupling*

In vitro studies are mostly carried out under water (immersion) with the transducers either placed a fixed distance apart or in contact with the sample. For airborne measurements at low ultrasonic frequencies ($\sim$50 kHz), Strelitzki *et al* have suggested direct contact on defatted samples [24].

14.12.2.   Transducers

Broadband ultrasound transducers are mostly used (see section 14.5.3). Focused transducers are preferred for scanning because of the improved spatial resolution, but unfocused transducers are generally used for fixed-site measurements. Highly-damped broadband transducers with centre frequencies from 0.5 to 1 MHz are most commonly employed.

14.12.3.   Transit time velocity measurements

The majority of velocity measurements in cancellous bone to date have been made using transit time measurements, utilizing the transmission technique (section 14.8.2). The time taken for a broadband ultrasonic pulse to traverse a specimen is determined using some characteristic feature (marker) on the waveform; velocity is calculated knowing the distance the wave has travelled [25]. The marker may be placed at the first apparent deviation of the rising edge [26], or a threshold may be applied [27], or a point at which the waveform crosses the time axis (zero-crossing) may be used [28] (figure 14.24). There is, however, no consensus on a standardized marker protocol for velocity determinations in bone.

Classically, two velocities are defined for wave motion: the phase velocity and the group velocity (section 14.2.5) [29]. The broadband pulse/transit time method does not explicitly yield either of these two classical parameters; rather it gives what can be termed signal velocities. In homogeneous solids, such as Perspex, aluminium or indeed cortical bone, where dispersion is minimal, the phase, group and signal velocities will be nearly identical. Where there is significant attenuation present, as in many biological tissues,

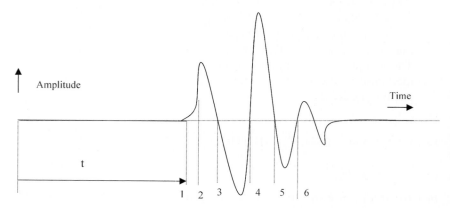

Figure 14.24. *The different transit time definitions used in velocity measurement. Marker 1 is the first arriving signal, marker 2 is the minimum threshold and marker 3 is the first zero crossing. No standardized protocol has been adopted for marker definition in bone measurement.*

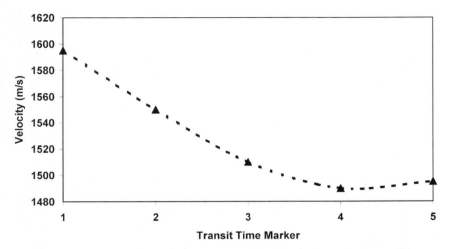

Figure 14.25. *The magnitude of the velocity measured will depend on the marker used, as demonstrated by these data obtained on human calcaneus measured in vitro (adapted by Strelitzki et al [31]).*

signal velocity will be influenced by the modification of the waveform by attenuation and dispersion [30]. In cancellous bone, frequency-dependent attenuation lowers the centre frequency and broadens the pulse, and hence signal velocity depends on the marker used to define transit time, as has been confirmed in measurements on calcanei *in vitro* (figure 14.25) [27, 31]. This pulse spreading is a function of BUA and the measured transit time velocity is consequently dependent on the attenuation [31]. One further consequence of such effects is that transit time velocity measurements, even if made using a standardized marker, will yield different results depending on the centre frequency of the transducers used [32].

14.12.4. Alternative velocity measurements

Improved methods for velocity measurement have been demonstrated *in vitro*. Transit time measurements made on single frequency tone-bursts [31, 33] have been attempted with some success, but may still be prone to errors associated with the transit time marker position. Cross-correlation can be used to derive a 'mean pulse velocity' which appears robust with respect to attenuation-related errors, but which is not readily related to the classical velocities and may be dependent on the original frequency content of the pulse [31, 34]. Phase spectral analysis of the broadband pulse can be used to derive phase velocity [27, 34, 35] and/or group velocity.

Methods which give phase and/or group velocity in cancellous bone have advantages in terms of accuracy and comparability, but their performance in terms of predicting clinically-relevant parameters has not been extensively validated. However, in one study [25], phase velocities in the calcaneus were shown to be very highly correlated with QCT-determined BMD ($r^2 = 0.88$), suggesting a predictive ability equal to or greater than conventional transit time velocities [35]. The reproducibility of phase velocity measurements has been shown to be comparable with that of transit time velocity measurements.

14.12.5.   Critical angle reflectometry

Antich and colleagues have described a different approach, ultrasound critical angle reflectometry (UCR), for velocity measurements in bone (see section 14.4.3) [36–39]. When ultrasound propagates across a water/solid interface, there is a critical angle of incidence at which total internal reflection occurs, and this is seen as a maximum in the reflected amplitude (figure 14.26b). This critical angle is directly related to the phase velocity of sound in the solid through Snell's law. Velocities measured by UCR correlate well with pulse transmission velocities, but are approximately 12% higher [36], a discrepancy which may be due to the different ways in which the two techniques are affected by anisotropy, heterogeneity and sample preparation [39]. UCR can be used to determine both longitudinal and shear wave velocities in cortical bone [40]. Velocities can be measured at all orientations in the plane of the bone surface, allowing the principal axes of elastic symmetry to be determined [36, 41]. One potential problem with UCR, however, is that it measures velocity at the surface, and the relationship between surface and bulk properties in the cortex is largely unknown.

14.12.6.   Attenuation

Attenuation is generally measured in cancellous bone as a function of frequency by comparing the power spectra of pulses propagating through a reference material (e.g. water) and through the bone specimen. Attenuation data can be quoted either as attenuation values (dB cm^{-1}) at specific frequencies, or, more commonly, as the slope of attenuation versus frequency (dB MHz^{-1} cm^{-1}) for a specified frequency range (i.e. BUA). The BUA (section 14.3.3) parameter assumes that attenuation is a linear function of frequency, but there is evidence, discussed below, that this is not always the case in cancellous bone.

14.12.7.   Error sources

There are a number of potential sources of error which are likely to affect attenuation and, to a lesser extent, velocity. These include diffraction, interface losses and phase cancellation. Historically, these have often been

(a)

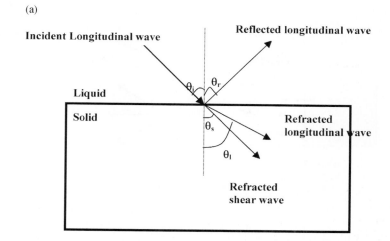

(b)

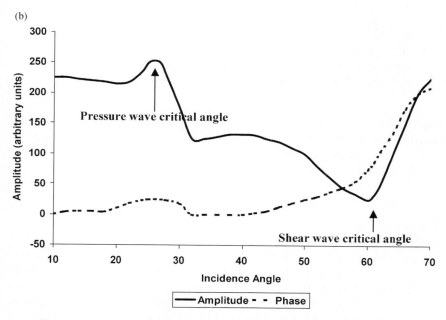

Figure 14.26. *Schematic representation of the ultrasound critical angle reflectometry (UCR). (a) There is reflection and refraction at the boundary between two media. For solid the refracted wave is partially converted into a shear wave. As the angle of incidence increases there comes a point where the longitudinal wave is no longer transmitted into the solid (total internal reflection). The corresponding incident angle is called the first critical angle. (b) A measure of the amplitude will reveal a peak at the critical angle. This is experimental data from bone sample (adapted from Antich et al [39]).*

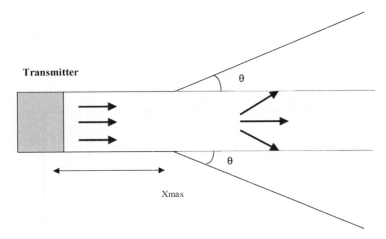

Figure 14.27. *In attenuation measurement diffraction is an error source: Close to the transmitter (near field) the wave approximates the parallel beam. At a transition X_{max}, the far-field begins and the beam starts to diverge at an angle θ. Both X_{max} and θ depend on frequency: at low frequencies X_{max} is small and θ is large (more beam spread); at higher frequencies D is large and θ small (less beam spread). Hence, the energy detected at a given frequency depends on the position of the receiver within the field.*

neglected in the context of clinical measurements, and existing evidence does suggest that their impact is limited. However, investigation of these effects is likely to be an important part of the search for improved accuracy and precision in clinical measurements.

14.12.7.1.   Diffraction

The attenuation measured using pulse transmission may include errors due to diffraction ('beam spread') (figure 14.27). When the immersion method is used and the sample velocity is close to that of water, diffraction losses can often be neglected [42]. With a contact method, diffraction errors may potentially be larger because the separation of the transducers changes to accommodate different specimen thicknesses. Analytical diffraction corrections can be applied, as is believed to be the case with some commercial devices (CUBA, McCue Ultrasonics, Winchester, UK), but there are no published data on their validation in the context of bone measurements. Theoretical work suggests possible diffraction related errors in BUA of $0.6\,\mathrm{dB\,MHz^{-1}}$ for immersion and $10\,\mathrm{dB\,MHz^{-1}}$ for contact measurements at the heel [21]. It should also be noted that velocity measurements can also be affected by diffraction, but that such effects are generally small [42].

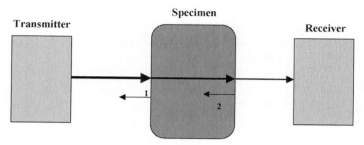

Figure 14.28. *At each interface, water/bone (1) and bone/water (2), a proportion of the incident signal will be reflected. Hence the attenuation recorded by the receiver represents the sum of the true energy losses within the specimen and the reflection losses at the two interfaces.*

14.12.7.2.   Interface losses

Interface losses refer to the loss of intensity of the transmitted beam due to partial reflection at interfaces between two different media (figure 14.28), and are generally assumed to be negligible in cancellous bone. There is some experimental evidence that such losses are of the order of 0.5 dB or less for a purely cancellous sample but are appreciably higher (up to 20 dB) when overlying cortical surfaces are present [43, 44]. Fry and Barger developed an analytical acoustic transmission model predicting a nonlinear variation in skull interface losses with frequency, in broad agreement with their experimental data [43]. In the human calcaneus, the frequency dependence of the interface losses has yet to be studied.

14.12.7.3.   Phase cancellation

Another potential source of error in attenuation measurements is phase cancellation. When ultrasound propagates through inhomogeneous media, components of the acoustic wave can arrive out of phase at the receiver, cancelling out and producing an erroneously high apparent attenuation (figure 14.29). This effect occurs only when the receiver is phase-sensitive, as is the case for the piezoelectric transducers used in all commercial bone ultrasound devices. In one study, it was claimed that phase-sensitive BUA measurements were 30% higher than phase-insensitive measurements [22]. However, another study found no difference in BUA measurements made on calcani *in vitro* using focused and unfocused transducers. Since a focused system will presumably be less affected by phase cancellation, this finding tends to contradict the earlier conclusion [32]. Measurements on the calcaneus using a multi-element receiver also suggest that phase cancellation effects on BUA in the calcaneus are small [45].

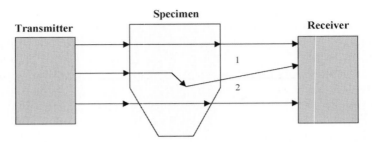

Figure 14.29. *Scattering within the specimen (1), sample thickness variation (2) and velocity variation (not shown) can change the propagation time of different components of the signals. Hence components arriving at the receiver out of phase will tend to cancel out, causing an artifactual increase in the measured attenuation (adapted from Nicholson and Njeh [53]).*

14.12.7.4.   Cortical surfaces

The cortex overlying the cancellous bone in the human calcaneus represents a potential source of error in ultrasonic measurements. Reflection losses will be greater at a water/cortical bone interface than a water/cancellous bone interface, since the acoustic impedance mismatch is higher. When the cortical surface is curved or irregular there are likely to be additional effects due to refraction and phase cancellation.

In measurements on intact calcanei and on purely cancellous, flat-sided, samples from the same specimens, the presence of the cortex produced a significant additional BUA of 3–5 dB MHz^{-1} cm^{-1}, but the two measurements were highly correlated ($r^2 = 0.86$–0.90) with a regression slope close to unity [46, 47]. These data suggest that the cortex produces a relatively constant offset in BUA. It has recently been suggested that this measurement artefact is due to phase cancellation [4], supported by computer and experimental simulations.

Similar studies have shown that removing the cortical faces reduced velocity in the calcaneus by approximately 20 m s^{-1} [48]. The velocity in the intact calcaneus correlated only moderately well with that in the cancellous bone alone ($r^2 = 0.64$–0.69), suggesting that the presence of the cortex degraded the ability to measure the velocity of cancellous bone within [48]. One explanation for this could be variations in cortical thickness between individuals, but further work is required to investigate this.

14.12.7.5.   Heel width

Some ultrasonic devices utilized clinically at the calaneus (e.g. Achilles, Lunar Corp, Madison, WI) do not measure and account for the width of the heel. With such machines, a pseudo-velocity termed speed of sound

(SOS) is calculated using the assumption of a constant heel width. SOS thus calculated will consequently be affected by bone size as well as bone material properties. Some have argued that the variation in heel width is small in practice and can safely be neglected [49]. In one cadaver study, SOS produced slightly better correlations with femoral strength than did true bone velocity in the calcaneus. This was explained by the independent positive correlation between calcaneal width and femur strength: larger people have stronger bones [50].

In commercial devices which do not measure heel width, the BUA value is given in $dB\,MHz^{-1}$ and is partially dependent on heel thickness. With the Achilles device, the final quoted 'BUA' value is obtained after additional arithmetic manipulation of the true BUA measurement. Where heel width is known, BUA can be normalized for thickness to give a value in $dB\,MHz^{-1}\,cm^{-1}$ which should theoretically reflect bone material properties alone. Wu *et al* [49] suggested that BUA was not a linear function of thickness in bovine bone, and that normalization for thickness was therefore inappropriate. This nonlinear behaviour is puzzling, but may represent effects associated with interface losses and/or diffraction in a specific experimental configuration. From the point of view of maximizing the information available from ultrasonic measurements, it would seem preferable to normalize ultrasonic measurements for bone thickness and thereby separate the influences of skeletal size and instrinsic acoustic properties.

14.12.7.6. Soft tissue

Soft tissue, including overlying skin, capillaries, fat layers and marrow fat within the trabecular spaces, may influence ultrasonic measurements. A large proportion (typically 70–95% in the calcaneus) of the volume of cancellous bone consists of fluid-filled pores, and the acoustic properties of the fluid can be expected to play a major role in sound propagation. Fat has a slightly lower density and acoustic velocity than water. However, in studies of calcanei measured at room temperature, removing the marrow fat and replacing it with water did not affect either BUA or velocity [47, 48]. The temperature-dependence of sound velocity in water is positive but is negative in fat, implying that differences between fat- and water-saturated bone may be larger at body temperature.

In relation to the overlying soft tissue, Kotzki *et al* [51], using the Achilles commercial device, observed that increasing the thickness of fat around cadaver heels decreased SOS but did not consistently affect BUA. However, *in vivo* studies have indicated that oedema in the soft tissues overlying the calcaneus significantly reduces measured velocity and BUA [52].

The composition and properties of marrow fat in the calcaneus, and their variation with age or disease, are not well understood. Similarly, the physical characteristics of the overlying soft tissues have not been studied in any depth. Without these data, it is difficult to assess the true clinical significance of soft tissue effects in ultrasonic bone measurements.

14.13.   *IN VITRO* EXPERIMENTAL FINDINGS

14.13.1.   QUS and bone density

14.13.1.1.   BUA and bone density

The relationship between QUS and bone density has been studied extensively *in vitro* and has been reviewed by Nicholson and Njeh [53]. Early studies by McCloskey *et al* [46] and McKelvie *et al* [54] found high correlation ($r = 0.85$ and 0.83, respectively) between BUA and apparent density in samples of cadaveric calcanei. Even higher correlations ($r = 0.97$) have been reported by Nicholson *et al* [55] in human vertebra samples measured in the anterior–posterior direction. Others have reported similar strong and positive relationships between density and BUA [47, 55–58]. However, in the more dense bovine cancellous bone, the relationship is much weaker or even completely absent, and both positive and negative regression slopes have been reported [56, 59–61]. These results suggest that the relationship between BUA and density over the large density range of trabecular bone is parabolic, with BUA rising to a maximum at a porosity of approximately 70%. Similar trends have also been reported in bone phantom materials [61–63], suggesting that this nonlinear behaviour is a general feature of porous, fluid-saturated solids [53]. The role of scattering and absorption in the attenuation of the ultrasound signal provides some form of qualitative explanation for the nonlinear relationship of BUA with density. Han *et al* [56] suggested that attenuation in low-density cancellous bone and in cortical bone may be due primarily to absorption, with scattering becoming important only in dense cancellous bone. In agreement with this, Duquette *et al* [58] found lower nBUA (BUA normalized for sample thickness) in samples with cortical bone and higher nBUA (51%) in trabecular bovine femur samples. Hodgskinson *et al* [61] suggested that the parabolic symmetry could be related to the scattering cross-sectional area between bone and marrow, being similar for low porosity (few pores) and high porosity (few trabeculae).

14.13.1.2.   Velocity and density

The nonlinearity that is observed for BUA in bovine trabecular bone is not present in the relationship between velocity and density. From equation (14.4), a power relationship should exist between velocity and density when $n \neq 3$. However, significant linear correlations (with r up to 0.94) have been reported between velocity and density [35, 55, 64]. The correlation is still significant for bovine samples over wide ranges of density [65]. However, studies on bone phantom materials suggest that a quadratic model may better represent the true relationship over a wide density range [63]. Since structure should affect velocity, especially in the extreme cases, the relationship with density (scalar quantity) may not be linear. Also when measurements are performed parallel to the direction of the principal trabecular orientation, two wave components may be observed, a fast and slow wave [18, 55].

14.13.2. *QUS and mechanical properties*

The ability of ultrasound to determine the mechanical properties of cancellous bone has been studied extensively [65–70] and recently reviewed by Nicholson and Njeh [53]. Early work includes that of Abendschein and Hyatt [71], who found a close correlation ($r = 0.91$) between the mechanically and ultrasonically determined elastic moduli of human cortical bone. Similar studies were carried out by Ashman and colleagues [67, 72, 73] using low frequencies (50 kHz) and defatted human and bovine cancellous specimens. Correlations between ultrasonically and mechanically measured moduli were excellent ($r^2 = 0.93$–0.96) and the magnitudes were closely comparable. These early studies have only looked at the association between velocity and mechanical properties [72, 74]. More recent studies have demonstrated that BUA can also be used to estimate the mechanical properties (Young's modulus and strength) of cancellous bone samples [47, 66, 75]. QUS can add predictive power beyond that of density for mechanical properties' estimation [68, 69, 76–79]. When BUA and SOS are used together, their predictive power is better than that of density for mechanical properties' estimation [80]. It has also been suggested that combining density and ultrasound velocity significantly improved the prediction of the ultimate bone strength [69, 74, 81, 82]. Grimm and colleagues [79] showed that BUA, but not UTV, was an independent predictor of Young's modulus when the correlation was adjusted for trabecular density both *in vitro* and *in situ*.

The ability of QUS to predict bone mechanical properties is diminished when measuring the bone *in situ* rather than in cancellous cubes or when predicting the strength of a bone at a remote location. BUA of the heel correlates moderately with the strength of the calcaneus itself ($r = 0.79$) [73] and the proximal femur ($r = 0.57$–0.71) [50, 83]. Femoral BMD is a significantly better predictor of femur strength (0.77–0.94) than heel QUS [50]. Also, calcaneal QUS is not as good as lumbar spine BMD in predicting vertebral strength [84, 85]. However, in contrast to these results, Lochmuller *et al* [86] found that calcaneal QUS correlates with failure load of the proximal femur similarly to femoral neck BMD. These mechanical and QUS measurements using whole samples are unreliable since they could be compounded by many error sources.

14.13.3.   QUS and bone structure

Since the introduction of QUS in the 1980s for the clinical assessment of bone status [87], it has been proposed that QUS provides information on bone structure in addition to density. Quite often reports in the scientific literature, which have described a poor association between QUS and BMD, have attributed this finding to the fact that QUS may be measuring structure. However, the poor association could be due to many factors including measurement errors and anatomical discordance [88]. Evidence of the structural dependence of QUS has come mainly from anisotropic, histo-morphometric and fractal analysis studies.

14.13.3.1.   Anisotropy in bone cubes

Since the late 1980s, researchers have been interested in exploring the relationship between ultrasound and bone structure. In the majority of cases, owing to the difficulties in obtaining meaningful parameters charac-terizing trabecular structure, these studies were essentially qualitative in nature. A number of workers reported on the acoustic anisotropy of cancellous bone cubes, from the equine metacarpel [89], the bovine femur [65] and radius [90], and the human vertebra [55] and femur [50]. Acoustic anisotropy implies that structure affects acoustic properties independently of density since the density of a given sample is constant. The approach was extended into a semi-quantitative analysis of the relationship between BUA and trabecular orientation on bovine bone by Glüer and colleagues [90]. Further qualitative evidence for relationships between structure and ultrasound came from the studies of Tavakoli and Evans [91], who demon-strated that when bovine cancellous samples were modified by demineraliza-tion and crushing, ultrasonic properties were affected by the structural changes, and further that BUA was more sensitive to these changes than was velocity.

14.13.3.2.   Quantitative structural measurements

A few studies have reported relationships between acoustic properties and quantitative micro-structural measurements. Langton *et al* [89] performed histomorphometry on equine cancellous bone but found no relationship with BUA. Glüer *et al* [92] reported a number of significant associations between structural parameters and ultrasound in bovine bone, even when adjusted for density. Hans *et al* [93] made histomorphometric measurements in a small number of human calcaneal specimens, and concluded that ultrasonic measurements reflected bone quantity rather than bone micro-architecture. The discrepancy between the latter two studies may be asso-ciated with the higher density of bovine bone and differences in trabecular structure compared with human cancellous bone.

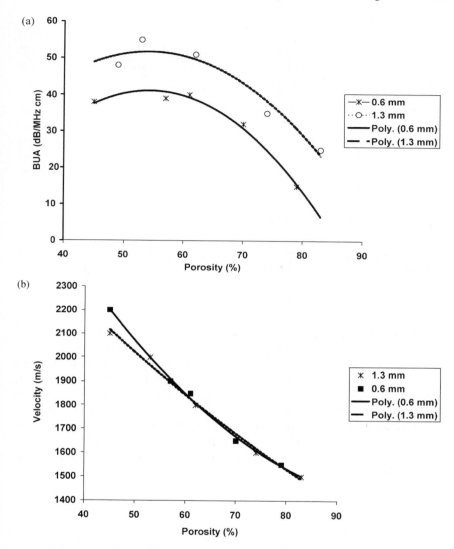

Figure 14.30. *Dependence of the ultrasonic properties on material porosity and on pore size. (a) BUA was greater for the phantom material with greater pore size. (b) Velocity was dependent on porosity but not on pore size. This is experiment data by Strelitzki et al [63] using the Leeds quantitative ultrasound phantom.*

Perhaps the most interesting work addressing this question is that of Strelitzki *et al* [64], who used a bone-mimicking material spanning a range of densities and with two distinct pore sizes. Velocity in the phantom material was independent of pore size (figure 14.30(a)), but BUA was greater for the larger pore size (figure 14.30(b)), indicating the sensitivity of BUA to a structural factor (pore size) independent of density. Further work along these lines would be valuable, and could include studies using phantom materials with a wider range of pore sizes, and the investigation of similar effects in cancellous bone using micro-computed tomography to assess pore size *in vitro*.

Whilst it is clear that structural effects on ultrasound can be demonstrated, it is by no means certain that these effects are significant in clinical measurements. As has been noted by a number of authors, the very strong correlations between density and acoustic properties measured mediolaterally in human cancellous bone *in vitro* imply only very limited room for structural factors to play a role [35, 47]. This, along with the other evidence cited above, suggests caution with regard to claims that clinical heel measurements are useful indicators of cancellous structure.

SECTION E: *IN VIVO* CLINICAL ASSESSMENT

14.14. COMMERCIAL SYSTEMS

Since the late 1980s, clinical QUS systems have become available. Currently there are a number of different machines on the market (table 14.1). These instruments have significant differences between them, including skeletal site of measurement, coupling method, velocity definition, transit time measurement and scanner design. Thus the readings obtained on different systems vary significantly [94].

Instead of describing the individual systems, the following sections will describe the general features. The individual systems have been described by Njeh *et al* [95]. Except for multi-site devices, all QUS systems use the pulse transmission technique.

14.14.1. Anatomical sites

Currently QUS assessment of the skeleton is limited to the peripheral sites. These sites include the calcaneus, phalanges, tibia, patella and radius. These sites vary in their bone composition (percentage of cortical to cancellous bone) and whether they are weight bearing or not. For osteoporosis assessment it has been advocated that for the best risk assessment the specific site of fracture must be measured.

14.14.1.1. Calcaneus

The calcaneus has 5 to 10 mm soft tissue covering it. In a study by Chappard *et al* [102], they found bone thickness in the range of 30.7 ± 2.7 mm and soft

Table 14.1.

Ultrasound system	Coupling medium	Parameter	Precision (CV)
Calcaneal single, fixed point QUS systems			
Lunar Achilles+	Water	BUA	0.8–2.5%
		SOS (TOF)	0.2–0.4%
		Stiffness	1.0–2.0%
McCue	Gel	BUA	1.5–4.0%
CUBAClinical		VOS (heel)	0.2–0.6%
Hologic Sahara	Gel	BUA	0.8–2.5%
		SOS (heel)	0.2–0.4%
		QUI	1.0–2.0%
Calcaneal imaging QUS systems			
DMS UBIS 3000	Water	BUA	0.8–2.5%
		SOS	0.2–0.4%
Osteometer DTU-1	Water	BUA	0.8–2.5%
		SOS	0.2–0.4%
Phalangeal QUS system			
IGEA	Gel	Ad-SOS	0.3–0.9%
DBM Sonic 1200		(cortex)	
Tibial QUS system (semi-reflection mode)			
Myriad	Gel	SOS	0.2–1.0%
Soundscan 2000		(cortex)	
Multi-site QUS system (semi-reflection mode)			
(calcaneum, tibia, phalanges, radius, ulna, trochanter, spine posterior process etc.)			
Sunlight	Gel	SOS	0.2–1.0%
Omnisense prototype			

tissue thickness of 8.8 ± 1.7 mm. Wu *et al* [49] using X-rays also measured bone thickness to be in the range of 29.6 ± 2.9 mm. Blake *et al* [182] also found bone width of 30.1 mm (range 22–40) with a standard deviation of 3.2 mm and precision of 1.1 mm.

The calcaneus is the most popular measurement site used for quantitative ultrasound for several reasons. The calcaneus is approximately 90% cancellous bone which, due to its high surface to volume ratio, has a higher metabolic turnover rate than cortical bone, and hence will manifest earlier metabolic changes [96]. The calcaneus is also easily accessible, and the medio-lateral surfaces are fairly flat and parallel. The choice of the calcaneus as the test site has been supported by Wasnich *et al* [97] and Black *et al* [98], who reported that the calcaneus appeared to be the optimal BMD measurement site for routine screening of perimenopausal women to

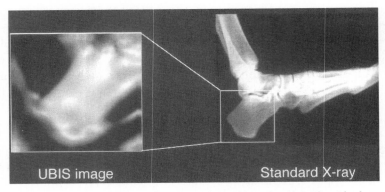

Figure 14.31. *A standard X-ray image of the calcaneus alongside with ultrasound generated image using the ultrasound bone imaging system (UBIS) 5000.*

predict the risk of any type of osteoporotic fracture. The calcaneus is measured in the mediolateral direction (see figure 14.31).

14.14.1.2. Phalanges

The phalanges are bones located in the fingers on the hand. Other significant bones in the hand are the metacarpals. The metacarpals were used as a measurement site in the early days of cortical index to measure bone status. For QUS application, the measurement site is the distal metaphysis of the first phalanx of the last four fingers. The medio-lateral surfaces are approximately parallel. In the metaphysis, both cortical and trabecular bone are present. Both types of bone tissue are extremely sensitive to age-related bone resorption. Cortical bone usually becomes more porous with advancing age. In addition, the cortices of long bone become thinner because the rate of endosteal resorption exceeds the rate of periosteal formation of bone. Taken together, the age-related losses of cortical and cancellous bone substantially increase the fragility of bone. In one study, the phalanges of elderly women had the highest deviation from peak adult bone mass compared with other techniques like spine DXA, spine QCT, femoral neck DXA and forearm DXA [99]. They are, therefore, appropriate to evaluate the risk of fracture. Depending on the technique, the phalanges can be measured in the mediolateral direction (DBM-sonic) or along the surface in the posterior–anterior direction.

14.14.1.3. Tibia

The mid-tibia has been chosen because of its long, straight and smooth surface. Also the overlying soft tissue is very thin, so minimizing errors in SOS measurement. Since 80% of the skeleton is cortical, and osteoporotic fracture may also involve cortical bone, it might be of clinical interest to measure a cortical bone such as the tibia. In addition, cortical bone loss may play an important role in determining whole bone strength. Measurement at this site is usually longitudinal ultrasound velocity along the anteromedial cortical border of the mid-tibia, thereby taking advantage of a site that is easily accessible in most individuals.

14.14.1.4.   Other sites

Measurement sites have been restricted to the peripheral skeleton due the highly attenuating nature of ultrasound. However, with the advent of axial transmission techniques, the accessible sites now include many others such as the femur, posterior process of the spine, radius and ulna.

14.14.2.   Methodology: coupling

As previously described in section 14.12.1, *in vitro* coupling can be achieved either using gel or water, or both.

In water coupled devices such as the Achilles+ and UBIS 5000, a temperature regulated water bath is used. The bath is maintained at 37 °C in the Achilles+ and 32 °C in the UBIS 5000. It is essential to regulate the water bath temperature since it has been established that temperature affects both SOS and BUA. Two water tanks are used: the reservoir to feed the bath and the other to drain the system. The transducers are mounted coaxially at a fixed separation of 95 mm for the Achilles and 100 mm for the UBIS. (See figure 14.32, water coupled devices.)

Most of the QUS devices now use gel for coupling. These include the AOS-100 (Aloka Co, Japan), CUBAClinical, Sahara, Osteospace, DBM-sonic and QUS II. In gel-coupled devices, the transducers are placed in direct contact with the heel. The approach to bringing the transducer into contact with the heel varies from device to device. Manual positioning is used in the A-100, spring action positioning in the DBM-sonic and motorized positioning in the CUBAClinical, Sahara and QUS II. With the Omnisense multi-site system, the transducers are manually moved over the region of interest. (See figure 14.33, gel coupled devices.)

For direct contact systems, different approaches have been used to tackle the problem with the non-uniformity of the calcaneus. The Sahara and CUBAclinical have soft elastomer (silicone) pads that are supposed to adapt to the heel shape. In the AOS-100, the transducers are encased in a rubber material that is soft enough to adapt to the shape of the calcaneus. Various water- or oil-based gels have been recommended for the different devices by the manufacturers.

A recent innovation to reduce the weight of water-coupled devices has been to encase the water in an inflatable balloon coupled to the heel via

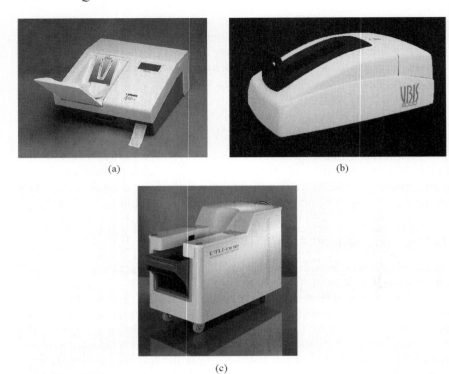

(a)　　　　　　　　　　　　(b)

(c)

Figure 14.32. *Some of the commercially available water-coupled QUS devices: (a) Achilles express (GE Lunar, Madison, Wisconson); (b) UBIS 5000 (DMS SA, Montpellier, France); (c) DTU-one (Osteometer, Meditech Inc, Hawthorne, CA).*

gel. This combined gel/water approach has been implemented in the Achilles Express and Paris systems.

14.14.3. Methodology: measurement variables

All QUS devices measure BUA and/or SOS. Some devices do report a composite index from a combination of SOS and BUA. However, the approaches to measure BUA and SOS vary significantly from device to device. In water-coupled devices, the time-of-flight approach is used to determine the speed of sound and is given by the equation

$$\text{SOS} = \frac{V_{\text{w}}x}{x - \Delta t V_{\text{w}}} \tag{14.12}$$

where x is the heel thickness, V_{w} is the velocity of ultrasound in water and Δt is the difference in transit time without and with the sample in the ultrasound

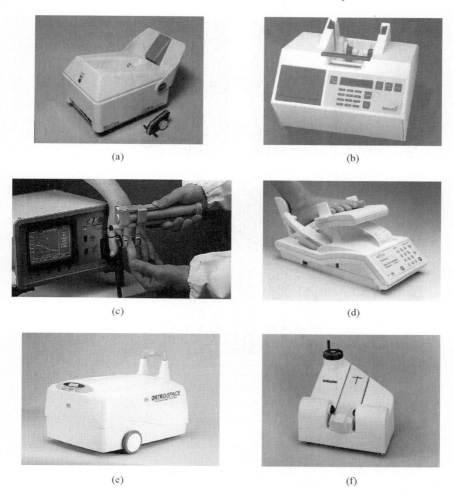

(a)

(b)

(c)

(d)

(e)

(f)

Figure 14.33. *Some of the commercially available gel-coupled QUS devices. From top left (a) CUBA Clinical (McCue Plc, UK), (b) Sahara Clinical bone sonographer (Hologic Inc, Bedford, MA, USA), (c) DMS sonic 1200 (Igea Srl, Carpi, Italy), (d) QUS-2 ultrasonometer (Quidel Corp, San Diego, CA, USA), (e) Osteospace (Medilink, Montpellier, France), (f) AOS-100 (acoustic osteo screener, Aloka, Co, Tokyo, Japan).*

path. Bone thickness (d) is assumed to be constant across all subjects (namely 40 mm for the Achilles+ and 29 mm for the UBIS 5000). For the contact systems, bone thickness is measured directly and equation (14.10) is used to determine the speed of sound (see figure 14.34).

Measurement of transit time is also problematic. As discussed in section 14.12.3, different markers can be used yielding different results. The

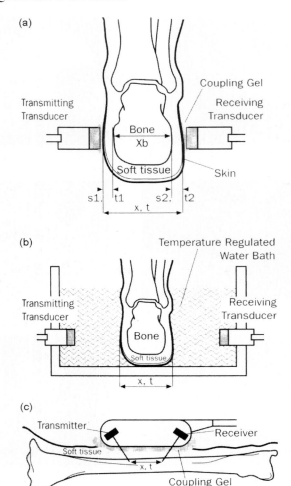

Figure 14.34. *Three schematic setup for velocity of ultrasound or (speed of sound) measurements as implemented in commercially available QUS devices. (a) Contact method where velocity is determined by measuring the heel thickness (x). More precise measurement could be obtained by subtracting soft tissue thickness (S1, S2) from heel thickness. (b) Time-of-flight or substitution technique, a constant bone thickness x is assumed. (c) Axial technique measures velocity along the bone surface. (Adapted from Njeh et al [105].)*

following approaches have been used in clinical QUS devices: first deviation from zero, zero crossing and threshold.

Attenuation is measured using the substitution method described in sections 14.3.3 and 14.12.6. However, the frequency range used to assess BUA varies slightly, being mostly calculated between 0.2 and 0.6 MHz.

14.14.4. *Quality assurance*

It is very important to apply a stringent QA program when using QUS in bone status assessment. This is mainly because changes seen in SOS and BUA due to disease or treatment are relatively small. Therefore, measurements of bone status changes have to be very precise because procedural errors, malfunctioning equipment or erroneous data analysis may cause substantial interference and hence mis-diagnosis, even if the data are erroneous by only a few percentage points. The QA protocols should be designed to ascertain that the equipment is functioning properly. Basic machine differences, lack of an absolute ultrasound bone phantom or of universally accepted cross-calibration procedures result in BUA and/or SOS variations when measuring the same subject on different systems. Hence this is not to be recommended.

The degree of complexity of the QA programme will depend on whether it is for an individual site or for a multi-centre clinical trial. At an individual site, for example, two types of testing could be carried out: acceptance testing and routine testing.

Acceptance testing at manufacture includes accuracy, *in vivo* and *in vitro* precision and power output measurement. *In vivo* measurement should cover a wide age and body status range. This is because precision has been reported to vary between normal and osteoporotic subjects.

Routine testing includes regular measurement of a QA phantom to detect any drift in the machine and periodic graphical analysis of QA data. Most manufacturers will provide system-specific phantoms. However, these system-specific phantoms are not anthropomorphic and their daily changes may not reflect what might happen *in vivo*. Non-manufacturer-led phantoms are the Vancouver phantoms, Leeds phantoms and CIRS phantom [62, 183], Clarke *et al*. The Leeds phantoms have both SOS and BUA values that fall within the biological range. For the Vancouver phantoms only BUA is within the biological range and hence it can only be used to monitor BUA stability of the device. Ultrasound measurement of these phantoms are highly influenced by temperature variations thus if used in the QA program, the temperature of the phantom must also be monitored. An alternative QA approach incorporates an electronic circuit to simulate the behaviour of ultrasound propagation through the calcaneus. Two commercial systems, the McCue CubaClinical and the DMS UBIS, have adopted this 'electronic phantom'.

Another aspect of QA is the proper training of the operator. Indeed, positioning error is one of the major sources of imprecision in QUS, being dependent on the working procedures of technicians.

14.14.5. *Cross-calibration*

The scientific community has not yet seriously addressed the issue of cross-calibration for different QUS devices. In many situations, the differences between the various QUS devices are not important. For example, in a clinic with only one machine on which all patients are measured, the relative differences between machines are not important. As long as the scanner used is stable, the analysis technique is consistent, and patients are not scanned on any other systems, any observed changes in ultrasound parameter over time can be assumed to be real and not due to machine artefact. However, investigators often do not have the luxury of using a single machine for the duration of a study. Also, many older QUS devices will be replaced in the coming years, and a hospital or clinic may choose to switch to another manufacturer's system. In either case, the continued monitoring of bone status in patients would be complicated by this change in equipment—even with newer versions of the same device, discrepancies can occur.

In clinical drug trials using ultrasonic information, multiple centres are typically used in order to quickly enrol the number of patients needed to establish the efficacy of a particular pharmaceutical. More often than not, the centres enrolled will be equipped with a variety of different scanners. Due to differences in ultrasound values, direct comparisons of centres with different equipment cannot be made without performing cross-calibration measurements. The determination of a 'true' ultrasound parameter becomes arbitrary at this point, as no ultrasound standard has yet been established.

Simple comparisons of patient ultrasound values with established normals are also complicated by the disparate results obtained on different scanners. Evaluations can only be performed properly by using normative data obtained on the same type of system, otherwise the comparisons will be invalid. By quantifying the differences between manufacturers, a set of universal normative data would be possible, compiled from the normative data results of all the manufacturers. The establishment of a single normative data set would unify the diagnostic results obtained on the various systems.

As ultrasound systems become widely used both for clinical diagnosis and drug trials the question of how to compare results from different machines arises. As described in the preceding sections, clinical systems have differences in the diameter of transducer used, the frequency range and the method of measuring velocity, all resulting in a disparity of the results obtained and different dynamic ranges. Causes of variation include the following.

- Transducer diameter, resulting in unequal volumes of the calcaneus being interrogated by the ultrasound wave.
- Use of focused and unfocused transducers.
- Region of interest (ROI) resulting in differences in the area of the calcaneus measured, knowing that the calcaneus is very heterogeneous.
- Definition of transit time, resulting in significant variation in calculated velocities. Nicholson *et al* [27] found that the zero crossing (as used by the

Achilles) resulted in the lowest velocity compared with the first deviation from zero (as used by the CUBAClinical) that resulted in the highest velocity.

- Definition of velocity also adds another complexity, as bone velocity, limb velocity and time-of-flight velocity are all frequently quoted [15].
- Use of fixed or assumed fixed heel width in the velocity calculation.
- Different algorithms for calculating BUA, resulting in a systematic difference in BUA between manufacturers. For example, the Achilles and DTU-one have a range of BUA values significantly narrower than those of the Sahara and CUBAClinical.

There is a need to accurately identify the source and magnitude of all the variations so that accurate cross-calibration can be achieved and measurements interchanged between machines.

Glüer *et al* [100] carried out a study on a group of 30 women using both the Walker Sonix UBA575 and the Lunar Achilles. They reported that the BUA readings obtained on the two machines differed substantially, and the correlation between UBA575 BUA and Achilles BUA was markedly lower than correlations between different bone densitometers measuring the same site. Therefore, BUA results cannot accurately be extrapolated from one device to the other. No large comparative assessment of velocity has been reported for these clinical systems. Strelitzki *et al* [101] have also addressed this problem, using phantoms on three different manufacturers' systems. They found intra- and inter-manufacturer variation in measured velocity values.

14.14.6.   *Artefacts and sources of errors*

There are two main sources of error: accuracy and precision. For clinical purposes, precision errors are considered more important than accuracy. Pertaining to QUS, some of the sources of accuracy errors have been discussed in section 14.12.7. *In vivo* precision is affected by coupling, temperature, positioning, *in vivo* soft tissue, heel thickness and oedema [102, 103]. Using the Walker Sonix UBA1001, Evans *et al* [30] carried out an extensive study of the factors that might affect precision. These included: immersion time of the foot in the water bath, water depth, water temperature, concentration of detergents and various rotations of the foot. They found that foot positioning was the major cause of measurement imprecision for BUA. This was due to the spatial variation of BUA in the calcaneus. They proposed that the optimum measuring temperature of the water bath should be $32 \pm 2\,^\circ C$, because of the demonstrable decrease in BUA with temperature (above $34\,^\circ C$). The thickness of the overlying soft tissue has also been reported to affect only velocity and not BUA, using the Achilles system [37]. One might expect that for the UBA575+ where soft tissue corrections are made and true bone velocity is measured, the effect will be negligible.

14.15.   *IN VIVO* APPLICATION OF ULTRASOUND

14.15.1.   In vivo studies

Several studies *in vivo* have evaluated the relationship between ultrasound parameters and the established ionizing radiation measurements of BMD, using simple linear regression analysis. This has been reviewed by Gregg *et al* [101, 106], Njeh *et al* [101, 105] and Prins *et al* [102, 104]. The correlation coefficients between BUA or velocity and BMD have ranged from 0.34 to 0.83. Improved but still poor correlations are observed between QUS and BMD measured at the calcaneus compared with BMD measured at the axial sites [107, 108]. For example, Greenspan *et al* [109] observed a correlation of 0.86 between stiffness and BMD of the calcaneus compared with 0.77, 0.80 and 0.68 for the trochanter, femoral neck and spine respectively. These correlations, although significant, are weak. It has been argued that this is due to a dependence upon other aspects of bone, in addition to density. One of the possible causes of low correlation between QUS and calcaneal BMD could be anatomical differences in location of the measurement site used with single-point QUS devices and DXA. The calcaneus is very heterogeneous in both density [110] and structure [111] and thus mismatched regions of interest (ROIs) could result in a poor correlation. An early study by Glüer *et al* [112] evaluated the impact of the differences in the measurement sites and showed that the correlation between BUA and BMD was not improved by matching the ROIs analysed by ultrasound and SXA. One could argue that the matching was not optimal in that particular study. This question has been further addressed using different QUS devices [107, 113–116]. Generally when BMD of the calcaneus was closely matched to the site of ultrasound measurement, an improved correlation of approximately $r = 0.9$ was observed.

Another factor that might impact the correlation between QUS and BMD is the reproducibility (precision) of each of the techniques. Apart from the study by Glüer *et al* [112], no other study has tried to compensate for the impact of precision on the correlation coefficients. They did not observe significant increase in correlation after correcting for precision errors [112]. Accuracy errors can also play a role in the degree of correlation between these two measurements of bone [117]. Regardless of the exact physical characteristics represented by these measurements, inaccuracies in their assessment will contribute to a poorer correlation between the two parameters. This is especially true when there is no positive correlation between the accuracy errors of QUS and DXA [118]. Accuracy errors in DXA are determined by comparison with a reference technique like ashing of bone [119–121]. It is known that the thickness and heterogeneity of surrounding soft tissue influences the accuracy of DXA [119]. Heel DXA has an accuracy of 3–6% [121, 122]. The accuracy of QUS has not been evaluated. This is primarily due to the lack of an accepted reference technique for the evaluation of QUS. Hassager *et al* [117] from theoretical analysis demonstrated that the correlation between forearm and spine BMD increased from 0.6 to 0.8–0.9 if one adjusted for accuracy errors. If one were to assume that the accuracy of QUS is no better than that of heel DXA, it may be possible to show significant improvements in correlations between BMD and QUS after adjustment for accuracy errors.

Other factors that could affect the correlations between QUS and ionizing radiation techniques include differing skeletal sites (biological variation), different attenuation mechanism for ultrasound and photon-based techniques, and the characteristics of the population studied (e.g. age range, gender, race, health status). The relationship between QUS and BMD, however, could possibly be explained by a power function rather than being simply linear.

14.15.2.   In vivo QUS measurement

Since the availability of clinical QUS, there has been a tendency to correlate it with the established ionizing radiation measurements of BMD. The correlation coefficients have ranged from 0.34 to 0.83 [105, 106]. These correlations, although significant, are moderate. It has been argued that this is due to QUS being dependent upon other qualities of bone in addition to density. The interpretation of the correlations between QUS and absorptiometry techniques is complicated by many factors, such as the surrounding soft tissue and accuracy of positioning. When error sources are reduced (considering an *in vitro* model, for example) the correlation approaches 0.9, suggesting that QUS is a measure principally of BMD. However, in current practice, it is generally accepted that ultrasound and densitometry interact differently with bone and that a high correlation between them should not be expected. Attention is now directed towards diagnostic accuracy of ultrasound rather than simple correlation with BMD, using statistical techniques such as receiver-operating characteristic (ROC) analysis, Z-scores, odds ratios and relative risk [121].

14.15.3.   Age-related change

Numerous cross-sectional studies of ultrasound normative data have been reported [124–127]. They all show that QUS parameters are inversely correlated with age, showing a significant decrease in both BUA and SOS, especially after the menopause. Before the menopause, both BUA and SOS are relatively stable, although two studies have shown a steady decline beginning at age 20 [127, 128]. Typical rates of change are 0.5–1.0 dB MHz^{-1} (0.5–1.0%) per year for BUA and 1–5 m s^{-1} (0.1–0.3%) per year for SOS at the heel, patella and tibia. A similar pattern is found at the phalanges with a rate of change between 0.3 and 0.5% per year for post-menopausal women $(7-12\,\mathrm{m\,s^{-1}})$ [129–132]. However, there are substantial differences between devices. The first longitudinal study was performed by Schott *et al* [133] on 140 healthy post-menopausal women measured at the calcaneus and confirmed by Krieg *et al* [134] in institutionalized elderly women. The decrease observed by Schott *et al* over two years was $-1.0\% \pm 4.3\%$ for BUA and $-0.8\% \pm 0.6\%$ for SOS. Mele *et al* [135] published the first longitudinal study for the phalanges. They found an average decrease of 2.9% for Ad-SOS over a period of three years ($\sim$1% per year) in post-menopausal women compared with 0.9% ($\sim$0.3% per year) in pre-menopausal women. The decrease in SOS was significantly greater soon after the menopause compared with earlier in life.

14.15.4. *Velocity diagnostic sensitivity*

Cross-sectional patient studies have demonstrated that QUS can discriminate normal from osteoporotic subjects as well as traditional bone densitometry techniques [106]. Early studies such as Heaney *et al* [136] and Rossman *et al* [137] showed discrimination between osteoporotic subjects compared with normal subjects. Recent studies [138–142] have demonstrated that ultrasound velocity discriminates between fracture and control groups. Turner *et al* [138] reported that velocity measured at the calcaneus had sensitivity comparable with femoral neck BMD and better than DXA spine BMD in hip fracture prediction. For hip fracture, they reported an area under the ROC curve of 0.85 for SOS, 0.78 for femoral neck BMD and 0.53 for spine BMD. Velocity measured at the calcaneus had similar diagnostic sensitivity to DXA spine BMD and femoral neck BMD for distinguishing patients with vertebral fractures from controls [138]. However, Gonnelli *et al* [140] observed that although SOS at the calcaneus was an independent predictor of vertebral fracture, BMD was a slightly better predictor. On the other hand, after examining the ultrasound of the phalanges, Scavalli *et al* [141] reported for spine fracture an area under the ROC curve of 0.85 for Ad-SOS (with an age-adjusted odds ratios of 1.8) versus 0.83 for spine BMD. These findings have also been confirmed by Hans *et al* [184] with an age-adjusted odds ratios of 2.3 (1.1–4.9) in women suffering vertebral fracture. The sensitivity and the specificity for Ad-SOS used in the diagnosis of established osteoporosis were 81.5 and 79.3%, respectively, when an appropriate cut-off point was selected. They concluded that Ad-SOS discriminates between normal post-menopausal women and patients with either low lumbar mineral density or prevalent fractures to the same extent as bone mineral density measurements do.

A prospective study of hip fractures measuring QUS and BMD reported by Hans *et al* [144] observed that velocity measured at the calcaneus had the

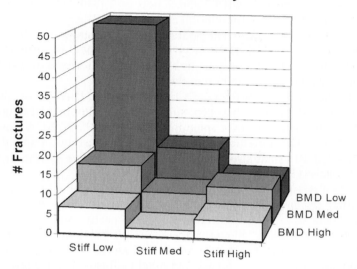

EPIDOS, # Fractures by Tertile

Figure 14.35. *In a large multi-centre prospective study called EPIDOS (Epidemiologie de l'osteoporose) quantitative ultrasound predicts fractures as well as bone mineral density. It is represented by tertile. Stiff = stiffness which is a combination of ultrasound velocity and broadband ultrasound attenuation (Hans et al [144]).*

same diagnostic sensitivity as femoral neck BMD in predicting hip fractures. The increased risk associated with a decrease of one standard deviation was estimated as 2.0 for both ultrasound velocity and DXA femoral neck BMD (see figure 14.35). Mele *et al* [135] undertook another prospective study, measuring instead at the phalanges in order to predict low energy (~osteoporotic) fracture (e.g. hip and forearm fractures). They estimated the increased risk associated with a decrease of one standard deviation as 1.5 (95% CI: 1.1–1.7) for Ad-SOS. The small number of fractures is a limitation of the study, but the prediction was found to be very significant ($p < 0.03$). Therefore, one could expect a stronger prediction power of when more fractures will occur.

14.15.5. *BUA diagnostic sensitivity*

The ability of BUA to discriminate between normal and osteoporotic subjects has been reported in the literature [105, 106, 145, 146]. The differences reported between normal and osteoporotic volunteers are statistically significant. In a retrospective study of a group of 50 women who had recently sustained a fracture of the hip, Stewart *et al* [147] demonstrated that BUA of the calcaneus had diagnostic sensitivity comparable with hip BMD but superior to spine BMD measured by DXA. These results have been supported by the work of Turner *et al* [137] and Schott *et al* [139]. They also demonstrated that BUA adjusted for femoral neck BMD still discriminated between fracture and control groups. The poor prediction of hip fracture by lumbar spine BMD measured by DXA may be because BMD is artificially but unpredictably elevated in the spine by aortic calcification and osteoarthritis, since these problems are common in the elderly [148].

With the exception of the report by Stewart *et al* [149], BUA has been reported to discriminate between subjects with vertebral fracture and controls as effectively as DXA spine BMD [138, 150–152]. BUA was still a statistically significant predictor of fracture after adjusting for spinal BMD [138, 150–152]. BUA has also been reported to be sensitive to other fractures such as Colle's [153, 154].

The first prospective study for hip fractures by Porter *et al* [155] demonstrated that calcaneal BUA predicts hip fracture risk in post-menopausal women. There have been two recent large prospective studies carried out by groups in France on 7598 very elderly women [144] and in the USA on 6183 elderly women [156]. Hans *et al* [144] observed that for one standard deviation decrease there was an age and weight adjusted relative hip fracture risk of 2.2 for BUA compared with 2.0 for femoral neck BMD. After adjusting BUA for femoral neck BMD, the logistic regression showed that BUA was still a significant independent predictor of hip fractures. They also demonstrated that combining the results of ultrasound and BMD significantly improved the prediction of hip fracture. Bauer *et al* [156] found an adjusted relative risk of 2.2, 2.5 and 3.0 for one standard deviation decrease in calcaneal BUA, calcaneal BMD and femoral neck BMD in hip fracture prediction respectively. However, contrary to Hans *et al* [144], they found that BUA was no longer significantly associated with hip fracture after adjusting for either hip or calcaneal BMD. It is worth noting that these two studies were carried out using two different ultrasound systems, the Lunar Achilles for the French study and the Walker Sonix UBA575 for the American study. The study population and the proportion of trochanteric and neck fractures could also contribute to the observed differences [157].

14.15.6. *QUS and longitudinal monitoring*

To date, few studies have reported on the effects of drugs or have addressed the usefulness of ultrasound to monitor the treatment of osteoporosis [158]. Hence further longitudinal studies are required. Four studies at the calcaneus with two-year follow-up measurements have shown positive changes in QUS parameters between controls and women treated with calcitonin (+4.2% for BUA and +0.8% for SOS) [158], hormone replacement therapy (HRT) (+3.6% for BUA and +0.7% for SOS) [159] and bisphosphonates (estimated

+18% for BUA) [160]. These changes were statistically significant for treatment with calcitonin and HRT but not with the bisphosphonate. The power of the bisphosphonate study was limited by its very small sample size ($n = 12$). Another study at the phalanges using the DBM Sonic 1200 with one-year follow-up measurements has shown positive changes in QUS parameters between controls (-0.7% for Ad-SOS) and women treated with HRT ($+0.7\%$ for Ad-SOS) [161]. Using the same device without any drug intervention, Ingle *et al* [380] concluded that finger quantitative ultrasound is sensitive enough to detect and monitor diffuse osteoporosis. These results suggest some utility for QUS to monitor patients longitudinally.

However, the current longitudinal precision of the measurement for all QUS devices will make evaluation of changes in the individual more difficult.

14.15.7.   *Paediatric application*

Different diseases such as cystic fibrosis [162], juvenile idiopathic arthritis (JIA) [163, 164], eating disorders [165] and renal transplantation [166] may directly or indirectly decrease the bone mineral content (BMC) in children. Also, some drugs like glucocorticoids [167] or anticonvulsants [168] will have deleterious effects on bone. Therefore, the choice of the right therapeutic strategy and the optimal follow-up of patients with disease that could affect their bone metabolism may be assisted by assessment of bone integrity. Also, bone mass accrued during childhood has been proposed as a determinant of an individual's susceptibility to osteoporotic fractures in adulthood [169]. Knowledge of the timing and magnitude of peak bone mass is important because it may be relevant to an individual's subsequent risk of osteoporosis, especially if a disease state or delayed sexual maturation is present. QUS has been shown to provide clinically useful information about bone status in adults [105], but limited information is available for children. There is now growing information on the use of QUS for paediatric measurements [170–175]. Currently fixed position QUS devices are not suitable for paediatric study because of the small size of the foot. Attempts have been made by using foot adaptors [176] or by reducing the size of the transducer [174]. However, imaging QUS system may be more suited since the ROI could be adjusted to suit the smaller calcaneus. In a recent study, Van den Bergh *et al* [175] found that, after correction for heel width and adjustment of the ROI size based on foot length, BUA and SOS were significantly associated with age in both boys ($r = 0.60$, $p < 0.001$ and 0.24, $p = 0.05$) and girls ($r = 0.73$ and 0.24, both $p = 0.001$).

14.15.8.   *Application to rheumatoid arthritis*

There is a growing interest in the use of phalangeal ultrasound [175] and calcaneal ultrasound [178–181] to assess bone status in rheumatoid arthritis

(RA) patients. Martin *et al* [179] demonstrated a significant decrease in both ultrasound attenuation (31.7%) and velocity (6.6%) measured at the calcaneus of RA patients compared with controls. In studies by Daens *et al* [180] and Blanckaert *et al* [181] QUS performs similarly to BMD measured at the spine and femur in evaluating glucocorticoid-induced osteoporosis.

REFERENCES

[1] Nicholson P H F, Lowet G, Langton C M, Dequeker J and Van de Perre G 1996 Comparison of time-domain and frequency-domain approaches to ultrasonic velocity measurement in trabecular bone *Phys. Med. Biol.* **41** 2421–35

[2] Wear K A and Garra B S 1998 Assessment of bone density using ultrasonic backscatter *Ultrasound Med. Biol.* **24**(5) 689–95

[3] Laugier P, Giat P, Chappard C, Roux C and Berger G 1997 *Clinical Assessment of the Backscatter Coefficient in Osteoporosis*

[4] Langton C M and Subhan M 2001 Computer and experimental simulation of a cortical end-plate phase cancellation artefact in the measurement of BUA at the calcaneus *Physiol. Meas.* **22**(3) 581–7

[5] Kaczmarek M, Pakula M and Kubik J 2000 Multiphase nature and structure of biomaterials studied by ultrasounds *Ultrasonics* **38** 703–7

[6] Wear K A 2000 Anisotropy of ultrasonic backscatter and attenuation from human calcaneus: implications for relative roles of absorption and scattering in determining attenuation *J. Acoust. Soc. Am.* **107**(6) 3474–9

[7] Wear K A 1999 Frequency dependence of ultrasonic backscatter from human trabecular bone: theory and experiment *J. Acoust. Soc. Am.* **106**(6) 3659–64

[8] McKelvie M L and Palmer S B 1991 The interaction of ultrasound with cancellous bone *Phys. Med. Biol.* **36**(10) 1331–40

[9] Strelitzki R, Nicholson P H and Paech V 1998 A model for ultrasonic scattering in cancellous bone based on velocity fluctuations in a binary mixture *Physiol. Meas.* **19**(2) 189–96

[10] Nicholson P H, Strelitzki R, Cleveland R O and Bouxsein M L 2000 Scattering of ultrasound in cancellous bone: predictions from a theoretical model *J. Biomech.* **33**(4) 503–6

[11] Biot M A 1956 Theory of propagation of elastic waves in a fluid-saturated solid. II. Higher frequency range *J. Acoust. Soc. Am.* **28** 179–91

[12] Schoenberg M 1984 Wave propagation in alternating solids and fluid layers *Wave Motion* **6** 303–20

[13] Haire T J and Langton C M 1999 Biot theory: a review of its application to ultrasound propagation through cancellous bone *Bone* **24**(4) 291–5

[14] Hughes E R, Leighton T G, Petley G W and White P R 1999 Ultrasonic propagation in cancellous bone: a new stratified model *Ultrasound Med. Biol.* **25**(5) 811–21

[15] Leclaire P, Kelders L, Lauriks W, Glorieux C and Thoen J 1997 Ultrasonic wave propagation in porous media: determination of acoustic parameters and high frequency limit of the classical models *Stud. Health. Technol. Inform.* **40** 139–55

[16] Williams J L, Grimm M J, Wehrli F W, Foster K R and Chung H W 1994 Prediction of frequency and pore size dependent of ultrasound in trabecular bone using Biot's

theory. In *Mechanics of Poroelasticity* ed A P S Selvadurai (Dordrecht: Kluwer) pp 263–71

[17] Williams J L 1992 Ultrasonic wave propagation in cancellous and cortical bone: prediction of some experimental results by Biot's theory *J. Acoust. Soc. Am.* **91**(2) 1106–12

[18] Hosokawa A and Otani T 1997 Ultrasonic wave propagation in bovine cancellous bone *J. Acoust. Soc. Am.* **101**(1) 558–62

[19] Wu K Y, Xue Q and Adler L 1990 Reflection and transmission of elastic waves from a fluid-saturated porous solid boundary *J. Acoust. Soc. Am.* **87**(6) 2349–58

[20] Xu W and Kaufman J J 1993 Diffraction correction methods for insertion ultrasound attenuation estimation *IEEE Trans. Biomed. Eng.* **40**(6) 563–70

[21] Strelitzki R and Evans J A 1998 Diffraction and interface losses in broadband ultrasound attenuation measurements of the calcaneum *Physiol. Meas.* **19**(2) 197–204

[22] Petley G W, Robins P A and Aindow J D 1995 Broadband ultrasonic attenuation: are current measurement techniques inherently inaccurate? *Br. J. Radiol.* **68**(815) 1212–4

[23] Brown T D and Ferguson A B 1980 Mechanical property distributions in the cancellous bone of the human proximal femur *Acta Orthop. Scand.* **51**(3) 429–37

[24] Strelitzki R, Paech V and Nicholson P H 1999 Measurement of airborne ultrasonic slow waves in calcaneal cancellous bone *Med. Eng Phys.* **21**(4) 215–23

[25] Rich C, Klinik E, Smith R and Graham B 1966 Measurement of bone mass from ultrasonic transmission time *Proc. Soc. Exp. Biol Med.* **123**(1) 282–5

[26] Langton C M, Ali A V, Riggs C M, Evans G P and Bonfield W 1990 A contact method for the assessment of ultrasonic velocity and broadband attenuation in cortical and cancellous bone. *Clin. Phys. Physiol. Meas.* **11**(3) 243–9

[27] Nicholson P H F, Lowet G, Langton C M, Dequeker J and Van der Perre G 1996 A comparison of time-domain and frequency-domain approaches to ultrasonic velocity measurement in trabecular bone *Phys. Med. Biol. (UK)* **41**(11) 2421–35

[28] Zagzebski J A, Rossman P J, Mesina C, Mazess R B and Madsen E L 1991 Ultrasound transmission measurements through the os calcis *Calcif. Tissue Intl.* **49**(2) 107–11

[29] Pain H J 1985 *The Physics of Vibrations and Waves* (Chichester: Wiley)

[30] Ragozzino M 1981 Analysis of the error in measurement of ultrasound speed in tissue due to waveform deformation by frequency-dependent attenuation *Ultrasonics* **19**(3) 135–8

[31] Strelitzki R, Clarke A J and Evans J A 1996 The measurement of the velocity of ultrasound in fixed trabecular bone using broadband pulses and single-frequency tone bursts *Phys. Med. Biol.* **41**(4) 743–53

[32] Nicholson P H F, Droin P and Laugier P 1997 Inter-system comparison of site-matched ultrasonic measurements of the calcaneus *Eur. J. Ultrasound* **5**(3) 191–202

[33] Pollintine P, Haddaway M J and Davie M W 2000 The use of tonebursts as an alternative to broadband signals in the measurement of speed of sound in human cancellous bone [In Process Citation] *Phys. Med. Biol.* **45**(7) 1941–51

[34] Strelitzki R and Evans J A 1996 On the measurement of the velocity of ultrasound in the os calcis using short pulses *Eur. J. Ultrasound* **4**(3) 205–13

[35] Laugier P, Droin P, Laval-Jeantet A M and Berger G 1997 In-vitro assessment of the relationship between acoustic properties and bone mass density of the calcaneus

by comparison of ultrasound parametric imaging and quantitative computed tomography *Bone* **20**(2) 157–65

[36] Antich P P, Anderson J A, Ashman R B, Dowdey J E, Gonzales J, Murry R C, Zerwekh J E and Pak C Y 1991 Measurement of mechanical properties of bone material in-vitro by ultrasound reflection: methodology and comparison with ultrasound transmission *J. Bone Min. Res.* **6**(4) 417–26

[37] Antich P P 1993 Ultrasound study of bone in-vitro *Calcif. Tissue Intl.* **53** Suppl 1 S157–61

[38] Antich P P, Pak C Y, Gonzales J, Anderson J, Sakhaee K and Rubin C 1993 Measurement of intrinsic bone quality in-vivo by reflection ultrasound: correction of impaired quality with slow-release sodium fluoride and calcium citrate *J. Bone Min. Res.* **8**(3) 301–11

[39] Antich P P and Mehta S 1997 Ultrasound critical-angle reflectometry (UCR): a new modality for functional elastometric imaging *Phys. Med. Biol.* **42**(9) 1763–77

[40] Mehta S and Antich P P 1997 Measurement of shear-wave velocity by ultrasound critical-angle reflectometry (UCR) *Ultrasound Med. Biol.* **23**(7) 1123–6

[41] Mehta S S, Oz O K and P. A P 1998 *Bone* elasticity and ultrasound velocity are affected by subtle changes in the organic matrix *J. Bone Min. Res.* **13** 114–21

[42] Verhoef W A, Cloostermans M J and Thijssen J M 1985 Diffraction and dispersion effects on the estimation of ultrasound attenuation and velocity in biological tissues *IEEE Trans. Biomed. Eng.* **32**(7) 521–9

[43] Fry F J and Barger J E 1978 Acoustical properties of the human skull *J. Acoust. Soc. Am.* **63**(5) 1576–90

[44] McKelvie M L and Palmer S B 1987 The interactions of ultrasound with cancellous bone. In *Physical Acoustic Group of the Institute of Physics and the Institute of Acoustics 1987* eds S B Palmer and C M Langton (Hull: Institute of Physics) pp 1–13

[45] Strelitzki R, Metcalfe S C, Nicholson P H, Evans J A and Paech V 1999 On the ultrasonic attenuation and its frequency dependence in the os calcis assessed with a multielement receiver *Ultrasound Med. Biol.* **25**(1) 133–41

[46] McCloskey E V, Murray S A, Charlesworth D, Miller C, Fordham J, Clifford K, Atkins R and Kanis J A 1990 Assessment of broadband ultrasound attenuation in the os calcis in-vitro *Clin. Sci. (Colch.)* **78**(2) 221–5

[47] Langton C M, Njeh C F, Hodgskinson R and Currey J D 1996 Prediction of mechanical properties of the human calcaneus by broadband ultrasonic attenuation *Bone* **18**(6) 495–503

[48] Njeh C F and Langton C M 1997 The effect of cortical endplates on ultrasound velocity through the calcaneus: an in-vitro study *Br. J. Radiol.* **70**(833) 504–10

[49] Wu C Y, Gluer C C, Jergas M, Bendavid E and Genant H K 1995 The impact of bone size on broadband ultrasound attenuation *Bone* **16**(1) 137–41

[50] Nicholson P H, Lowet G, Cheng X G, Boonen S, van der Perre G and Dequeker J 1997 Assessment of the strength of the proximal femur in-vitro: relationship with ultrasonic measurements of the calcaneus *Bone* **20**(3) 219–24

[51] Kotzki P O, Buyck D, Hans D, Thomas E, Bonnel F, Favier F, Meunier P J and Rossi M 1994 Influence of fat on ultrasound measurements of the os calcis *Calcif. Tissue Intl.* **54**(2) 91–5

[52] Johansen A and Stone M D 1997 The effect of ankle oedema on bone ultrasound assessment at the heel *Osteoporos. Intl.* **7**(1) 44–7

[53] Nicholson P H F and Njeh C F 1999 Ultrasonic studies of cancellous bone in-vitro. In *Quantitative Ultrasound: Assessment of Osteoporosis and Bone Status* eds C F Njeh, D Hans, T Fuerst, C C Gluer and H K Genant (London: Martin Dunitz) pp 195–220

[54] McKelvie M L, Fordham J, Clifford C and Palmer S B 1989 In-vitro comparison of quantitative computed tomography and broadband ultrasonic attenuation of trabecular bone *Bone* **10**(2) 101–4

[55] Nicholson P H F, Haddaway M J and Davie M W J 1994 The dependence of ultrasonic properties on orientation in human vertebral bone *Phys. Med. Biol. (UK)* **39**(6) 1013–24

[56] Han S, Rho J, Medige J and Ziv I 1996 Ultrasound velocity and broadband attenuation over a wide range of bone mineral density *Osteoporos. Intl.* **6**(4) 291–6

[57] Serpe L and Rho J Y 1996 The nonlinear transition period of broadband ultrasound attenuation as bone density varies *J. Biomech.* **29**(7) 963–6

[58] Duquette J, Honeyman T, Hoffman A, Ahmadi S and Baran D 1997 Effect of bovine bone constituents on broadband ultrasound attenuation measurements *Bone* **21**(3) 289–94

[59] Evans J A and Tavakoli M B 1990 Ultrasonic attenuation and velocity in bone *Phys. Med. Biol.* **35**(10) 1387–96

[60] Alves J M, Xu W, Lin D, Siffert R S, Ryaby J T and Kaufman J J 1996 Ultrasonic assessment of human and bovine trabecular bone: a comparison study *IEEE Trans. Biomed. Eng.* **43**(3) 249–58

[61] Hodgskinson R, Njeh C F, Whitehead M A and Langton C M 1996 The non-linear relationship between BUA and porosity in cancellous bone *Phys. Med. Biol. (UK)* **41**(11) 2411–20

[62] Clarke A J, Evans J A, Truscott J G, Milner R and Smith M A 1994 A phantom for quantitative ultrasound of trabecular bone *Phys. Med. Biol. (UK)* **39**(10) 1677–87

[63] Strelitzki R, Evans J A and Clarke A J 1997 The influence of porosity and pore size on the ultrasonic properties of bone investigated using a phantom material *Osteoporos. Intl.* **7**(4) 370–5

[64] Njeh C F, Kuo C W, Langton C M, Atrah H I and Boivin C M 1997 Prediction of human femoral bone strength using ultrasound velocity and BM D: an in-vitro study *Osteoporos. Intl.* **7**(5) 471–7

[65] Njeh C F, Hodgskinson R, Currey J D and Langton C M 1996 Orthogonal relationships between ultrasonic velocity and material properties of bovine cancellous bone *Med. Eng Phys.* **18**(5) 373–81

[66] Rho J Y 1998 Ultrasonic characterisation in determining elastic modulus of trabecular bone material *Med. Biol. Eng. Comput.* **36**(1) 57–9

[67] Ashman R B, Corin J D and Turner C H 1987 Elastic properties of cancellous bone: measurement by an ultrasonic technique *J. Biomech.* **20**(10) 979–86

[68] Bouxsein M L and Radloff S E 1997 Quantitative ultrasound of the calcaneus reflects the mechanical properties of calcaneal trabecular bone *J. Bone Min. Res.* **12**(5) 839–46

[69] van den Bergh J P, van Lenthe G H, Hermus A R, Corstens F H, Smals A G and Huiskes R 2000 Speed of sound reflects Young's modulus as assessed by microstructural finite element analysis *Bone* **26**(5) 519–24

[70] Wu C, Hans D, He Y, Fan B, Njeh C F, Augat P, Richards J and Genant H K 2000 Prediction of bone strength of distal forearm using radius bone mineral density and phalangeal speed of sound *Bone* **26**(5) 529–33

[71] Abendschein W F and Hyatt G W 1972 *Ultrasonics* and physical properties of healing bone *J. Trauma* **12**(4) 297–301

[72] Ashman R B and Rho J Y 1988 Elastic modulus of trabecular bone material *J. Biomech.* **21**(3) 177–81

[73] Rho J Y 1996 An ultrasonic method for measuring the elastic properties of human tibial cortical and cancellous bone *Ultrasonics* **34**(8) 777–83

[74] Turner C H and Eich M 1991 Ultrasonic velocity as a predictor of strength in bovine cancellous bone *Calcif. Tissue Intl.* **49**(2) 116–9

[75] Han S U, Medige J and Ziv I 1996 Combined models of ultrasound velocity and attenuation for predicting trabecular bone strength and mineral density *Clin. Biomech.* **11**(6) 348–53

[76] Hodgskinson R, Njeh C F, Currey J D and Langton C M 1997 The ability of ultrasound velocity to predict the stiffness of cancellous bone in-vitro *Bone* **21**(2) 183–90

[77] Han S, Medige J, Davis J, Fishkin Z, Mihalko W and Ziv I 1997 Ultrasound velocity and broadband attenuation as predictors of load-bearing capacities of human calcanei *Calcif. Tissue Intl.* **60** 21–5

[78] Bouxsein M L, Coan B S and Lee S C 1999 Prediction of the strength of the elderly proximal femur by bone mineral density and quantitative ultrasound measurements of the heel and tibia *Bone* **25**(1) 49–54

[79] Grimm M J and Williams J L 1997 Assessment of bone quantity and 'quality' by ultrasound attenuation and velocity in the heel *Clin. Biomech.* **12**(5) 281–5

[80] Lochmuller E M, Eckstein F, Zeller J B, Steldinger R and Putz R 1999 Comparison of quantitative ultrasound in the human calcaneus with mechanical failure loads of the hip and spine *Ultrasound Obstet. Gynecol.* **14**(2) 125–33

[81] Toyras J, Kroger H and Jurvelin J S 1999 *Bone* properties as estimated by mineral density, ultrasound attenuation, and velocity *Bone* **25**(6) 725–31

[82] Nicholson P H and Strelitzki R 1999 On the prediction of Young's modulus in calcaneal cancellous bone by ultrasonic bulk and bar velocity measurements *Clin. Rheumatol.* **18**(1) 10–6

[83] Bouxsein M L, Courtney A C and Hayes W C 1995 Ultrasound and densitometry of the calcaneus correlate with the failure loads of cadaveric femurs *Calcif. Tissue Intl.* **56**(2) 99–103

[84] Cheng X G, Nicholson P H, Boonen S, Lowet G, Brys P, Aerssens J, Van der Perre G and Dequeker J 1997 Prediction of vertebral strength in-vitro by spinal bone densitometry and calcaneal ultrasound *J. Bone Min. Res.* **12**(10) 1721–8

[85] Lochmuller E M, Eckstein F, Kaiser D, Zeller J B, Landgraf J, Putz R and Steldinger R 1998 Prediction of vertebral failure loads from spinal and femoral dual-energy X-ray absorptiometry, and calcaneal ultrasound: an in situ analysis with intact soft tissues *Bone* **23**(5) 417–24

[86] Lochmuller E M, Zeller J B, Kaiser D, Eckstein F, Landgraf J, Putz R and Steldinger R 1998 Correlation of femoral and lumbar DXA and calcaneal ultrasound, measured in situ with intact soft tissues, with the in-vitro failure loads of the proximal femur *Osteoporos. Intl.* **8**(6) 591–8

[87] Langton C M, Palmer S B and Porter R W 1984 The measurement of broadband ultrasonic attenuation in cancellous bone *Eng. Med.* **13**(2) 89–91

[88] Miller P D, Bonnick S L, Johnston C C Jr, Kleerekoper M, Lindsay R L, Sherwood L M and Siris E S 1998 The challenges of peripheral bone density testing: Which

patients need additional central density skeletal measurements *J. Clin. Densitom.* **1**(3) 211–7

[89] Langton C M, Evans G P, Hodgskinson R and Riggs C M 1990 Ultrasonic, elastic and structural properties of cancellous bone. In *Current Research in Osteoporosis and Bone Mineral Measurements 1990* ed E F G Ring (Bath: British Institute of Radiology) pp 10–1

[90] Glüer C C, Wu C Y and Genant H K 1993 Broadband ultrasound attenuation signals depend on trabecular orientation: an in-vitro study *Osteoporos. Intl.* **3**(4) 185–91

[91] Tavakoli M B and Evans J A 1992 The effect of bone structure on ultrasonic attenuation and velocity *Ultrasonics* **30**(6) 389–95

[92] Glüer C C, Wu C Y, Jergas M, Goldstein S A and Genant H K 1994 Three quantitative ultrasound parameters reflect bone structure *Calcif. Tissue Intl.* **55**(1) 46–52

[93] Hans D, Arlot M E, Schott A M, Roux J P, Kotzki P O and Meunier P J 1995 Do ultrasound measurements on the os calcis reflect more the bone microarchitecture than the bone mass?: a two-dimensional histomorphometric study *Bone* **16**(3) 295–300

[94] Njeh C F, Hans D, Li J, Fan B, Fuerst T, He Y Q, Tsuda-Futami E, Lu Y, Wu C Y and Genant H K 2000 Comparison of six calcaneal quantitative ultrasound devices: precision and hip fracture discrimination *Osteoporos. Intl.* **11**(12) 1051–62

[95] Njeh C F, Hans D, Fuerst T, Gluer C-C and Genant H K 1999 *Quantitative Ultrasound: Assessment of Osteoporosis and Bone Status* (London: Martin Dunitz)

[96] Vogel J M, Wasnich R D and Ross P D 1988 The clinical relevance of calcaneus bone mineral measurements: a review *Bone Miner.* **5**(1) 35–58

[97] Wasnich R D, Ross P D, Heilbrun L K and Vogel J M 1987 Selection of the optimal skeletal site for fracture risk prediction *Clin. Orthop.* **216** 262–9

[98] Black D M, Cummings S R, Genant H K, Nevitt M C, Palermo L and Browner W 1992 Axial and appendicular bone density predict fractures in older women *J. Bone Min. Res.* **7**(6) 633–8

[99] Kleerekoper M, Nelson D A, Flynn M J, Pawluszka A S, Jacobsen G and Peterson E L 1994 Comparison of radiographic absorptiometry with dual-energy X-ray absorptiometry and quantitative computed tomography in normal older white and black women *J. Bone Min. Res.* **9**(11) 1745–9

[100] Glüer C C, C W and Genant H K 1993 Disparity of different BUA approaches *Calcif. Tissue Intl.* **52** S171

[101] Strelitzki R and Truscott J G 1998 An evaluation of the reproducibility and responsiveness of four 'state-of-the-art' ultrasonic heel bone measurement systems using phantoms *Osteoporos. Intl.* **8**(2) 104–9

[102] Chappard C, Camus E, Lefebvre F, Guillot G, Bittoun J, Berger G and Laugier P 2000 Evaluation of error bounds on calcaneal speed of sound caused by surrounding soft tissue *J. Clin. Densitom.* **3**(2) 121–31

[103] Chappard C, Berger G, Roux C and Laugier P 1999 Ultrasound measurement on the calcaneus: influence of immersion time and rotation of the foot *Osteoporos. Intl.* **9**(4) 318–26

[104] Prins S H, Jorgensen H L, Jorgensen L V and Hassager C 1998 The role of quantitative ultrasound in the assessment of bone: a review *Clin. Physiol.* **18**(1) 3–17

[105] Njeh C F, Boivin C M and Langton C M 1997 The role of ultrasound in the assessment of osteoporosis: a review *Osteoporos. Intl.* **7**(1) 7–22

[106] Gregg E W, Kriska A M, Salamone L M, Roberts M M, Anderson S J, Ferrell R E, Kuller L H and Cauley J A 1997 The epidemiology of quantitative ultrasound: a review of the relationships with bone mass, osteoporosis and fracture risk *Osteoporos. Intl.* **7**(2) 89–99

[107] Kang C and Speller R 1998 Comparison of ultrasound and dual energy X-ray absorptiometry measurements in the calcaneus *Br. J. Radiol.* **71**(848) 861–7

[107] Salamone L M, Krall E A, Harris S and Dawson-Hughes B 1994 Comparison of broadband ultrasound attenuation to single X-ray absorptiometry measurements at the calcaneus in postmenopausal women *Calcif. Tissue Intl.* **54**(2) 87–90

[109] Greenspan S L, Bouxsein M L, Melton M E, Kolodny A H, Clair J H, Delucca P T, Stek M Jr, Faulkner K G and Orwoll E S 1997 Precision and discriminatory ability of calcaneal bone assessment technologies *J. Bone Min. Res.* **12**(8) 1303–13

[110] Burston B, McNally D S and Nicholson H D 1998 Determination of a standard site for the measurement of bone mineral density of the human calcaneus *J. Anatomy* **193**(Pt 3)(1) 449–56

[111] Lin J C, Amling M, Newitt D C, Selby K, Srivastav S K, Delling G, Genant H K and Majumdar S 1998 Heterogeneity of trabecular bone structure in the calcaneus using magnetic resonance imaging *Osteoporos. Intl.* **8**(1) 16–24

[112] Glüer C C, Vahlensieck M, Faulkner K G, Engelke K, Black D and Genant H K 1992 Site-matched calcaneal measurements of broad-band ultrasound attenuation and single X-ray absorptiometry: do they measure different skeletal properties? *J. Bone Min. Res.* **7**(9) 1071–9

[113] Jorgensen H L and Hassager C 1997 Improved reproducibility of broadband ultrasound attenuation of the os calcis by using specific region of interest *Bone* **21**(1) 109–12

[114] Damilakis J, Perisinakis K, Vagios E, Tsinikas D and Gourtsoyiannis N 1998 Effect of region of interest location on ultrasound measurements of the calcaneus *Calcif. Tissue Intl.* **63**(4) 300–5

[115] Roux C, Fournier B, Laugier P, Chappard C, Kolta S, Dougados M and Berger G 1996 Broadband ultrasound attenuation imaging: a new imaging method in osteoporosis *J. Bone Min. Res.* **11**(8) 1112–8

[116] Langton C M and Langton D K 2000 Comparison of bone mineral density and quantitative ultrasound of the calcaneus: site-matched correlation and discrimination of axial BM D status [in Process Citation] *Br. J. Radiol.* **73**(865) 31–5

[117] Hassager C, Jensen S B, Gotfredsen A and Christiansen C 1991 The impact of measurement errors on the diagnostic value of bone mass measurements: theoretical considerations *Osteoporos. Intl.* **1**(4) 250–6

[118] Rifkin R D 1995 Effects of correlated and uncorrelated measurement error on linear regression and correlation in medical method comparison studies *Stat. Med.* **14**(8) 789–98

[119] Svendsen O L, Hassager C, Skodt V and Christiansen C 1995 Impact of soft tissue on in-vivo accuracy of bone mineral measurements in the spine, hip, and forearm: a human cadaver study *J. Bone Min. Res.* **10**(6) 868–73

[120] Hagiwara S, Engelke K, Yang S O, Dhillon M S, Guglielmi G, Nelson D L and Genant H K 1994 Dual X-ray absorptiometry forearm software: accuracy and intermachine relationship *J. Bone Min. Res.* **9**(9) 1425–7

[121] Yamada M, Ito M, Hayashi K and Nakamura T 1993 Calcaneus as a site for assessment of bone mineral density: evaluation in cadavers and healthy volunteers. *Am. J. Roentgenol.* **161**(3) 621–7

[122] Szucs J, Jonson R, Granhed H and Hansson T 1992 Accuracy, precision, and homogeneity effects in the determination of the bone mineral content with dual photon absorptiometry in the heel bone *Bone* **13**(2) 179–83

[123] Lu Y and Glüer C-C 1999 Statistical tools in quantitative ultrasound applications. In *Quantitative Ultrasound: Assessment of Osteoporosis and Bone Status* eds C F Njeh, D Hans, T Fuerst, C-C Glüer and H K Genant (London: Martin Dunitz) pp 77–100

[124] Rosenthall L, Tenenhouse A and Caminis J 1995 A correlative study of ultrasound calcaneal and dual-energy X-ray absorptiometry bone measurements of the lumbar spine and femur in 1000 women *Eur. J. Nucl. Med.* **22**(5) 402–6

[125] van Daele P L, Burger H, Algra D, Hofman A, Grobbee D E, Birkenhager J C and Pols H A 1994 Age-associated changes in ultrasound measurements of the calcaneus in men and women: the Rotterdam Study *J. Bone Min. Res.* **9**(11) 1751–7

[126] Funck C, Wuster C, Alenfeld F E, Pereira-Lima J F S, Fritz T, Meeder P J, Gotz M and Ziegler R 1996 Ultrasound velocity of the tibia in normal German women and hip fracture patients *Calcif. Tissue Intl.* **58**(6) 390–4

[127] Schott A M, Hans D, Sornay-Rendu E, Delmas P D and Meunier P J 1993 Ultrasound measurements on os calcis: precision and age-related changes in a normal female population *Osteoporos. Intl.* **3**(5) 249–54

[128] Moris M, Peretz A, Tjeka R, Negaban N, Wouters M and Bergmann P 1995 Quantitative ultrasound bone measurements: normal values and comparison with bone mineral density by dual X-ray absorptiometry *Calcif. Tissue Intl.* **57**(1) 6–10

[129] Duboeuf F, Hans D, Schott A M, Giraud S, Delmas P D and Meunier P J 1996 Ultrasound velocity measured at the proximal phalanges: precision and age-related changes in normal females *Rev. Rhum. Engl. Ed.* **63**(6) 427–34

[130] Ventura V, Mauloni M, Mura M, Paltrinieri F and de Aloysio D 1996 Ultrasound velocity changes at the proximal phalanxes of the hand in pre-, peri- and post-menopausal women *Osteoporos. Intl.* **6**(5) 368–75

[131] Murgia C, Cagnacci A, Paoletti A M, Pilia I, Meloni A and Melis G B 1996 Comparison between a new ultrasound densitometer and single-photon absorptiometry *Menopause: J. North Am. Menopause Soc.* **3**(3) 149–53

[132] Aguado F, Revilla M, Hernandez E R, Villa L F and Rico H 1997 Ultrasound bone velocity on proximal phalanges in premenopausal, perimenopausal, and post-menopausal healthy women *Invest. Radiol.* **32**(1) 66–70

[133] Schott A M, Hans D, Garnero P, Sornay-Rendu E, Delmas P D and Meunier P J 1995 Age-related changes in Os calcis ultrasonic indices: a 2-year prospective study *Osteoporos. Intl.* **5**(6) 478–83

[134] Krieg M A, Thiebaud D and Burckhardt P 1996 Quantitative ultrasound of bone in institutionalized elderly women: a cross-sectional and longitudinal study *Osteoporos. Intl.* **6**(3) 189–95

[135] Mele R, Masci G, Ventura V, de Aloysio D, Bicocchi M and Cadossi R 1997 Three-year longitudinal study with quantitative ultrasound at the hand phalanx in a female population *Osteoporos. Intl.* **7**(6) 550–7

[136] Heaney R P, Avioli L V, Chesnut C H D, Lappe J, Recker R R and Brandenburger G H 1989 Osteoporotic bone fragility. Detection by ultrasound transmission velocity *J. Am. Med. Assoc.* **261**(20) 2986–90

[137] Rossman P, Zagzebski J, Mesina C, Sorenson J and Mazess R 1989 Comparison of speed of sound and ultrasound attenuation in the os calcis to bone density of the radius, femur and lumbar spine *Clin. Phys. Physiol. Meas.* **10**(4) 353–60

[138] Turner C H, Peacock M, Timmerman L, Neal J M and Johnson C C Jr 1995 Calcaneal ultrasonic measurements discriminate hip fracture independently of bone mass *Osteoporos. Intl.* **5**(2) 130–5

[139] Schott A M, Weill-Engerer S, Hans D, Duboeuf F, Delmas P D and Meunier P J 1995 Ultrasound discriminates patients with hip fracture equally well as dual energy X-ray absorptiometry and independently of bone mineral density *J. Bone Min. Res.* **10**(2) 243–9

[140] Gonnelli S, Cepollaro C, Agnusdei D, Palmieri R, Rossi S and Gennari C 1995 Diagnostic value of ultrasound analysis and bone densitometry as predictors of vertebral deformity in postmenopausal women *Osteoporos. Intl.* **5**(6) 413–8

[141] Stegman M R, Heaney R P and Recker R R 1995 Comparison of speed of sound ultrasound with single photon absorptiometry for determining fracture odds ratios *J. Bone Min. Res.* **10**(3) 346–52

[142] Heaney R P, Avioli L V, Chesnut C H R, Lappe J, Recker R R and Brandenburger G H 1995 Ultrasound velocity through bone predicts incident vertebral deformity *J. Bone Min. Res.* **10**(3) 341–5

[143] Scavalli A S, Marini M, Spadaro A, Messineo D, Cremona A, Sensi F, Riccieri V and Taccari E 1997 Ultrasound transmission velocity of the proximal phalanxes of the non-dominant hand in the study of osteoporosis *Clin. Rheumatol.* **16**(4) 396–403

[144] Hans D, Dargent-Molina P, Schott A M, Sebert J L, Cormier C, Kotzki P O, Delmas P D, Pouilles J M, Breart G and Meunier P J 1996 Ultrasonographic heel measurements to predict hip fracture in elderly women: the EPIDOS prospective study *Lancet* **348**(9026) 511–4

[145] Hans D, Schott A M and Meunier P J 1993 Ultrasonic assessment of bone: a review. *Eur. J. Med.* **2**(3) 157–63

[146] Kaufman J J and Einhorn T A 1993 Ultrasound assessment of bone *J. Bone Min. Res.* **8**(5) 517–25

[147] Stewart A, Reid D M and Porter R W 1994 Broadband ultrasound attenuation and dual energy X-ray absorptiometry in patients with hip fractures: which technique discriminates fracture risk *Calcif. Tissue Intl.* **54**(6) 466–9

[148] Orwoll E S, Oviatt S K and Mann T 1990 The impact of osteophytic and vascular calcifications on vertebral mineral density measurements in men *J. Clin. Endocrinol. Metab.* **70**(4) 1202–7

[149] Stewart A, Felsenberg D, Kalidis L and Reid D M 1995 Vertebral fractures in men and women: how discriminative are bone mass measurements? *Br. J. Radiol.* **68**(810) 614–20

[150] Ross P, Huang C, Davis J, Imose K, Yates J, Vogel J and Wasnich R 1995 Predicting vertebral deformity using bone densitometry at various skeletal sites and calcaneus ultrasound *Bone* **16**(3) 325–32

[151] Funke M, Kopka L, Vosshenrich R, Fischer U, Ueberschaer A, Oestmann J W and Grabbe E 1995 Broadband ultrasound attenuation in the diagnosis of osteoporosis: correlation with osteodensitometry and fracture *Radiology* **194**(1) 77–81

[152] Bauer D C, Gluer C C, Genant H K and Stone K 1995 Quantitative ultrasound and vertebral fracture in postmenopausal women. Fracture Intervention Trial Research Group *J. Bone Min. Res.* **10**(3) 353–8

[153] Dretakis E C, Kontakis G M, Steriopoulos C A and Dretakis C E 1994 Decreased broadband ultrasound attenuation of the calcaneus in women with fragility fracture.

85 Colles' and hip fracture cases versus 77 normal women *Acta Orthop. Scand.* **65**(3) 305–8

[154] Kroger H, Jurvelin J, Arnala I, Penttila K, Rask A, Vainio P and Alhava E 1995 Ultrasound attenuation of the calcaneus in normal subjects and in patients with wrist fracture *Acta Orthop. Scand.* **66**(1) 47–52

[155] Porter R W, Miller C G, Grainger D and Palmer S B 1990 Prediction of hip fracture in elderly women: a prospective study [see comments] *Br. Med J.* **301**(6753) 638–41

[156] Bauer D C, Gluer C C, Cauley J A, Vogt T M, Ensrud K E, Genant H K and Black D M 1997 Broadband ultrasound attenuation predicts fractures strongly and independently of densitometry in older women. A prospective study. Study of Osteoporotic Fractures Research Group. *Arch. Intern. Med.* **157**(6) 629–34

[157] Hans D, Dargent P, Schott A M, Breart G, Meunier P J and Group ATE 1996 Ultrasound parameters are better predictors of trochanteric than cervical hip fractures: the EPIDOS prospective study. In *World Congress on Osteoporosis 1996* eds S E Papapoulos, P Lips, H A P Pols, C C Johnston and P D Delmas (Amsterdam: Elsevier) pp 161–6

[158] Gonnelli S, Cepollaro C, Pondrelli C, Martini S, Rossi S and Gennari C 1996 Ultrasound parameters in osteoporotic patients treated with salmon calcitonin: a longitudinal study *Osteoporos. Intl.* **6**(4) 303–7

[159] Giorgino R, Lorusso D and Paparella P 1996 Ultrasound bone densitometry and 2-year hormonal replacement therapy efficacy in the prevention of early postmenopausal bone loss *Osteoporos. Intl.* **6**(suppl 1) S341

[160] Ryan P, Herd R, Blake G C and Fogelman I 1996 Calcaneal BUA changes in a 2 year placebo controlled study of pamidronate in post menopausal osteoporosis *Osteoporos. Intl.* **6**(suppl 1) PMO 516

[161] de Aloysio D, Rovati L C, Cadossi R, Paltrinieri F, Mauloni M, Mura M, Penacchioni P and Ventura V 1997 Bone effects of transdermal hormone replacement therapy in postmenopausal women as evaluated by means of ultrasound: an open one-year prospective study *Maturitas* **27**(1) 61–8

[162] Gibbens D T, Gilsanz V, Boechat M I, Dufer D, Carlson M E and Wang C I 1988 Osteoporosis in cystic fibrosis *J. Pediatrics* **113**(2) 295–300

[163] Cetin A, Celiker R, Dinçer F and Ariyürek M 1998 Bone mineral density in children with juvenile chronic arthritis *Clin. Rheumatol.* **17**(6) 551–3

[164] Falcini F, Trapani S, Civinini R, Capone A, Ermini M and Bartolozzi G 1996 The primary role of steroids on the osteoporosis in juvenile rheumatoid patients evaluated by dual energy X-ray absorptiometry *J. Endocrinol. Invest.* **19**(3) 165–9

[165] Mazess R B, Barden H S and Ohlrich E S 1990 Skeletal and body-composition effects of anorexia-nervosa *Am. J. Clinical Nutrition* **52**(3) 438–41

[166] Chesney R W, Rose P G and Mazess R B 1984 Persistence of diminished bone mineral content following renal transplantation in childhood *Pediatrics* **73**(4) 459–66

[167] Chesney R W, Mazess R B, Rose P and Jax D K 1978 Effect of prednisone on growth and bone mineral content in childhood glomerular disease *Am. J. Dis. Child.* **132**(8) 768–72

[168] Timperlake R W, Cook S D, Thomas K A, Harding A F, Bennett J T, Haller J S and Anderson R M 1988 Effects of anticonvulsant drug therapy on bone mineral density in a pediatric population *J. Pediatr. Orthop.* **8**(4) 467–70

[169] Seeman E, Young N, Szmukler G, Tsalamandris C and Hopper J L 1993 Risk factors for osteoporosis *Osteoporos. Intl.* **3** Suppl 1 40–3

[170] Sundberg M, Gardsell P, Johnell O, Ornstein E and Sernbo I 1998 Comparison of quantitative ultrasound measurements in calcaneus with DXA and SXA at other skeletal sites: a population-based study on 280 children aged 11–16 years *Osteoporos. Intl.* **8**(5) 410–7

[171] Mughal M Z, Langton C M, Utretch G, Morrison J and Specker B L 1996 Comparison between broad-band ultrasound attenuation of the calcaneum and total body bone mineral density in children *Acta Paediatr.* **85**(6) 663–5

[172] Halaba Z and Pluskiewicz W 1997 The assessment of development of bone mass in children by quantitative ultrasound through the proximal phalanxes of the hand *Ultrasound Med. Biol.* **23**(9) 1331–5

[173] Lappe J M, Recker R R and Weidenbusch D 1998 Influence of activity level on patellar ultrasound transmission velocity in children *Osteoporos. Intl.* **8**(1) 39–46

[174] Jaworski M, Lebiedowski M, Lorenc R S and Trempe J 1995 Ultrasound bone measurement in pediatric subjects *Calcif. Tissue Intl.* **56**(5) 368–71

[175] van den Bergh J P, Noordam C, Ozyilmaz A, Hermus A R, Smals A G and Otten B J 2000 Calcaneal ultrasound imaging in healthy children and adolescents: relation of the ultrasound parameters BUA and SOS to age, body weight, height, foot dimensions and pubertal stage *Osteoporos. Intl.* **11**(11) 967–76

[176] Mughal M Z, Ward K, Qayyum N and Langton C M 1997 Assessment of bone status using the contact ultrasound bone analyser *Arch. Dis. Child.* **76**(6) 535–6

[177] Njeh C F, Boivin C M, Gardner-Medwin J M, Shaw N J and Southwood T R 1997 The use of finger ultrasound velocity to monitor bone status in juvenile chronic arthritis: A pilot study (abstract) *Osteoporos. Intl.* **7**(3) 300

[178] Madsen O R, Egsmose C, Hansen B and Sorensen O H 1998 Soft tissue composition, quadriceps strength, bone quality and bone mass in rheumatoid arthritis *Clin. Exp. Rheumatol.* **16**(1) 27–32

[179] Martin J C, Munro R, Campbell M K and Reid D M 1997 Effects of disease and corticosteroids on appendicular bone mass in postmenopausal women with rheumatoid arthritis: comparison with axial measurements *Br. J. Rheumatol.* **36**(1) 43–9

[180] Daens S, Peretz A, de Maertelaer V, Moris M and Bergmann P 1999 Efficiency of quantitative ultrasound measurements as compared with dual-energy X-ray absorptiometry in the assessment of corticosteroid-induced bone impairment *Osteoporos. Intl.* **10**(4) 278–83

[181] Blanckaert F, Cortet B, Coquerelle P, Flipo R M, Duquesnoy B, Marchandise X and Delcambre B 1997 Contribution of calcaneal ultrasonic assessment to the evaluation of postmenopausal and glucocorticoid-induced osteoporosis *Revue du Rhumatisme (English Edn)* **64**(5) 305–13

[182] Blake G M, Herd R J, Miller C G and Fogelman I 1994 Should broadband ultrasonic attenuation be normalized for the width of the calcaneus? *Br. J. Radiol.* **67**(804) 1206–9

[183] Fuerst T, Njeh C F and Hans D 1999 Quality assurance and quality control in quantitative ultrasound. In *Quantitative Ultrasound: Assessment of Osteoporosis and Bone Status* ed H K Genant (London: Martin Dunitz) pp 163–75

[184] Hans D, Njeh C F, Genant H K and Meunier P J 1998 Quantitative ultrasound in bone status assessment *Rev. Rhum. Engl. Ed.* **65**(7–9) 489–98

Chapter 15

Finite element modelling

Bert van Rietbergen

15.1. INTRODUCTION

This chapter discusses the application of the finite element (FE) method for the analysis of bone. Since the principal function of bone is a mechanical one, the emphasis of this chapter will be on the use of the FE method for the determination of bone mechanical properties. The possibilities and requirements for FE analysis of bone are largely determined by the structural level of organization that is considered and for this reason a distinction is made between different levels of structural organization of bone ranging from the whole-bone level down to the trabecular, tissue and finally ultra-structural level. A structural and mechanical characterization relevant for finite element analysis is given for each of these levels. In the remainder of the chapter, recent applications and results are described for the FE analysis of bone at the whole-bone and trabecular level.

SECTION A: FINITE ELEMENT ANALYSIS OF BONE: GENERAL CONSIDERATIONS

15.2. FUNDAMENTALS OF FE ANALYSIS

The FE method is a numerical approximation method for the solution of boundary value cases that can arise in a wide number of physical situations, for example, when resolving the temperature distribution in a heated body, the deformation of a solid structure due to external forces acting on it, the velocity profile of a pressurized fluid etc. With the FE method, a two-dimensional or three-dimensional domain is subdivided into a finite number of sub-domains with a simple geometry, called elements, connected to each other at their corner points, called nodes. The differential equations describing the underlying problem are then integrated per element. The finite element discretization leads to a set of coupled linear algebraic equations (usually very large) with the unknown parameter defined at the nodal points that can be determined by solving the set of equations. After solving, the distribution of the parameter can be determined throughout the entire domain by using a suitable set of interpolation functions (usually linear or higher order polynomials).

The focus of this chapter is on the application of the FE method for the solution of problems as they appear in structural solids mechanics. In such cases, the domain represents the geometry of the structure. In the two-dimensional case this domain is usually subdivided into triangular or 4-node elements, in the three-dimensional case into tetrahedron or hexahedron elements. Within the elements a continuum assumption is adopted, i.e. it is assumed that the material distribution, the stresses and strains within an infinitesimal material neighbourhood can be regarded as essentially uniform. In most of the problems considered in this chapter, the parameter of interest is the deformation field of the body. By solving the FE equations, it is possible to determine this deformation field as a function of the constitutive behaviour of the material and the applied external and internal loads. From the calculated deformation field, derived parameters such as strain and stress fields can be calculated as well.

A detailed description of the fundamentals of the FE method is outside the scope of this chapter but can be found in several text books [e.g. 1, 2]. In the remainder of this section, some practical aspects and considerations of the finite element method relevant for solid structural analysis of bone are discussed.

15.3.   FE ANALYSIS APPLIED TO BONE

The principal task of our skeleton is to withstand forces acting upon the body. As such, the main task of bone tissue is a mechanical one, and it is the mechanical parameters that determine its functional behaviour and quality. For this reason, the most common goal of FE analysis of bone is to obtain a quantitative evaluation of mechanical parameters such as stresses and strains in the bone tissue due to forces applied to the bone. The principal aspects of this type of FE analysis are discussed below.

15.3.1   *Structural and solid mechanics FE analysis*

15.3.1.1.   *Linear FE analysis*

For forces in the physiological range, the deformations in the bone tissue are relatively small (strains are usually less than 0.5%). This implies that for physiological loading conditions, geometrically linear analyses, where the geometry is defined only in the undeformed situation, are suitable. Bone tissue is also known to exhibit linear elastic constitutive behaviour for strains in the physiological range. Consequently, when analysing bone under physiological loading conditions, linear FE analyses are usually adequate.

Material properties for linear elastic materials, as they need to be specified for FE analyses, are fully described by the components of a fourth-rank stiffness or compliance tensor in the generalized Hook's law [3, 4]. The stiffness and compliance tensors are usually represented by symmetric six-by-six matrices. In its most general form, these matrices hold 21 independent components that must be determined from experiments. If planes of elastic symmetry exist in the material, some of these coefficients are interdependent or zero when measured in a coordinate system aligned with the normals to the symmetry planes. In the case of orthotropy, with three orthogonal planes of elastic symmetry, nine independent elastic coefficients remain to be determined [5]. The number of independent engineering constants is reduced further to five for the case of transverse isotropy when each plane through a longitudinal axis is a plane of elastic symmetry. In the case of isotropy only two independent engineering constants remain. Bone displays significant anisotropy [6], but the elastic symmetries are dependent on the level of structural organization. The anisotropic models that best describe the elastic behaviour at each of these levels will be described in the next section.

Boundary conditions for FE models of bones consist of external forces and prescribed or suppressed displacements. Force boundary conditions can represent, for example, contact forces at the joints, muscle forces or forces applied in an experimental setting. Displacement boundary conditions are needed to avoid rigid-body motion and can represent, for example, prescribed displacements in experiments. Internal forces due to the weight of the bone can usually be neglected since the loads acting on most bones far exceed the loads due to the weight of the bone itself.

15.3.1.2. Nonlinear FE analysis

If strains within the bone tissue exceed physiological values, the small-deformation and linear elasticity assumption no longer holds and a nonlinear finite element approach is required to account for deformation-dependent changes in geometry or material properties. With nonlinear finite element analyses, the loading region is subdivided into a number of smaller load increments in which the deformations and material properties do not change significantly. The behaviour during each of these increments can then be considered as being linear and the nonlinear problem is solved as a piece-wise linear problem, with a geometry and material properties that are updated iteratively based on the actual stress and strain state.

Except in the case of large deformations, nonlinear FE analyses may also be required in the case of changing interface settings or boundary conditions. Typical examples are contact problems, in which the region of contact is dependent on the applied force, thus requiring an iterative procedure to find the correct contact regions. Nonlinear FE analyses are also required when the analysis includes soft tissues (with large deformations even for physiological loading conditions), materials which display essentially nonlinear behaviour (seen for many soft tissues), or when the time-dependent behaviour (creep behaviour, stress relaxation) of the (bone-) tissue plays an important role. Typical examples of the latter are poroelastic FE analyses, as discussed in section 15.3.2. Finally, in the case of bone (and other biological tissues) nonlinear analyses may also be required to account for active biological behaviour of the tissue. A typical example is bone remodelling resulting in changes in bone mass and morphology as discussed in the last sections.

15.3.2.   *Poroelastic FE analysis*

In vivo, bone is saturated with fluid (blood, marrow and fat). Linear elastic FE analysis for structures and solids as described above do not explicitly account for the water content. In certain situations, however, the bone fluid can play an important role in the mechanical behaviour of bone tissue. The theory of poroelasticity has been developed to model the interaction of deformation and fluid flow in a fluid-saturated porous medium. There are several different approaches to the development of the equations for the theory of poroelasticity, but all approaches lead to the same set of equations (for a discussion of the different theories see [7]). The theory used most often for the poroelastic analysis of bone is the effective medium approach based on the work of Biot [8]. In this theory, the poroelastic field variables are the total stress, pore pressure, strain in the solid phase and variation in fluid content. The theory can also be applied to soft tissues (cartilage and fibrous-like tissues), but in these situations mixture-based theories such as the small strain incompressible poroelastic formulation by Mow *et al* [9] (biphasic theory) are more commonly used.

As discussed earlier, analyses of poroelastic behaviour require nonlinear finite element analyses. Poroelastic FE analyses require the measurement of additional material parameters, such as the (anisotropic) porosity and permeability of the solid phase, and the bulk modulus and viscosity of the fluid. In addition, extra boundary conditions are needed to define the permeability of the boundary.

15.3.3.   *Other types of FE analysis*

As mentioned earlier, the FE approach can be used to solve a very wide number of boundary value cases. With regard to bone, the FE method has been used to analyse various cases such as those related to the temperature distribution in bones due to friction at the head of a femoral implant [10] or different ambient conditions [11] and the effect of shock wave treatment to destroy cement around femoral prostheses [12]. Although the underlying equations to be solved for such problems are different from those for structural solid mechanics, the FE approach is largely the same and many general-purpose FE packages can solve these types of problems. Obviously, the parameters that are needed to define the problem are different and dependent upon it (e.g. temperature coefficients for temperature analyses).

15.4. GENERATION OF FE MODELS

The first step in the finite element approach is usually the definition and quantification of the domain geometry and the subsequent division of this domain into finite elements (meshing). The domain can be two-dimensional, two-dimensional axisymmetric or three-dimensional. The choice of the domain dimensionality depends on the shape of the object to be analysed, the goal of the analysis, the available geometry data and the available (computer) time.

Presently, the geometry of FE models of bone is usually based on digitized images. These can be projected (two-dimensional) images such as radiographs, or images obtained from more advanced medical imaging techniques such as computed tomography (CT) and magnetic resonance (MR) scanners that can provide a large number of sequential cross-sectional images. There are several possibilities of utilizing these images for the generation of an FE model. First, it is possible to measure contours from the images and import the measured points into some finite element pre-processing program. This pre-processing program can be used to further define the domain and to subdivide it into elements. When using quantitative CT or other calibrated imaging techniques, the pixel grey-values in the images in fact represent a measure of the local density of the bone. These image grey-values can be used to assign proper density values to the elements in the FE model from which, in turn, (elastic) properties of the element can be derived [13]. The whole process of mesh generation usually requires a substantial amount of user input and this can inhibit the analysis of large numbers or very complicated structures.

Recently, a new method for the fully automated generation of three-dimensional FE models from images has been developed. With this technique, a large number of sequential digitized cross-sectional images are stacked in a computer [14]. The stack of images then represents a digitized reconstruction of the geometry of the bone in a voxel (three-dimensional pixel) grid. Two methods have been developed for the conversion of these voxel data to finite element models. With the first of these techniques, the

Generation of 3-D FE models

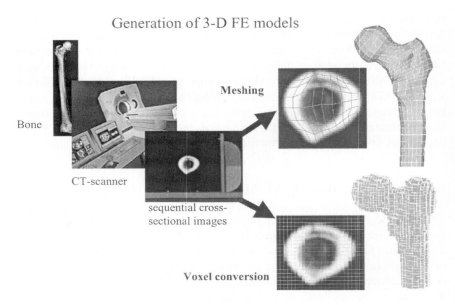

Figure 15.1. *Two contemporary methods for creating three-dimensional FE models. With both methods, the geometry of the bone (in vitro or in vivo) is based on a large number of sequential cross-sectional images obtained from a whole-body CT scanner. With the meshing approach (e.g. [13]) a two-dimensional element grid is projected on the images and the nodal points of the elements are adjusted to the contours. After combining two two-dimensional meshes with a three-dimensional mesh build of tetrahedron or hexahedron elements, a three-dimensional model of the bone results. With this approach, the density of each element can be determined from the CT data. With the voxel conversion approach ([15]) the voxels in the CT scans are directly converted to equally-shaped hexahedron elements in an FE model. An advantage of this approach is the fact that it can be done in a fully automated way. Disadvantages are the large number of elements and the 'jagged boundaries' that are generated.*

'voxel-conversion' technique, the voxels representing bone tissue are directly converted to equally shaped brick elements in a finite element model [15, 16]. A major advantage of this approach is the fact that all elements in the FE model will have the same geometry, orientation and size, enabling the use of very fast special-purpose solvers that can compensate for the large number of elements that are usually created with this technique. A disadvantage is that the created models will have 'jagged' surfaces. It has been shown that this can lead to oscillating values for the calculated stresses and strains, in particular at the bone surfaces [17, 18] (figure 15.1).

The second fully automated meshing technique is based on the 'marching cubes' algorithm [19, 20]. With this approach, voxels representing bone tissue are converted to FE models built of tetrahedron elements, enabling the formation of smooth surfaces. A disadvantage of this technique is the lower accuracy of tetrahedron elements and the fact that no purpose-built fast solvers exist for these models.

15.5.   EQUIPMENT AND SOFTWARE

The only equipment needed for FE analysis itself is a (fast) computer and FE software. Several general-purpose software packages are commercially available for FE analysis. All packages suitable for structural solids analyses support linear elastic analyses and other relatively simple analyses such as temperature problems; however, support for nonlinear analyses is sometimes limited. Although many packages implemented large-displacement options to account for geometrically nonlinear analyses and material nonlinear behaviour (to account for, for example, plastic behaviour), poroelastic analyses are only supported by a few packages. Packages that support poroelastic analyses with both compressible and incompressible constituents are, for example, Abaqus [21] and Diana [22]. Other packages (e.g. Marc [23]) do support poroelastic analyses, but only for incompressible matrix materials. Although this assumption is reasonable for most soft tissues (and cartilage) it is not for the analysis of bone.

General-purpose FE solvers that implement iterative solvers can usually solve problems of the size of 10^5 elements. For larger problems, e.g. those arising from FE analysis of trabecular structures (as discussed in section D), special-purpose finite element solvers are usually required. Some of these codes are now commercially available [24, 25].

SECTION B: BONE MECHANICAL CHARACTERIZATION AND FE MODELLING AT DIFFERENT LEVELS OF STRUCTURAL ORGANIZATION

The possibilities and requirements for FE analysis of bone are largely determined by the structural level of organization that is considered. In this section, four commonly distinguished levels of bone structural organization are described (figure 15.2), and a characterization of relevant parameters and analysis types is given for each of these levels.

Levels of structural organization

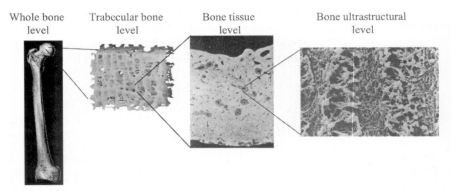

Figure 15.2. *The four levels of structural organization that are commonly distinguished. At the whole-bone level the structural organization is determined by the bone external and internal geometry, the bone density distribution and the bone anisotropy. At the trabecular bone level, it is determined by the trabecular organization of the bone. Farther down, at the bone tissue level, structural organization is merely determined by the porosity of the bone due to Haversian canals, lacunae and canaliculi and finally at the ultrastructural level, by the organization of the apatite and collagen that form the constituents of the bone tissue (ultrastructural level figure taken from [51]).*

15.6.   THE WHOLE BONE (APPARENT) LEVEL

15.6.1.   *Structural characterization*

At the bone organ level, bone is considered as a continuous material with material properties that represent the average properties of a representative bone volume. At this level, two types of bone are distinguished: cortical bone and cancellous bone. The latter type is the porous type of bone which is often referred to as trabecular bone. In this chapter, however, the term cancellous bone is used when the continuum level properties of this type of bone are considered, whereas the term trabecular bone will be used when this type of bone is considered as a discontinuous porous structure. For cortical bone, a representative volume can be rather small ($\sim$1 mm^3). For cancellous bone regions, however, a representative volume must hold a large enough number of trabeculae in order to provide sufficiently averaged continuum quantities. It has been stated that a representative volume of cancellous bone should be at least five inter-trabecular lengths in size (approximately 3–5 mm) [26]. It is not possible to define meaningful continuum-level elastic properties for cancellous bone volumes smaller than this minimum size. A typical length scale associated with this level thus is of the size of a few millimetres.

At this level the cancellous bone structural properties are first characterized by the bone volume fraction (often represented by the parameter BV/TV: bone volume over total volume), which is the volume of bone per tissue unit of volume and is dimensionless. A similar scalar quality is the total mass per unit volume of bone, called the structural density or apparent density, which measures the degree of mineralization as well (dimensions: $kg\,m^{-3}$). Since the bone volume fraction does not provide any information about the actual structure of the bone (other than its density), other parameters have been developed. Among these are scalar parameters that quantify the average geometry of the trabeculae (e.g. mean trabecular thickness and spacing), the connectivity and the fractal dimension of the trabecular network. In order to quantify the directionality and the anisotropy of the structure, tensor parameters have been introduced. The one most commonly used is the mean intercept length (MIL) [144, 145]. In MIL measurements, a grid of parallel lines is projected over a cross-sectional image of the bone structure. The number of bone–marrow interfaces is then counted for a large number of grid angles. When representing the total line length over the number of bone–marrow interfaces in a polar plot, an ellipse-shaped figure results (an ellipsoid in three dimensions), with its largest principal axis indicating the principal or axial direction of the material. The ratio between the lengths of the principal axes is an indication of the anisotropy of the structure. It should be emphasized, however, that no unique definition exists for the structural anisotropy of bone; other tensor measures have been developed as well [27].

15.6.2. Mechanical characterization

For cortical bone, elastic behaviour is usually best described as transversally isotropic, with the symmetry axis aligned with the bone longitudinal axis. The longitudinal modulus of cortical bone is typically in the range of 15–20 GPa whereas transversal moduli are somewhat less. A detailed overview of human and animal cortical bone elastic properties can be found in the literature [e.g. 28, 29].

The elastic modulus of cancellous bone can vary in a wide range (roughly speaking from 0.1 to 2000 MPa) depending on the bone volume fraction. The stiffness is also dependent on the direction in which the bone is loaded, demonstrating the anisotropic behaviour of cancellous bone. For cancellous bone, it has long been assumed that at least three orthogonal planes of elastic symmetry exist (orthotropic elastic behaviour), but it is only recently that this could be proven [30]. Depending on the site, cancellous bone can also display transversally isotropic or isotropic behaviour. Many earlier studies have documented relationships between the cancellous bone axial modulus (the stiffness of the bone in its stiffest direction) and volume fraction. In an often-cited study by Carter and Hayes [31], a cubic relationship was found between the axial modulus and apparent density for a set of human and bovine cancellous and cortical bone specimens. Later studies have indicated that a square relationship will yield more accurate results for human cancellous bone alone [32]. The application of these relationships requires the measurement of bone density (or volume fraction). This can be done using quantitative imaging techniques such as dual energy X-ray absorptiometry (DXA), or quantitative CT or MR techniques if three-dimensional information is needed [33, 34]. More recent studies have documented detailed relationships between (a) all orthotropic elastic constants of cancellous bone and bone volume fraction [30], (b) a combination of bone volume fraction and a mean intercept length, or (c) other tensors that describe the average trabecular orientation [35]. With the use of such relationships, it becomes possible to quantify the site-specific anisotropic elastic behaviour of bone from structural measurements. Since such structural measurements are possible with high-resolution imaging methods, these relationships enable a detailed mechanical characterization of bone even *in vivo*.

15.6.3.   FE modelling

The geometry of the FE model at this level describes the external contours of the bone. The external geometry of the bone can be obtained in the case of cadaver bones, from direct measurements, but in most studies it is obtained by measuring the geometry from radiographs or cross-sectional images obtained from whole-body or peripheral CT or MR scanners.

Voxel conversion techniques for the fully automated generation of FE models have also been used at this level [15]. This has enabled the fully automated generation of FE models and assignment of CT-derived material properties. A disadvantage of this technique is the limited resolution of the images that can be obtained with clinical CT scanners at this level, which is of the order of 1 mm, resulting in models with severely 'jagged' surfaces. Validation studies, however, have demonstrated that strains predicted by FE models generated with this approach correlated very well with strains measured in an experiment using the same bone [36].

Depending on the type of analysis, forces applied to the bone can represent physiological joint forces measured *in vivo* using telemetric devices, muscle forces calculated from mathematical models or non-physiological forces applied by a test machine. Extensive and complete data about bone loading conditions are now available from telemetry measurements for the hip joint [37] and the spine [38], and good estimates of forces acting in other joints and forces due to non-physiological loading conditions (e.g. a fall) can also be found in the literature [142, 143].

15.7. THE TRABECULAR BONE LEVEL

15.7.1. *Structural characterization*

At this level, the trabecular architecture of cancellous bone is explicitly accounted for. Cancellous bone has a very complex internal geometry, built of many struts and plates. A typical length scale for the trabecular thickness is about 100 μm. The trabecular tissue that forms the trabeculae is built of collagen fibres oriented in the trabecular direction and reinforced with apatite crystals. At this level, the bone tissue itself is considered as a continuum.

15.7.2. *Mechanical characterization*

Bone tissue elastic properties and strength have been measured using standard engineering test methods such as tensile tests, three- or four-point bending tests and buckling tests. Values found for the tissue modulus range from 0.76 to 10 GPa when using tensile tests, from 3.2 to 5.4 GPa for three- or four-point bending tests and from 8.7 to 14 GPa for buckling tests (for an overview see Rho *et al* [39]). A major problem when using standard engineering test methods for the determination of bone tissue properties is the relatively small size of trabeculae (thickness 100–200 μm, length 1–2 mm), resulting in inaccuracies in the displacement measurements and thus in the calculation of moduli. Another problem is the irregular shape of trabeculae, where standard engineering tests require standardized specimen geometry. To overcome the latter problem, some studies have used machined specimens [29, 40], but it is unclear to what extent machining artefacts can affect the stiffness of the specimens. In other studies, ultrasound was used for the measurement of tissue elastic properties. Values found with this method are generally higher than those obtained from the standard tests, 11–15 GPa [41], approaching the stiffness measured for cortical bone.

Based on its microstructural composition, it is generally assumed that the tissue material will behave transversally isotropic with its symmetry axis in the trabecular longitudinal direction. Recent studies, however, have indicated that the modelling of this tissue as an isotropic material with an 'effective isotropic' modulus can be sufficient for most FE analyses at this level [42].

15.7.3. *FE modelling*

The geometry of the trabecular architecture is extremely complex, making it difficult, if not impossible, to create three-dimensional models of it with traditional meshing techniques based on the measurement of bone contours. For this reason, the generation of three-dimensional FE models at this level

micro-FE approach: voxel conversion

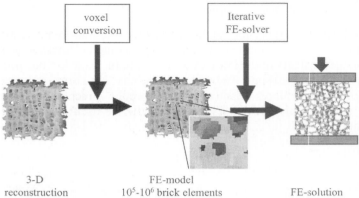

Figure 15.3. *Schematic overview of the voxel conversion technique for the generation of µFE models. A large number of sequential µCT images (or other high-resolution images) are used to create a three-dimensional reconstruction of the original bone specimen in a computer. The models thus generated consist of a very large number of voxels representing the bone tissue. Voxels representing the bone tissue are converted to equally shaped 8-node brick elements, resulting in FE models with about 10^5 to 10^6 elements per cm^3. Using special-purpose iterative FE solvers, such models can be solved within reasonable amounts of computer time [16]. The result of such µFE analyses can be used to calculate the stiffness of a bone specimen as a function of its architecture or to determine tissue-level stresses and strains.*

has been the exclusive domain of fully automated meshing techniques that make use of digitized images (figure 15.3). The resolution of these images should be good enough to resolve the trabecular structure, i.e. of the order of 100 µm or better. Several imaging methods exist to create such high-resolution images. These methods may be destructive (e.g. serial sectioning and serial milling techniques [43, 44]) or non-destructive (e.g. micro-CT (µCT) and micro-MR (µMR) imaging [45–48]). In both cases, a resolution of 50 µm or better could be obtained for cancellous bone regions of approximately 1 cm in size. Although the voxel conversion technique [16] is the most common technique to create FE models from voxel data, the 'marching cubes' algorithm has also been used in several studies [20, 49].

Material properties assigned to the elements are usually isotropic and homogeneous, with the same stiffness applied to all elements. The main reason for this choice is the fact that, to date, the micro-imaging methods used at this level can provide no reliable data about bone tissue density or anisotropy. As discussed above, however, the choice of an 'effective isotropic' modulus will be sufficient in most cases.

15.8. BONE TISSUE AND ULTRASTRUCTURAL LEVEL

15.8.1. Structural characterization

At the tissue level the bone tissue porosity due to Haversian canals, cells and canaliculi is the main structural feature. At the ultrastructural level, it is the collagen fibres and apatite crystals that form about 90% of the tissue, with the remainder being made up by porosity [50, 51]. A typical length scale at these levels is in the range 1–10 μm. As such, most imaging methods used at these levels are based on images from microscopes, requiring the sectioning and preparation of specimens. Non-destructive methods for cross-sectional images based on synchrotron radiation computed tomography have also been developed. These methods enable a high spatial resolution down to the micron level and a high density resolution allowing a local quantification of bone mineralization and three-dimensional reconstruction of the bone tissue [52].

15.8.2. Mechanical characterization

The tissue elastic properties at the tissue level have been measured using nano-indentation tests [53]. The stiffness measured at this level was in the range 15–20 GPa, and thus closely resembles the Young's modulus of cortical bone measured at a much higher level (e.g. from standard tensile tests).

Acoustic microscopy techniques have been used to measure the elastic properties of single mineral apatite crystals at angular intervals [54, 55]. The elastic properties of synthetic hydroxyapatite mineral was found to be isotropic with a Young's modulus of 120 GPa and a Poisson's ratio of 0.27, but it should be noted that the actual modulus of the apatite in bone is likely to be less due to impurities. It was also found that the orientation of mineral crystals is the primary determinant of bone anisotropy, and the collagen matrix within osteonal bone has little directional orientation [56].

15.8.3. FE modelling

At the bone tissue level, the geometry of FE models represents the external contours of trabeculae in the case of trabecular bone or osteons in the case of cortical bone. The structure of bone associated with bone fluid and their interfaces and the bone poroelastic parameter values have been well documented [7]. Poroelastic models have been developed to calculate load-induced fluid flow or strain generated potential using mathematical models [57, 58], finite difference models [59] or finite element models [60].

At the ultrastructural level, where the actual constituents of the bone tissue are considered, the geometry of FE models represents the spatial organization of the bone tissue mineral phase and an organic matrix usually simplified as a repetitive structure. Models at this level resemble those of short-fibre composites and can be used to determine mechanical properties of the bone tissue as a function of its composition and the material properties of its constituents [61]. An extensive overview of different models used at this level has been given by Lucchinetti [146, 147].

SECTION C: FE ANALYSIS OF BONE AND BONES AT THE ORGAN LEVEL: CONTEMPORARY APPLICATIONS AND RESULTS

The FE method has been used extensively in tissue mechanics and orthopaedic implant design studies. The first part of this section summarizes some of the major applications and results of the FE analysis as a tool for the analysis of bone mechanical behaviour and loading. FE analyses that focused on the application of this technique in a clinical setting are discussed in the second part. Rather than just as an analysis tool, however, the FE method has also been used to enable the simulation of mechanically-induced adaptive processes in bone. This is discussed in the third part of this section.

15.9.   ANALYSIS OF BONE MECHANICAL PROPERTIES AND LOADING

15.9.1.   *Bone failure load*

FE analyses have been used to predict bone fracture risk due to either bone diseases (metastatic defects, osteoporosis) or non-physiological loading conditions (a fall). The predictive value of FE analysis for the evaluation of bone strength reduction due to metastatic cortical defects in long bones was investigated by comparing FE results with strength measurements of defective bones [62, 63]. In these studies it was found that nonlinear FE analyses which accounted for both material plasticity and the stress concentration effects of the cortical defects yielded good correlation with the experimental data. Linear models, however, proved insufficient for the prediction of failure loads. Similar results were obtained in other studies where a linear FE approach to predict metastatic strength reduction in the femoral neck resulted in a considerable underestimation of the percentage reduction in the *in vitro* fracture strength [64]. In this study, it was suggested that this finding may reflect a fundamental inability of a linear, continuum-based analysis to predict accurately the fracture strength of a bone structure as complex as the human femur.

A similar approach, combining experimental strength measurements and FE analyses, was used to investigate whether linear or nonlinear FE analyses could predict bone strength for high-impact loading conditions such as a fall. Good results were obtained for the fracture prediction of the intact proximal human femur [65, 66]. In these studies, it was found that there was excellent agreement between *in vitro* failure data and the results of the linear model, especially when using a Von Mises effective strain failure criterion. The use of nonlinear models further improved the accuracy of the prediction and demonstrated different fracture mechanisms for different loading configurations.

Finally, FE models were used to investigate the increase in stress distribution due to osteoporosis and the effect of the impact direction of a fall [67, 68]. It was concluded that both factors could affect the failure load to a similar extent [69].

15.9.2. Bone fracture healing and tissue differentiation analysis

It is well known that the mechanical environment is of importance for the tissue differentiation processes that play a role during the healing and repair of bone fractures. Several hypothetical models have been developed to explain tissue differentiation as a result of the local stress and strain state quantified by stress and strain tensors or their invariants [70, 71]. FE analysis has been used to evaluate the hypothetical models by analysing these mechanical parameters in specific situations where tissue differentiation leads to bone formation. A typical situation analysed in a number of studies is the initial state of fracture healing. Here, the domain of interest is the bone–gap interface. Since the initial stiffness of the gap region is much lower than that of bone, nonlinear elastic analyses are required to analyse stresses and strains in the callus region. When analysing later stages of the fracture healing process, or situations where strains are largely reduced due to external fixators, however, linear FE analysis could be used without comprising the results. In one of the earlier studies, DiGioia *et al* [72] found that in a general nonlinear material, nonlinear analysis is necessary to model an osteotomy gap subjected to a maximum longitudinal strain of 100%. Restricting the maximum longitudinal strain to 10%, however, allowed the use of a linear geometric formulation. In this and later studies it was found that the local stress and strain distribution can explain the initial formation of different tissues during fracture healing as well as during distraction osteogenesis as the result of the tensile and hydrostatic stress distribution [73–75]. In recent studies [76, 77], including animal experiments, cell culture studies and FE analysis (figure 15.4), intramembraneous bone formation was found for strains smaller than 5% and small hydrostatic pressure (<0.15 MPa). Larger strains led to the formation of connective tissue.

Tissue differentiation processes also play a role during the fixation of implants and bone osteogenesis in the embryonic state. Prendergast *et al* [78] used a biphasic finite element model to study the mechanical environment

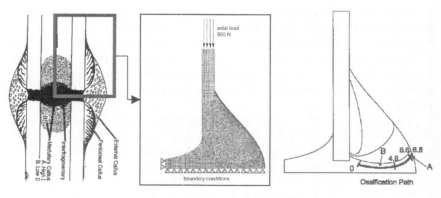

Figure 15.4. *FE analysis of callus formation after bone fracture (left). An axisymmetric FE model (middle) represents bone and callus tissue at one side of the fracture. Material properties of the callus tissue are adapted to represent different stages of the healing process as observed in complementary animal experiments. By comparing the stresses and strains calculated at the ossification path at each stage (right) it is possible to relate mechanical conditions and types of tissue in a fracture callus (adapted from Claes and Heigele [77]).*

for tissues around a micromotion device implanted in dogs. By analysing the state of stress in the tissues at four different stages during the differentiation process they were able to propose a 'mechano-regulatory pathway' that can be used to describe the interaction between biophysical stimuli and tissue phenotype.

Tanck *et al* [79] used poroelastic FE analyses to explain the typical shape of the bone mineralization front in mouse metatarsal bones as the result of local mechanical loading conditions. They found that this shape could not be explained by the fluid pressure, but it could be explained by distortional strain and the authors thus concluded that strain rather than pressure modulates the rate of the mineralization process.

15.9.3. Consequences of orthopaedic implants and interventions

One of the primary roles of FE analysis in orthopaedic research has been to quantify changes in bone mechanical loading conditions caused by orthopaedic implants and interventions. Due to the large difference in stiffness, the placement of implants can lead to a reduction of loads in the surrounding bone, a phenomenon often referred to as 'stress shielding' (figure 15.5). It is well known that bone responds to changes in loading by adding bone in overloaded areas and removing bone in low-loaded areas ('Wolff's law'). Stress-shielding effects can thus result in the resorption of bone, which can jeopardize the treatment and complicate revision operations. FE analysis has been used to quantify this stress-shielding effect after hip arthroplasty

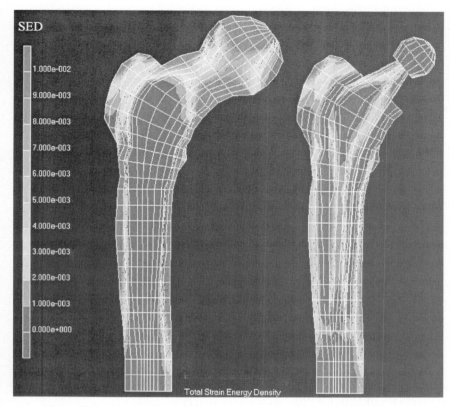

SED

1.000e-002

9.000e-003

8.000e-003

7.000e-003

6.000e-003

5.000e-003

4.000e-003

3.000e-003

2.000e-003

1.000e-003

0.000e+000

Total Strain Energy Density

Figure 15.5. *Contour plots of the strain energy density (SED) distribution at the midfrontal plane for a healthy femur (left) and for a femur with hip implant (right) calculated from three-dimensional FE models. Red corresponds to high values, dark green to low values. Note that the SED values in the proximal region of the bone are significantly reduced after the placement of the implant since the implant will carry a part of the load ('stress shielding') (adapted from Huiskes et al [109]).*

[80–83], humeral joint arthroplasties [84–86], knee arthroplasty [87, 88] or the placement of fixation plates [89–91]. FE analysis has thus enabled a fast and relatively cheap pre-clinical evaluation of the effect of different implant design variables (geometry, stiffness, coating etc.) on the stresses in the surrounding or underlying bone.

For more information on this topic, the interested reader is directed to overview papers that focus on this topic [92–94].

It should be noted that in most of the studies mentioned previously, the purpose was to compare stresses in the healthy and operated situation or between different implant designs. For this purpose, the modelling of the bone itself is not very critical, and in most of these studies it was simply modelled as a linear elastic and isotropic continuum material with, in some of the studies, a Young's modulus that is dependent on the local density measured from radiographs or CT scans.

15.10. CLINICAL ASSESSMENT OF BONE MECHANICAL PROPERTIES

A clinical patient-specific application of FE analysis requires the use of automated meshing methods based on images obtained from modalities available in the clinic. Keyak *et al* [15] introduced the voxel conversion technique for the generation of FE models of the human femur in a fully automated way. The images used were obtained from a whole-body CT scanner, enabling clinical use of this approach. Although the authors initially found good agreement between calculated principal stresses and values reported in the literature, they later concluded that the prediction of femoral fracture load from FE analyses was about as good as a prediction based on the bone density data, such that the extra effort and radiation dose needed for this approach could not be justified at that time [95]. Using a similar approach, however, Cody *et al* [96] recently found that the femoral strength of bones is better predicted by FE analyses than by QCT or DXA measurements. The same authors also found that the short-term *in vivo* precision of proximal femur finite element fracture load prediction was about 6%, and concluded that this could be a suitable method for serial research studies on changes in femoral bone strength *in vivo* [97].

Using models generated from CT images, FE analysis for the assessment of bone fracture load was also applied to other bones, such as the skull [98] and vertebral bodies [99, 100]. As for the femur, it was concluded in these studies that CT-based FE analysis could be used successfully to predict both global and local failure behaviour of simplified skeletal structures.

15.11. SIMULATION OF MECHANICALLY INDUCED BIOLOGICAL PROCESSES

In the previous sections it was described how FE analysis has been used to analyse the state of mechanical parameters (stress, strain) in bone. A single FE analysis, however, is not suitable to predict the outcome of mechanically induced biological processes such as load adaptive bone remodelling or tissue differentiation. This is because bone and other tissues will change their mechanical properties during the process, leading to a redistribution of the mechanical parameters when the process evolves. A correct prediction of the outcome of such processes requires the formulation and solving of an initial value problem with a differential equation describing the rate of change of a parameter (e.g. bone density). The initial state represents the bone just after the orthopaedic intervention took place or at the start of the healing process. By solving the initial value problem, the changes in bone density or tissue material properties are obtained as a function of time. Solving methods for initial value problems include simple explicit (e.g. forward Euler) and more advanced (e.g. Runge–Kutta) methods. With such methods, the time scale is subdivided in a limited number of increments at which the solution is obtained. In this way, the evolvement of the mechanically induced adaptive process can be simulated and changes in the mechanical properties of the tissues can be predicted as a function of time. Since the actual solving of this initial value problem requires the evaluation of the boundary value problem (i.e. at least one FE analysis) at each time increment to obtain the state of the mechanical parameter, these analyses can be computationally expensive.

15.11.1. Bone remodelling

A consistent mathematical theory for load adaptive bone remodelling in accordance with 'Wolff's law' was developed by Cowin [101]. According to this theory, known as the theory of adaptive elasticity, bone has a characteristic equilibrium configuration represented by a particular shape and density distribution. It is assumed that for normal external loading conditions a typical equilibrium state of stress and strain in the bone is obtained. A deviation from this equilibrium state (for example after the placement of an implant) is the driving force for an adaptation in shape (external remodelling) or bone density (internal remodelling) back to the equilibrium state. The mathematical implementation of this theory results in differential equations of the form

$$\frac{\mathrm{d}\rho(\vec{x})}{\mathrm{d}t} = c(S(\vec{x}) - k) \tag{15.1}$$

where $\rho(\vec{x})$ is a scalar describing the external geometry (external remodelling) or the density of the bone (internal remodelling) as a function of the spatial coordinate $\vec{x}$, $S(\vec{x})$ is a local mechanical signal, k is a reference value representing the equilibrium state which can be a function of location ('site-specific' remodelling) or a constant ('non-site specific' remodelling) [102] and c is a rate constant. In agreement with Wolff's law, this equation will predict bone formation ($\mathrm{d}\rho/\mathrm{d}t > 0$) if the signal S exceeds the reference value k, and bone resorption if S is less than k. According to the theories of Frost [103] and Carter [104] it is assumed in many studies that a 'lazy zone' or 'dead zone' exist where $\mathrm{d}\rho/\mathrm{d}t = 0$, when S is in the neighbourhood of k. FE analysis is used to calculate the spatially distributed mechanical signal $S(\vec{x})$ and a simple time-stepping method (forward Euler) is usually implemented

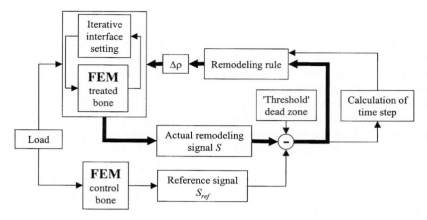

Figure 15.6. *Overview of a 'site-specific' bone remodelling adaptation model incorporating FE models for the calculation of the actual site-specific remodelling signal S in a model with implant and the reference signal S_{ref} in an intact bone. Depending on the difference between both signals, the density ρ and material properties in the FE model with implant are adapted according to the chosen remodelling rule. This process continues until a converged state is reached (adapted from van Rietbergen et al [110]).*

to solve the resulting initial value problem (figure 15.6). At present, however, there is no consensus with regard to the choice for the actual parameter that should be used as the mechanical signal $S(\vec{x})$ to which bone responds. In many studies it is assumed that $S(\vec{x})$ is the elastic strain energy density [81, 102, 105] sometimes normalized by the density, but other stress- and strain-based formulations have also been used. Prendergast *et al* [106] developed a formulation based on damage accumulation in the bone. In their formulation the deviation of the mechanical signal $S(\vec{x})$ from k is replaced by a time integral of the deviation of the rate of damage production from the rate of damage production at remodelling equilibrium.

Several authors have used bone remodelling simulation models to predict bone resorption patterns around orthopaedic implants. In the earlier of these studies, two-dimensional FE models were used. These studies have demonstrated the important effect of implant material properties (steel, titanium), fixation method (cemented, uncemented) and coating placement on the predicted bone resorption patterns [81, 107–109] (figure 15.7). Later, more realistic three-dimensional models were used which enabled the validation of the results of such simulation studies relative to animal experiments. The similarity between resorption patterns around hip implants seen after two years in animal experiments, and those predicted by the simulation model was excellent [82]. Using this validated model it was possible to accurately predict resorption patterns around other types of prostheses used in similar animal experiments [110].

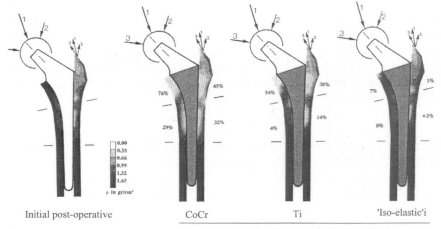

Initial post-operative   CoCr   Ti   'Iso-elastic'i

Bone loss predicted by the remodeling simulation

Figure 15.7. *Results of a bone remodelling simulation study. The grey levels represent the bone density in the initial post-operative situation (left) and the predictions by the simulation model for different stem materials (right). In agreement with clinical observations, the model predicts more bone loss for stiffer stems (CoCr) than for Ti or iso-elastic stems (adapted from Weinans et al [106]).*

Recently, Kerner *et al* [111] compared the amount of bone loss measured in patients with the amount of bone loss predicted by the simulation model using patient-specific FE models. Patterns of predicted bone loss corresponded very well with the DXA measurements. In agreement with DXA measurements, their model also predicted that the amount of bone loss is inversely correlated with the initial bone mineral content, and they concluded that this correlation could be explained by mechanically induced remodelling.

For further information on the topic of FE-based bone remodelling simulation, the reader is redirected to overview papers by Hart and Fritton [112], Huiskes and Hollister [94] and Prendergast [92].

15.11.2. *Tissue differentiation and fracture healing*

To date, most of the studies that investigated mechanically induced tissue differentiation and fracture healing analysed the mechanical state in the tissues only at a limited number of stages [76, 78]. These studies, discussed previously, have focused on understanding the relationships between mechanical loading and tissue differentiation rather than on simulating the actual process.

Based on results obtained from these studies, La Croix *et al* [113] recently developed a simulation model to predict the time-dependent fracture healing process in bone. In their study, an osteotomy of a long bone was analysed using an axisymmetric FE model in which all tissues were modelled as biphasic tissues. Material properties were updated every iteration depending on the strain state of the collagenous phase and fluid velocity of the interstitial fluid phase of the different tissue. It was found that this simulation model could explain the typical ossification patterns seen during the development of the fracture healing process.

SECTION D: FE ANALYSIS AT THE BONE TRABECULAR LEVEL: RECENT APPLICATIONS AND RESULTS

As discussed earlier, the voxel conversion technique has enabled the fully automated meshing of very complex structures such as the trabecular architecture. For such complex structures, however, the voxel conversion technique will result in the generation of FE models with a very large number of elements ($\sim 10^5$–10^6 elements per cm^3). With the development of purpose-built FE solvers that make use of the special features of FE models generated in this way (all elements have the same geometry, size and orientation) and the increase in computer performance, it recently became possible to solve such large FE models within reasonable limits of computation time [16]. In the following this FE approach at the trabecular level will be referred to as the micro-FE (µFE) approach.

Over the past decade, the µFE technique has mainly been applied to reach three distinct goals. First, to obtain an accurate and complete characterization of cancellous bone mechanical behaviour in relation to its morphology. Second, to quantify physiological bone tissue loading conditions. Third, to analyse cancellous bone mechanical properties *in vivo* in order to improve the prediction of bone fracture risk or to investigate the efficacy of drugs for the treatment of bone diseases. In this section, these applications and some recent results will be briefly discussed.

15.12. ANALYSIS OF BONE MECHANICAL PROPERTIES AND LOADING

15.12.1. *Elastic properties*

Micro-FE techniques have provided unique possibilities to obtain a complete characterization of the apparent anisotropic elastic properties of bone specimens as they relate to its morphology [114, 115]. This is an important advantage over experimental tests, from which, at best, the orthotropic elastic constants can be derived. Another advantage over real tests is the fact that, by specifying isotropic and homogenous tissue properties in the model, the µFE results are determined by the bone structure only, whereas with real tests the effect of bone tissue properties and structure on apparent level properties cannot be separated. Complete elastic properties have been determined for large sets of bone specimens [35, 116, 117]. The results obtained from these µFE analyses have led to a number of new conclusions with regard to the elastic behaviour of cancellous bone. First, it was found that the anisotropic behaviour of cancellous bone is the result of its micro-structure alone [42, 118]. This implies that the anisotropic elastic properties of a bone specimen can be calculated from µFE models with an 'effective' isotropic tissue modulus. In fact, the determination of a tissue Young's modulus by comparing experimental tests and µFE results has been an important purpose in itself in a number of studies [118, 119]. It is interesting to note that, when using accurate experiments, the values found with this approach are generally higher than those obtained from experiments on single trabeculae, and closely approach values common for cortical bone. Second, this anisotropic elastic behaviour of cancellous bone can be described as orthotropic [30]. Consequently, only the nine orthotropic elastic constants or the nine common engineering constants (three longitudinal moduli, three shear moduli and three Poisson's ratios) are needed to describe the elastic behaviour of cancellous bone. Third, it was found that there is a very strong stochastic relationship between these orthotropic elastic constants of cancellous bone with its fabric and density [35]. Using such relationships, it is possible to accurately predict the orthotropic elastic constants from measurements of bone fabric and density. Fourth, a strong correlation between bone anisotropic properties and volume fraction alone exists as well [30]. The predictive value of these relationships is similar to those based on density and fabric, but obviously, relationships based on density alone cannot predict the direction of the orthotropic symmetry axes. Finally, it was found that fabric and orthotropic principal directions of cancellous bone are closely related [120]. Although the coincidence of fabric and orthotropic principal axes as well as the orthotropic elastic behaviour have long been generally assumed, it was not possible to prove these assumptions earlier for cancellous bone.

15.12.2. *Strength and yield properties*

Whereas linear-elastic µFE analyses for the calculation of the elastic behaviour of cancellous bone has been the subject of many studies, there are only a few studies to date that used this technique to investigate the failure behaviour of cancellous bone. For the simulation of bone failure behaviour, where strain-dependent changes in material properties and large deformation can play an important role, a nonlinear µFE approach is needed. Consequently, such analyses are much more complex and computationally expensive than linear analyses. The results of the few studies that have used this approach have nevertheless shown that the simulation of bone failure mechanisms as a function of tissue level stresses and strains and of tissue level failure parameters is feasible [121, 122]. In a recent study, it was shown that such finite element analyses, implementing a bilinear constitutive model with asymmetric tissue yield strains in tension and compression, could capture the apparent strength behaviour of bovine bone specimens to an outstanding level of accuracy [123]. In that study, it was proposed that such computational models have reached a level of fidelity that qualifies them as surrogates for destructive mechanical testing of real specimens. Although it remains to be demonstrated that similar accurate results can be obtained for human (osteoporotic) bone, these results indicate that the accurate prediction of bone failure with these techniques is possible.

15.12.3. Assessment of physiological bone tissue loading

The ability to calculate stresses and strains at the bone tissue level is a unique feature of the μFE technique. In many studies, tissue stresses and strains were calculated for extracted bone specimens loaded in compression [115, 124, 125]. For the determination of *physiological* tissue loading conditions, however, the analyses of whole bones rather than bone test specimens is required, since natural loading conditions are only known at the whole-bone level. The computational demands for such analyses are huge: the analysis of whole human bones ($\sim$100 cm^3), with elements small enough to accurately represent trabeculae, requires μFE models with approximately 100 times more elements than those representing bone test specimens ($\sim$1 cm^3). The amount of required computer memory and time would increase by a similar factor. To reduce the required computational resources to an acceptable value, the first μFE study which aimed to analyse physiological tissue stresses and strains used a small canine proximal femur (considerably smaller than a human femur) [126]. In this study, the μFE model, which consisted of over seven million elements, was solved for external loading conditions representing the walking phase, and tissue level stresses and strains were calculated (figure 15.8).

   With the recent introduction of μCT scanners that can hold larger pieces of bone and the availability of large parallel supercomputers, the application of this technique for the analysis of human bones has become feasible. Van Rietbergen *et al* [127] recently solved μFE models representing the proximal part of human femurs, containing of the order of 100 million elements. By comparing the tissue strain distribution calculated in a μFE model of a healthy human femur with that calculated in an osteoporotic human femur, they were able to demonstrate that the tissue strain distribution in a healthy situation is more uniformly distributed than that in the osteoporotic situation. This result was, in fact, predicted long ago [128], but it is only with the introduction of these μFE techniques that it is possible to actually prove

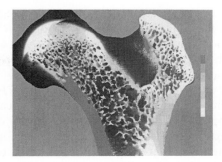

Figure 15.8. *Results of a μFE analysis of the proximal part of a dog femur. The μFE model, built of over seven million brick elements, can represent the trabecular architecture of the bone in great detail. Colours represent the calculated bone tissue strain energy density (SED). Red areas indicate high values whereas in the white areas the SED is close to zero (van Rietbergen et al [126]).*

such concepts. It should be noted, however, that solving such large μFE models is still a challenge, requiring several weeks even when using large parallel computer systems.

15.13. CLINICAL ASSESSMENT OF BONE MECHANICAL PROPERTIES

The application of the μFE technique to analyse bone *in vivo* is complicated by the limited resolution of the images that can be obtained from bone *in vivo*. Presently, only pQCT and MR imaging techniques can produce high-resolution images of bone *in vivo*, but the resolution that can be obtained (~150 μm) is less than is generally used for μFE models. In a number of validation studies, it was found that μFE analyses with models based on pQCT and MR images can provide accurate results when care is taken that elements in the resulting FE model are well connected to each other and that the load-carrying bone mass is preserved [49, 129, 130].

Using a pQCT scanner, Ulrich *et al* [131] were able to generate a μFE model of the distal radius of a young, healthy volunteer. After solving this model for external loading conditions representing a fall, they found that regions for which high bone tissue loading was calculated corresponded well with the regions in which typical Colles' type fractures occur. In a follow-up study, Pistoia *et al* used the same pQCT scanner and the same FE approach to analyse the distal radii of 58 cadaver arms [132]. After imaging, these cadaver arms were compressed in a test machine and the load at which the first signs of fracture occurred were measured. They found that a good correlation existed between the fracture load measured in experiments for each of the 58 cadaver arms, and a predicted fracture load that was based on the μFE calculated tissue loading ($R^2 = 0.76$). The fact that this correlation was much higher than the correlation between fracture load and bone density ($R^2 = 0.41$) indicates that such FE analyses might improve the prediction of bone fracture risk.

Recently, μFE analysis based on *in vivo* MR images has also been introduced in longitudinal clinical trials to investigate the efficacy of drugs for the treatment of osteoporosis [133]. In this study, a total of 56 patients was divided into groups that either received a new drug for the treatment of osteoporosis (idoxifene) or were untreated. High-resolution MR images of the calcaneus were made at baseline and after one year. FE analyses were made to analyse elastic properties of cubic volumes of interest at the centre of the calcaneus. The results of these FE analyses revealed significant changes in elastic properties between baseline and one-year results, whereas there were no significant changes in volume fraction or DXA measurements. It was therefore suggested that μFE analyses could potentially enhance the sensitivity of these longitudinal studies.

15.14. SIMULATION OF MECHANICALLY INDUCED BIOLOGICAL PROCESSES

Since mechanically induced biological processes in bone are regulated by cells, knowledge about the mechanical loading at the level of the bone cells could provide a better understanding of these processes. The introduction of μFE models of trabecular architecture has provided new tools to unravel the processes that result in bone adaptation and tissue differentiation seen at the whole-bone level.

15.14.1. *Bone remodelling*

As described earlier, bone remodelling simulation models, incorporating continuum FE models, can explain local bone mass adaptation as the result of local loading demands. Weinans *et al* [134], however, discovered that the density distribution predicted by these models in fact yields a discontinuous result, a result that was hidden in earlier studies due to the interpolation of the post-processing part of the FE codes. They proposed that the bone re-modelling process leads to a self-organizational control process that can produce discontinuous and even trabecular-like structures. In this and several later studies, it was shown that these discontinuous, trabecular like structures result only for certain conditions of the parameters used whereas other combinations of parameters would yield a smooth density distribution [135, 136].

The idea that trabeculae are formed due to the existence of a local biological control process of bone regulation was proposed over 100 years ago

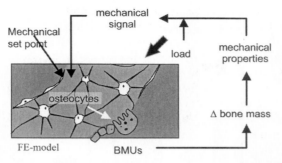

Figure 15.9. *Schematic representation of the hypothetical non-site-specific bone remodelling mechanism according to Mullender et al [137]. A finite element model representing trabecular structures is used to calculate the stresses and strains in the bone tissues. It is assumed that osteocyte cells in the tissue can sense this local mechanical loading. It is further assumed that osteocytes can activate basic multicellular units (BMUs), to add or remove bone. Changes in bone mass are implemented as changes in bone elastic properties in the FE model and the mechanical signal is re-calculated. This process continues until a converged state is reached.*

by Roux [128] and Wolff [137]. It is now generally assumed that mechano-sensor cells exist in the bone tissue that can evaluate the local mechanical loading in the bone tissue. It is further assumed that the mechanosensors will activate osteoblast cells to add bone in regions of high loading or activate osteoclast cells in regions of low loading, and thus produce a trabecular archi-tecture that is somehow optimized for the mechanical loading conditions.

Mullender *et al* [138] implemented such a hypothetical bone remodelling model combined with FE models for the calculation of tissue-level loading to investigate if it could explain morphogenesis and adaptation of trabecular bone architecture (figure 15.9). They proposed a regulatory mechanism in which osteocytes located within the bone sense mechanical signals and mediate osteoclasts and osteoblasts in their vicinity to adapt bone mass [139]. In this study, they demonstrated that this relatively simple regulatory model could produce trabecular-like structures aligned with external loading in a two-dimensional FE model. In later studies, they extended this model to three dimensions and demonstrated that the model could be used to investigate changes in trabecular architecture due to a disturbance in the regulatory mechanism [140] (figure 15.10).

In a recent study Huiskes *et al* [141] further refined this computational model and demonstrated that the emergence and maintenance of trabecular architecture as an optimal mechanical structure, as well as its adaptation to alternative external loads, can be explained by a computational model in which the coupling between osteoclast and osteoblast activity is governed by feedback from mechanical load transfer alone.

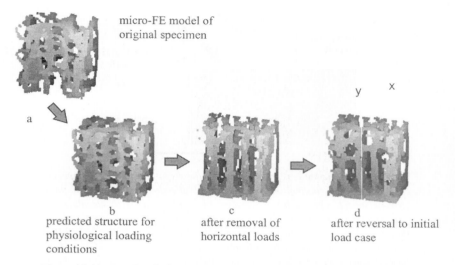

micro-FE model of
original specimen

a

y   x

b
predicted structure for
physiological loading
conditions

c
after removal of
horizontal loads

d
after reversal to initial
load case

Figure 15.10. *Results of a bone remodelling simulation study using μFE models
of trabecular bone. A physiological external load is applied to a μFE model
representing a bone specimen (a). After reaching a converged state (b) the
horizontal components of the external loads are removed, leading to loss of
horizontal trabeculae (c). After the initial loading situation is re-applied, the
lost trabeculae are not restored, but the remaining horizontal trabeculae get
thicker (adapted from Mullender et al [138]).*

15.15. SUMMARIZING CONCLUSION

Finite element analysis of structural solids is a flexible and versatile approach
for the quantitative assessment of bone mechanical properties and parameters.
Combined with high-resolution imaging techniques to derive bone geometry
and density distribution, it is now possible to mechanically analyse realistic
bone structures at different levels of structural organization, ranging from
the whole-bone down to the trabecular or even ultrastructural level. Computational resources now enable the analysis of very large finite element problems
with many millions of degrees of freedom as well as complicated geometrically
or materially nonlinear and poroelastic analyses. In combination with
(hypothetical) progression models, it can also be used to simulate the development of load-adaptive biological processes (such as bone remodelling and
tissue differentiation). As described in this chapter, such analyses and simulations have provided unique information about bone mechanical behaviour at
all levels of organization of bone. Presently, its clinical application is merely
limited by the imaging methods available. Using contemporary CT or MR
imaging techniques, it is now possible to create continuum finite element
models of whole bones *in-vivo* that can be used for strength calculation, pre-
operative planning or the prediction of patient specific bone remodelling

patterns. With the introduction of imaging devices with an improved resolution, it will soon be possible to create finite element models that account for the trabecular architecture as well, which will enable the assessment of changes in bone mechanical properties (e.g. strength) that are related to (subtle) structural changes. As discussed in this chapter, first efforts into this direction have been made in the laboratory with encouraging results.

REFERENCES

[1] Zienkiewicz O C and Morgan K 1989 *The Finite Element Method—Basic Concepts and Linear Applications* vol 1 4th edn (London: McGraw-Hill)

[2] Hughes T J R 1987 *The Finite Element Method: Linear Static and Dynamic Finite Element Analysis* (Englewood Cliffs, NJ: Prentice Hall)

[3] Nye J F 1957 *Physical Properties of Crystals: their Representation by Tensors and Matrices* (Oxford: Clarendon Press)

[4] Bunge H J 1982 *Texture Analysis in Material Science* (London: Butterworths)

[5] Cowin S C and Mehrabadi M 1987 On the identification of material symmetry for anisotropic elastic materials, *Q. J. Mech. Appl. Math.* **40** 451

[6] Cowin S C and Mehrabadi M 1989 Identification of the elastic symmetry of bone and other materials *J. Biomech.* **22** 503

[7] Cowin S C 1999 Bone poroelasticity *J. Biomech.* **32** 217–38

[8] Biot M A 1941 General theory of three-dimensional consolidation *J. Appl. Phys.* **12** 155–64

[9] Mow V C, Kuei S C, Lai W L and Armstrong C 1980 Biphasic creep and stress relaxation of articular cartilage in compression theory and experiments *J. Biomech. Eng.* **102** 73–84

[10] Bergmann G, Graichen F, Rohlmann A, Verdonschot N and van Lenthe G H 2001 Frictional heating of total hip implants. Part 2: finite element study *J. Biomech.* **34** 429–35

[11] Barnett S S, Smolinski P and Vorp D A 2000 A three-dimensional finite element analysis of heat transfer in the forearm *Comput. Meth. Biomech. Biomed. Eng.* **3** 287–96

[12] May T C, Krause W R, Preslar A J, Smith M J, Beaudoin A J and Cardea J A 1990 Use of high-energy shock waves for bone cement removal *J. Arthroplasty* **5** 19–27

[13] Huiskes R and van Rietbergen B 1995 Preclinical testing of total hip stems *Clin. Orthop.* **319** 64–76

[14] Odgaard A, Andersen K, Ullerup R, Frich L H and Melsen F 1994 Three-dimensional reconstruction of entire vertebral bodies *Bone* **15** 335–42

[15] Keyak J H, Meagher J M, Skinner H B and Mote C D Jr 1990 Automated three-dimensional finite element modelling of bone: a new method *J. Biomed Eng.* **12** 389–97

[16] van Rietbergen B, Weinans H, Polman B J W and Huiskes R 1996 Computational strategies for iterative solutions of large FEM applications employing voxel data, *Int. J. Num. Meth. Eng.* **39** 2743–67

[17] Jacobs C R, Mandell J A and Beaupre G S 1993 A comparative study of automatic finite element mesh generation techniques in orthopaedic biomechanics *ASME/BED* **24** 512–4

[18] Guldberg R E, Hollister S J and Charras G T 1998 The accuracy of digital image-based finite element models *J. Biomech Eng.* **120** 289–95

[19] Frey P, Sarter B and Gautherie M 1994 Fully automated mesh generation for 3-D domains based upon voxel sets, *Int. J. Num. Meth. Eng.* **37** 2735

[20] Müller R and Rüegsegger P 1995 Three-dimensional finite element modelling of non-invasively assessed trabecular bone structures *Med. Eng. Phys.* **17** 126

[21] Hibbit, Karlson & Sorensen Inc, Pawtucket, Rhode Island, USA

[22] Diana Analysis BV, Delft, The Netherlands

[23] Marc Analysis Research Corporation, Palo Alto, CA

[24] Voxelcon, Version 2.0, User's Manual, Psiphics Technologies Inc, Ann Arbor, Michigan, USA, 1997

[25] Scanco Medical AG, Bassersdorf, Switzerland

[26] Harrigan T P, Jasty M, Mann R W and Harris W H 1988 Limitations of the continuum assumption in cancellous bone *J. Biomech.* **21** 269–75

[27] Odgaard A 1997 Three-dimensional methods for quantification of cancellous bone architecture *Bone* **20** 315–28

[28] Abé H, Hayashi K and Sato M 1996 *Data Book on Mechanical Properties of Living Cells, Tissues, and Organs* (Tokyo: Springer)

[29] Choi K, Kuhn J L, Ciarelli M J and Goldstein S A 1990 The elastic moduli of human subchondral, trabecular, and cortical bone tissue and the size-dependency of cortical bone modulus *J. Biomech.* **23** 1103–13

[30] Yang G, Kabel J, van Rietbergen B, Odgaard A, Huiskes R and Cowin S C 1999 The anisotropic Hooke's law for cancellous bone and wood *J. Elasticit.* **53** 125

[31] Carter D R and Hayes W C 1977 The compressive behaviour of bone as a two-phase structure *J. Bone Joint. Surg. Am.* **59** 954

[32] Rice J C, Cowin S C and Bowman J A 1988 On the dependence of the elasticity and strength of cancellous bone on apparent density *J. Biomech.* **21** 155

[33] Laib A and Ruegsegger P 1999 Calibration of trabecular bone structure measurements of in vivo three-dimensional peripheral quantitative computed tomography with 28-micron-resolution microcomputed tomography *Bone* **24** 35–9

[34] Lang T, Augat P, Majumdar S, Ouyang X and Genant H K 1998 Noninvasive assessment of bone density and structure using computed tomography and magnetic resonance *Bone* **22** S149–53

[35] Kabel J, van Rietbergen B, Odgaard A and Huiskes R 1999 The constitutive relationships of fabric, density and elastic properties in cancellous bone architecture *Bone* **25** 481

[36] Keyak J H, Fourkas M G, Meagher J M and Skinner H B 1993 Validation of an automated method of three-dimensional finite element modelling of bone *J. Biomed. Eng.* **15** 505–9

[37] Bergmann G, Graichen F and Rohlmann A 1993 Hip joint loading during walking and running, measured in two patients *J. Biomech.* **26** 969–90

[38] Rohlmann A, Graichen F, Weber U and Bergmann G 2000 Monitoring in vivo implant loads with a telemeterized internal spinal fixation device *Spine* **25** 2981–6

[39] Rho J Y, Ashman R B and Turner C H 1993 Young's modulus of trabecular and cortical bone material: ultrasound and microtensile measurements *J. Biomech.* **26** 111

[40] Choi K and Goldstein S A 1992 A comparison of the fatigue behaviour of human trabecular and cortical bone tissue *J. Biomech.* **25** 1371

[41] Ashman R B and Rho J Y 1998 Elastic modulus of trabecular bone material *J. Biomech.* **21** 177

[42] Kabel J, van Rietbergen B, Dalstra M, Odgaard A and Huiskes R 1999 The role of an effective isotropic tissue modulus in the elastic properties of cancellous bone *J. Biomech.* **32** 673–80

[43] Odgaard A, Andersen K, Melsen F and Gundersen H J G 1990 A direct method for fast three-dimensional serial reconstruction *J. Microscopy* **159** 335–42

[44] Beck J D, Canfield B L, Haddock S M, Chen T J, Kothari M and Keaveny T M 1997 Three-dimensional imaging of trabecular bone using the computer numerically controlled milling technique *Bone* **21** 281

[45] Feldkamp L A, Goldstein S A, Parfitt A M, Jesion G and Kleerekoper M 1989 The direct examination of three-dimensional bone architecture in vitro by computed tomography *J. Bone Min. Res.* **4** 3

[46] Ruegsegger P, Koller B and Muller R 1996 A microtomographic system for the nondestructive evaluation of bone architecture *Calcif. Tissue Intl.* **58** 24

[47] Bonse U, Busch F, Gunnewig O, Beckmann F, Pahl R, Delling G, Hahn M and Graeff W 1994 3D computed X-ray tomography of human cancellous bone at 8 microns spatial and 10(-4) energy resolution *Bone Min.* **25** 25

[48] Hipp J A, Jansujwicz A, Simmons C A and Snyder B D 1996 Trabecular bone morphology from micro-magnetic resonance imaging *J. Bone Min. Res.* **11** 286

[49] Ulrich D, van Rietbergen B, Weinans H and Rüegsegger P 1998 Finite element analysis of trabecular bone structure: a comparison of image-based meshing techniques *J. Biomech.* **31** 1187–92

[50] Gong J K, Arnold J S and Cohn S H 1964 Composition of trabecular and cortical bone *Anat. Rec.* **149** 325–32

[51] Marotti G 1996 The structure of bone tissues and the cellular control of their deposition *Ital. J. Anat. Embryol.* **101** 25–79

[52] Peyrin F, Salome M, Nuzzo S, Cloetens P, Laval-Jeantet A M and Baruchel J 2000 Perspectives in three-dimensional analysis of bone samples using synchrotron radiation microtomography *Cell. Mol. Biol.* **46** 1089–102

[53] Rho J Y, Roy M E, Tsui T Y and Pharr G M 1997 Young's modulus and hardness of trabecular and cortical bone in the various directions determined by nano-indentation *Proc. 43rd Ann. Meeting ORS* **43** 811

[54] Katz J L and Ukraincik K 1971 On the anisotropic elastic properties of hydroxy-apatite *J. Biomech.* **4** 221–7

[55] Gardner T N, Elliott J C, Sklar Z and Briggs G A 1992 Acoustic microscope study of the elastic properties of fluorapatite and hydroxyapatite, tooth enamel and bone *J. Biomech.* **25** 1265–77

[56] Hasegawa K, Turner C H and Burr D B 1994 Contribution of collagen and mineral to the elastic anisotropy of bone *Calcif. Tissue Intl.* **55** 381–6

[57] Weinbaum S, Cowin S C and Zeng Y 1994 A model for the excitation of osteocytes by mechanical loading-induced bone fluid shear stresses *J. Biomech.* **27** 339–60

[58] Wang L, Fritton S P, Cowin S C and Weinbaum S 1999 Fluid pressure relaxation depends upon osteonal microstructure: modelling an oscillatory bending experiment *J. Biomech.* **32** 663–72

[59] Steck R, Niederer P and Knothe Tate M L 2000 A finite difference model of load-induced fluid displacements within bone under mechanical loading *Med. Eng. Phys.* **22** 117–25

[60] Mak A F and Zhang J D 2001 Numerical simulation of streaming potentials due to deformation-induced hierarchical flows in cortical bone *J. Biomech Eng.* **123** 66–70

[61] Wagner H D and Weiner S 1992 On the relationship between the microstructure of bone and its mechanical stiffness *J. Biomech.* **25** 1311–20

[62] McBroom R J, Cheal E J and Hayes W C 1988 Strength reductions from metastatic defects in long bones *J. Orthop. Res.* **6** 369–78

[63] Hipp J A, McBroom R J, Cheal E J and Hayes W C 1989 Structural consequences of endosteal metatstatic lesions in long bones *J. Orthop. Res.* **7** 828–37

[64] Cheal E J, Hipp J A and Hayes W C 1993 Evaluation of finite element analysis for the prediction of the strength reduction due to metastatic lesions in the femoral neck *J. Biomech.* **26** 251–64

[65] Lotz J C, Cheal E J and Hayes W C 1991 Fracture prediction for the proximal femur using finite element models. Part I—Linear analysis *J. Biomech. Eng.* **113** 353–60

[66] Lotz J C, Cheal E J and Hayes W C 1991 Fracture prediction for the proximal femur using finite element models. Part II—Nonlinear analysis, *J. Biomech. Eng.* **113** 361–5

[67] Lotz J C, Cheal E J and Hayes W C 1995 Stress distribution within the proximal femur during gait and falls: implications for osteoporotic fracture *Osteoporos. Int.* **5** 252–61

[68] Ford C M, Keaveny T M and Hayes W C 1996 The effect of impact direction on the structural capacity of the proximal femur during falls *J. Bone Min. Res.* **11** 377–83

[69] Pinilla T P, Boardman K C, Bouxsein M L, Myers E R and Hayes W C 1996 Impact direction from a fall influences the failure load of the proximal femur as much as age-related bone loss *Calcif. Tissue Int.* **58** 231–5

[70] Pauwels F 1980 *Biomechanics of Locomotion Apparatus* (Berlin: Springer)

[71] Carter D R, Blenman P R and Beaupre G S 1988 Correlations between mechanical stress history and tissue differentiation in initial fracture healing *J. Orthop. Res.* **6** 736–48

[72] DiGioia A M, Cheal E J and Hayes W C 1986 Three dimensional strain fields in a uniform osteotomy gap *J. Biomech. Eng.* **108** 273–80

[73] Meroi E A and Natali A N 1989 A numerical approach to the biomechanica; analysis of bone fracture healing *J. Biomed. Eng.* **11** 390–7

[74] Cheal E J, Mansmann K A, DiGioia A M, Hayes W C and Perren S M 1991 Role of interfragmentary strain in fracture healing: ovine model of a healing osteotomy *J. Orthop. Res.* **9** 131–42

[75] Blenman P R, Carter D R and Beaupre G S 1989 Role of mechanical loading in the progressive ossification of a fracture callus *J. Orthop. Res.* **7** 398

[76] Claes L E, Heigele C A, Neidlinger-Wilke C, Kaspar D, Seidl W, Margevicus K J and Augat P 1998 Effects of mechanical factors on the fracture healing process *Clin. Orthop.* **355** (Suppl) S132–47

[77] Claes L E and Heigele C A 1999 Magnitudes of local stress and strain along bony surfaces predict the course and type of fracture healing *J. Biomech.* **32** 255–66

[78] Prendergast P J, Huiskes R and Soballe K 1997 Biophysical stimuli on cells during tissue differentiation at implant interfaces *J. Biomech.* **30** 539–48

[79] Tanck E, Blankevoort L, Haaijman A, Burger E H and Huiskes R 2000 Influence of muscular activity on local mineralization patterns in metatarsals of the embryonic mouse *J. Orthop. Res.* **18** 613–9

[80] Cook S D, Skinner H B, Weinstein A M and Haddad R J 1982 Stress distribution in the proximal femur after surface replacement effects of prosthesis and surgical techniques *Biomater. Med. Devices Artif. Organ* **10** 85–102

[81] Huiskes R, Weinans H, Grootenboer H J, Dalstra M, Fudala B and Slooff T J 1987 Adaptive bone-remodelling theory applied to prosthetic-design analysis *J. Biomech.* **20** 1135–50

[82] Weinans H, Huiskes R, van Rietbergen B, Sumner D R, Turner T M and Galante J O 1993 Adaptive bone remodelling around bonded noncemented total hip arthroplasty: a comparison between animal experiments and computer simulation *J. Orthop. Res.* **11** 500–13

[83] Skinner H B, Kim A S, Keyak J H and Mote C D 1994 Femoral prosthesis implantation induces changes in bone stress that depend on the extent of porous coating *J. Orthop. Res.* **12** 553–63

[84] Orr T E and Carter D R 1985 Stress analyses of joint arthroplasty in the proximal humerus *J. Orthop. Res.* **3** 360–71

[85] Pressel T, Lengsfeld M, Leppek R and Schmitt J 2000 Bone remodelling in humeral arthroplasty: follow-up using CT imaging and finite element modelling—an in vivo case study *Arch. Orthop. Trauma Surg.* **120** 333–5

[86] Stone K D, Grabowski J J, Cofield R H, Morrey B F and An K N 1999 Stress analyses of glenoid components in total shoulder arthroplasty *J. Shoulder Elbow Surg.* **8** 151–8

[87] Vasu R, Carter D R, Schurman D J and Beaupre G S 1986 Epiphyseal-based designs for tibial plateau components—I. Stress analysis in the frontal plane *J. Biomech.* **19** 647–62

[88] Tissakht M, Ahmed A M and Chan K C 1996 Calculated stress-shielding in the distal femur after total knee replacement corresponds to the reported location of bone loss *J. Orthop. Res.* **14** 778–85

[89] Woo S L, Simon B R, Akeson W H, Gomez M A and Seguchi Y 1983 A new approach to the design of internal fixation plates *J. Biomed. Mater. Res.* **17** 427–39

[90] Woo S L, Lothringer K S, Akeson W H, Coutts R D, Woo Y K Simon, B R and Gomez M A 1984 Less rigid internal fixation plates: historical perspectives and new concepts *J. Orthop. Res.* **1** 431–49

[91] Beaupre G S, Carter D R, Orr T E and Csongradi J 1988 Stresses in plated long-bones: the role of screw tightness and interface slipping *J. Orthop. Res.* **6** 39–50

[92] Prendergast P J 1997 Finite element models in tissue mechanics and orthopaedic implant design *Clin. Biomech.* **12** 43–366

[93] Huiskes R and Chao E Y 1983 A survey of finite element analysis in orthopedic biomechanics: the first decade *J. Biomech.* **16** 385–409

[94] Huiskes R and Hollister S J 1993 From structure to process, from organ to cell: recent developments of FE-analysis in orthopaedic biomechanics *J. Biomech. Eng.* **115** 520–7

[95] Keyak J H, Ross, S A, Jones K A and Skinner H B 1998 Prediction of femoral fracture load using automated finite element modelling *J. Biomech.* **31** 125–33

[96] Cody D D, Gross G J, Hou F J, Spencer H J, Goldstein S A and Fyhrie D P 1999 Femoral strength is better predicted by finite element models than QCT and DXA *J. Biomech.* **32** 1013–20

[97] Cody D D, Hou F J, Divine G W and Fyhrie D P 2000 Short term in vivo precision of proximal femoral finite element modelling *Ann. Biomed Eng.* **28** 408–14

[98] Bandak F A, Vander Vorst M J, Stuhmiller L M, Mlakar P F, Chilton W E and Stuhmiller J H 1995 An imaging-based computational and experimental study of skull fracture: finite element development *J. Neurotrauma*, **12** 679–88

[99] Bozic K J, Keyak J H, Skinne, H B, Bueff H U and Bradford D S 1994 Three-dimensional finite element modelling of a cervical vertebrae: an investigation of burst fracture mechanism *J. Spinal Disord.* **7** 102–10

[100] Silva M J, Keaveny T M and Hayes W C 1998 Computed tomography-based finite element analysis predicts failure loads and fracture patterns for vertebral sections *J. Orthop. Res.* **16** 300–8

[101] Cowin S C and Hegedus D H 1976 Bone remodelling I: A theory of adaptive elasticity *J. Elasticity* **6** 313–26

[102] Beaupré G S, Orr T E and Carter D R 1990 An approach for time-dependent bone modelling and remodelling—theoretical development *J. Orthop. Res.* **8** 651–61

[103] Frost H M 1983 A determinant of bone architecture: the minimum effective strain *Clin. Orth. Rel. Res.* **175** 286–92

[104] Carter D R 1984 Mechanical loading histories and cortical bone remodelling *Calcif. Tissue Int.* **36** S19–24

[105] Carter D R 1987 Mechanical loading history and skeletal biology *J. Biomech.* **20** 1095–109

[106] Prendergast P J and Taylor D 1994 Prediction of bone adaptation using damage accumulation *J. Biomech.* **27** 1067–76

[107] Weinans H, Huiskes R and Grootenboer H J 1992 Effects of material properties of femoral hip components on bone remodelling *J. Orthop. Res.* **10** 845–53

[108] Orr T E, Beaupre G S, Carter, D R and Schurman D J 1990 Computer predictions of bone remodelling around porous-coated implants *J. Arthroplasty* **5** 191–200

[109] Huiskes R 1995 Bone remodelling around implants can be explained as an effect of mechanical adaptation. In *Total Hip Revision Surgery* eds J O Galante, A G Rosenberg and J J Callaghan (New York: Raven Press)

[110] van Rietbergen B, Huiskes R, Weinans H, Sumner D R, Turner T M and Galante J O 1993 ESB Research Award 1992. The mechanism of bone remodelling and resorption around press-fitted THA stems *J. Biomech.* **26** 369–82

[111] Kerner J, Huiskes R, van Lenthe G H, Weinans H, van Rietbergen B, Engh C A and Amis A A 1999 Correlation between pre-operative periprosthetic bone density and post-operative bone loss in THA can be explained by strain-adaptive remodelling *J. Biomech.* **32** 695–703

[112] Hart R T and Fritton S P 1997 Introduction to finite element based simulation of functional adaptation of cancellous bone *Forma* **12** 277–99

[113] Lacroix D and Prendergast P J 2002 A mechano-regulation mode for tissue differentiation during fracture healing: analysis of gap size and loading *J. Biomech.* **35**(9) 1163–71

[114] Hollister S J, Brennan J M and Kikuchi N 1994 A homogenization sampling procedure for calculating trabecular bone effective stiffness and tissue level stress *J. Biomech.* **27** 433

[115] van Rietbergen B, Odgaard A, Kabel J and Huiskes R 1996 Direct mechanics assessment of mechanical symmetries and properties of trabecular bone architecture *J. Biomech.* **29** 1653–7

[116] Ulrich D, van Rietbergen B, Laib A and Rüegsegger P 1999 The ability of three-dimensional structural indices to reflect mechanical aspects of trabecular bone *Bone* **25** 55–60

[117] van Rietbergen B, Ulrich D, Kabel J and Rüegsegger P 1998 Estimates in normal versus osteoportic bone *Proc. 14th Europ. Soc. Biomat.*

[118] van Rietbergen B, Kabel J, Odgaard A and Huiskes R 1997 Determination of trabecular bone tissue elastic properties by comparison of experimental and finite element results. In *Material Identification Using Mixed Numerical Experimental Methods* eds H Sol and C W J Oomens (Darmstadt: Kluwer) pp 183–92

[119] Jacobs C R, Davis B R, Rieger C J, Francis J J, Saad M and Fyhrie D P 1999 The impact of boundary conditions and mesh size on the accuracy of cancellous bone tissue modulus determination using large-scale finite-element modelling *J. Biomech.* **32** 1159–64

[120] Odgaard A, Kabel J, van Rietbergen B, Dalstra M and Huiskes R 1997 Fabric and elastic principal directions of cancellous bone are closely related *J. Biomech.* **30** 487–95

[121] van Rietbergen B, Ulrich D, Pistoia W, Huiskes R and Rüegsegger P 2000 Prediction of trabecular bone failure parameters using a tissue failure criterion *Comp. Sim.* **1**(2) 98–101

[122] Niebur G L, Yuen J C and Keaveny T M 2000 Role of hard tissue quality in trabecular bone failure mechanisms *Trans. 46th ORS* 11

[123] Niebur G L, Feldstein M J, Yuen J C, Chen T J and Keaveny T M 2000 High-resolution finite element models with tissue strength asymmetry accurately predict failure of trabecular bone *J. Biomech.* **33** 1575–83

[124] Hollister S J and Kikuchi N 1992 Direct analysis of trabecular bone stiffness and tissue level mechanics using an element-by-element homogenization method *Trans. 38th A Meeting Orthop. Res. Soc.* 559

[125] Fyhrie D P and Hamid M S 1993 The probability distribution of trabecular level strains for vertebral cancellous bone *Trans. 39th A Meeting Orthop. Res. Soc.* 175

[126] van Rietbergen B, Müller R, Ulrich D, Rüegsegger P and Huiskes R 1999 Tissue stresses and strain in trabeculae of a canine proximal femur can be quantified from computer reconstructions *J. Biomech.* **32** 443–51

[127] van Rietbergen B, Huiskes R, Eckstein F and Rüegsegger P 2003 Trabecular bone tissue strains in the healthy and osteoporotic human femur *JBMR* **18**(10)

[128] Roux W 1881 *Der Kampf der Theile im Organismus* (Leipzig: Engelmann)

[129] Pistoia W, van Rietbergen B, Laib A and Rüegsegger P 2001 High-resolution three-dimensional-pQCT image can be an adequate basis for *in-vivo* micro-FE analysis of bone *J. Biomech. Eng.* **123**(2) 176–83

[130] van Rietbergen B, Majumdar W, Pistoia D C, Newitt M, Kothari A, Laib A and Rüegsegger P 1998 Assessment of cancellous bone mechanical properties from micro-FE models based on micro-CT, pQCT and MR images *Techn. Health Care* **6** 413–20

[131] Ulrich D, van Rietbergen B, Laib A and Rüegsegger P 1999 Load transfer analysis of the distal radius from in-vivo high-resolution CT-imaging *J. Biomech.* **32** 821–8

[132] Pistoia W, van Rietbergen B, Lochmüller E M, Lill C, Eckstein F and Rüegsegger P 2002 Estimation of distal radius failure load with micro-finite element analysis models based on three-dimensional peripheral quantitative computed tomography images *Bone* **30**(6) 842–8

[133] van Rietbergen B, Majumdar S, Newitt D and McDonald B 2002 High-resolution MRI and micro-FE for the evaluation of changes in bone mechanical properties during longitudinal clinical trials: application to calcaneal bone in postmenopausal women after one year of idoxifene treatment *Clin. Biomech.* **17**(2) 81–8

[134] Weinans H, Huiskes R and Grootenboer H J 1992 The behaviour of adaptive bone-remodelling simulation models *J. Biomech.* **25** 1425–41

[135] Harrigan T P and Hamilton J J 1992 An analytical and numerical study of the stability of bone remodelling theories: dependence on microstructural stimulus *J. Biomech.* **25** 477–88.

[136] Cowin S C, Luo G M, Sadegh A M and Harrigan T P 1994 On the sufficiency conditions for the stability of bone remodelling equilibrium *J. Biomech.* **27** 183–6

[137] Wolff J 1892 *Das Gesetz der Transformation der Knochen* (Berlin: A Hirchwild). 1986 Translated as *The Law of Bone Remodelling* by P Maquet and R Furlong (Berlin: Springer)

[138] Mullender M G and Huiskes R 1995 Proposal for the regulatory mechanism of Wolff's law *J. Orthop. Res.* **13** 503–12

[139] Cowin S C, Moss-Salentijn L and Moss L 1991 Candidates for the mechanosensory system in bone *J. Biomech. Eng.* **113** 191–7

[140] Mullender M, van Rietbergen B, Ruegsegger P and Huiskes R 1998 Effect of mechanical set point of bone cells on mechanical control of trabecular bone architecture *Bone* **22** 125–31

[141] Huiskes R, Ruimerman R, van Lenthe G H and Janssen J D 2000 Effects of mechanical forces on maintenance and adaptation of form in trabecular bone *Nature* **405** 704–6

[142] Robinovitch S N, Hayes W C and McMahon T A 1991 Prediction of femoral impact forces in falls on the hip *J. Biomech. Eng.* **113** 366–74

[143] Chiu J and Robinovitch S N 1998 Prediction of upper extremity impact forces during falls on the outstretched hand *J. Biomech.* **31** 1169–76

[144] Whitehouse W J 1974 The quantitative morphology of anisotropic trabecular bone *J. Microsc.* **101** 153–68

[145] Harrigan T P and Mann R W 1984 Characterization of microstructural anisotropy in orthotropic materials using a second rank tensor *J. Mater. Sci.* **19** 761

[146] Lucchinetti 2001 Composite models of bone properties *Bone Mechanics Handbook* 2nd edn, ed S C Cowin (New York: CRC)

[147] Lucchinetti 2001 Dense bone tissues as a molecular composite *Bone Mechanics Handbook* 2nd edn, ed S C Cowin (New York: CRC)

Chapter 16

Vibration analysis

James L Cunningham

SECTION A: INTRODUCTION

The study of vibration and vibrational waves is a major topic in the physical sciences as the effects of vibration often have an undesirable or unpleasant effect on both structures and individuals [1, 2]. It is important to know how structures will respond to vibration as this will have consequences for the design of the structure, an obvious example of this being in the design of buildings in earthquake prone areas.

Vibration can also be used as a method of non-destructive testing and this may be applied to the study of biological structures such as bones, the discussion of which will be the aim of this chapter. However, before commencing directly on this, it is perhaps worth considering some applications of vibration in engineering to set this in context.

16.1. CONDITION MONITORING OF MACHINERY

Rotating machines produce vibration during operation and by monitoring the vibration their running condition can be assessed. The simplest analysis technique is to measure the overall level of vibration and relate this to how roughly or smoothly the machine is running. A more sensitive analysis can be obtained by a spectral analysis of the vibration signal produced by the machine [3]. This can be used to identify particular frequencies which generate the largest signals, and hence high vibration levels, which can then be compared with the signals which could be expected to arise from various parts of the machine. Signal averaging can also be used in which the vibration around a particular frequency (corresponding to a particular component) is monitored, changes in vibration behaviour at this frequency then being related to potential problems with that component [3].

16.2.   MODAL ANALYSIS

Modal analysis is widely used in engineering and provides an understanding of the structural characteristics, operating conditions and performance of structures subject to vibration. Modal analysis determines the fundamental vibration mode shapes and the frequencies at which these occur. This can involve a relatively straightforward analysis for basic components of a simple system or an extremely complicated analysis when examining a complex device or structure exposed to time varying loading. Modal analysis is widely used in the design of many types of structure, including automotive structures, aircraft structures, spacecraft and of sports equipment such as hockey sticks, golf clubs and tennis racquets.

16.3.   NON-DESTRUCTIVE TESTING

Vibration analysis has been used in non-destructive testing (NDT) for many years; indeed one of the oldest non-destructive tests involved hitting the component to be tested (e.g. railway wheels) with a hammer and listening to the sound produced, a flawed component producing a dull tone which diminishes rapidly in intensity [4]. Using modern instrumentation, the accuracy of this basic approach can be greatly enhanced and both changes in resonant frequency and damping have been used as methods for NDT of engineering components. A recent example of the use of vibration is in the NDT of composite materials, for example, in determining whether de-bonding has occurred between joined composite structures (e.g. the host structure and the composite repair patch) [5].

16.3.1.   *Transverse (flexural) vibration methodology*

The natural frequency at which a component will vibrate is a function of the shape of the component, the material from which it is constructed and the support (or boundary) conditions. For specimens of a simple shape, it is possible to derive a relationship between these various parameters and the frequency of the different modes of vibration. For example, for a beam undergoing free (i.e. unconstrained) transverse vibration, the natural frequencies are given by

$$f = \frac{\alpha}{L^2} \sqrt{\frac{EI}{A\rho}} \qquad (16.1)$$

where f is the frequency in Hz, α is the constant which depends on the support conditions and mode number, L is the length of beam, E is Young's modulus, I is the second moment of area, A is the cross-sectional area, and ρ is the density.

If the length of the beam (L) is known and the support conditions can be controlled, the natural frequency of vibration is seen to be a function of the stiffness (EI) and the mass/unit length $(A\rho)$ of the beam. Hence, through measurements of resonant frequency, it is possible to infer the material and structural properties of the component under test.

16.4. VIBRATIONAL MEASUREMENTS APPLIED TO BONE

As noted above, vibration testing has been used for many years in engineering applications, but its potential in medical diagnostics is only now beginning to emerge. The unique relationship between resonant frequencies and material properties has led to the use of vibration being used as a method of non-invasively determining the stiffness of long bones to assess the effect that osteoporosis may have on the mechanical properties [6]. It has also been used to diagnose the occurrence of a bone fracture, determine the rate of healing of fractures and detect when a prosthetic hip has loosened [7, 8].

Each of these applications of vibration to the testing of bone will be discussed in turn with reference to previously published work, and a summary regarding the potential usefulness of the application of vibration analysis to each of these areas will be given.

SECTION B: MATERIAL PROPERTIES OF WHOLE LONG BONES

16.5. FREQUENCY RESPONSE MEASUREMENTS

As suggested above, the response of a bone to a vibrational input can be used to infer the material and structural properties of a component. This has obvious advantages when applied to a bone *in vivo*, as it offers the possibility of a simple, non-invasive method of assessing bone quality without the use of ionizing radiation. A number of techniques have been developed to do this and these are described below, including, where these have been measured, the results obtained *in vivo*.

16.5.1. Early studies

Jurist [9] presented a theoretical basis and a measurement method for determining the resonant frequency of the ulna *in vivo*. Vibration was applied

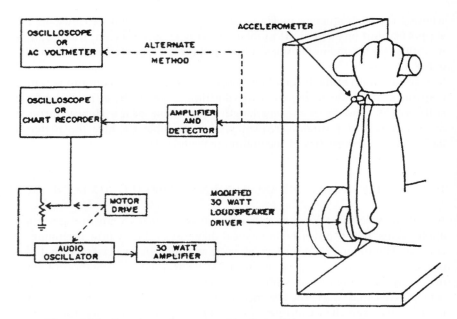

Figure 16.1. *Experimental set-up used by Jurist [9] to measure the resonant frequency of the ulna.*

to the olecranon process through a modified loudspeaker driver and the response was measured at the distal end of the ulna by an accelerometer (figure 16.1). The arm of the subject was supported vertically in a specially designed frame which aimed to hold the boundary conditions constant. Resonant frequencies were found by scanning through the frequency range from 200 to 1000 Hz, and recording the frequency response. The product fL, where f is the resonant frequency and L is the length, was calculated in order to obtain a quantity proportional to $\sqrt{E/\rho}$, where E is Young's modulus and ρ is the average density, by assuming the relationship $fL = K\sqrt{E/\rho}$, where K is a constant, for a vibrating bar. (Note that according to equation (16.1), a more appropriate expression might be $fL^2 = K\sqrt{E/\rho}$.) For the lowest resonant frequency ($\sim$250 Hz) the standard deviation (SD) of repeated measurements was quoted as 2–3%. Using this technique Jurist then went on to measure the resonant frequency of the ulna in normal, osteoporotic and diabetic subjects [10]. The product fL was shown to decrease with age. For asymptomatic women over 45, a bimodal distribution was found suggesting that approximately 35% showed reduced values of this product, indicating early development of osteoporosis. For women with symptomatic osteoporosis, fL values were approximately 44% below those of age-matched controls, and for the diabetic subjects a reduction of approximately 25% was found.

Doherty *et al* [11] carried out measurements on three excised tibiae (two normal and one osteoporotic) using a forced steady-state vibration in which both the acceleration and the force applied to the bone at the point of excitation were measured. From the results obtained, they suggested that both stiffness (flexural rigidity) and dynamic mass are more sensitive to changes in the physical state of the bone than is the resonant frequency.

Subsequent studies have standardized on three main techniques. These all excite vibration in the bone to be measured. Two of the techniques, impulse frequency response (IFR) and bone resonance analysis (BRA), determine the resonant frequencies over a frequency range of approximately 50–1000 Hz, the third, mechanical response tissue analysis (MRTA), computes the bending stiffness of the bone under test from mechanical impedance measurements in the frequency range up to 1500 Hz. These techniques will now be described and the results obtained using them discussed.

16.5.2. Impulse frequency response (IFR) technique

Vandecasteele *et al* [12] used modal analysis to identify the modes of vibration of isolated tibiae (both dry and fresh). Excitation was provided via a hammer impact and the resulting vibration was measured at different positions along the tibia with an accelerometer. This technique is known as the IFR technique. Both the input signal from the hammer impact $F(t)$ and the measured acceleration $A(t)$ were determined as functions of time and, following Fourier transformation, as functions of frequency, $F(f)$ and $A(f)$. The transfer function, $A(f)/F(f)$, was then plotted as a function of frequency, discrete peaks in this plot occurring at the natural frequencies of the measured bone. The magnitude of the transfer function at a specific resonant frequency and location are then used to generate the mode shape at that frequency (figure 16.4). For the dry tibia, three modes were found, the first two corresponding to single bending about the two principal axes (I_1 and I_2). The first mode, which was found in a plane normal to the medial face (min. I) occurred at a frequency between 430 and 700 Hz depending on the condition of the tibia (dry or fresh). The second, in a plane parallel to the medial face, occurred at a frequency between 640 and 930 Hz. The third mode was identified at ± 780 Hz and corresponded to double bending in the plane of minimum stiffness.

The IFR technique was subsequently used to obtain *in vivo* measurements on the tibiae of 20 males aged 23 ± 2 years [12]. For these measurements, it was noted that the frequency peaks in the plot of transfer function against frequency were very diffuse in comparison with those obtained on dry tibiae and they attributed this to pronounced damping of the soft tissues. This caused some problems in the exact identification of resonant frequencies and mode shapes. The low resonances (<300 Hz) were interpreted as rigid body modes and the high resonances (>300 Hz) as single bending modes.

16.5.3.   Bone resonance analysis (BRA) technique

A driving point impedance technique (often referred to as bone resonance analysis or BRA), similar to that described by Young *et al* [13], was used by Borgwardt Christensen *et al* [14] to determine the lowest resonant frequency in the human tibia. The technique consisted of an electromechanical shaker on which an impedance head had been mounted. For the *in vivo* studies the subject lay in a supine position and the limb was supported in a splint (figure 16.2). The impedance head was placed on the anterior tibial margin and the excitation frequency was scanned from 30 to 1000 Hz. The frequency of the lowest resonant frequency was obtained by tuning the input signal to the maximum velocity response from the impedance head.

Evans *et al* [15], using a similar technique but using white noise excitation, determined the frequency response up to 6 kHz at various points along the length and circumference of two isolated human ulnas under free–free vibration conditions. Similar patterns of multiple resonant frequencies were found for each bone, two fundamental modes occurring in mutually perpendicular planes and higher resonances appearing at higher frequencies in the same planes as the fundamental frequencies. Marked coupling between adjacent resonances was observed which could vary the observed resonant frequency by up to 10%.

Collier and Donarski [16] presented a similar method to BRA which, by including mass compensation, overcame some of the artefact caused by the

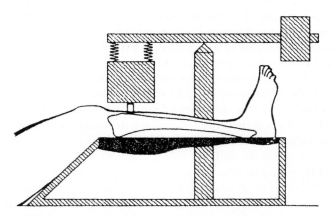

Figure 16.2. *Bone resonance analysis (BRA) method in which the leg is supported in a splint and the impedance head is placed on the anterior tibial margin. Note that the impedance head is counter-weighted to allow for different pre-loads (from Borgwardt-Christensen et al [14]).*

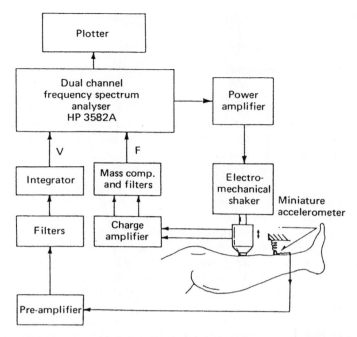

Figure 16.3. *Measurement system used by Collier and Donarski [16] in which an electromechanical shaker applies a sinusoidal driving force via an impedance head, the resulting vibration being measured with a low-mass accelerometer. A further accelerometer is used to correct the measurement by removing the effect of coupling of the skin.*

overlying soft tissues. The measurement system used is shown in figure 16.3, and consisted of an electromechanical shaker which applied a sinusoidal driving force to the bone via an impedance head, a small static force being used as a pre-load. The resulting vibration was measured with a low-mass accelerometer. A further accelerometer was used to correct the measurement by removing the effect of coupling of the skin. The results obtained showed that the effect of using mass coupling was to significantly reduce the dependence on static pre-load of the shaker and improve the signal-to-noise ratio at frequencies >300 Hz.

16.5.4. *Comparison of IFR and BRA techniques*

The techniques of BRA (as previously described by Borgwardt-Christensen *et al* [14]) and impulse frequency response (IFR—figure 16.4) as described by Vandecasteele *et al* [12] were compared by Christensen *et al* [18]. Resonant frequencies and mode shapes of a fresh excised tibia and an amputation specimen were determined using both BRA and IFR techniques. Each of

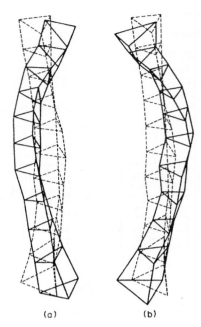

                    (a)   (b)

Figure 16.4. *Single bending modes of a dry tibia in the free–free condition obtained by modal analysis (from Van der Perre and Cornelissen [17]). The broken line shows the undeformed shape. (a) Bending about the principal x axis (motion normal to the medial face); (b) bending about the principal y axis (motion parallel to the medial face).*

these techniques has its own inherent support conditions which were found to have a major effect on both the mode shapes and resonant frequencies obtained. The differences in the excitation and signal processing used in each technique were found to result only in minor differences in the resonant frequencies. For the BRA technique, a rigid body mode of 165 Hz in the saggital plane and a single bending mode of 315 Hz close to the saggital plane were found. For the IFR technique, a similar rigid body mode to that found with the BRA technique was obtained at 167 Hz, and two single bending modes, corresponding to free–free support conditions, were found at 303 Hz close to the frontal plane and 470 Hz in the saggital plane.

16.5.5. *Mechanical response tissue analysis (MRTA)*

Young *et al* [19] compared measurements obtained using an impedance technique (shown in figure 16.5) on five subjects. The electromechanical shaker used to apply the vibration was placed at the end of a balance beam to permit the application of a variable pre-load and impedance was measured

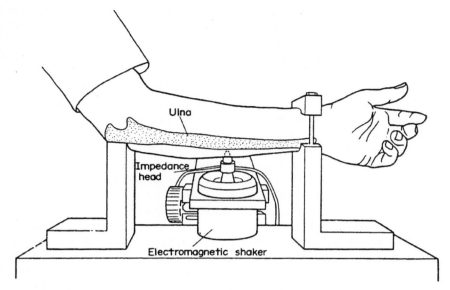

Figure 16.5. *Impedance probe technique for determining the impedance of the ulna (from Orne [20]).*

at discrete frequencies between 65 and 1000 Hz. The forearm was placed in a support rig which clamped the distal end of the ulna and bone mineral content (BMC) was measured at a section of the ulna 30 cm from the distal end. The results were compared with a mathematical model of the system and demonstrated generally good agreement between the predicted and experimental results up to 350 Hz, for a pre-load of 4.9 N.

In a subsequent study, Young *et al* [13] used a similar impedance probe to measure the *in vivo* bending rigidity of simian ulnae and tibiae. The mechanical impedance was used to determine bending rigidity through the use of a lumped parameter model in which the bone and surrounding musculature were treated as a spring/mass/dashpot system and the skin was considered as a simple spring [20, 21]. The effective stiffness of the bone was the product of individual values of frequency and impedance over the frequency range 100–250 Hz. Repeatability of rigidity measurements was 4%. Rigidity was found to be correlated with BMC (measured using single photon absorptiometry) of the cross-section ($r = 0.899$). Prolonged restraint reduced the measured rigidity by 12–22% and the BMC by 7–24%.

This technique was subsequently developed and named mechanical response tissue analysis (MRTA) by Steele *et al* [22]. A point mechanical impedance technique was used, similar to that of Young *et al* [13], utilizing a specially designed probe for good skin contact and a 'proper' support of the proximal and distal ends of the bone to obtain a best approximation to simple supports. A seven-parameter model for the mechanical response (including first anterior–posterior (AP) mode of bending, the damping and spring effect of soft tissues and the damping of musculature) was developed to determine the bending stiffness and critical buckling load from the vibration measurements. Eighty human subjects were evaluated using this system and also measured using single photon absorptiometry (SPA) to compare bending stiffness and mineral content. Good correlations were obtained between EI and BMC ($r = 0.81$) and between critical buckling load and BMC ($r = 0.89$). The authors also introduced a parameter called the 'bone sufficiency parameter' (i.e. critical buckling load/body weight) which they demonstrated to be closely related to activity level of the subject.

Using the MRTA technique McCabe *et al* [23] measured bending stiffness (EI) and BMC in the ulnae of 48 healthy women, 25 in an older group and 23 in a younger group. Both BMC and EI were found to be greater ($p < 0.005$) in the younger group. In the younger group correlations of BMC and body weight (BW) were significant. BW did not change significantly with age and ulnar width and the ratio of ulnar width to length were similar in both groups. Both BMC and ulnar width were significantly greater on the dominant arm for the young women only. In the younger group, using multiple regression analysis, bone width was found to be the only independent predictor of stiffness. There was an age-related decrease in ulnar bending stiffness but not in overall ulnar width or length. Hence, in the younger group, EI was predicted by the bone width and in the older group EI was predicted by the BMC.

16.5.6.   *Effect of soft tissue on frequency*

In order to try to ascertain the effect of the soft tissues, previously noted in the *in vivo* measurements of Vandecasteele *et al* [12], Van der Perre *et al* [24] subsequently determined natural frequencies, mode shapes and damping ratios of dry and fresh excised tibiae. The measurements were performed on *in vivo* tibia and on an amputation specimen (tibia) at different stages of dissection. Using a hanging leg protocol, with hammer impact to the medial malleolus and accelerations measured at defined points along the length of the tibia, the tibia was found to exhibit free–free (i.e. unconstrained) boundary conditions. Two single bending modes, corresponding to vibration about each principal bending axis, were identified at 270 and 340 Hz, although it was noted that a certain coupling or overlap could occur which affected the determination of the resonant frequency for each mode. The main differences between the frequencies observed on the isolated tibia and those observed *in vivo* were found (as a result of the experiment on the amputation specimen) to result almost completely from the damping effect of the surrounding musculature. The main difference (a decrease $\sim$200 Hz in resonant frequency) between the dry and fresh isolated tibiae was found to result from the presence of bone marrow in the fresh specimens.

For both the BRA and IFR techniques, Cornelissen *et al* [25] assessed the influenced of the joints and soft tissues on the vibration behaviour of the lower limb by gradually dissecting amputated lower limbs. For both techniques, the soft tissues were found to exert a similar influence on the vibration of the tibia. The skin exerted only a very small influence (<5%) on the resonant frequencies while the musculature markedly decreased the resonant frequencies (due to its mass) and increased the damping. The fibula was found to exert a stiffening effect on the tibia, resulting in a decrease in resonant frequencies of 5–11%. The joints were found to have a small influence for the IFR technique while the supports used for the BRA method were found to exert a profound influence on the boundary conditions and therefore the modes observed. However, throughout the dissection, the vibration behaviour of the tibia was not found to change markedly.

Tsuchikane *et al* [26] carried out measurements on five cadaveric specimens to determine the effect of skin, soft tissue and fibula on the vibration parameters of resonant frequency and damping ratio. The technique used was similar to the IFR technique previously described [12], and used a hammer excitation to the medial malleolus, the resulting vibration being measured with an accelerometer placed proximal to this. Similar to Cornelissen *et al* [25] the skin was found to have very little effect whereas the foot and surrounding musculature had a significant influence on the resonant frequencies measured. The resonant frequency was found to be primarily dependent on the weight of the limb but it was also influenced by the individual knee ligaments. Resonant frequency increased with the removal of skin, muscles and foot and then decreased with the removal of the femur and fibula. The damping ratio decreased throughout the dissection. By modal analysis, the tibia was found to have a primary single bending mode and the shape of this mode was not influenced by the knee joint, ankle joint and fibula if the leg was suspended. The authors therefore concluded that for the limb suspended or hanging free, changes in the resonant frequency of the tibia represent a change in the condition of the tibia itself.

16.6. LONGITUDINAL WAVE PROPAGATION

The transmission of longitudinal waves along a bone have been used to assess bone quality. An impulse is provided at one end of the bone and the time taken for the longitudinal wave to travel a known distance is measured. Here it is the velocity of a travelling wave which is of interest and not the resonant frequencies generated within the bone. This is analogous to the longitudinal propagation of ultrasound, although using low frequency mechanical waves. Two techniques have been developed, one in which the time delay between the impact and the wave being received at a single trans-ducer (accelerometer) is used to calculate the velocity (one-point method), the other in which the time delay is measured between two accelerometers (two-point method).

16.6.1. *One-point method*

Pelker and Saha [27] examined the characteristics of a travelling wave generated by a compressive pulse produced by the impact of a steel ball bearing on the proximal end of human long bones (femora, tibiae and humerii). Wave propa-gation was measured using strain gauges bonded to the bone. The velocity of the travelling wave and its attenuation and dispersion coefficients were deter-mined from the output from the strain gauges. These parameters were corre-lated with the mineral density, porosity and cross-sectional area of the bones tested. Statistically significant correlations were found between the travelling wave characteristics and the measured bone properties. Of particular note were the parabolic relationships found between the attenuation coefficient and the bone porosity and cross-sectional area, which suggested a potential use of this technique in the detection of osteoporosis. Using a similar technique (illustrated in figure 16.6), Wong *et al* [28] measured transverse wave propa-gation on embalmed human tibiae and on 17 male and 13 female volunteers. Wave velocity was determined from the distance between the measurement points and the time delay between impact and the received stress wave. Significant negative correlations were found between the wave velocity and age in both males and females, and two subjects with bone disease were found to have velocities well below that of age matched controls. *In vitro* studies of embalmed tibiae demonstrated a negative correlation of wave velocity with osteoporosis index and a positive correlation with mass/unit length [28].

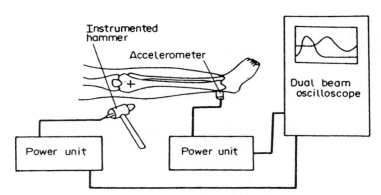

Figure 16.6. *Experimental set-up used by Wong et al [28] to measure the velocity of flexural waves in the tibia.*

A mathematical model of stress wave propagation in bone was presented by Chen and Saha [29]. The bone was assumed to be a thick walled cylinder filled with fluid. Using this model they demonstrated the sensitivity of phase and group velocity to changes in cortical thickness and density and suggested that these could potentially be used to differentiate between groups of male and female subjects of different age ranges.

16.6.1.1.   Effect of pre-load on phase velocity

Nokes *et al* [30] assessed the effect of accelerometer pre-load on the results obtained from the measurement of the transmission of a stress wave along the bone. It was found that excessive pre-load resulted in a distortion of the high-frequency component of the received signal and that the pre-load required was proportional to the soft tissue thickness. For any given soft tissue thickness, however, the range of pre-load to give an acceptable output signal was within the range 3.8–5.4 N.

16.6.2.   Two-point method

Stussi [31, 32] presented a two-point method (in contrast to the one-point method of Pelker and Saha [27]) of measuring the phase velocity of flexural waves in the tibia. An electromechanical hammer applied an impulse to the medial tibial condyle. Two accelerometers mounted in an adapter to hold them a set distance apart were attached to the medial face of the tibia with elastic straps. The impact was in a direction parallel to the axes of the accelerometers. From the output acceleration signals, the phase velocity (range 250–650 m/s, similar to that observed by Wong *et al* [28]) as a function of frequency was derived from the phase spectra, as

$$c(\omega) = \frac{\omega d}{\phi(\omega) + 2\pi N} \tag{16.2}$$

where $c(\omega)$ is the phase velocity, ω is the frequency, d is the distance between accelerometers, $\phi(\omega)$ is the phase spectra of the cross-correlation and N is an integer $(\pm 1, \pm 2, \pm 3, \ldots)$.

In 43 subjects (11 normal, 32 with renal disease), phase velocity was found to increase with increasing frequency or decreasing wavelength. It was also shown to be positively correlated with BMC/bone width; different dispersion curves were also observed for different values of BMC/bone width.

Cheng *et al* [33] measured vibrational wave propagation in the tibia of 56 female volunteers. An impact force was produced by a force-instrumented hammer and the response measured with two accelerometers. Parameters measured were the peak amplitude of the measured accelerations, the velocity of the vibrational wave and the damping time. The results obtained for vibrational wave propagation were of a similar magnitude (300–500 m/s) to those obtained previously by Wong *et al* [28] and Stussi and Fah [31]. Significant negative correlations were demonstrated with age for peak amplitude and velocity; damping time was also found to decrease with increasing age.

16.7. ASSOCIATION OF RESONANT FREQUENCY WITH TORSIONAL AND BENDING STIFFNESS

Lowet *et al* [34] developed a simple beam model for the prediction of torsional stiffness from resonant frequency, bone mass and length which was also developed from equation (16.1):

$$f = \alpha \sqrt{\frac{EI}{ML^3}} \qquad (16.3)$$

where M is the mass of the beam ($= \rho LA$) and the other symbols are as for equation (16.1).

For a beam of circular cross-section, the torsional stiffness is given by

$$S = \frac{GJ}{L_i} \qquad (16.4)$$

where G is the shear modulus, J is the polar moment of area and L_i is the length of the beam over which the torsional stiffness was measured ($\neq$ total length of the beam).

As noted earlier, the first two natural frequency modes for the tibia (f_1 and f_2) correspond to bending about two principal axes (I_1 and I_2). By combining equations (16.3) and (16.4), it can be shown that

$$S = \frac{ML^3(f_1^2 + f_2^2)}{2\alpha L_i(1 + \nu)}. \qquad (16.5)$$

Assuming constant values of α and ν, the torsional stiffness

$$S \propto \frac{ML^3(f_1^2 + f_2^2)}{L_i}. \qquad (16.6)$$

Therefore the torsional stiffness should be a function of the mass, the first two resonant frequencies and the length (total length and the length over which the torsional stiffness was measured). To test this relationship Lowet *et al* [34] obtained resonant frequency data from 142 excised long bones from different anatomical locations (tibia, femur and radius) and different species (sheep, dog and monkey). A hammer impact was used for excitation and a microphone for vibration measurement to minimize any mass loading effect for these small, light bones. Torsional stiffness was obtained from mechanical testing. A good linear correlation was obtained between predicted and measured torsional stiffness, both for individual groupings of bones and for all of the bones tested between resonant frequency and torsional stiffness, implying that the model developed is valid for all of the different bones tested. For *in vivo* application of the model, total bone mineral content (BMC) was substituted for the mass and again a good correlation was obtained between the predicted and measured torsional stiffness.

A model relating bending stiffness (*EI*) and resonant frequency was developed by Van der Perre and Lowet [35] from equation (16.1) assuming the bone to consist of a hollow cylindrical shell filled with marrow, the resonant frequency being given by

$$f = \frac{\alpha}{L^2} \sqrt{\frac{EI}{(A_1 \rho_1 + A_2 \rho_2)}} \tag{16.7}$$

where the subscripts 1 and 2 refer to the marrow and the bone respectively. From this equation, the bending rigidity can be approximated by

$$EI \approx f^2 (A_1 \rho_1 + A_2 \rho_2) L^4. \tag{16.8}$$

The quantity $(A_1 \rho_1 + A_2 \rho_2)L$ is the total wet bone mass. Lowet *et al* [34] had previously demonstrated a good linear correlation between the total bone mineral content (TBMC) and the total wet bone mass in a large number of different animal bones. Hence knowing TBMC, the bending rigidity can be estimated from

$$EI \approx f^2 (\text{TBMC}) L^3. \tag{16.9}$$

The model was verified experimentally on the same population of long bones as has been used previously by Lowet *et al* [34], and a population of normal controls and osteoporotic subjects was also measured. The calculated bending rigidity was found to decrease in the osteoporotic group compared with the controls. It was concluded that a combination of BMC with resonant frequency data offers significant advantages over BMC alone, as this only gives information on the amount of bone present and not on its structural stiffness.

16.8. Use of vibration to monitor treatment effect

In a simian model, Geusens *et al* [36] investigated the effect of increasing doses of Tiludronate (a bisphosphonate which inhibits bone resorption) on changes in BMD and the torsional and vibrational properties of the radii. Measurements were made with the radii supported on foam to give free–free vibration conditions, vibration excitation being performed with hammer impact and response measured with a microphone. Transverse stiffness was calculated from $f^2 m$, where f is the resonant frequency and m is the total wet mass, and buckling strength calculated from $f^2 m l$, where l is the length. Reproducibility of the resonant frequency measurements was

0.2%. Torsional stiffness was found to increase after treatment but no significant differences were found between experimental and control bones one year after treatment had stopped. Transverse bending stiffness and buckling strength calculated from resonant frequency measurements indicated increased transverse stiffness after treatment and increased transverse bending stiffness and buckling strength one year after treatment.

16.9.   VIBRATION MODELLING STUDIES

Many modelling studies of vibration long bones have been performed usually in parallel with experimental studies. The principal purpose of these studies was to assess the influence of the principal parameters affecting the vibration and to enable predictions of material and structural properties to be made from vibration data. A summary of the modelling studies performed and their main conclusions is given below.

16.9.1.   Ulnar model

Jurist and Kianian [37] evaluated three models for the prediction of the resonant frequency of the ulna as a function of length, width and mineral content at a specific location, this being one-third of the length from the proximal end. The ulna was modelled as a homogeneous isotropic cylindrical tube, the differences in the models being the boundary conditions used. For all three models the resonant frequency predicted was within 20% of that measured in 118 subjects.

A model of a simply supported viscoelastic beam subjected to a harmonically varying force close to its mid-span was developed by Orne [20]. The soft tissue between the impedance head and the ulna was modelled as a three-parameter viscous solid. The theoretical impedance curves obtained using this model were in better agreement with the previous experimental data than an earlier lumped parameter model. Refinements to this model, to include the resistance to lateral vibration produced by the surrounding musculature, were produced by Orne and Mandke [21]. This was represented as a continuous series of damped oscillators attached to the ulna and characterized by specific mass spring and damping coefficients. The incorporation of physically plausible parameters for the musculature in the model gave closer agreement with previous experimental results over a wider range of forcing frequency than the earlier model.

Three models for the prediction of ulnar resonant frequency from length, mineral content and ulnar width were evaluated by Spiegel and Jurist [38]. The ulna was modelled as a homogeneous isotropic cylindrical tube and the effect of varying the support conditions was assessed. The most useful model and that which gave the best agreement with experimental data was a model with rigid supports (hinged–hinged) incorporating bone marrow.

The effects of pre-twist, non-uniformities in mass and flexural stiffness, rotary inertia and shear deformation on the natural frequency of intact bones were determined using a three-dimensional linear elastic finite element (FE) model by Orne and Young [39]. Variations in mass and flexural stiffness along the length of the bone strongly influenced the first three natural frequencies and they concluded that a model which included these but that neglected pre-twist, rotary inertia and shear deformation was sufficient to model the vibrational behaviour of a long bone.

16.9.2. Tibia model

Hight *et al* [40] modelled the tibia with beam-type finite elements to predict the effect of geometry, mass distribution and boundary conditions on the natural frequencies. The boundary conditions chosen were found to have an important influence on the natural frequencies, due not only to the boundary conditions themselves but also to the coupled effects of the curvature and twist of the bone with the boundary conditions. It was felt to be important that an accurate estimate of the boundary conditions (and their stiffness) was made in order for the natural frequency to be predicted.

Collier *et al* [41] carried out measurements on an isolated tibia and presented theoretical models for the three main vibration modes in long bones with a view to being able to identify the various resonances and to show the types of modes which exist *in vitro*. The model was composed of a homogeneous hollow beam with a constant cross-section of an isosceles triangle. Using this model, they were able to explain the two main types of transverse mode occurring and the predicted resonances obtained were less than 6% in error from the measured values although the error increased to 24% for torsional modes. Good agreement was obtained between the predicted and measured resonant frequencies, although in addition to two single bending modes (in approximately the medio-lateral and anterior–posterior directions) they also noted the presence of a triple resonance at each of these modal frequencies. In a response to this paper, Van der Perre and Cornelissen [17] noted that for a beam with uniform cross-sectional properties along the longitudinal axis, there are only two single bending modes (figure 16.7), these being associated with bending about the principal x and y axes:

$$f_x = \alpha \sqrt{\frac{EI_{xx}}{\mu L^4}} \tag{16.10}$$

$$f_y = \alpha \sqrt{\frac{EI_{yy}}{\mu L^4}} \tag{16.11}$$

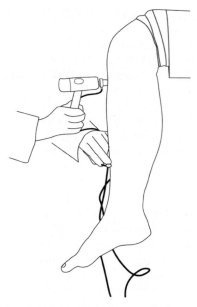

Figure 16.7. *Impulse frequency response (IFR) technique in which the vibration is applied by hammer impact incorporating a force transducer and measured with a hand-held accelerometer pressed against the bone (from Christensen et al [18]).*

where E is Young's modulus, I is the second moment of area of the cross-section about the axis considered, μ is the mass/unit length $= \rho A$, L is the length and α is the dimensionless constant depending on the boundary conditions and the mode. Hence the lower the I value, the lower will be the resonant frequency about that plane. They also noted that the triple resonances previously noted by Collier *et al* [41] may be artefactual and could result from coupling effects between the vibrator head and the bone.

A theoretical and experimental study of the free–free vibration of an excised tibia was described by Thomsen [42]. Seven natural frequencies in the range 0–3 kHz were identified experimentally by structural transfer functions determined through impulse excitation and vibrational response measured with an accelerometer and dual channel fast Fourier transform (FFT) analysis. The frequencies observed appeared to correspond to three pairs of flexural modes and one torsional mode. In the theoretical model, the tibia was modelled by finite element beam elements composed of two isotropic linear elastic materials (cortical and cancellous bone) and one perfectly flexible material (bone marrow). The unknown parameters in this model were varied until maximum agreement with the experimental observations was obtained. The optimized model was then used to predict higher resonances in the frequency range 3–5 kHz. A sensitivity analysis of the FE model suggested that, in a

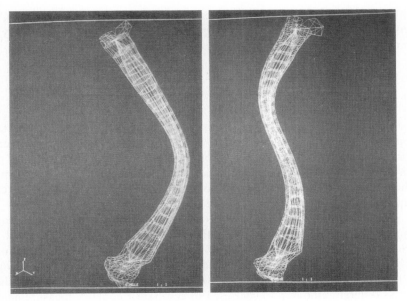

Figure 16.8. *Finite element model of Hobatho et al [43] showing the mode shapes of the first two bending modes in the sagittal plane.*

vibrational sense, the tibia was more uniform than its complex geometry might otherwise suggest and that flexural vibrations are significantly influenced by shear deformations. This suggested that simple models of the tibia which include shear deformations could be considered as adequate as the more detailed FE model, particularly for the lower vibration modes.

CT data were used to construct a geometrically accurate three-dimensional FE model of a human tibia by Hobatho *et al* [43]. An experimental modal analysis was used to optimize the FE model (figure 16.8). With the optimized model the predicted resonant frequencies were found to be within 10% of those determined experimentally, this reducing to 3% if the inhomogeneity of the tibia was considered.

Roberts *et al* [44] presented a refined model from their previous study [22], in which a more advanced model of the theoretical behaviour of the tissue was incorporated. The results obtained using this model were used to interpret both *in vivo* and *in vitro* results obtained previously using MRTA [22]. Compared with the previous model, marked improvements between predicted and measured values of bone stiffness and strength were obtained.

16.9.3.   Femur model

Couteau *et al* [45] developed a three-dimensional FE model of a human femur from CT images. The results obtained using this model were compared with a modal analysis of the same femur and good agreement (within 4%) was obtained. The model was subsequently used to assess the sensitivity of changes in the mechanical properties on the vibrational properties.

16.10. SUMMARY

It is apparent that vibrational measurements have much potential when applied to the measurement of bone. The advantage of these measurements is that they can be related, often directly, to the mechanical properties of the bone, and they thus serve as a potentially very useful method of determining when a reduction in the ability of bone to carry load has occurred, as can happen with various metabolic diseases.

The two main techniques used (resonant frequency measurement and longitudinal wave propagation) both give results which can be used to detect absolute and relative changes in bone properties. The measurement of longitudinal wave propagation is not dependent on the support conditions of the bone and therefore the results obtained could be more easily applied across different studies and centres. For resonant frequency measurements, which depend on standardized and repeatable support conditions, interpretation of the measurements between different study designs and centres can be more difficult unless a strict protocol is adhered to.

The principal disadvantages of these measurements is the requirement for the measured bone to be accessible to vibration through a thin layer of soft tissue and to be of a reasonable length so that the vibration modes are simple, or to allow a reasonable length for a longitudinal wave to propagate. This limits these measurements to such long bones as the tibia and ulna, which are not commonly the site of fractures associated with osteoporosis. However, these long bones will be affected by the process of osteoporosis (as is the os-calcis from which ultrasound velocity and attenuation have been measured and used to indicate fracture risk in osteoporotic subjects) and therefore could be used to infer changes in bone properties throughout the appendicular skeleton.

SECTION C: THE USE OF VIBRATION IN THE MONITORING OF FRACTURE HEALING

16.11. INTRODUCTION

The assessment of fracture healing remains largely a matter of clinical judgement based upon a subjective impression of fracture stiffness made by manual loading, and an assessment of radiological union by observing the presence of bridging callus or the obscuration of the fracture. These assessments have however been previously shown to be unreliable in determining functional union, manual loading being insensitive above relatively low values of stiffness [46], and plain radiology has been shown to be unreliable in determining union [47, 48]. In many instances fracture union progresses satisfactorily and these somewhat crude clinical assessments remain satisfactory. However, in certain long bone fractures (the tibial shaft remains a prime example) fracture union may not progress satisfactorily and, despite modern methods of treatment, the rate of delayed union (19% at 20 weeks post-fracture) or non-union (4%) remains high [49].

The development of reliable and accurate quantitative measurements of fracture healing enables objective assessments of both the rate of fracture healing and the time to union to be made. These measurements will be related, either directly or indirectly, to the strength and stiffness of the healing fracture, and therefore should provide a measurement which is superior to the usual clinical assessments in that it is quantitative and related to the end point of fracture healing, i.e. the restoration of skeletal integrity and strength. During the early stages of healing, when clinical and radiological assessments of healing do not give much information on the progression of healing, such quantitative measurements provide invaluable information on the progression of healing and can discriminate between fractures which are healing normally and those in which healing is delayed.

Auscultatory percussion using a stethoscope to listen to the sound conducted along a bone has been used as a means of diagnosing a fracture and assessing fracture healing [50–52]. The sound can be produced by tapping or placing a vibrating tuning fork on a bony prominence adjacent to the fracture and listening to the sound at the opposite end of the fractured bone or an adjacent articulating bone. The progress of healing can be assessed in terms of increasing sound transmission across the fracture, the intact contralateral bone being used as a control. This method has particular value in developing countries when radiographs are not readily available.

Such qualitative measurements can be quantified by the use of instrumentation. Broadly speaking these measurements can be split into two types: those measuring the transmission and attenuation of a vibratory signal across the fracture site (low frequency wave propagation) and those measuring the resonant frequency of the bone (resonant frequency measurement).

16.12.   LOW FREQUENCY WAVE PROPAGATION

16.12.1.   *Propagation and measurement of low frequency waves ('stress waves')*

Stress waves are generated by providing an impulse to one end of a fractured bone, in a similar manner to that used to generate phase velocities as described in section 16.6. The impulse used can be provided by variety of sources including hammer impact [53, 54] and the impact of a steel ball dropping from a set height [27, 55]. Measurements of the transmitted vibration signal are made with accelerometers placed at a known distance from the impact source.

16.12.2.   In vitro results

Stress wave propagation was examined in embalmed and fresh human tibiae and humerii by Pelker and Saha [27]. A stress pulse was generated by the impact of a steel ball bearing on the proximal end of the bone. Stress waves were monitored with semiconductor strain gauges bonded to the bone and with a magnetic velocitometer. Healing was simulated in reverse by gradually cutting through the bone from the intact state until the bone was completely transected. A statistically significant relationship was found between the relative size of the discontinuity and the transmission coefficient (defined as the ratio of the strain before cut to strain after cut divided by the transmission coefficient for the uncut bone). The transmission coefficient was found to increase with normalized area by a relationship of the form $y = 1 - e^{-x}$ (figure 16.9) and a transmission coefficient of unity was achieved when the normalized area was around 0.5. Both relative pulse width (the dispersion resulting from the wave passing across the fracture) and the delay in transit time decreased linearly with increasing normalized area, again reaching unity when a normalized area of 0.5, or 50% of the intact cross-sectional area, was reached.

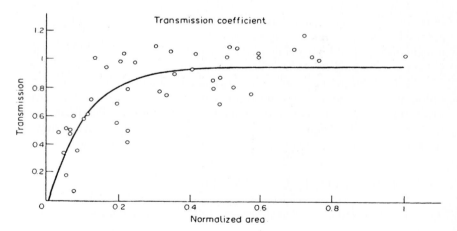

Figure 16.9. *Results obtained by Pelker and Saha [27] from an in vitro model of fracture healing showing change in transmission time with increased normalized area.*

16.12.3. In vivo results

The transmission of stress waves across the fracture in a canine model of bone healing was measured by Sonstegard and Matthews [53]. Hypodermic needles were placed in the cortical surface of the bone at right angles to the longitudinal axis on either side of the fracture. A third needle was placed at 45° to the longitudinal axis, and the excitation was produced through this needle with a hammer impact (figure 16.10). With increasing healing, the amplitude of the received signal, relative to the signal on the same side of the fracture as the impact excitation, was found to increase and the time delay of the signal across the fracture site was found to decrease. Limited clinical results from 11 patients with various fractures (tibial, humeral, femoral) were also presented and these suggested the usefulness of this technique in diagnosing non-union.

Sekiguchi and Hirayama [54] measured the vibrational signal transmitted across a healing tibial fracture. An impulse was provided by a tapping device at the medial malleolus and the resulting transmitted vibration signal was measured at the tibial tuberosity. This is a similar technique to the auscultatory percussion technique described above but with the transmitted vibration signal being recorded. Differences in the received signal from that obtained on intact bones were used to infer the state of healing. With increasing healing, the waveform from the fractured bone approached that of intact bone, whereas in delayed union the received waveform was not found to change with time.

Nokes *et al* [55] described a simple quantitative method for the assessment of fracture healing using an experimental setup similar to that of Pelker and Saha [27]. An impulse was again applied to the tibial tubercle with a steel ball dropped through a Perspex tube from a known height and transverse vibration waves were measured at the proximal and distal ends of the bone using pre-loaded accelerometers. Twenty patients with unilateral diaphyseal fractures were measured at various stages of healing. The attenuation factor (maximum amplitude of signal from distal

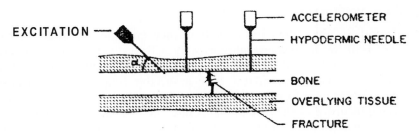

Figure 16.10. *Measurement technique utilized by Sonstegard and Matthews [53] in which hypodermic needles were used to provide direct coupling to the bone.*

accelerometer divided by maximum amplitude of signal from proximal accelerometer) was found to increase with time towards the range found for normal tibiae (mean 0.6) by a relationship of the form $y = A(1 - e^{bx})$. Note the similarity of this relationship obtained from healing fractures *in vivo* to that given by Pelker and Saha [27] for their *in vitro* simulation of fracture healing. They also noted that with an established non-union, the attenuation factor remained low (at around 0.1). Subsequent work by Flint *et al* [56] employing this same technique [55], measured seven patients with mid-shaft tibial fractures treated with unilateral external fixation. The fixator was removed prior to the measurements being taken and the accelerometers were attached to the pins of the external fixator on either side of the fracture. In only two of the patients studied was an increase in attenuation factor with time post-fracture demonstrated and the authors suggest that monitoring of propagation of the transverse wave may not be as sensitive to healing as monitoring the longitudinal wave.

16.13.   RESONANT FREQUENCY MEASUREMENT

In an inverse manner to that used to determine if a flaw exists in a metal structure, resonant frequency measurements can be used to monitor the course of healing. In simple terms, the principle of such measurements is that a fractured bone will resonate at significantly different frequencies to a similar intact one (usually the contralateral bone is used for this purpose). As the fracture heals and the two broken ends of the bone become joined by a stiff calcified callus, the resonant frequencies of the bone approach those of the intact bone thus providing a method of quantifying healing. Two methods have been used to excite resonant frequencies, one in which the bone is resonated by applying swept sinusoidal vibration in the frequency range 50–1000 Hz and one in which the bone is excited by a single impulse (impulse-response method).

16.13.1.   Swept sinusoidal vibration

The hypothesis that tibial resonant frequency and the amplitude of the resonant peak would increase progressively with healing was put forward by Markey and Jurist [57]. In a clinical study of 24 patients with tibial fractures, a mechanical oscillator, with a variable frequency drive (range 80–500 Hz), was used to vibrate the tibial tuberosity, the resulting acceleration being recorded at the medial malleolus. The resonant frequency of the fractured tibia relative to the control tibia was determined and was found to increase in a parabolic manner with the time post-fracture. This parameter was also found to be closely correlated to the radiographic appearance of the healing fractures.

Borgwardt Christensen *et al* [14] used a driving point impedance technique (often referred to as bone resonance analysis or BRA), similar to that of Markey and Jurist [57], to determine the lowest resonant frequency in the human tibia. A fracture was simulated by gradually cutting through the bone at the mid-shaft, the resonant frequency being measured at each stage of bone removal. The resonant frequency was found to decrease with the depth of cut, the most marked decrease being for depth of cut greater than 70% of the bone depth. Clinical measurements from four patients with tibial fractures was also presented which showed distinct differences in resonances between the fractured and non-fractured tibiae at the time of measurement. No follow-up measurements were made so changes in resonant frequency with healing were not determined.

The technique described earlier by Collier and Donarski [16] for determining the *in vivo* resonant frequencies of the tibia was subsequently applied to the measurement of healing tibial fractures [58]. It was found that the resonant frequency of the fractured tibia was often as low as half that of the intact bone but that as the fracture heals, increases in the resonant frequency occurred. No coupling of the fracture fragments was found above 200 Hz in the very early stages soon after fracture. As bridging of the fracture with callus occurs, there is a significant increase in the high frequencies transmitted across the fracture and it was suggested that this high-frequency signal could be a useful indicator of the healing process.

Lowet [59] reported on clinical measurements of fracture healing in seven patients using the BRA technique but using a modified experimental protocol. In most cases a linear relationship was found between the resonant frequencies (rigid body and single bending) and healing of the fractured tibia with an increase in resonant frequency of 1–2% per week. The average standard deviation from repeated measurements on both intact and fractured tibia was in the range 1.6–7.3%.

16.13.2. *Impulse-response method*

Lewis [60] carried out measurements on a healing tibial fracture by applying a standardized mechanical impulse to one end of the bone and measuring the response with an accelerometer at the other end. With increasing healing, an increase in the amplitude of the high frequency content of the received signal was obtained. However, he concluded that monitoring changes in the maximum acceleration was unlikely to be an adequate measure of healing (except possibly in the very early stages), and that a harmonic analysis of the accelerometer signal would be required if quantitative data on changes in stiffness with healing were to be obtained.

Cornelissen *et al* [25] carried out experiments on one above-knee amputation specimen in which the tibia was gradually transected. Two support conditions were investigated: limb hanging down with knee flexed at 90° and the

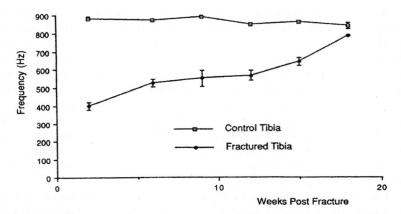

Figure 16.11. *Changes in resonant frequencies with time for the double bending mode of a fractured tibia compared with the un-fractured contralateral (from Cunningham et al [61]).*

limb supported in a specially designed splint with the knee flexed at 45°. The single bending mode was found to be the most sensitive to weakening of the bone, particularly with free–free support conditions, the reduction in resonant frequency being most marked at greater amounts of transection.

Nokes et al [55] carried out measurements on 11 patients with mid-shaft fractures of the tibia. An impulse was applied to the tibial tubercle and the response measured with skin-mounted accelerometers on either side of the fracture. The mean frequency of the distal fragment was found to be less than that of the proximal fragment, these frequencies reducing and converging as healing progressed. No control measurements of the non-fractured contralateral were made.

Cunningham et al [61] used the IFR technique described earlier by Vandecasteele et al [12] to determine healing in tibial fractures treated by cast, the cast having been removed prior to the measurement. They found that the frequencies of the different modes (rigid body, single bending and double bending) increased with time and approached those of the intact contralateral (the modal frequencies of which remained constant throughout healing) as healing progressed (figure 16.11). Where there was no or minimal convergence of modal frequencies, delayed union was found. Based on their preliminary results it was suggested that for clinical fracture healing the frequency of the fractured tibia should be greater than 90% of that of the intact contralateral.

Tower et al [62] performed resonant frequency analyses on 74 tibiae with a healing fracture or osteotomy. The fractures were treated with a variety of fixation methods and measurements made throughout healing. The IFR technique was again used, with the limb to be tested hanging free and hammer excitation being applied to the medial malleolus, the resulting response measured with an accelerometer pressed against the anterior proximal medial tibial plateau. FFT was used to obtain frequency spectra from which resonant frequencies were obtained. The tibial stiffness index (TSI—ratio of resonant frequency of the fractured tibia to that of the intact contralateral) was determined from the resonant frequency measurements and compared with clinical and radiological indices of healing. A significant positive correlation was found between the TSI and the indices for fractures treated by cast, unreamed unlocked intramedullary nails and external fixation; for fractures treated by unreamed locked intramedullary nailing no such correlation was found. Fractures were considered clinically healed when the TSI exceeded 0.8.

Benirschke *et al* [63] examined the correlation between bending rigidity and resonant frequency in model of fracture healing in which a neoprene rubber insert was inserted into a defect made in cadaveric tibiae. Fracture healing was simulated in reverse by freezing and gradually thawing the tibia with insert, to produce a decrease in 'callus' stiffness with time. Frequency was found to be highly correlated with flexural rigidity to the power 0.582, with a coefficient of determination r^2, of 0.815. A measurement protocol was developed which consisted of the leg hanging freely over the edge of a table, an impulse being applied with an electronic hammer to the anteriomedial face of the tibia, halfway between the mid-shaft and the distal end, and the resulting acceleration measured by a hand held accelerometer placed against the anteriomedial surface, below the tibial plateau. This protocol was found to give highly reproducible results ($\pm 3\%$) which were independent of any left/right bias. In 14 patients with a tibial fracture, the ratio of the frequency for the first bending mode of the fractured tibia to the non-fractured contralateral was determined with time. A trend of increasing frequency ratio with percentage healing time (healing being defined as a frequency ratio of 0.88 corresponding to 75% of the intact flexural rigidity from the *in vitro* tests) was observed (figure 16.12), which approximated to a second-order polynomial.

Nakatsuchi *et al* [64, 65] carried out both an *in vitro* experimental study and a clinical study in 67 patients with tibial fractures. The experimental study [64], consisted of an isolated tibia and fibula supported on a sheet of rubber to simulate free–free vibration conditions. Two separate vibrations were found, probably corresponding to vibration about the planes of maximal and minimal bending stiffness, with one vibrating strongly in the lateral direction, the other vibrating weakly in the AP direction, the modal frequencies being higher in the AP direction than in the lateral direction. Consolidation of the fracture callus was simulated by the hardening of an adhesive in a created transverse fracture of 2 mm width fixed with a metal plate. Changes in the hardness of the adhesive with time were measured simultaneously with vibration measurements. The model fractures were also fixed with Ender's nails, an intramedullary

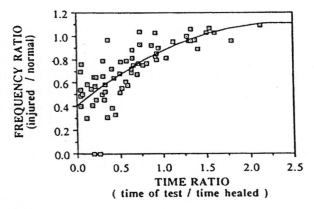

Figure 16.12. *Clinical data from 14 patients with tibial fractures showing increases in the frequency ratio (the ratio of the frequency for the first bending mode of the fractured tibia to the non-fractured contralateral) with increasing time ratio (from Benirschke et al [63]). Note that 'healing' was defined when a frequency ratio of 0.88 was reached.*

nail and an external fixator. Resonant frequency was found to increase with the hardness of the adhesive, most of the increase in frequency occurring up to a hardness of 40% of the final adhesive hardness, after which the frequency did not change significantly. Similar results were obtained for the model fractures treated with the plate, Ender's nails and intramedullary nail but both the initial frequency and the final frequency, after adhesive hardening, were notably higher for the external fixator.

In the clinical study [65], the IFR method, with a similar experimental protocol to that of Benirschke *et al* [63], was used to quantitatively assess healing in 67 tibial fractures treated with a wide variety of treatment methods. A temporary decrease in resonant frequency was found early in treatment in just over half of the cases examined; it was also found to decrease on removal of the fixation device. The resonant frequency of the single bending mode reached within 20 Hz (or exceeded the value obtained from the contralateral side) in 30% of the patients studied or did not reach within 20 Hz in 70%. The authors suggested that removal of the fixation device can be considered when the resonant frequency is within 30 Hz (or approximately 10%) of that of the contralateral; note that this is in agreement with the recommendation of Cunningham *et al* [61].

Akkus *et al* [66] measured vibration of excised specimens from an experimental model of healing consisting of a mid-shaft osteotomy in the lapine tibia stabilized with intramedullary fixation. By plotting the ratio of amplitudes, RA (maximum amplitude of vibration of fractured tibia versus maximum amplitude of vibration of non-fractured control), they were able to demonstrate a progression with time post-fracture towards unity. Changes in the flexural rigidity of the excised bones seemed to parallel changes in the RA for the first and second bending modes.

16.14. MODELLING OF THE EFFECT OF HEALING

A number of simulation studies on the effects of healing on bone resonant frequency have been performed.

Lewis [67] modelled a healing fracture as two elastic beams joined by a compliant insert. This model was capable of both bending and shear deformation. By varying the modulus of the insert, the effect of increasing healing on the vibrational behaviour could be modelled. The mode shapes calculated clearly demonstrated a change from where the two halves of the fracture are vibrating separately (at low insert modulus) to where the model vibrates as a single beam. The model also demonstrated that the natural frequency of the first six modes of vibration increased with increasing insert modulus, becoming asymptotic to the modal frequency of the solid bar as the modulus of the insert approaches that of the beam material.

Laura *et al* [68] describe a simple two-dimensional FE model corresponding to the human ulna. Healing was simulated by increasing the value of Young's modulus in the central element, and the vibration frequency of the first four modes determined. Frequency for all modes was observed to increase with increasing Young's modulus, the trend of this increase being similar to that observed experimentally with time post-fracture [57, 63].

An analytical model of a healing tibia subjected to both lateral and axial vibration was developed by Nikiforidis *et al* [69]. The tibia was modelled as a beam with a small local zone exhibiting different elastic properties in order to determine the sensitivity of the resonant frequency and the phase shift to local changes in stiffness at the fracture site. Lateral vibration gave the most clearly demonstrable changes in frequency with healing although axial vibration was found to be useful when the lateral vibration frequency spectrum was 'crowded'. Experimental results using osteotomized cadaveric tibiae with simulated fractures at different stages of repair (obtained by joining the two halves of the tibia with materials of increasing stiffness) demonstrated marked shifts in natural frequencies with healing as had been suggested by the analytical model. Limited clinical results on one patient with a delayed union were performed which showed that the natural frequencies of the fractured tibia were lower than those of the contralateral.

Lowet *et al* [70] constructed two FE models of a fractured tibia with a healing callus; in the first model the callus was in the middle of the diaphysis and in the second model it was located at the junction of the middle and distal third. Resonant frequencies were found to increase with increasing callus stiffness. Single bending modes were found to be more sensitive to the callus in the middle of the diaphysis while double bending modes were

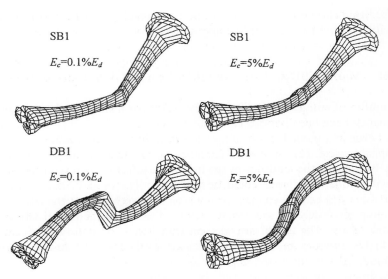

Figure 16.13. *Mode shapes of the first single bending and double bending modes of a healing tibia obtained by FE analysis (Lowet et al [70]) for different values of callus Young's modulus.*

found to be most sensitive to the callus at the junction of the middle and distal third. Similar mode shapes to the intact bone were obtained when the stiffness of the callus was equal to or greater than 5% of the intact diaphyseal bone stiffness (figure 16.13).

Roberts and Steele [71] carried out a numerical simulation to determine the effects of fracture length and location, end conditions and callus stiffness on the mechanical properties (resonant frequency, bending stiffness) of the healing bone. The mechanical properties were found to increase rapidly up to 30–70% of the intact stiffness; this increase then tapered off dramatically. Bending stiffness was found to be most sensitive to changes in callus properties for a larger proportion of the healing than was the torsional stiffness or the resonant frequency. They suggested that a measurement of the mechanical properties of a healing bone could provide insights into callus strength up to half of the intact callus strength. Bending stiffness and resonant frequency were also found to be highly sensitive to initial changes in callus strength but insensitive beyond 15–35% of intact callus strength.

16.15. OTHER MEASUREMENTS

Campbell and Jurist [72] performed a preliminary study of the feasibility of determining the amount of fracture healing of femoral neck fractures by the measurement of mechanical impedance and reactance. The reactance and impedance of the femur was measured as a function of frequency between 20 and 4000 Hz for the femur intact and at various stages of bone removal to simulate a femoral neck fracture. Large changes in reactance were found between 380 and 700 Hz following removal of a wedge of bone and removal of the femoral head.

Collier *et al* [73] and Collier and Ntui [74] examined the use of the phase delay of a mechanical vibration travelling across the fracture site (defined as the time delay of the vibration wave expressed in terms of the fraction of one complete cycle of the vibration). This method has the advantage that, unlike measurement of resonant frequencies, there are no problems in measuring specific frequencies or in correctly identifying a resonant mode. For an intact tibia, the phase delay was shown to be reasonably small (10^0) over a length of 10–15 cm. This compares with a fractured tibia where the phase delay over a similar length can be more than 100^0. In measurements carried out on healing fractures, the phase delay across the fracture site was found to decrease with increasing healing.

16.16. SUMMARY

Despite the substantial amount of research carried out into vibrational measurements of fracture healing, and the promising results obtained, this technique has never gained wide acceptance, and remains very much a research tool. In terms of quantifying healing, good results have been obtained with both of the main techniques employed (wave propagation and resonant frequency analysis), although measurements of resonant frequency are better researched and seem to more reliable (i.e. repeatable across different centres). Healing appears to be best quantified in terms of the non-fractured contralateral limb and there is some agreement between different studies as to what constitutes a 'healed' fracture in vibrational terms. With both techniques, the protocol employed can have a significant influence on the results and care should be taken to standardize this between measurements and between centres for a multi-centre study.

SECTION D: THE USE OF VIBRATION IN THE DIAGNOSIS OF PROSTHESIS LOOSENING

The idea of using vibration as a diagnostic test for hip prostheses was first suggested by Chung *et al* [75] and Poss *et al* [76]. They explored the phenomenon of shifting resonant frequencies with fixation of the prosthesis and investigated the changes in resonant frequency which occur as the bone-cement hardens. A similar principle could be applied to the changes in resonant frequency which will occur as the prosthesis loosens, but an initial measurement would need to be obtained when the prosthesis is not loose.

Van der Perre [6] noted that in addition to a shift in the resonant frequency with loosening, the measured output was 'noisy'. This phenomenon was explored by Rosenstein *et al* [77], who demonstrated that an implant that is loose will not transmit mechanical vibration free from distortion. In this study, a sine wave was applied to the medial femoral condyle using a vibrator and the output measured at the greater trochanter using a low mass accelerometer. The transmitted waveform was analysed using an oscilloscope and spectrum analyser. Since a loose femoral implant will not move uniformly in all directions, a flattening of the sinusoidal vibration in a particular direction will occur (figure 16.14). This was found to produce harmonics of the original sine wave which can be measured or discretized using a spectrum analyser.

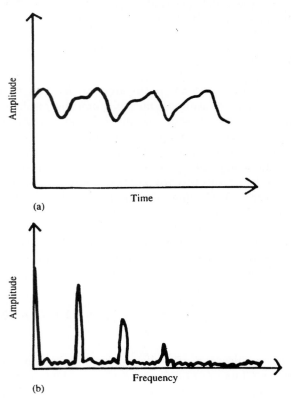

Figure 16.14. *Output waveform and frequency spectrum obtained from a loose prosthesis showing distortion to the wave and the presence of multiple harmonics (from Rosenstein et al [77]).*

Li *et al* [78, 79] used the same technique [77] in an *in vitro* model and found that early loosening of the prosthesis, defined as stable prosthesis/ cement and cement/bone interfaces but without penetration of the bone cement into the trabecular spaces (macrolock without microlock), was only found in 40% of the specimens tested using the vibrational technique.

Rosenstein *et al* [77] also carried out limited clinical tests on seven patients prior to revision surgery. In five of the seven cases the vibration test prior to surgery was indicative of a loose prosthesis which was subsequently confirmed at surgery. In the two remaining patients, the vibration test suggested that the prosthesis was secure, and although plain radiographs were suggestive of loosening, both prostheses were found to be securely fixed during surgery.

In a recent clinical study, Georgiou and Cunningham [80] collected vibration data from 23 patients admitted to hospital for a revision total hip replacement (THR). Prosthesis stability was determined intra-operatively one day later, by a surgeon unaware of the findings of the vibration test. Control data were also collected from 10 patients who had undergone a primary THR several months earlier. Three 'blind' clinicians assessed the stability of each patient's prosthesis based solely on the patient's radiographs, in order to allow comparison of the two techniques for the same patients.

Vibration testing was found to be 81% sensitive and 89% specific in the diagnosis of prosthesis loosening, but was unable to produce a definitive diagnosis in 10% of patients. The positive predictive value (PPV) was 93% and the negative predictive value (NPV) was 73%. Radiographs from the same patients showed them to have a similar accuracy. A correct diagnosis of loosening was obtained with vibration testing in 100% of loose femoral components, 50% of loose acetabular components and 88% of cases where both components were loose.

16.17. SUMMARY

It would seem that the vibrational technique described above is sensitive to the presence of a loose prosthesis both *in vitro* and *in vivo* and is at least as accurate as radiographs in determining loosening in a clinical setting. The potential usefulness of this technique lies in its ability to screen out patients with a loose prosthesis from those presenting with hip pain from an uniden-tifiable source. Further work will, however, be required to determine the potential of this technique in a larger clinical setting.

REFERENCES

[1] Bishop R E D 1979 *Vibration* 2nd edn (Cambridge University Press)
[2] Chaffin D B and Andersson G B J 1991 *Occupational Biomechanics* 2nd edn (Chichester: Wiley)

[3] Neal M J 1988 A review of condition monitoring techniques. In *Non-Destructive Testing* vol 1 eds J M Farley and R W Nichols (Oxford: Pergamon) pp 134–8

[4] Halmshaw R 1991 *Non-Destructive Testing* 2nd edn (Edward Arnold)

[5] Cawley P and Adams R D 1988 The mechanics of the coin-tap method of non-destructive testing *J. Sound Vibr.* **122** 299–316

[6] Van der Perre G 1984 Dynamic analysis of human bones. In *Functional Behaviour of Orthopaedic Biomaterials* eds P Ducheyne and G W Hastings (Boca Raton, FL: CRC Press) pp 99–159

[7] Nokes L D M and Thorne G C 1988 Vibrations in orthopaedics *CRC Crit. Rev. Biomed. Eng.* **15** 309–49

[8] Nokes L D M 1999 The use of low frequency vibration measurement in orthopaedics *Proc. Inst. Mech. Eng.* [H] **213** 271–90

[9] Jurist J M 1970 *In-vivo* determination of the elastic response of bone. I. Method of ulnar resonant frequency determination *Phys. Med. Biol.* **15** 417–26

[10] Jurist J M 1970 *In-vivo* determination of the elastic response of bone. II. Ulnar resonant frequency in osteoporotic, diabetic and normal subjects *Phys. Med. Biol.* **15** 427–34

[11] Doherty W P, Bovill E G and Wilson E L 1974 Evaluation of the use of resonant frequencies to characterize physical properties of human long bones *J. Biomech.* **7** 559–61

[12] Vandecasteele J, Van der Perre G, Van Audekercke R and Martens M 1981 Evaluation of bone strength and integrity by vibration methods: identification of *in-vivo* excited modes. In *Mechanical Factors and the Skeleton* ed. I A F Stokes (London: John Libbey) pp 98–105

[13] Young D R, Howard W H, Cann C and Steele C R 1979 Noninvasive measures of bone bending rigidity in the monkey (*M. memestrina*) *Calcif. Tissue Int.* **27** 109–15

[14] Borgwardt Christensen A B, Tougaard L, Dyrbye C and Vibe-Hansen H 1982 Resonance of the human tibia. Method, reproducibility and effect of transection *Acta Orthop. Scand.* **53** 867–74

[15] Evans E J 1985 Vibratory properties and resonances of the isolated human ulna *J. Biomed. Eng.* **7** 144–8

[16] Collier R J and Donarski R J 1987 Non-invasive method of measuring resonant frequency of a human tibia *in-vivo*. Part 1 *J. Biomed. Eng.* **9** 321–8

[17] Van der Perre G and Cornelissen P 1983 On the mechanical resonances of a human tibia *in-vitro*. *J. Biomech.* **16** 549–52

[18] Christensen A B, Ammitzboll F, Dyrbye C, Cornelissen M, Cornelissen P and Van der Perre G 1986 Assessment of tibial stiffness by vibration testing *in situ*—I. Identification of mode shapes in different supporting conditions *J. Biomech.* **19** 53–60

[19] Young D R, Thompson G A and Orne D 1976 *In-vivo* determination of mechanical properties of the human ulna by means of mechanical impedance tests: experimental results and improved mathematical model *Med. Biol. Eng.* **14** 253–62

[20] Orne D 1974 The *in-vivo*, driving-point impedance of the human ulna-a viscoelastic beam model *J. Biomech.* **7** 249–57

[21] Orne D and Mandke J 1975 The influence of musculature on the mechanical impedance of the human ulna, an *in-vivo* simulated study *J. Biomech.* **8** 143–9

[22] Steele C R, Zhou L J, Guido D, Marcus R, Heinrichs W L and Cheema C 1988 Noninvasive determination of ulnar stiffness from mechanical response-*in-vivo*

comparison of stiffness and bone mineral content in humans *J. Biomech. Eng.* **110** 87–96

[23] McCabe F, Zhou L J, Steele C R and Marcus R 1991 Noninvasive assessment of ulnar bending stiffness in women *J. Bone Min. Res.* **6** 53–9

[24] Van der Perre G, Van Audekercke R, Martens M and Mulier J C 1983 Identification of *in-vivo* vibration modes of human tibiae by modal analysis *J. Biomech. Eng.* **105** 244–8

[25] Cornelissen M, Cornelissen P, Van der Perre G, Christensen A B, Ammitzboll F and Dyrbye C 1987 Assessment of tibial stiffness by vibration testing *in-situ*—III. Sensitivity of different modes and interpretation of vibration measurements *J. Biomech.* **20** 333–42

[26] Tsuchikane A, Nakatsuchi Y and Nomura A 1995 The influence of joints and soft tissue on the natural frequency of the human tibia using the impulse response method. *Proc. Inst. Mech. Eng.* [Part H] **209** 149–55

[27] Pelker R R and Saha S 1985 Wave propagation across a bony discontinuity simulating a healing fracture *J. Biomech.* **18** 745–53

[28] Wong F Y, Pal S and Saha S 1983 The assessment of in vivo bone condition in humans by impact response measurement *J. Biomech.* **16** 849–56

[29] Chen I I and Saha S 1987 Wave propagation characteristics in long bones to diagnose osteoporosis. *J. Biomech.* **20** 523–7

[30] Nokes L D M, Mintowt-Czyz W J, Fairclough J A, Mackie I, Howard C and Williams J 1984 Natural frequency of fracture fragments in the assessment of tibial fracture healing *J. Biomed. Eng.* **6** 227–9

[31] Stussi E and Fah D 1988 Assessment of bone mineral content by *in vivo* measurement of flexural wave velocities *Med. Biol. Eng. Comput.* **26** 349–54

[32] Fah D and Stussi E 1988 Phase velocity measurement of flexural waves in human tibia *J. Biomech.* **21** 975–83

[33] Cheng S, Komi P V, Kyröläinen H, Kim D H and Häkkinen K 1989 *In vivo* vibrational wave propagation in human tibae at different ages *Eur. J. App. Physiol.* **59** 128–30

[34] Lowet G, Van Audekercke R, Van der Perre G, Geusens P, Dequeker J and Lammens J 1993 The relation between resonant frequencies and torsional stiffness of long bones in vitro. Validation of a simple beam model. *J. Biomech.* **26** 689–96

[35] Van der Perre G and Lowet G 1994 Vibration, sonic and ultrasonic wave propagation analysis for the detection of osteoporosis *Clin. Rheumatol.* **13** (Suppl. 1) 45–53

[36] Geusens P, Nijs J, Van der Perre G, Van Audekercke R, Lowet G, Goovaerts S, Barbier A, Lacheretz F, Remandet B, Jiang Y *et al* 1992 Longitudinal effect of tiludronate on bone mineral density, resonant frequency, and strength in monkeys *J. Bone Min. Res.* **7** 599–609

[37] Jurist M and Kianian K 1973 Three models of the vibrating ulna *J. Biomech.* **6** 331–42

[38] Spiegl P V and Jurist J M 1975 Prediction of ulnar resonant frequency *J. Biomech.* **8** 213–7

[39] Orne D and Young D R 1976 The effects of variable mass and geometry, pre-twist, shear deformation and rotatory inertia on the resonant frequencies of intact long bones: a finite element model analysis *J. Biomech.* **9** 763–70

[40] Hight T K, Piziali R L and Nagel D A 1980 Natural frequency analysis of a human tibia *J. Biomech.* **13** 139–47

[41] Collier R J, Nadav O and Thomas T G 1982 The mechanical resonances of a human tibia: part I—*in-vitro. J. Biomech.* **15** 545–53
[42] Thomsen J J 1990 Modelling human tibia structural vibrations *J. Biomech.* **23** 215–28
[43] Hobatho M C, Darmana R, Pastor P, Barrau J J, Laroze S and Morucci J P 1991 Development of a three-dimensional finite element model of a human tibia using experimental modal analysis *J. Biomech.* **24** 371–83
[44] Roberts S G, Hutchinson T M, Arnaud S B, Kiratli B J, Martin R B and Steele C R 1996 Noninvasive determination of bone mechanical properties using vibration response: a refined model and validation in vivo *J. Biomech.* **29** 91–8
[45] Couteau B, Hobatho M C, Darmana R, Brignola J C and Arlaud J Y 1998 Finite element modelling of the vibrational behaviour of the human femur using CT-based individualized geometrical and material properties *J. Biomech.* **31** 383–6
[46] Webb J, Herling G, Gardner T, Kenwright J and Simpson A H R W 1996 Manual assessment of fracture stiffness *Injury* **27** 319–20
[47] Nichols P J, Beng E, Bliven F E and King J M 1979 X-ray diagnosis of fracture healing in rabbits *Clin. Orthop. Rel. Res.* **142** 234–6
[48] Hammer R R R, Hammerby R and Lindholm B 1985 Accuracy of radiological assessment of tibial shaft fracture union in humans *Clin. Orthop. Rel. Res.* **199** 233–7
[49] Oni O O A, Hui A and Gregg P J 1988 The healing of closed tibial shaft fractures *J. Bone Joint Surg.* **70-B** 787–90
[50] Lowet G 1993 Vibration and ultrasound wave propagation analysis in long bones. Application to monitoring of fracture healing and detection of osteoporosis. Doctoraat Thesis, Katholieke Universiteit Leuven
[51] Misurya R K, Khare A, Mallick A, Sural A and Vishwakarma G K 1987 Use of tuning fork in diagnostic auscultation of fractures *Injury* **18** 63–4
[52] Siffert R S and Kaufman J J 1996 Acoustic assessment of fracture healing. Capabilities and limitations of 'a lost art' *Am. J. Orthop.* **25** 614–8
[53] Sonstegard D A and Matthews L S 1976 Sonic diagnosis of bone fracture healing—a preliminary study *J. Biomech.* **9** 689–94
[54] Sekiguchi T and Hirayama T 1979 Assessment of fracture healing by vibration *Acta Orthop. Scand.* **50** 391–8
[55] Nokes L, Mintowt-Czyz W J, Fairclough J A, Mackie I and Williams J 1985 Vibration analysis in the assessment of conservatively managed tibial fractures *J. Biomed. Eng.* **7** 40–4
[56] Flint A J, Nokes L D and Macheson M 1994 Determination of fracture healing by transverse vibration measurement: a preliminary report *J. Med. Eng. Technol.* **18** 205–7
[57] Markey E L and Jurist J M 1974 Tibial resonant frequency measurements as an index of the strength of fracture union *Wis. Med. J.* **73** S62–5
[58] Collier R J and Donarski R J 1987 Non-invasive method of measuring the resonant frequency of a human tibia *in-vivo*. Part 2 *J. Biomed. Eng.* **9** 329–31
[59] Peltier L F 1977 The diagnosis of fractures of the hip and femur by auscultatory percussion *Clin. Orthop.* **123** 9–11
[60] Lewis J L 1975 A dynamic model of a healing fractured long bone *J. Biomech.* **8** 17–25
[61] Cunningham J L, Kenwright J and Kershaw C J 1990 Biomechanical measurements of fracture repair *J. Med. Eng. and Technol.* **14** 92–101

[62] Tower S S, Beals R K and Duwelius P J 1993 Resonant frequency analysis of the tibia as a measure of fracture healing *J. Orthop. Trauma* **7** 552–7

[63] Benirschke S K, Mirels H, Jones D and Tencer A F 1993 The use of resonant frequency measurements for the noninvasive assessment of mechanical stiffness of the healing tibia *J. Orthop. Trauma*. **7** 64–71

[64] Nakatsuchi Y, Tsuchikane A and Nomura A 1996 The vibrational mode of the tibia and assessment of bone union in experimental fracture healing using the impulse response method *Med. Eng. Phys.* **18** 575–83

[65] Nakatsuchi Y, Tsuchikane A and Nomura A 1996 Assessment of fracture healing in the tibia using the impulse response method *J. Orthop. Trauma* **10** 50–62

[66] Akkus O, Korkusuz F, Akin S and Akkas N 1998 Relation between mechanical stiffness and vibration transmission of fracture callus: an experimental study on rabbit tibia *Proc. Inst. Mech. Eng.* [Part H] **212** 327–36

[67] Lewis J L 1975 A dynamic model of a healing fractured long bone *J. Biomech.* **8** 17–25

[68] Laura P A, Rossi R E and Maurizi M J 1990 Dynamic analysis of a simplified bone model during the process of fracture healing *J. Biomed. Eng.* **12** 157–60

[69] Nikiforidis G, Bezerianos A, Dimarogonas A and Sutherland C 1990 Monitoring of fracture healing by lateral and axial vibration analysis *J. Biomech.* **23** 323–30

[70] Lowet G, Dayuan X and Van der Perre G 1996 Study of the vibrational behaviour of a healing tibia using finite element modelling *J. Biomech.* **29** 1003–10

[71] Roberts S G and Steele C R 2000 Efficacy of monitoring long-bone fracture healing by measurement of either bone stiffness or resonant frequency: numerical simulation *J. Orthop. Res.* **18** 691–7

[72] Campbell J N and Jurist J M 1971 Mechanical impedance of the femur: a preliminary report *J. Biomech.* **4** 319–22

[73] Collier R J, Donarski R J, Worley A J and Lay A 1993 The use of externally applied mechanical vibrations to assess both fractures and hip prostheses. In *Micromovement in Orthopaedics* ed A R Turner-Smith (Oxford: Clarendon Press) pp 151–63

[74] Collier R J and Ntui J A 1994 *In-vivo* measurements of the phase constants of transverse mechanical waves in a human tibia from 100 to 1000 Hz *Med. Eng. Phys.* **16** 379–83

[75] Chung J K, Pratt G W, Babyn P S, Poss R and Brigham R B 1979 A new diagnostic technique for the evaluation of prosthetic fixation. *Proc. of the first annual conference of the IEEE/Engineering in Medicine and Biological Sciences*, New York pp 158–60

[76] Poss R, Pratt G W and Chung J K 1984 An evaluation of total hip replacement cementing technique using sonic resonance *Eng. Med.* **13** 191–6

[77] Rosenstein A D, McCoy G F, Bulstrode C J, McLardy-Smith P D, Cunningham J L and Turner-Smith A R 1989 The differentiation of loose and secure femoral implants in total hip replacement using a vibrational technique: an anatomical and pilot clinical study *Proc. Inst. Mech. Eng.* [Part H] **203** 77–81

[78] Li P S, Jones N B and Gregg P J 1995 Loosening of total hip arthroplasty. Diagnosis by vibration analysis *J. Bone Joint Surg.* **77-B** 640–4

[79] Li P S, Jones N B and Gregg P J 1996 Vibration analysis in the detection of total hip prosthetic loosening *Med. Eng. Phys.* **18** 596–600

[80] Georgiou A P and Cunningham J L 2001 Accurate diagnosis of hip prosthesis loosening using a vibrational technique *Clin. Biomech.* **16** 315–23

SECTION 5

CLINICAL APPLICATIONS OF PHYSICAL MEASUREMENT OF BONE

Chapter 17

Human studies

Christopher F Njeh, John Shepherd and Harry K Genant

17.1. INTRODUCTION

Although conventional radiographs are readily available and fairly inexpensive, estimation of bone mineral density (BMD) from appearance is insensitive and inaccurate, since the subjective assessment is influenced by radiographic exposure factors, patient size and film processing techniques [1, 2]. BMD must decline by as much as 40% before it can be detected reliably on radiographs [3]. These factors have supported the need for objective, non-invasive methods of bone density measurements. These methods should be accurate, precise (reproducible), sensitive, inexpensive and involve minimal exposure to ionizing radiation. Current techniques include radiographic absorptiometry (RA), single X-ray absorptiometry (SXA), dual X-ray absorptiometry (DXA), quantitative computed tomography (QCT) and quantitative ultrasound. These techniques have been extensively described in chapters 8–14, varying in precision, accuracy, fracture discrimination and fundamental methodology, along with their clinical and research utility (table 17.1). This chapter discusses the clinical utility of bone densitometry in evaluating bone status. Although primary osteoporosis is the main use of bone densitometry, it is also used to assess bone status in paediatrics, subjects with secondary osteoporosis and arthritis.

SECTION A: PRESENTATION OF BMD

17.2. UNITS OF MEASURE

Bone fragility and health can be quantified by measuring the bone mass in regions of interest (ROIs) throughout the body. Since bone mass, measured

Table 17.1. *Comparison of available modalities for bone status assessment (Adapted from Genant et al [27] and Scheiber [50]).*

Technique	Precision error (%)	Accuracy error (%)	Effective dose* (μSv)	Advantages	
RA				Low cost per test/ equipment	Limited to the phalanges
Phalanx/metacarpal	1–2	5	~5	Equipment mobile	
SXA/pDXA				Low cost per test/ equipment	Limited to wrist or heel Area density
Radius/calcaneus	1–2	4–6	<1	Equipment mobile Low radiation dose	Limited correlation to spine/hip
DXA				Multiple sites capability	Limited mobility
PA spine	1–1.5	4–10	~1	Low radiation exposure	Areal density
Lateral spine	2–3	5–15	~3		Moderate cost of equipment
Proximal femur	1.5–3	6	~1		
Forearm	~1	5	<1		
Whole body	~1	3	~3		
QCT				Volumetric density	Higher radiation exposure
Spine trabecular	2–4	5–15	~50		Recalibration between test
Spine integral	2–4	4–8	~50		Difficulty measuring the hip
pQCT				Volumetric density	Limited to wrist
Radius trabecular	1–2	?	~1	Equipment mobile	Limited correlation with spine/hip
Radius total	1–2	2–8	~1	Low radiation dose	
QUS				Radiation free	Limited to peripheral sites
SOS calcaneus/tibia	0.3–1.2	?	0	Low cost per test/ equipment	Limited correlation to the spine/hip
BUA calcaneus	1.3–3.8	?	0	Equipment mobile	

* Dose for annual background ~2000 μSv, for abdominal radiograph ~500 μSv and for abdominal CT ~4000 μSv [34]).

in grams, is dependent on the size of the bone and the size of the region of interest, the bone mass is typically normalized to bone area to create an areal bone density in grams per unit area (mg/cm^2) or to bone volume and represented as a volumetric bone density in grams per unit volume (mg/cm^3). For anterior–posterior (AP) spine and total femur BMD, the units of measure have been standardized across manufacturers such that the BMD measured on different devices and different manufacturers is on average the same. To distinguish this standardized BMD (sBMD) from the manufacturer-specific BMD, sBMD is reported in units of milligrams per centimetre squared (mg/cm^2). Both areal bone density and volumetric bone density are commonly referred to as bone mineral density (BMD). The most common regions of interest to diagnose osteoporosis are areas where fragility fractures are likely to occur, namely, the spine, proximal femur and the distal forearm. All of the regions can be measured using projectional dual X-ray absorptiometry (DXA) techniques for areal density or quantitative computed tomography for volumetric density. In the rest of this chapter, we make no distinction between areal or volumetric density except where noted since the clinical information is applicable in general to all clinically available measurement sites.

17.3.   REFERENCE POPULATION

There is no consensus on how bone mineral density (BMD) measurements should be presented. To be clinically useful, BMD results for individual patients must be related to similar values obtained from a healthy reference population, yet there are no universal standards for reference data. However, it is typical to use, when available, a population that most closely describes the patient in terms of sex, race and country of origin. For many combinations, exact matches are not available and the clinician has to use the reference data available and be aware of possible differences. As an example, Huang *et al* [4] found that native Japanese women had an average BMD 6% lower than Japanese American women living in Hawaii. Reference populations are usually described in terms of the mean BMD and standard deviation of the population as a function of age, sex and race, and in some cases nationality. Figure 17.1 shows the sex and race mean BMD values for the total femur ROI from the National Health and Nutrition Examination Survey (NHANES) III [5]. As can be seen, what is considered normal bone density can vary by as much as 25% between black males and Caucasian females of the same age.

17.4.   T-SCORES

T-scores are comparisons of the patient's BMD with a population's peak reference values and can be represented in units of population standard

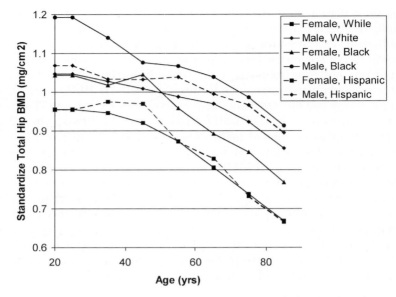

Figure 17.1. *Reference data from the NHANES III study [5]. BMD is shown in standardized BMD units.*

deviations (SD) or as a percentage of the reference population's peak value. The T-score$_{sd}$ represents the number of SD a BMD measurement is above or below the mean peak bone mass of a young normal population matched for sex and race. In units of standard deviations, T-score$_{sd}$ is defined as

$$\text{T-score}_{sd} = \frac{(\text{BMD}_{\text{patient}} - \text{BMD}_{\text{peak}})}{\text{Std.Dev}_{\text{peak}}} \qquad (17.1)$$

where BMD$_{\text{patient}}$ is the patient's BMD value, BMD$_{\text{peak}}$ is the reference peak BMD value, and Std.Dev$_{\text{peak}}$ is the population standard deviation at that peak value. As a percentage of the population's peak value, T-score$_{\%}$ is defined as

$$\text{T-score}_{\%} = \frac{\text{BMD}_{\text{patient}}}{\text{BMD}_{\text{peak}}} \times 100. \qquad (17.2)$$

17.5.   Z-SCORES

Z-scores are comparisons of the patient's BMD to a population's age-matched reference values and can also be represented in units of population standard deviations or as a percentage of the reference population's peak value. The

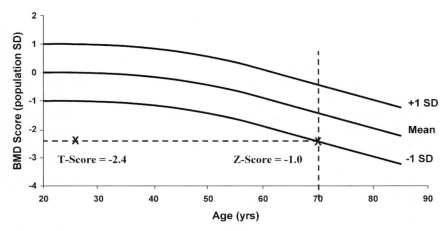

Figure 17.2. *Calculating T- and Z-scores. Graph is for the total femur from the NHANES III study for female Caucasian. The patient shown is a 60-year-old Caucasian woman with a BMD of 0.64 mg/cm². Her Z-score is −1.0 while her T-score is −2.4.*

Z-score$_{sd}$ expresses the number of SD a subject differs from the mean value for an age, sex and race matched reference population. Z-scores should not be used to define osteoporosis or fracture risk since their use would result in the apparent prevalence and risk of the disease not increasing with age. The Z-score, however, is quite useful clinically in assessing a patient's skeletal status relative to his or her peers. In units of standard deviations, Z-score$_{sd}$ is defined as

$$\text{Z-score}_{sd} = \frac{(\text{BMD}_{patient} - \text{BMD}_{age\text{-}matched})}{\text{Std.Dev}_{age\text{-}matched}} \qquad (17.3)$$

where $\text{BMD}_{patient}$ is the patient's BMD value, $\text{BMD}_{age\text{-}matched}$ is the reference BMD value at the same age as the patient, and $\text{Std.Dev}_{age\text{-}matched}$ is the population standard deviation at the patient's age. As a percentage of the population's age-matched BMD value, Z-score$_\%$ is defined as

$$\text{Z-score}_\% = \frac{\text{BMD}_{patient}}{\text{BMD}_{age\text{-}matched}} \times 100. \qquad (17.4)$$

A sample calculation of the T- and Z-score is shown graphically in figure 17.2. A 70-year-old female patient has a BMD that is 2.4 population standard deviations below the peak mean BMD value of a young reference population and 1 population standard deviation below the mean BMD value for women her age in the reference population. From figure 17.2, we see that her Z-score$_{sd}$ is −1.0 while her T-score$_{sd}$ is −2.4.

SECTION B: INTERPRETATION OF BMD RESULTS

17.6. WHO CRITERIA

In 1994, the World Health Organization established a definition for osteoporosis and its milder form, osteopenia based on either bone mineral content (BMC) or bone mineral density (BMD) alone [36] (see table 17.2). The criteria define patients with osteoporosis as having a BMC or BMD value that is more than 2.5 SD below the mean of normal peak bone mass. For osteopenia, the criterion is a BMC or BMD value between 1 SD and 2.5 SD below the mean of normal peak bone mass. These recommendations, originally intended for population-based studies, have been widely used as individual diagnostic and treatment thresholds in the absence of other factors. This cutoff approach is certainly easy to apply. However, there are inherent problems to such a simplistic threshold approach. For one, it is heavily dependent on estimates of young adult reference means and standard deviations; different patterns of bone loss at specific sites are not taken into consideration [6]. In addition, it ignores the continuous increase of risk with decreasing BMD. These are general guidelines for diagnosis but are not intended to require or restrict therapy for individual patients. Rather, the physician and the patient should use the BMD information in conjunction with knowledge of the patient's specific medical and personal history to determine the best course of action for each individual. This definition is particularly important for those who have osteopenia, since a person whose bone density is 1 SD below the young normal mean might not have pathologically low bone mass today but still could be at sufficiently high risk of fracture over their remaining lifetime [7]. The density information may help people to start early preventive

Table 17.2. *WHO definitions of osteoporosis based on BMD or BMC value [36].*

Classification	Description
Normal	A value of BMD greater than 1 SD below the average value of a young adult ($T > -1$)
Low bone mass (osteopenia)	A value of BMD more than 1 SD below the young adult average but not more than 2.5 SD below ($-2.5 < T \leq -1$)
Osteoporosis	A value of BMD more than 2.5 SD below the young adult average value ($T \leq -2.5$)
Severe (established) osteoporosis	A value for BMD more than 2.5 SD below the young adult average and there has been one or more osteoporotic fracture

treatments, thus reducing the probability of fracture occurrence in the future.

17.7. LIMITATIONS OF WHO CRITERIA

The WHO's definition of osteoporosis was based on the notion that bio-mechanical competence of the skeleton is optimal among young adults but is degraded as bone loss occurs with ageing [8]. However, there may not be a unique age for peak bone mass that applies consistently to all skeletal sites. Evidence from previous studies shows that bone loss from different bone sites has different timings. The proximal femur begins to lose bone mass in the early 20s [9]; the spine starts to lose after the 30s [6] and the distal forearm starts to lose at the time of menopause [10]. In addition, there is no unequivocal histological dividing line between osteoporosis and normal. BMD is a normally distributed (continuum) variable in the population and there is a significant overlap between BMD in normal and fracture populations. For example, a lumbar spine BMD that is 2 SD below the young adult mean has a sensitivity and specificity for fracture of only around 60% [11]. Hence specific threshold-based diagnosis with T-scores results in substantial misclassification from one site or technique to the next.

The applicability of the WHO criteria to groups other than white women is also not certain. For instance, Asian women have a much lower BMD than white women but have a lower rate of hip fracture than white women, which warrants separate differential criteria [12, 13]. Moreover, when based on female BMD cutoff values, 1.4% of men have osteoporosis and 15–33% have osteopenia; when the male BMD cutoff values are used, it was found that 3–6% of men have osteoporosis and 28–47% have osteopenia [51].

Another limitation resulting from defining osteoporosis as a state of bone mass is that it does not include information about the geometry and quality of bone, which is also an important component of fracture risk. The occurrence of fracture is a consequence not only of low bone mass but also change of bone quality. For example, two people may have the same low mineral density. In one, the bone may reflect diffusely thin trabeculae with the maintenance of normal trabecular connection and, in the other, the bone may exhibit thicker trabeculae with reduced trabecular connectivity. The first person may have a trivial fracture risk, whereas the bones of the second person may be in real jeopardy [14]. Current densitometric tools for the clinical assessment of bone mass would not distinguish between these two situations. Criteria based on BMC or BMD measurements, therefore, may not be adequate for diagnostic and therapeutic decision-making in an individual patient. While the WHO's definition may be of limited effectiveness, it nevertheless avoids the need to restrict treatment to end-stage disease [15, 16] and to date there is no better alternative.

17.8.   NOF RECOMMENDATIONS

The National Osteoporosis Foundation (NOF) is a non-profit US health organization dedicated to promoting bone health to reduce the widespread prevalence of osteoporosis and associated fractures. It has made treatment recommendations based on BMD and other risk factors [17].

> *'Initiate therapy to reduce fracture risk in women with BMD T-scores below −2 in the absence of risk factors and in women with T-scores below −1.5 if other risk factors are present.'*

This guideline is an enhancement of the WHO criteria that includes other risks factors in addition to just the BMD T-score. The precise threshold values for treatment mentioned are currently under review.

17.9.   FRACTURE RISK ASSESSMENT

In a clinical setting, the essential role of bone densitometry is to identify patients at risk of osteoporotic fracture. There have been several large epidemiological studies to estimate fracture risk from BMD measurement at various anatomical sites. The Study of Fractures (SOF) quantified hip fracture risks on 8134 Caucasian women aged from 65 to more than 85 years from BMD measurements of the proximal hip, lumbar spine, forearm and the calcaneus [18]. Figure 17.3 shows the relative fracture risk as a function of Z-score using the SOF results. Thus, a post-menopausal Caucasian woman with a Z-score of −2 at the total hip has an approximate increase of fracture risk of the hip by 5.4 times that of women her age with a Z-score of 0.

Melton *et al* [19] introduced the concept of lifetime fracture risk. This model uses clinical risk factors to determine the risk of fracture in the patient's remaining lifetime. The concept is that, for a given proximal femur BMD, the lifetime fracture risk decreases with increasing age and decreasing life expectancy. For example, a woman with a high relative risk of fracture at age 85 may have a lower lifetime fracture risk than a woman with a lower relative fracture risk at age 70 because of the longer life expectancy of the latter. Although fracture risk data are available on select populations, such as Caucasian women, there is no universally accepted method to clinically utilize fracture risk for diagnosis. Wasnich [20] proposed the concept of remaining lifetime fracture probability (RLFP). RLFP is a reporting tool that determines absolute fracture probabilities calculated from long-term epidemiological studies. The model allows for input from a variety of BMD

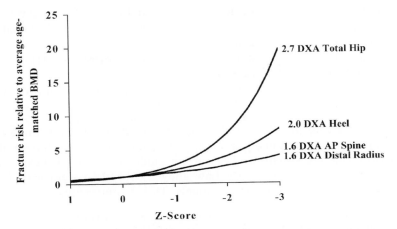

Figure 17.3. *Fracture risk associated with the BMD measurement from different measurement sites taken from the SOF study. The risk of fracture relative to women at the mean BMD is shown as a function of increasing Z-score.*

measurement sites in specific BMD units from a variety of manufacturers, ethnicities and gender. The physician can choose specific treatment strategies for individual patients and view how the RLFP changes over the short term (the next year) and the long term (remaining life expectancy).

SECTION C: UTILITY OF BMD

17.10. WHO SHOULD BE TESTED?

There are general guidelines for diagnosis, which are not intended to require or restrict therapy for individual patients. Rather, the physician and the patient should use the BMD information in conjunction with knowledge of the patient's specific medical and personal history to determine the best course of action for each individual.

Given the current awareness of osteoporosis, bone densitometry is becoming widely used in routine medical practice, yet BMD should principally be assessed after consideration of the individual's risk profile, and testing is generally not indicated unless the results could influence a treatment decision. A consensus is forming as to when a bone density scan is appropriate in the assessment of osteoporosis. The following are the recommendations made by the National Osteoporosis Foundation [17]. BMD testing should be performed on the following.

- All post-menopausal women under age 65 who have one or more additional risk factors for osteoporotic fracture (besides menopause).

- All women aged 65 and older regardless of additional risk factors.
- Post-menopausal women who present with fractures (to confirm diagnosis and determine disease severity).
- Women who are considering therapy for osteoporosis, if BMD testing would facilitate the decision.
- Women who have been on hormone replacement therapy for prolonged periods.

However, the guidelines for reimbursement from insurance companies and government coverage in many cases drive the patient's and clinician's ability to perform the measurement. The current guidelines from Medicare were established by a 1998 Medicare Law allowing for the coverage of bone density tests for five groups of individuals. These groups include:

1. Oestrogen-deficient women at clinical risk of osteoporosis.
2. Individuals with vertebral abnormalities as demonstrated by X-ray to be indicative of osteoporosis, low-bone mass or vertebral fracture.
3. An individual receiving long-term glucocorticoid (steroid) therapy.
4. An individual with primary hyperparathyroidism.
5. Individuals being monitored to assess the response to or efficacy of an FDA-approved osteoporosis drug therapy.

The guidelines state that these individuals will be covered for BMD tests at a frequency of every two years. The exception to the two-year rule are individuals receiving high steroid doses (i.e. ≥ 7.5 mg/day of prednisone), who qualify for a six-month or yearly measurement, and individuals who are first diagnosed on a peripheral BMD site and an axial baseline BMD is desired.

17.11.   HOW TO APPLY BMD

It should be recognized that the best use of any bone densitometry technique depends on the nature of the clinical problem and the age of the patient, as well as technical factors [21]. The primary purpose of measuring bone status should be to assess fracture risk in individual patients in order to make clinical decisions about intervention to minimize that risk. In fact, a meta-analysis demonstrated that BMD is a better predictor of the risk of fracture than cholesterol is of coronary disease or blood pressure is of a stroke [22]. Bone densitometry is appropriate in the evaluation of perimenopausal women if the result will influence subsequent clinical decisions. As an example, a patient with severe symptoms of oestrogen deficiency would be appropriately treated with oestrogen regardless of the bone density. However, if the density value will affect a decision to undertake HRT (or an alternative such as alendronate or raloxifene), then testing is clearly defensible. An international panel has recently formulated guidelines on the clinical use and interpretation of BMD [23]. The main consensus statements are listed below.

- Bone mass measurements predict a patient's future risk of fracture.
- Osteoporosis can be diagnosed on the basis of bone mass measurements even in the absence of prevalent fractures.
- Bone mass measurements provide information that can affect the management of patients.
- The choice of the appropriate skeletal measurement site(s) may vary depending on the specific circumstances of the patient.
- The technique chosen for bone mass measurements should be based on an understanding of the strength and limitations of the different techniques.
- Bone mass data should be accompanied by a clinical interpretation.

17.12. DIAGNOSTIC ALGORITHMS

Several examples of diagnostic algorithm have been promoted for osteo-porosis evaluation. The NOF has released recommendations that are based heavily on the initial willingness of the patient to consider a treatment option [17]. In this method, the BMD itself is not available to the patient if, after counselling with their physician, the patient is not considering the use of a drug treatment. An alternative approach is shown in figure 17.4. The patient is assessed through a clinical interview to determine if he/she is at risk of fracture. If not, he/she is then reassessed at a later date and, if found to have risk factors, then a quantitative risk assessment is performed. This may be any one of the following: BMD assessment of the hip and spine; peripheral BMD site such as the forearm, heel or hand; quantitative ultra-sound assessment of the heel, forearm or hand; or X-ray examination of the spine to identify prevalent but unknown fractures. If the quantitative assessment does not show an elevated risk, then the patient is reassessed at a later date.

If the patient is at an elevated risk, then the clinician must try to evaluate if the osteoporosis is due to secondary causes such as inflammatory disorders (RA), bone marrow disorders, disorders associated with low-body weight, defects of connective tissue synthesis (OI), malabsorption disorders, endocrinological disorders (hyperparathyroidism), hypogonadism or immobilization. If any of these conditions are found, they should be addressed. Conversely, when a patient with a known metabolic bone disorder seeks medical attention, a BMD measurement may be obtained to determine the magnitude of the deleterious effect of the disorder on the skeleton and, in this setting, the use of the Z-score is particularly helpful, independent of the patient's T-score or fracture risk.

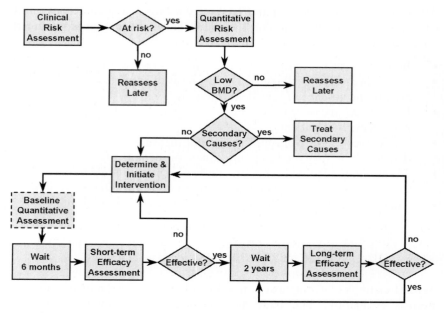

Figure 17.4. *Diagnostic algorithm for the diagnosis of osteoporosis.*

With no secondary causes present, the clinician and patient would discuss and initiate intervention strategies. Once chosen, and if the previous quantitative assessment would not be adequately sensitive to monitor drug therapy, then a baseline BMD assessment should be performed with the measurement device and site that is most responsive to therapy such as the spine or hip. A biochemical marker test may also be useful to determine short-term efficacy of the treatment plan (i.e. was there a change in the absorption and/or resorption bone markers? Is the patient tolerating taking the medication? etc). If the patient is proceeding, then a delay of one to three years is appropriate before quantitative long-term efficacy assessment. If the patient is not showing signs of a decrease in fracture risk, the intervention method being used should be re-evaluated and the algorithm started over again. This method accounts for using different technologies and measurement sites to manage the patient.

SECTION D: WHICH SITE TO MEASURE

17.13. AVAILABLE SITES

Different skeletal sites contain different proportions of trabecular and cortical bone, which differ in their rates of bone loss. The spine measured

AP or PA by DXA contains approximately 66% trabecular bone, while the femoral neck contains only about 25% trabecular bone. The proximal forearm consists of predominantly cortical bone (95%), while the distal and ultradistal contain about 50% cortical bone [24, 25]. However, osteoporosis is a systemic disease and loss of bone occurs at all sites albeit at different rates. Hence, one could assume that for diagnostic purposes bone density could be measured at any site, except for the lumbar spine in the very elderly (in whom degenerative disease falsely elevates DXA-derived BMD) [18, 26]. Low bone density at the many different anatomical sites is significantly associated with the risk of osteoporotic fracture [18, 23]. However, BMD assessed in areas remote from the fracture site generally has weaker predictive power.

In the previously mentioned meta-analysis [22] for the prediction of fracture occurring at any site, the risk ratios for different measurement sites were similar, implying that any bone density measurement could be used. However, for site-specific fracture prediction, that site should be measured (i.e. measure the hip when predicting hip fracture) [21].

Chapters 8–14 in this book described the various techniques available for the assessment of bone status at different anatomical sites. Some prospective studies [22, 27, 28] have shown that these techniques are comparable in their ability to identify patients at risk of fracture. Although these techniques were able to identify different groups of subjects, none to date have provided absolute discrimination between fracture and non-fracture subjects [29, 30]. Once a measurement site has been chosen, the best technique to measure at that site will depend upon at least the precision, accuracy, availability and cost [21]. It is important to recognize that bone densitometry provides only an estimate of the patient's fracture risk. The clinician should interpret the scan findings along with other clinical risk factors (table 17.2) to determine the best management for the individual patient.

17.14. LIMITATIONS

The clinician should be aware of the limitations of BMD, such as the confounding effect of other diseases or medication upon a particular anatomical site. For example, changes in BMD that occur in the immediate post-menopausal period or as a result of osteoporosis treatment are often more marked, and can be detected earlier, in the spine than at the hip or wrist. In the elderly, where degenerative disease may be prevalent, BMD assessed by spinal QCT or at an appendicular site such as the femur may be more effective for risk assessment. Alternatively, lateral DXA scanning of the lumbar spine reduces the influence of degenerative changes and may be a more suitable measurement in the elderly and perhaps in younger subjects as well [31].

17.15. COMBINING SITES TO INCREASE DIAGNOSTIC POWER

The question of whether more information would be gained by making multiple measurements has also been addressed. Several studies have documented the discordance of diagnosis from measurements at different sites and the possibility of frequent misdiagnosis of women if BMD is measured at a single site [32]. Other reports suggest that, for hip fracture, BMD at the hip is the superior measurement and that additional measurement at the spine, calcaneus or radius adds little new information [27, 33]. In contrast, Wasnich and colleagues [32] found that combining BMC measurements from multiple sites improved the prediction of incident vertebral fracture. Lu *et al* [49] showed that combining the four sites of the hip increased the relative fracture risk from 2.4 (if only one of the four sites was osteoporotic) to 8.6 (if all four sites were osteoporotic). However, there are no current guidelines as to how to combine the results from different regions of interest or different sites for any particular manufacturer. Further investigation is required to clarify the role of multiple measurements with particular attention to the type of fracture, the characteristics of the population studied, and the statistical techniques employed. Until this is resolved, it is advisable to scan several sites if practical, or at least consider all of the regional measures, when testing for osteoporosis.

SECTION E: TREATMENT CONSIDERATIONS

Serial measurements are very useful to monitor the natural history of the condition or to assess response to therapy. The choice of anatomical site for longitudinal assessment depends on two variables: the rate of change of bone mass within the skeleton itself and the precision of bone densitometry measurement. In general, the measured change in bone density should be 2.8 times the long-term precision error for the measured variable in order to be statistically significant [34, 35]. The precision of serial DXA scans is approximately 1–2% when performed by an experienced technologist. Thus a change of bone mass of at least 3–6% at the spine and 6–8% at the femoral neck is required to be considered significant. The rate of bone loss as assessed by DXA of the spine or hip in early menopause is on average 2% per year, but may range from less than 1% by radial SXA to more than 5% per year by spinal QCT [36]. Hence, measurements at about one- to two-year intervals are often sufficient to document the natural history of bone change [37]. Testing too infrequently may mean that patients who are rapidly losing bone may be missed and go untreated. For a patient with borderline bone mass measurements, repeated measurements or alternative measurement sites may help to clarify decisions on intervention. For patients with secondary conditions producing osteoporosis, such as corticosteroid therapy, a rapid rate of bone loss can be expected and more frequent measurements such as every six months may be useful [38].

Repeated measurements are also useful for assessing the response to therapy. The majority of patients respond well to HRT and bisphosphonates. Measurement of the spine at one- to two-year intervals will detect these significant improvements. Generally, follow-up measurements should be used to establish that the therapeutic intervention is effective, as well as to provide incentive for the patient to maintain treatment. It is difficult for individuals to continue long-term treatment programmes, especially when the disease is asymptomatic, when there is no direct impact on their sense of wellbeing. In applying serial measures however, one must be aware of over-interpreting modest changes and of the phenomenon of 'regression to the mean', which can confound interpretation of serial changes [39].

Poor compliance can impact the long-term preventive strategy of osteoporosis management. Poor compliance is particularly observed in patients prescribed hormone replacement therapy (HRT). One study in the UK revealed that almost 40% of women prescribed HRT were not taking treatment after an average of eight months [40]. Compliance with treatment for osteoporosis can be enhanced through regular doctor/nurse follow-up, including education about the benefits of treatment and which side effects to expect. There is some evidence to suggest that knowledge of bone density enhances the acceptance of prescribed lifestyle changes and therapeutic intervention [41, 42]. Monitoring could also be used to follow patients after cessation of therapy [43].

Another important reason for bone densitometry measurement in treatment decision-making has recently been established. It has been shown that in subjects with prevalent vertebral fractures, those individuals with low BMD may benefit more from anti-resorptive treatment than those with high BMD [44]. Similarly, it has been reported that among those who have not yet suffered a vertebral fracture, only individuals who are in the osteoporotic range will benefit from anti-resorptive treatment, at least with regard to short-term fracture reduction [45]. Subjects with a normal BMD, however, did not show fracture reduction.

SECTION F: MEASUREMENT ERRORS

The subject of measurement errors has been discussed in great detail in chapter 4. These will be discussed here briefly for completeness. Measurement errors are inherent in quantitative bone densitometry measurements, there being two types: accuracy error and precision error [46]. Accuracy error (also termed 'validity') refers to the ability of a technique to measure the true value. To evaluate accuracy errors we need to know the true values of the measured parameters. For most BMD techniques, accuracy has generally been defined as the degree to which bone densitometry is able to estimate the calcium content. The measured BMC or BMD is compared directly with the ash or volumetric weight of the measured bone. One should note that for clinical applications only the fraction of the accuracy error that varies from patient to patient in an unknown fashion is of relevance. The other fraction, i.e. the systematic, constant inaccuracy, can be averaged across subjects and is not important for two reasons. First, for diagnostic use of the technique, the reference data will be affected by the same systematic error and thus the T- and Z-scores remain unaffected. The relative relationship between healthy and osteoporotic bone is maintained. Second, when monitoring a patient's bone change, the systematic error is present in both the baseline and follow-up measurements and does not contribute to the measured change [47]. For these reasons, small systematic accuracy errors are of little clinical significance provided they remain constant.

Precision error (also termed reliability) reflects the reproducibility of a technique. It measures the ability of a technique to measure a parameter

Table 17.3. *Factors contributing to osteoporosis (adapted from Kanis et al [48]).*

Genetic
Ethnicity: Caucasian or Asian
Family history of fracture
Small body stature

Lifestyle and nutritional
Cigarette smoking
Excessive alcohol consumption
Sedentary or prolonged immobilization

Medical conditions
Anorexia nervosa
Rheumatoid arthritis
Amenorrhoea
Hyperthyroidism
Primary Hyperparathyroidism
Multiple myeloma
Transplantation
chronic diseases—kidney, lung or gastrointestinal

Drugs
Chronic corticosteriod therapy
Excessive thyroid therapy
Anticoagulants
Chemotherapy
Gonadotropin-releasing hormone agonist or antagonist
Anticonvulsant

consistently over multiple measurements. In the clinical context of monitoring, precision is more important than accuracy. There are two major sources of precision error: equipment and technologist/patient. Precision errors can be further separated into short-term and long-term precision errors. Short-term precision errors characterize the reproducibility of a technique and are useful in describing the limitations of measuring changes in skeletal status. Long-term precision errors are used to evaluate instrument stability.

17.16. CONCLUSIONS

There is a growing awareness of osteoporosis and its economic and social cost. The value of bone status assessment in fracture risk prediction is as good as or better than the ability of blood pressure to predict a stroke or serum cholesterol to predict a heart attack. Currently, DXA is the most widely used technique for bone mineral assessment. There is an increased momentum to establish the role of peripheral techniques such as QUS and pDXA due to their low costs and potentially wide availability. This expanded access is also driven by the availability of effective and acceptable therapies for osteoporosis. Clinical questions, availability and cost should govern the choice of site, number of sites and the technique. BMD or ultrasound should not be used in isolation in the assessment of osteoporosis, but should be combined with risk factors (see table 17.3), medical history, physical examination, and other laboratory tests in overall patient management.

REFERENCES

[1] Williamson M R, Boyd C M and Williamson S L 1990 Osteoporosis: diagnosis by plain chest film versus dual photon bone densitometry *Skeletal Radiol.* **19**(1) 27–30

[2] Masud T, Mootoosamy I, McCloskey E V, O'Sullivan M P, Whitby E P, King D *et al* 1996 Assessment of osteopenia from spine radiographs using two different methods: the Chingford Study *Br. J. Radiol.* **69**(821) 451–6

[3] Jergas M, Uffmann M, Escher H, Glüer C-C, Young K C, Grampp S *et al* 1994 Interobserver variation in the detection of osteopenia by radiography and comparison with dual X-ray absorptiometry (DXA) of the lumbar spine *Skeletal Radiol.* **23**(3) 195–9

[4] Huang C, Ross P D, Fujiwara S, Davis J W, Epstein R S, Kodaama K *et al* 1996 Determinants of vertebral fracture prevalence among native Japanese women and women of Japanese descent living in Hawaii *Bone* **18** 5

[5] Looker A C, Wahner H W, Dunn W L, Calvo M S, Harris T B, Heyse S P *et al* 1995 Proximal femur bone mineral levels of US adults *Osteoporos. Intl.* **5** 389–409

[6] Slosman D O, Rizzoli R, Pichard C, Donath A and Bonjour J P 1994 Longitudinal measurement of regional and whole body bone mass in young healthy adults *Osteoporos. Intl.* **4**(4) 185–90

[7] Johnston C C Jr, Melton L J III, Lindsay R and Eddy D M 1989 Clinical indications for bone mass measurements *J. Bone Min. Res.* **4** (Suppl. 2) 1–28

[8] Melton L J I, Thamer M, Ray N F, Chan J K, Chesnut C H R, Einhorn T A *et al* 1997 Fractures attributable to osteoporosis: report from the National Osteoporosis Foundation *J. Bone Min. Res.* **12**(1) 16–23

[9] Slemenda C, Longcope C, Peacock M, Hui S and Johnston C C 1996 Sex steroids, bone mass, and bone loss. A prospective study of pre-, peri-, and postmenopausal women *J. Clin. Invest.* **97**(1) 14–21

[10] Matkovic V, Jelic T, Wardlaw G, Ilich J Z, Goel P K, Wright J K *et al* 1994 Timing of peak bone mass in Caucasian females and its implications for the prevention of osteoporosis *J. Clin. Invest.* **93** 799–808

[11] Overgaard K, Hansen M A, Riis B J and Christiansen C 1992 Discriminatory ability of bone mass measurements (SPA and DEXA) for fractures in elderly postmenopausal women *Calcif. Tissue Intl.* **50** 30–5

[12] Ross P D, Wasnich R D, Davis J W and Vogel J M 1991 Vertebral dimension differences between Caucasian populations, and between Caucasians and Japanese *Bone* **12**(2) 107–12

[13] Ross P D, Norimatsu H, Davis J W, Yano K, Wasnich R D, Fujiwara S *et al* 1991 A comparison of hip fracture incidence among native Japanese, Japanese Americans, and American Caucasians *Am. J. Epidemiol.* **133**(8) 801–9

[14] Marcus R 1991 Understanding osteoporosis *West J. Med.* **155**(1) 53–60

[15] Melton L J I 1996 Epidemiology of osteoporosis and fractures. In: *Osteoporosis Diagnosis and Treatment* ed D J Sartoris (New York: Marcel Dekker) pp 57–78

[16] Melton L J I 1996 Epidemiology of hip fractures: implications of the exponential increase with age *Bone* **18** (Suppl. 3) S121–5

[17] National Osteoporosis Foundation 1998 *Physician's Guide to Prevention and Treatment of Osteoporosis* (Washington, DC: National Osteoporosis Foundation)

[18] Cummings S R, Black D M, Nevitt M C, Browner W, Cauley J, Ensrud K *et al* 1993 Bone density at various sites for prediction of hip fractures. The Study of Osteoporotic Fractures Research Group *Lancet* **341**(8837) 72–5

[19] Melton L J D, Kan S H, Wahner H W and Riggs B L 1988 Lifetime fracture risk: an approach to hip fracture risk assessment based on bone mineral density and age *J. Clin. Epidemiol.* **41**(10) 985–94

[20] Wasnich R D 1996 A new, standardized approach to fracture risk interpretation *Hawaii Med. J.* **55**(8) 141–3

[21] Baran D T, Faulkner K G, Genant H K, Miller P D and Pacifici R 1997 Diagnosis and management of osteoporosis: Guidelines for the utilization of bone densitometry *Calcif. Tissue Intl.* **61** 433–40

[22] Marshall D, Johnell O and Wedel H 1996 Meta-analysis of how well measures of bone mineral density predict occurrence of osteoporotic fractures [see comments] *Br. Med. J.* (Clinical Research edn) **312**(7041) 1254–9

[23] Miller P D, Bonnick S L, Rosen C J, Altman R D, Avioli L V, Dequeker J *et al* 1996 Clinical utility of bone mass measurements in adults: consensus of an international panel. The Society for Clinical Densitometry *Semin. Arthritis Rheum.* **25**(6) 361–72

[24] Schlenker R A, Oltman B G and Kotek T J 1976 Proceedings: Bone mineral mass and width in normal white women and men *Am. J. Roentgenol.* **126**(6) 1282–3

[25] Schlenker R A 1976 Proceedings: Percentages of cortical and trabecular bone mineral mass in the radius and ulna *Am. J. Roentgenol.* **126**(6) 1309–12

[26] Melton L J I, Atkinson E J, O'Fallon W M, Wahner H W and Riggs B L 1993 Long-term fracture prediction by bone mineral assessed at different skeletal sites *J. Bone Min. Res.* **8**(10) 1227–33

[27] Genant H K, Engelke K, Fuerst T, Glüer C-C, Grampp S, Harris S T *et al* 1996 Noninvasive assessment of bone mineral and structure: state of the art *J. Bone Min. Res.* **11**(6) 707–30

[28] Gregg E W, Kriska A M, Salamone L M, Roberts M M, Anderson S J, Ferrell R E *et al* 1997 The epidemiology of quantitative ultrasound: a review of the relationships with bone mass, osteoporosis and fracture risk *Osteoporos. Intl.* **7**(2) 89–99

[29] Grampp S, Genant H K, Mathur A, Lang P, Jergas M, Takada M *et al* 1997 Comparisons of noninvasive bone mineral measurements in assessing age-related loss, fracture discrimination, and diagnostic classification *J. Bone Min. Res.* **12**(5) 697–711

[30] Kleerekoper M and Nelson D A 1997 Which bone density measurement? [editorial]. *J. Bone Min. Res.* **12**(5) 712–4

[31] Yu W, Glüer C-C, Grampp S, Jergas M, Fuerst T, Wu C Y *et al* 1995 Spinal bone mineral assessment in postmenopausal women: a comparison between dual X-ray absorptiometry and quantitative computed tomography *Osteoporos. Intl.* **5**(6) 433–9

[32] Davis J W, Ross P D and Wasnich R D 1994 Evidence for both generalized and regional low bone mass among elderly women *J. Bone Min. Res.* **9**(3) 305–9

[33] Black D M, Palermo L, Nevitt M C, Genant H K, Epstein R, San Valentin R *et al* 1995 Comparison of methods for defining prevalent vertebral deformities: the Study of Osteoporotic Fractures *J. Bone Min. Res.* **10**(6) 890–902

[34] Blake G M and Fogelman I 1997 Technical principles of dual energy X-ray absorptiometry *Semin. Nucl. Med.* **27**(3) 210–28

[35] Hassager C, Jensen S B, Gotfredsen A and Christiansen C 1991 The impact of measurement errors on the diagnostic value of bone mass measurements: theoretical considerations *Osteoporos. Intl.* **1**(4) 250–6

[36] World Health Organization 1994 *Assessment of Fracture Risk and its Application to Screening for Postmenopausal Osteoporosis* (Geneva: WHO) Technical Report Series No. 843

[3] He Y F, Ross P D, Davis J W, Epstein R S, Vogel J M and Wasnich R D 1994 When should bone density measurements be repeated? *Calcif. Tissue Intl.* **55**(4) 243–8

[38] Eastell R 1995 Management of corticosteroid-induced osteoporosis. UK Consensus Group Meeting on Osteoporosis *J. Intern. Med.* **237**(5) 439–47

[39] Cummings S R, Palermo L, Browner W, Marcus R, Wallace R, Pearson J *et al* 2000 Monitoring osteoporosis therapy with bone densitometry: misleading changes and regression to the mean. Fracture Intervention Trial Research Group *J. Am. Med. Assoc.* **283**(10) 1318–21

[40] Ryan P J, Harrison R, Blake G M and Fogelman I 1992 Compliance with hormone replacement therapy (HRT) after screening for post menopausal osteoporosis [see comments] *Br. J. Obstetrics Gynaecol.* **99**(4) 325–8

[41] Rubin S and Cummings S 1992 Results of bone densitometry affect women's decisions about taking measures to prevent fractures *Ann. Intl. Med.* **116** 990–5

[42] Phillipov G, Mos E, Scinto S and Phillips P J 1997 Initiation of hormone replacement therapy after diagnosis of osteoporosis by bone densitometry *Osteoporos. Intl.* **7**(2) 162–4

[43] Stock J L, Bell N H, Chesnut C H III, Ensrud K E, Genant H K, Harris S T *et al* 1997 Increments in bone mineral density of the lumbar spine and hip and suppression of bone turnover are maintained after discontinuation of alendronate in postmenopausal women *Am. J. Med.* **103**(4) 291–7

[44] Hochberg M C, Ross P D, Black D, Cummings S R, Genant H K, Nevitt M C *et al* 1999 Larger increases in bone mineral density during alendronate therapy are associated with a lower risk of new vertebral fractures in women with postmenopausal osteoporosis. Fracture Intervention Trial Research Group *Arthritis Rheum.* **42**(6) 1246–54

[45] Cummings S R, Black D M, Thompson D E, Applegate W B, Barrett-Connor E, Musliner T A *et al* 1998 Effect of alendronate on risk of fracture in women with low bone density but without vertebral fractures: results from the Fracture Intervention Trial *J. Am. Med. Assoc.* **280**(24) 2077–82

[46] Glüer C C, Blake G, Lu Y, Blunt B A, Jergas M and Genant H K 1995 Accurate assessment of precision errors: how to measure the reproducibility of bone densitometry techniques *Osteoporos. Intl.* **5**(4) 262–70

[47] Lu Y and Glüer C-C 1999 Statistical tools in Quantitative ultrasound applications. In: *Quantitative Ultrasound: Assessment of Osteoporosis and Bone Status* eds C F Njeh, D Hans, T Fuerst, C-C Glüer and H K Genant (London: Martin Dunitz) pp 77–100

[48] Kanis J A, Delmas P, Burckhardt P, Cooper C and Torgerson D 1997 Guidelines for diagnosis and management of osteoporosis *Osteoporos. Intl.* **7**(4) 390–406

[49] Lu Y, Genant H K, Shepherd J, Zhao S, Mathur A, Fuerst T P and Cummings S R 2001 Classification of osteoporosis based on bone mineral densities *J. Bone Miner. Res.* **16**(5) 901–10

[50] Scheiber L B 2nd and Torregrosa 1998 Evaluation and treatment of postmenopausal osteoporosis *Semin. Arthritis Rheum.* **27**(4) 245–61

[51] Looker A C, Orwoll E S, Johnston C C Jr, Lindsay R L, Wahner H W, Dunn W L, Calvo M S, Harris T B and Heyse S P 1997 Prevalence of low femoral bone density in older US adults from NHANES III *J. Bone Miner. Res.* **12**(11) 1761–8

Chapter 18

Animal studies

Jenny Zhao, Yebin Jiang, Christopher F Njeh,
Roger Bouillon, Piet Geusens and Harry K Genant

18.1. INTRODUCTION

To study the etiology and pathophysiology and to improve the management of
certain diseases, it is necessary to establish suitable animal models for validating
the safety and efficacy of therapy. Various mammalian species such as rats,
dogs, sheep, pigs, rabbits, guinea pigs and monkeys have been used to model
osteoporosis. Each of these animal models provides certain advantages and
disadvantages [1–3] (table 18.1). Osteoporosis has been induced by ovariectomy
(OVX), immobilization or space flight, low calcium diet, orchidectomy etc.
[4–6]. According to the US Food and Drug Administration Guidelines for
Preclinical and Clinical Evaluation of Agents Used in the Prevention or
Treatment of Post-menopausal Osteoporosis [7], the OVX rat and primate
are the most commonly used models for osteoporotic research, as their reactions closely mimic those of humans, and prediction of the human reaction is
the most important criteria for selecting an animal model. Various techniques
can be used to assess osteoporotic animal models and their treatment, including
dual X-ray absorptiometry (DXA) and peripheral quantitative computed
tomography (pQCT) for bone mineral and geometric measurements; histomorphometry and advanced imaging such as micro-CT (μCT) and magnetic
resonance imaging (MRI) microscopy for bone structure analyses; and
biomechanical testing for bone strength determination (see previous chapters).

18.2. ANIMALS MODELS

18.2.1. Introduction

To select an appropriate animal model for osteoporosis, certain criteria must
be met. These include [1, 8]:

Table 18.1. *Characteristics of five animal models used for the study of osteoporosis (adapted from Thorndike and Turner [1]).*

Animal	Sheep	Mongrel dog	Beagle	Swine	Rodents
Cost (US$)	165	250	350	350	12
Maintenance cost/day	1.50	2.20	2.31	2.31	0.28
Handling/ trained tech	Docile/no	No	No	Aggressive/ yes	Some training required
Social concerns	Low	High	High	Low	Low
Oestrus cycle	Seasonal polyoestrus	Dioestrus	Dioestrus	Polyoestrus	Polyoestrus 4–5 days
Physiology	Ruminant/ herbivore	Monogastric	Monogastric	Monogastric/ omnivore	Monogastric

1. appropriateness as a model for oestrogen deficiency;
2. genetic homogeneity of organisms where applicable;
3. background knowledge of biological properties;
4. cost and availability;
5. ease of experimental manipulation;
6. ecological consideration;
7. ethical and societal implications; and
8. generalizability of the results.

It is of great importance that the model does not add too many variables to an already complex problem [1]. Ideally, the model should closely mimic human disease in its induction, progression and pathology. It is apparent that no animal model can meet all of these requirements, so compromises must be made and the best possible model chosen. Before proceeding to a discussion of the various animal models that have been used to study osteoporosis, a brief discussion of how osteoporosis is induced in these models is appropriate.

18.2.2. Modelling osteoporosis in animals

There are a variety of endocrine excess or deficiency conditions that induce osteopenic reactions. These include ovariectomy (OVX), orchidectomy, hypophysectomy, hyperparathyroidism, hyperthyroidism, prolactinaemia, induced diabetes, induced renal failures, increased glucocorticosteroids and adrenalectomy [9].

18.2.2.1.   Ovariectomy

Even though most of the animals used in these models do not experience a natural menopause, it can be induced with some reliability by ovariectomy (OVX). This is the process whereby the ovaries are surgically removed under total anaesthesia. OVX is the most popular model for studying osteoporosis. In considering which species to use, the oestrus or menstrual cycles should be considered. The rat, for example, under laboratory conditions will maintain continuous oestrus cycles of 4–5 days in duration. The rat skeleton will undergo regular fluctuations in gonadal steroids and thus be more sensitive to the loss of ovarian hormones. Other species, e.g. dogs, are generally seasonal breeders entering oestrus only once or twice per year, thus their skeleton would not be exposed to levels of gonadal hormones as would occur in other species [9].

18.2.2.2.   Orchidectomy

Similar to females, bone loss can be induced in male rats by surgical removal of the testes. Older orchidectomized male rats lose metaphyseal cancellous bone, particularly evident in the long bones [10]. In contrast to OVX rats, orchidectomized rats have decreased body weight gain and increased cortical porosity. Cancellous bone osteopenia following orchidectomy appears to be due to decreased growth (unlike OVX) and a reduced rate of bone turnover, with resorption exceeding formation [9].

18.2.3.   Rat as a model for osteoporosis

OVX rats have been extensively used to study peri- and post-menopausal skeletal changes in humans [4, 5, 11–13]. The same mechanisms control gains in bone mass (longitudinal bone growth and modelling drifts) and losses (basic multicellular units-based remodelling) in humans as in rats [5]. Rats and humans respond similarly to hormones, drugs and other agents, and mechanical influences [5]. Rats have no significant spontaneous bone loss until after 24 months of age [14]. OVX in rats results in increased bone turnover after 2 weeks [15] and induces an imbalance between bone resorption and formation. Bone resorption apparently exceeds bone formation so that net bone loss occurs, as reported in early post-menopausal women [11, 13, 16].

Bone length, weight, density and calcium content of the female Wistar rat were found to increase at a rapid rate between the ages of 1 and 3 months [4]. In older rats, longitudinal growth becomes trivial or stops [4, 5]. The apparently non-remodelling cortical bone in the rat has a latent remodelling capacity [4]. Even in young adult rats an appropriate stimulus such as NaF, PGE2 administration, low dietary calcium or castration can activate intracortical bone remodelling [17–20].

To ensure comparable data between research groups, two standardized rat models for osteoporosis research have been proposed: a 'mature rat model' and an 'aged rat model' aged 3 and 12 months at the beginning of the study, respectively. They should be OVX at the beginning of the study. A baseline group should be included to be certain that the bone loss induced by OVX below that of the baseline control represents true bone loss, and is not complicated by relative differences in the rate of bone growth [4].

There are other advantages of the rat model: rats are easy to house and to work with and their use does not suffer from constraints, such as the increasing cost of purchasing and maintaining the animals, and the decreasing availability as a result of social pressures that are encountered in working with other potential models such as dogs and nonhuman primates. The short life span of rats also facilitates studies on the effects of ageing on bone.

There are, however, some disadvantages to the use of rats as a model of osteoporosis. Rodents do not experience a natural menopause, but OVX can produce an artificial menopause. Although aged rodents have Haversian systems and OVX results in significant bone loss, they have limited naturally occurring basic multicellular unit-based remodelling. Rats also have lamellar bone (although most is 'fine-fibred'). Another limitation is the absence of impaired osteoblast function during the late stages of oestrogen deficiency which may be due to the decrease in bone fatigue experienced by small quadruped animals such as the rodent [21]. Rats have a different loading pattern than humans, with very little intracortical bone remodelling and pronounced bone modelling throughout life [22].

18.2.4. *Sheep as a model of osteoporosis*

It would be useful to model osteoporosis in large animal models with proven intracortical bone remodelling. This is supported by FDA guidelines [7]. The most efficient large animal model has yet to be determined, but primates, minipigs, dogs and sheep are all being investigated [22]. Oestrogen-deficient dogs [23] and baboons exhibit high turnover osteopenia [3, 24].

Sheep are a promising model for various reasons: they are docile, easy to handle and house, relatively inexpensive, available in large numbers, ovulate spontaneously and have hormone profiles similar to women [3, 25]. Ovariectomy results in a slight loss of bone from the ovine iliac crest, and biochemical markers such as osteocalcin are well characterized. Physiological disadvantages are that they lack a natural menopause, that normal oestrus cycles are restricted to autumn and winter and that they have a different gastrointestinal system. Sheep have cortical bone that is plexiform in structure although Haversian remodelling is seen in older animals. Whether biomechanical incompetence of bone follows OVX is presently unknown.

There is no single ideal model for the study of post-menopausal osteoporosis; all have advantages and disadvantages. Researchers in this field must recognize the limitations of the model they choose, and select the one that is most appropriate to their needs.

18.3.   BONE STATUS MEASUREMENTS

18.3.1.   Introduction

Bone status assessment is conducted in animals to evaluate the efficacy of drugs. The important bone variables are bone density (section 18.4), bone structure (section 18.5) and bone strength (section 18.6). These variables have been discussed in the preceding chapters. However, they will be discussed briefly in this chapter to set the stage for the techniques used to evaluate these variables in animal models.

18.3.2.   Bone density

Bone density is usually defined as the mass per unit volume. However, as evident from previous chapters, the definition is dependent on the technique used to measure density. The fundamental variable is the bone mineral content (BMC). The BMC can be divided by the area to give 'areal' density reported by the densitometric techniques, or by the volume to give volumetric density reported by QCT.

18.3.3.   Bone structure

Bone is a matrix of mineral and organic materials with a certain architecture and texture. The biomechanical competence of bone depends on its architecture and material properties. Many studies have found that bone strength is only partially explained by bone mineral and that bone architecture has a significant influence on bone strength. Just as with bridges and buildings, the biomechanical competence of bone depends not only on the amount of material but also on its spatial distribution, its structure, its degree of primary and secondary mineralization and its architecture [26].

The deterioration of trabecular bone structure is characterized by a change from plate elements to rod elements. Consequently the terms 'rod-like' and 'plate-like' are frequently used for a subjective classification of cancellous bone. A new morphometric parameter called the structure model index (SMI) makes it possible to quantify the characteristic form of a three-dimensional structure in terms of the amount of plates and rods composing the structure [27]. The SMI is calculated by three-dimensional image analysis based on a differential analysis of the triangulated bone surface. The SMI value is 0 for an ideal plate structure and 3 for an ideal rod structure, independent of the physical dimensions. For a structure with both plates and rods of equal thickness the value lies between 0 and 3, depending on the volume ratio of rods and plates. The geometrical degree of anisotropy (DA) is defined as the ratio between the maximal and the minimal radius of the mean intercept length (MIL) ellipsoid [28]. The MIL distribution is calculated by superimposing parallel test lines in different directions on the three-dimensional image. The directional MIL is the total length of the test lines in one direction divided by the number of intersections with the bone marrow interface of the test lines in the same direction. The MIL ellipsoid is calculated by fitting the directional MIL to a directed ellipsoid using a least-squares fit. Also, the relationship of these parameters to *in vitro* measures of strength and their application to microfinite element modelling has been shown [29].

18.3.4. *Bone biomechanical properties*

Bone biomechanical properties of animal models can be determined using biomechanical testing, as discussed in chapter 3. For long bone consisting of cortical bone, three-point bending and torsion testing are usually used, while the and compression of the vertebral body are most commonly used. In biomechanical testing, a curve of applied load with resultant deformation is obtained for structural properties, and stress and strain can be determined with parameters of geometric properties to obtain material properties of the bone specimens tested.

18.4. TECHNIQUES FOR MEASURING BONE DENSITY

18.4.1. *Dual X-ray absorptiometry (DXA)*

Bone mineral in osteoporotic animal models can be determined using X-ray based densitometry, e.g. SPA in bone specimens [30], DXA in bone with and without soft tissue, and pQCT (see chapters 7–10). They are non-invasive measures and may allow repeated measurements in longitudinal studies.

DXA (a device applying alternating energy levels) or filtering the wide spectrum of X-ray at two different energy levels are widely used in animal (preclinical) studies. DXA was initially developed for human studies, but software has been developed for measuring BMD in small laboratory animals. An ultrahigh-resolution mode is used for rats and other small animals such as rabbits and cats. Also a lower mA current is required to avoid saturating the detectors. In larger animals such as dogs, pigs and sheep, software used in humans has been adapted to measure BMC/BMD and body composition [31]. The human spine and left and right hip protocols are adapted easily to animals of this size, and the software for body composition has been adapted to dogs.

Precision of DXA in the non-invasive measurement of BMD in animals is dependent on positioning and the ability of the operator to define the same region of interest using clearly defined anatomical landmarks on the scan image, and the intrinsic nature of the equipment [32]. These are essential requirements for successful densitometry in animals. While measurement of the vertebral spine is relatively standardized, there is a wide variety in the methods of measuring the hind limb. Some researchers have used isolated bones immersed in different materials such as saline or the whole hind limb *in vitro* [33].

DXA is fast, precise and accurate, and can measure any skeletal site of the body *in vivo* and on *in situ*, *ex vivo* specimens. An excellent correlation ($r = 0.99$) was found between rat femoral BMC and ash weight [34], and a CV of 0.2–0.7% was found in repeated measurements of rat distal femoral metaphysis, femoral midshaft and lumbar spine [19]. The measured values are given as BMD (BMD is merely a BMC corrected for the number of pixels in which it is measured). DXA can also measure the composition of soft tissue (lean body mass and fat mass).

DXA can detect accelerated trabecular bone loss in the region of the distal femoral metaphysis, in the femoral neck and in the lumbar spine in OVX rats. Oestrogen replacement therapy prevents this trabecular bone loss, and naproxen, a prostaglandin inhibitor, shows a modest bone sparing effect [19].

Because rat cortical bone may not mature until 7.5 months of age, the proposed 3-month-old mature rat model might only be suitable for trabecular bone study. For a mature rat model, it might be more appropriate to delay OVX until the rats reach peak bone mass. OVX induces an increase in rat cortical BMC and bone expansion if performed before the cortical bone has matured [19]. Like OVX, low calcium diets can greatly weaken cortical bone mass; a low calcium diet will accelerate OVX-induced osteopenia.

DXA has been shown to detect the bone sparing effects of various agents such as bisphosphonate [92] in rats, dogs and monkeys, and the preventive effect of calcitonin in OVX sheep [6].

18.4.2.   *Peripheral dual X-ray absorptiometry*

DXA densitometers using rectilinear scanning require as much as 30 min for a total body acquisition with relatively poor precision (>5%) and spatial resolution (>0.5 mm). The long process requires careful sedation techniques and often endangers animal safety. Peripheral DXA is another modality for use in animal bone studies. A new transportable pDXA unit (PIXImus, Lunar/GE) that is designed specifically for small animal research has recently been introduced. The point resolution of the scanner is 0.186 mm (pixel size of 0.18×0.18). The system uses dual kVp (35/80) at 500 µA to scan the animals. The PIXImus X-ray energy is therefore lower than that used for peripheral densitometry in humans resulting in better contrast in the extremely low density bone. Good precision and accuracy have been reported for studies in guinea pigs [35].

18.4.3.   Peripheral quantitative computed tomography (pQCT)

pQCT scanners specifically designed for use on animal bones have been used in the study of osteoporotic animal models, such as mice [2] but mostly in rats, with osteopenia induced by OVX [36, 37], orchidectomy [38] immobilization [39] and corticosteroid administration [40]. It has also been employed to evaluate effects of pharmaceutical intervention in osteopenic models using anti-resorptive [37] or anabolic agents [39]. pQCT can determine cortical volumetric BMD and geometry [39] which contribute significantly to the estimation of cortical bone biomechanical strength [37, 40], and trabecular volumetric BMD [36, 38, 39]. Longitudinal analysis of the proximal tibia *in vivo* showed a significant reduction of 17% in BMD 31 days after OVX [36]. It has been reported that pQCT volumetric BMD is more sensitive than DXA to bone loss in the rat [36]. pQCT was used in a study that found that a low dose of a Chinese herbal preparation containing epimedium leptorhizum was effective in preventing osteoporosis development in weight bearing proximal femur and tibia but not in the lumbar spine in ovariectomized rats [41]. pQCT and μCT examinations showed that the distribution of the vertebral mineral into cortical and trabecular compartments is regulated genetically in 12BXH recombinant inbred mice [82]. The CVs for bone mineral measurements in rats are in the range 1.6–5.9% *in vivo*, with repositioning [42, 60]. There was excellent agreement between trabecular BMD measurements *in vivo* and *ex vivo* ($r = 0.91$) [42].

18.5.   TECHNIQUES FOR MEASURING BONE STRUCTURE

18.5.1.   Introduction

Several non-invasive/non-destructive techniques have been developed for measuring bone structure. Radiography has been used for many years to detect and quantify bone changes. Much progress has been made in developing μCT and MRI microscopy for non-invasive and/or non-destructive assessment of three-dimensional trabecular structure and connectivity. The availability of three-dimensional measuring techniques and three-dimensional image processing methods allows direct quantification of unbiased morphometric parameters, such as direct volume and surface determination [43], model independent assessment of thickness [44] and three-dimensional connectivity estimation [45]. Assessment of three-dimensional trabecular structural characteristics may further improve our ability to understand the pathophysiology and progression of osteoporosis and other bone disorders. Three-dimensional structure is typically inferred indirectly from histomorphometry and stereology on a limited number of two-dimensional sections based on a parallel plate model [46]. With age and disease, trabecular plates are perforated and connecting rods are dissolved, with a continuous shift from one structural type to the other. The traditional histomorphometric measurements based on a fixed model type will lead to questionable results [47]. The introduction of three-dimensional measuring techniques in bone research makes it possible to capture the true trabecular architecture without assumptions of the structure type.

18.5.2. Radiography, microradiography and radiogrammetry

Conventional radiography using high-detail films and small spot X-ray tubes is a non-invasive means of visualizing animal bone structure. It also provides qualitative information on bone density at the organ and tissue level. The decrease in density is a result of a decrease in both mineralized bone volume and total calcium (mineralization). The amount of calcium per unit of mineralized bone volume remains constant in osteoporosis and osteopenia, resulting in decreased absorption of the X-ray beam and changes in the bone structure on radiographs. With improvements in micro-focus radiography [48], details of the bone structure of small animals can be clearly displayed. Micro-focus radiography and specimen slab radiography have been employed in preclinical studies [18, 49] and in toxicological and safety studies, e.g. to investigate effects of on/off treatments of high doses of a bisphosphonate (an inhibitor of osteoclastic activities and bone resorption) on bone modelling and bone remodelling in growing non-human primates and rodents. Radiological changes are very closely correlated with corresponding histopathological findings (figure 18.1). The radiologic manifestations in baboons are very similar to changes in humans treated with pamidronate [50]. Sophisticated image processing techniques can be used to segment and extract trabecular morphometric parameters.

Radiogrammetry can measure cortical thickness and bone size [17]. Morphometric radiogrammetry measurements of the long tubular bones on conventional radiographs include bone length (L), the outer diameter or periosteal width (D) and inner diameter or medullary space (d) of the cortical bone at 50% of the length. The cortical thickness can be obtained by simple subtraction ($D - d$). Since the long tubular bones are approximately circular at midshaft and the medullary cavity is nearly centred in the tubular bone cylinder at that point, $D^2 - d^2$ can be regarded as an index of the cortical cross-sectional area after omitting the constant $\pi/4$; $(D - d)/D$, the ratio of cortical thickness to the midshaft width; and $(D^2 - d^2)/D^2$, the ratio of cortical cross-sectional area to the total midshaft cross-sectional area. The reproducibility as CV of six determinations with

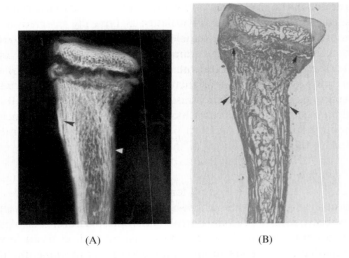

(A)   (B)

Figure 18.1. *A baboon treated with a high dose of a bisphosphonate. Slab contact radiograph (A) of the proximal end of the radius shows irregularity of the growth plate with dense dots, increased density in the spongiosa, intracortical lucent lines and periosteal reaction (arrowhead), which corresponds in the histological macrosection (B, HE staining) to cartilage calcification in the growth plate, increased number of trabeculae in the metaphysis, enlargement of the Haversian canals in the cortical bone and periosteal woven bone formation (arrowhead).*

repositioning of the radiographs at each measurement is $L = 0.2\%$, $D = 1.2\%$, $d = 2.1\%$ for conventional radiographs of the radius of baboons and $L = 0.07\%$, $D = 0.65\%$, $d = 1\%$ for microradiographs ($\times 100$, with focal spot 5 μm) of rat femurs. High correlations are found in both baboons and rats between midshaft periosteal diameter (D) and torsional strength and stiffness ($r = 0.7$); between cortical area index ($D^2 - d^2$) and stiffness ($r = 0.6$); and among D, $D^2 - d^2$ and DXA measurement ($0.7 \leq r \leq 0.9$), indicating that radiogrammetric measurements may be useful for estimating bone mineral and strength in preclinical studies [51].

Quantitative microradiography using an aluminium calibration stepwedge, similar to quantitative backscattered electron imaging and small-angle X-ray scattering imaging [52, 53], can measure the focal degree of mineralization and secondary mineral apposition. Mineralization of the trabecular bone consists of primary and secondary mineral apposition. Primary mineral apposition is mineralization of osteoid, which can be detected using traditional histomorphometry that measures the amount of bone matrix or bone mass. Secondary mineral apposition is a slow and gradual maturation of the mineral component, including an increase in the amount of crystals and/or an augmentation of crystal size towards their maximum dimensions. Increased bone turnover, as in post-menopausal osteoporosis, results in hypomineralization, because resorption begins prematurely, before mineralization is complete. The least mineralized bone is highly radiotranslucent and is the most recently formed, or has just achieved its primary mineralization, and represents about 75% of the complete mineralization. It contrasts with the adjacent fully mineralized interstitial bone, with low radiotranslucency.

18.5.3. *Peripheral quantitative computed tomography (pQCT)*

pQCT can be used to determine aspects of cortical geometry such as area, thickness and cross-sectional moments of inertia which contribute significantly to the estimation of cortical bone biomechanical strength. Cross-sectional moments of inertia (CSMI) describe the geometric configuration or distribution of the bone mass around the central axis.

In bending or torsion, the cross-sectional area of a structure is more important in resisting loads than is its mass or density [54]. In humans, whole bone biomechanical behaviour is largely influenced by differences in geometrical properties, and much less by differences in material properties [55]. Similarly in the rat, bone geometric changes have a much greater effect on the whole bone biomechanical behaviour than do the material changes [56, 57]. CSMI describes the geometric configuration or distribution of the bone mass around the central axis. The magnitude of the CSMI depends on both the cross-sectional area and the squared distance of each unit area from the sectional neutral axis that always intersects the centroid, or centre of the area of the section. This is an important parameter because it is possible for the cross-sectional area of two bones to be the same while the distribution of bone is vastly different. In torsion, deformation would be resisted more efficiently if bone were distributed farther away from the neutral torsional axis. This property is known as the 'polar moment of inertia'. Ideally, in bending or torsion, bone should be distributed as far away from the neutral axis of the load as possible. The maximum and minimum CSMIs, here simply designated I_{max} and I_{min}, indicate the relative magnitudes of greatest and least bending rigidity of a section, respectively. The principal axes of a section define the directions of greatest and least bending rigidity (major axis—greatest, minor axis—least). The sum of the I_{max} and I_{min}, or the polar CSMI (J), reflects torsional strength and rigidity in non-circular sections. Changes in geometry of the skeleton can offset the effects of altered bone density [58].

With *ex vivo* specimen studies, it might be convenient to examine the bones after dissection and exposure to air by densitometry, as pQCT can comparably and reproducibly quantify both volumetric trabecular BMD in either water or in air [59]. It is difficult to determine the trabecular structure in preclinical studies because of limited spatial resolution and increased slice thickness in the pQCT images, though apparent trabecular structural parameters can be derived from pQCT images of human cubic bone specimens using special imaging processing techniques [59]. Should the spatial resolution and the signal-to-noise ratio be improved, pQCT could be useful for trabecular bone structural assessment. In OVX rats, longitudinal analysis of the proximal tibia *in vivo* showed a significant reduction of 17% in BMD 31 days post-OVX, and the images showed a progressive loss of trabecular bone, widening of marrow spaces and thinning of the cortical bone [36]. Thus, pQCT may be a reasonable surrogate for static histomorphometric measurement in preclinical studies [38, 39, 60].

18.5.4.   Micro-computed tomography (μCT)

The μCT system was introduced using a micro-focus X-ray tube as a source, an image intensifier as a two-dimensional detector, and a cone-beam reconstruction to create a three-dimensional object, for detecting small structural defects in ceramic materials [61, 62]. Instead of rotating the X-ray source and detectors during data collection as in clinical CT, the specimen is rotated at various angles. X-rays are partially attenuated as the specimen rotates in equal steps in a full circle about a single axis. At each rotational position, the surviving X-ray photons are detected by a planar two-dimensional array. A three-dimensional reconstruction array is created directly in place of a series of two-dimensional slices.

The discrete elements in digital images of μCT are called pixels for two-dimensional and voxels for three-dimensional. From a physical perspective, spatial resolution is typically given in line pairs per millimetre or as a 5 or 10% value of the modulation transfer function, requiring measurement of the modulation transfer function or the point spread function of the imaging system. For simplicity, pixel or voxel dimensions are used which may result in a 'pseudo' resolution that typically overestimates the physically 'correct' resolution by a factor of 2–4. Therefore, to image 50 μm thick structures the most 'appropriate' resolution must be at least 50 μm, requiring pixel/voxel size of 25 μm or smaller according to Nyquist's theorem [63].

The segmentation algorithm is important to accurately and reproducibly classify pixels from images in a fast, objective, non-user-specific manner. Segmentation is usually threshold- or edge-based. The threshold depends on the absolute grey value of the pixel, i.e. bone and marrow densities. Global thresholds apply the same threshold to the whole image, which works well in high-resolution and high-contrast images, while local thresholds adapt this value to a neighbourhood of a selectable size. Edge-based methods apply the first- or second-order derivative and therefore detect changes in grey values [63]. Although some decisions have to be made by the operator, such as placement of the region of interest (ROI), the important advance is that the algorithm is implemented in a uniform manner across all images.

An *in vitro* study showed that precision error as coefficient of variation (CV) and standardized CV of trabecular structural parameters was <5% [59].

Morphological measures such as the Euler characteristic quantify connectivity, a measure of the maximum number of branches which can be removed before the structure is divided into multiple pieces. The connectivity density (CD) of a two-component system such as bone and marrow can be derived directly from the Euler characteristic, E, by $CD = (1 - E)/TV$ [45], when all the trabeculae and bone marrow cavities were connected without isolated marrow cavities inside the bone. The measure of the maximum number of branches which can be removed before the structure is divided into multiple pieces, E, was determined in the μCT data set without prior skeletonization [45, 61]. It is defined as $E = \beta_0 - \beta_1 + \beta_2$. β_0 is the number of bone pieces, usually assumed to be one piece, and β_1 is the connectivity, or the maximum number of connections that must be broken to split the structure into two pieces, while β_2 is the number of marrow cavities surrounded completely by bone [45]. Discrepant results of the connectivity measurements, ranging from highly linear to non-linear relationships with bone volume, have been reported [45, 61, 64, 65], which might be due to different specimen and sample sizes studied. Goulet [66] utilized images of bone cubes and related these image-based parameters to Young's modulus, a measure of elasticity of bone. Based on data sets from Feldkamp's μCT, Engelke [67] developed a three-dimensional digital model of trabecular bone that could be used to compare two- and three-dimensional structural analysis methods and to investigate the effect of spatial resolution and image processing techniques on the extraction of structural parameters.

The three-dimensional cone-beam μCT imaged the trabecular bone architecture in small samples of human tibias and vertebrae, *ex vivo*, with a spatial resolution of 60 μm [61]. A resolution of 60 μm, although acceptable for characterizing the connectivity of human trabeculae, may be insufficient for studies in small animals like the rat, where the trabecular widths average about 50 μm and trabecular separations average 150 μm or less [64]. Furthermore, Smith and Smith [68] have reported that three-dimensional images from cone-beam scanners are inevitably distorted away from the central slice because the single-orbit cone-beam geometry does not provide a complete data set. These distortions and associated loss of spatial resolution have been particularly evident in samples containing plate-like structures, even when the cone-beam angle is less than 6.5° [68]. A recent study with direct comparison of fan-beam (obtained on the central plane) and cone-beam (obtained from a divergent section near the periphery of the volume) techniques over a full cone angle of 9°, using a 100 μm micro-focus X-ray tube and isotropic 33 μm voxels, showed that the bone volume fraction based on a grey-scale threshold in the excised lumbar vertebrae from normal adult rats was not adversely affected by cone-beam acquisition geometry for cone angles typically used in μCT [69].

18.5.5.   Synchrotron radiation μCT

The μCT method was further enhanced by resorting to synchrotron radiation with a spatial resolution of 2 μm [70]. The X-ray intensity of synchrotron radiation is higher in magnitude than in X-ray tubes. When scanning time is important, and for resolution less than 1–5 μm, a synchrotron radiation X-ray source is a better choice than X-ray tubes. Adaptation of the X-ray energy to the sample can be optimized by using monochromatic radiation because of the continuous X-ray spectrum of synchrotron radiation, which can minimize radiation exposure when examining small animals *in vivo*. It uses parallel-beam imaging geometry, avoids the distortions and loss of resolution inherent in cone-beam methods, and can make distortionless images using a CT at a synchrotron electron storage ring [70].

The use of a synchrotron radiation X-ray source was first suggested by Grodzins [71] for high-resolution micro-tomography (μT) of small samples. It provides a continuous energy spectrum with a high photon flux. The optimum energy for a given sample can be selected from the synchrotron radiation white beam with a small energy bandwidth (0.1–0.001%) using a crystal monochromator, while at the same time keeping the photon flux rate high enough for efficient imaging. The monochromaticity of the beam is very important, while conventional polychromatic X-ray sources result in beam hardening artefacts in the reconstructed images due to the stronger attenuation of the soft X-ray in the sample. The monochromaticity of the beam is especially important to perform accurate density measurement. The high photon flux available and small angular source size from synchrotron radiation X-ray sources lead to negligible geometrical blur, making it possible to obtain images with high spatial resolution and high signal-to-noise ratio.

Synchrotron μCT at 23 μm/voxel has been applied to living rats for longitudinal study in the proximal tibial metaphysis [72–74]. Recently, μCT using high intensity and tight collimation synchrotron radiation that achieves a spatial resolution of 1–2 μm has provided the capability to assess additional features such as resorption cavities [75–77].

In addition, synchrotron radiation X-ray μT using new X-ray optic components, has been designed to assess the ultrastructure of individual trabeculae with a resolution of 1 μm, and to describe microscopic variations in mineral loading within the bone material of an individual trabecular rod. Artefacts from X-ray refraction and diffraction require methods different from those used for other μCT techniques. Delicate and minimal individual trabecular specimen handling and no microtome cutting preserve the specimen geometry and internal micro-fractures. The histological features of the mineral ultrastructure can be evaluated using volumetric viewing. The volume, shape and orientation of osteocyte lacunae and major cannaliculae can be observed. Quantitative measures of trabecular ultrastructure are now being considered including BMU (Basic Multicellular Unit of bone remodelling with activation–resorption–formation sequence) volume, lamellar thickness and density gradients [78].

The hardware for synchrotron radiation µCT, however, is not readily accessible. Electron storage rings are stationary and cannot be operated in a small laboratory and only a few synchrotron radiation centres are available worldwide. Ruegsegger *et al* [79] developed a µCT device dedicated to the study of bone specimens, without synchrotron radiation, which has been used extensively in laboratory investigations [79].

18.5.6.   *µCT three-dimensional assessment*

Image processing algorithms, free from the model assumptions used in two-dimensional histomorphometry, have been developed to segment and directly quantify three-dimensional trabecular bone structure [27, 44, 47]. Trabecular thickness is determined by filling maximal spheres in the structure with the distance transformation, then calculating the average thickness of all bone voxels. Trabecular separation is calculated with the same procedure, but the voxels representing non-bone parts are filled with maximal spheres. Separation is the thickness of the marrow cavities. Trabecular number is taken as the inverse of the mean distance between the mid-axes of the observed structure. The mid-axes of the structure are assessed from the binary three-dimensional image using the three-dimensional distance transformation and extracting the centre points of non-redundant spheres which fill the structure completely. Then the mean distance between the mid-axes is determined similarly to the separation calculation, i.e. the separation between the mid-axes is assessed.

The early uses of three-dimensional µCT focused on the technical and methodological aspects of the systems, but recent developments emphasize the practical aspects of µT imaging. µCT has been used to measure trabecular bone structure in rats. Most studies have focused on the trabecular bone in the proximal tibial metaphysis, but the trabecular bone in rat vertebrae is of interest because of its similarity to the human fracture site, and because biomechanical testing is practical [19]. µCT with an isotropic resolution of $11 \, \mu m^3$ has been used to examine the three-dimensional trabecular bone structure of the vertebral body in ovariectomized rats treated with oestrogen replacement therapy (figure 18.2) [19]. µCT three-dimensionally determined trabecular parameters show greater percentage changes than those observed with DXA, and they show better correlation with biomechanical properties. Combining trabecular bone volume with trabecular structural parameters provides better prediction of biomechanical properties than either parameter alone. A study of the anabolic effects of low-dose (5 ppm in drinking water) long-term (9 months) sodium fluoride (NaF) treatment in intact and OVX rats shows that NaF treatment increases trabecular bone volume, possibly

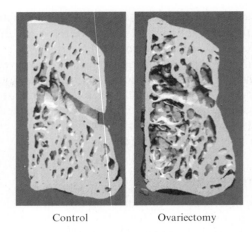

Control   Ovariectomy

Figure 18.2. *μCT three-dimensional images of rat vertebral body with isotropic resolution of 11 μm³. Compared with age-matched sham-operated control (left), ovariectomy results in a remarkable decrease in the trabecular bone volume, thickness, number, and a conspicuous increase in trabecular separation. The different trabecular patterns are noticeable, i.e. plate-like trabeculae in the control and rod-like trabeculae in the ovariectomized rat. The cortical thickness is also decreased after ovariectomy.*

by increasing trabecular thickness through increasing bone formation on existing trabeculae, rather than by increasing trabecular number. NaF in sham-operated rats increases trabecular structural parameters and bone mineral, but decreases compressive stress in the vertebral body. NaF partially protects against OVX-induced changes in bone mineral and structure but this protection does not translate into a corresponding protection of bone biomechanical properties [80, 81].

The wide availability of genetically altered mice has increased the usefulness of the murine model for investigating osteoporosis and other skeletal disorders. Independent genetic regulation of three-dimensional vertebral trabecular microstructure in 12BXH recombinant inbred mice as measured by μCT contributed information regarding the variation in biomechanical properties among the strains [82]. Mice homozygous for a null mutation of the PTHrP gene die at birth with severe skeletal deformities. Heterozygotes survive and by 3 months of age develop osteopenia characterized by decreased trabecular bone volume and increased bone marrow adiposity. PTHrP wild type and heterozygous-null mice were ovariectomized at 4 months of age and sacrificed at 5 weeks. Three-dimensional μCT was use to examine the trabecular structure of the mice, with an isotropic resolution of 9 μm³. Bone specimens from mice heterozygous for the PTHrP null allele demonstrate significant changes as compared with wild-type litter mates in

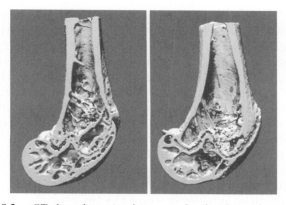

Figure 18.3. *μCT three-dimensional images of trabecular and cortical bone structure of mice with isotropic resolution of $9\,\mu m^3$. Compared with age-matched sham-operated control (left), ovariectomy results in a dramatic loss of trabecular bone volume and other trabecular bone microstructure (right).*

most parameters examined. However, measurements of trabecular number and trabecular thickness were not significantly different between the two groups. These findings support the notion that PTHrP haplo-insufficiency leads to abnormal bone formation in the adult mouse skeleton.

OVX induces short-term high-turnover accelerated deterioration of three-dimensional trabecular structure in mice. In a study using 3 month old Swiss Webster mice 5 and 13 weeks after OVX, three-dimensional μCT trabecular structure was measured in the secondary spongiosa of the distal femur with an isotropic resolution of $9\,\mu m^3$. The trabeculae become more rod-like and more isotropic, thinner and more widely-separated after OVX (figure 18.3) [83]. HRT prevented OVX-induced bone loss. Percentage changes in pQCT volumetric BMD were similar to the changes measured by μCT but less pronounced.

18.5.7. Magnetic resonance imaging (MRI) microscopy

MR microscopy has advantages similar to those of μCT: it can show the three-dimensional structure in arbitrary orientations, it is non-destructive and non-invasive and, unlike μCT, it is non-ionizing (see chapter 13). MRI is a complex technology based on the application of high magnetic fields, transmission of radiofrequency (RF) waves, and detection of radiofrequency signals from excited hydrogen protons. MRI can delineate bone so well because bone mineral lacks free protons and generates no MR signal while adjacent soft tissue and marrow contain abundant free protons and give strong signals.

In MR the appearance of the image is affected by many factors beyond spatial resolution, including the field strength—since signal is related to the field strength, the specific pulse sequence used, the echo time and the signal-to-noise achieved. Acquisition, analysis and interpretation of MR images are more complicated than for X-ray based images of CT. Both spin–echo [84] and gradient–echo [85] sequences can be used to generate high-resolution MRI of bone structure. Gradient–echo techniques require less sampling time. A 180° RF pulse is applied to obtain an echo signal for generating an image in spin–echo sequences, while a reversal of the magnetic field gradient is used for such an echo in gradient–echo sequences. The apparent trabecular size depicted in both spin–echo and gradient–echo images may differ from their true dimensions [84]. The overestimation of trabecular dimension is more pronounced in gradient–echo images, especially when echo time (TE) is increased. An increase in the repetition time (TR) leads to an increase in signal-to-noise ratio, but also an increase in the total scan time. The higher the bandwidth or the total time in which the MR signal is sampled, the shorter the achievable echo time, and the lower the signal-to-noise ratio. All these parameters need to be carefully weighed and adjusted as they interact in multiple ways and affect resolution and quality of the image [85].

Segmentation to derive binary images for structural measurement can be challenging, since adaptive threshold- and edge-based methods tend to amplify susceptibility differences in the bone marrow, resulting in mis-classification of marrow as bone. Global thresholds based on the histogram of the grey-value distribution [85], or local thresholds using internal calibration based on fat, air, tendon and cortical bone [86], have been used. Hipp and colleagues [87] examined bovine cubes in a small bore micro-imaging spectrometer at $60 \, \mu m^3$ resolution and found three-dimensional results heavily dependent on the threshold and image processing algorithm. Considerable resolution dependence was observed for traditional stereo-logical parameters, some of which could be modulated by appropriate thresholds and image processing techniques. As results can be affected by operator-adjustable parameters in the threshold-based methods, it is recommended that the same segmentation scheme be used for all specimens in a specific project. Application of this technology to small animal bones is more demanding, though high-resolution MR has been used successfully for *in vitro* quantitative evaluation of human trabecular bone, as resolution requirements are more stringent because of the considerably smaller trabecular size. The need for higher resolution, dictated by the thinner trabeculae, entails a significant penalty for signal-to-noise ratio and acquisition time. Recently, the ability of MRI microscopy to assess osteoporosis in animal models has been well explored. Using a small high-efficiency coil in a high-field imager, MRI microscopy can be performed at resolutions sufficient to discriminate individual trabeculae. MRI of trabecular structure in the distal radius of rat shows trabecular bone loss after OVX (figure 18.4). µMRI of a rat tail showed three-dimensional cortical bone, trabecular

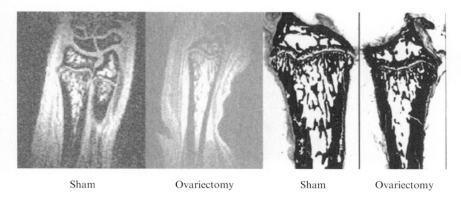

Sham   Ovariectomy   Sham   Ovariectomy

Figure 18.4. *μMRI using a solenoid coil and a 2 T imager (GE/Bruker) with isotropic resolution of 78 μm³, and their corresponding histology of the distal radius show loss in trabecular structure of rats after ovariectomy.*

network, and other soft tissue. μMRI shows increased bone mass in the distal femoral metaphysis of rats treated with a bisphosphonate, increased cartilage thickness in the growth plate, and corresponding recovery changes after withdrawal of treatment. The trabecular structure in the femoral neck of a ewe can be clearly shown on MRI. In this study, with appropriate choices, it was possible to image trabecular bone in rats *in vivo* and *in vitro*. Segmenting trabecular bone from adjacent tissues has been a useful technique in the quantification of trabecular bone in MRI images. In a study of OVX in rats, analysis of MRI demonstrated differences in rat trabecular bone that were not detected by DXA measurements [88].

MRI showed OVX-induced losses in trabecular bone volume fraction and structure that oestrogen replacement therapy (ERT) prevented [89]. There are excellent correlations between MRI with resolution up to $24 \times 24 \times 250$ μm and histological assessment of intact rat tibiae and vertebrae [89]. It has been reported that rat tibiae were imaged at 9.4 T *in vitro* with isotropic resolution of 46 μm³. It has also been shown that alendronate maintains trabecular bone volume and structure about midway between intact and OVX, whereas prostaglandin E_2 returned them to intact levels [90].

MRI shows three-dimensional bone structure and some other tissues at the same time. In the rabbit knee, MRI shows trabecular structure and cartilage. In an osteoarthritis model induced by menisectomy [91] or anterior cruciate ligament transection [92], MRI shows subchondral osteosclerosis and decreased cartilage thickness. MRI also shows osteophytes in a rabbit osteoarthritis model. However, radiographs show only subchondral osteosclerosis, while osteophytes could not be found in a rabbit osteoarthritis model.

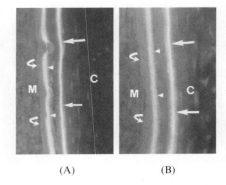

(A)   (B)

Figure 18.5. *Fluorescent double labels as seen under epifluorescent microscopy in the endocortical surface of a longitudinal section of the proximal tibial metaphysis of a sham ovariectomized rat (A) and an ovariectomized rat (B). The incorporation of the labels into bone as time markers permits the dynamic description of current remodelling activity. The thickness of bone between the first (straight arrows) and the second labels (arrowheads) which has been formed over a period of 10 days is greater in an OVX than in a sham rat. This thickness divided by the labelling interval (10 days) yields the mineral apposition rate (MAR). Osteoid is stained pink (curved arrows), located between the second label and bone marrow (m). Fluorescent micrographs, 20 μm thick undecalcified ground sections, Villanueva bone staining (10 × 25).*

The usefulness of this powerful tool is balanced by its disadvantages: it is expensive to obtain, operate and maintain.

18.5.8.   Histomorphometry

Histomorphometry, a well-established method, is used extensively in the assessment of two-dimensional trabecular structure of osteoporotic animal models and the various treatment effects (see chapter 6). In addition to the static trabecular structural parameters, the bone formation rate, mineral apposition rate, osteoid and mineralization lag time, osteoblastic and osteoclastic activities and other dynamic histomorphometric parameters can be measured (figures 18.1 and 18.5). Histomorphometry has been used to show that bisphosphonate and oestrogen and its analogues can suppress OVX-induced bone turnover and prevent OVX-induced bone loss, while anabolic agents such as PTH, prostaglandins, NaF, growth factors etc. can restore lost bone in various animal models. Microvasculature is useful for investigation of cortical and trabecular bone modelling and remodelling. There are several approaches to micro-angiography. One is to inject a contrast medium, such as barium sulphate, into the micro-vasculature, followed by μCT or radiography. Another approach is to stain the micro-vasculature, using for example Chinese ink, or a mixture of contrast media and Chinese ink. For bone micro-angiography, μCT images or contact radiographs can be taken after injection of the contrast media, with or without decalcification. Higher contrast and more detailed structure of the micro-vasculature when the specimen has been thoroughly decalcified. Chinese ink is a good staining agent for micro-angiography, as it stays firmly in the micro-vasculature, and will not influence histopathological observation of the structure of other cells in bones, joints or muscles [109, 110].

18.6.   BONE STRENGTH MEASUREMENT

Densitometric techniques such as DXA do not furnish sufficient information about the quality of bone. An important endpoint when using animal models of osteoporosis is the fragility of bone as a result of OVX or consequent treatment. Thus one of the important steps in the characterization of the model is the determination of the biomechanical changes of the bone.

Bone strength of the osteoporotic animal models can be determined using biomechanical testing as discussed in chapter 5. Among various testing methods, torsion or three-point bending of the long bones and compression of the vertebral body are most commonly used. In biomechanical testing, the following structural mechanical parameters can be derived from the curve of applied loads and the resulting deformations: maximum strength (load) at fracture; maximum deformation; stiffness, as the ratio of applied load to resultant deformation in the linear region of the curve; and strain energy absorption before fracture or toughness as the area under the curve.

Normally, loads acting on the human vertebral bodies are compressive. Thus, the most suitable biomechanical test for the vertebral body is compression [58]. Cancellous bone cores can be removed from lumbar vertebral bodies of rats for compressive tests [22]. Compression of the whole bone can also be performed. It has been shown that both OVX and a low calcium diet can decrease vertebral compressive strength, with OVX causing more loss than a low calcium diet. Oestrogen depletion causes more changes in trabecular bone than in the cortical bone. As the long bones in rats are much more subject to loading events than the vertebral bodies, the vertebral bodies might show a clearer pattern of bone response to oestrogen depletion [93].

A torsion test can be used to measure the mechanical properties of bone in shear stress. When a torque or twisting movement is applied to a circular specimen, shear stress varies from zero at the centre of the specimen to a maximum at the surface. Torsional loading subjects the whole bone to equal torque at every cross-section along its length, enabling the identification of the weakest section of the structure. Cortical bone is the major component of bone mass in the mammalian long bones and ensures their mechanical competence [94]. Small changes in the amount of compact bone can make a big difference in bone strength. It has been shown that a low calcium diet causes a greater loss of biomechanical competence in cortical bone than does OVX. OVX results in improved femoral geometric properties and strength in growing adult rats. Though cortical bone BMD in OVX rats is less than in sham-OVX rats, OVX results in an increase in femoral geometric properties, such as cross-sectional moments of inertia, because OVX increases modelling-dependent bone gain on the periosteal envelope [95–97]. The increase in periosteal bone formation offsets the increase in endocortical resorption. Therefore, the bone material is distributed farther from the neutral axis and improves its resistance to torsional loads, though the net effect is an expansion of the marrow cavity and cortical thinning with decreases in cortical width and cortical area. Moreover, OVX of rats induces an increase in body weight, leading to increased mechanical loading to the bones, which may also contribute to the increased strength of the whole bone structure.

In addition, biomechanical tests of the femoral neck and the tibia [98] show decreased biomechanical parameters in OVX, orchidectomized and immobilized rats, with the greater decrease in the femoral neck [99, 100]. Compressive testing of the trabecular bone of the distal femoral metaphysis shows decreased biomechanical competence after OVX and increased strength after PTH treatment [101].

In humans, BMD is the single best predictor of trabecular bone strength, and alone can explain 49–93% of the variance of the biomechanical properties such as the modulus of elasticity or compressive strength [102]. There is an extensive overlap in bone mass values between normal and osteoporotic persons. Measuring both bone mineral and structure can provide better prediction of bone biomechanical properties than either parameter alone [59, 103]. In animal studies, an increase in bone mass and geometry is not always necessarily followed by an increase in bone biomechanical strength. For example, oestrogen fully, and naproxen partially, prevent OVX-induced loss in both bone mass and strength. However, a low dose long-term sodium fluoride treatment increases bone mass and geometry of vertebrae and femur, but also increases intracortical bone remodelling and porosity and does not increase vertebral strength nor proportionally improve femoral strength. In both rats and baboons treated with a bisphosphonate, low doses improve bone mass and strength, while high doses beyond the therapeutic window result in modelling and mineralization defects, a decrease in stiffness and an increase in deformation, despite increased bone mass.

18.7.   SUMMARY AND PERSPECTIVES

The ultimate therapeutic benefit for osteoporotic patients is resistance to fracture. Accordingly, the most important aim of any therapy in animal models of osteoporosis is to improve the load-bearing capacity of the skeleton or bone biomechanical properties that depend on bone geometry and material. The shape, size, primary and secondary mineralization and structure determine the geometric variables that are related to bone modelling. Material properties relate to remodelling, which affects collagen fibre orientation and mineralization etc.

Animal models will continue to play very important roles in investigating etiology, prevention and treatment of osteoporosis, and in understanding the contribution of specific genes to the establishment of optimal bone architecture, peak bone mass, and the genetic basis for a predisposition towards accelerated bone loss in the presence of co-morbidity factors such as oestrogen deficiency and deprivation. It is very important to keep in mind that the enormous potential of laboratory animals as models for osteoporosis can only be realized if great care is taken in the choice of an appropriate species, age, experimental design and measurements. Poor choices will results in misinterpretation of results which ultimately can bring harm to patients who suffer from osteoporosis by delaying advancement of knowledge [2].

Bone mineral determined by densitometry [104, 105] significantly correlates with bone strength. Densitometric measurement is simple and fast, and is suitable for monitoring development of animal models and for screening the efficacy and safety of new drugs. However, an increase in the quantity of bone mineral is not necessarily followed by an increase in bone quality. Therefore, direct bone biomechanical testing should be performed to evaluate bone quality.

For assessment of bone structure, imaging techniques are unbiased and free from the model assumptions used in two-dimensional histomorphometry. Since a true three-dimensional assessment of the trabecular bone structure is possible, rod or plate model assumptions are no longer necessary. They are able to directly measure three-dimensional structure and connectivity in arbitrary orientations in a highly automated, fast, objective, non-user-specific manner, with little sample preparation, allowing greater numbers of samples for unbiased comparisons between controls and subjects. They can have large sample sizes and therefore less sampling error. They are non-destructive (which allows multiple tests such as biomechanical testing and chemical analysis on the same sample) and non-invasive (which permits longitudinal studies and may reduce the number of subjects in each study) and they may provide a complementary technique to dynamic histomorphometry. These methods also have weaknesses: they require robust image processing algorithms to segment and quantify bone structure, and may have limitations in spatial resolution for certain structures. They cannot provide information on cellular activities and on dynamic mineralization processes. Imaging might have better precision that is better for monitoring disease progression and drug efficacy, while histology might have better accuracy which is essential for reliable diagnosis. Histology is labour-consuming and subject to sectioning artefacts.

Rather than replacing bone histomorphometry, these imaging methods provide additional and valuable information and are a useful complement to traditional techniques in the evaluation of osteoporosis and other bone disorders.

Finally, extrapolation of the results of study in animals directly to humans should be done with caution since the loading pattern of the quadruped animals is different from bipedal humans. Though there is evidence of trabecular and cortical bone remodelling [5, 106–108], rats show less bone remodelling than is seen in larger animals and humans.

REFERENCES

[1] Thorndike E A and Turner A S 1998 In search of an animal model for postmenopausal diseases *Front. Biosci.* **3** c17–26

[2] Turner R T, Maran A, Lotinun S, Hefferan T, Evans G L, Zhang M *et al* 2001 Animal models for osteoporosis *Rev. Endocr. Metab. Disord.* **2**(1) 117–27

[3] Bellino F L 2000 Nonprimate animal models of menopause: workshop report *Menopause* **7**(1) 14–24

[4] Kalu D N 1991 The ovariectomized rat model of postmenopausal bone loss *Bone Miner.* **15**(3) 175–91

[5] Frost H M and Jee W S 1992 On the rat model of human osteopenias and osteoporoses *Bone Miner.* **18**(3) 227–36

[6] Geusens P, Boonen S, Nijs J, Jiang Y, Lowet G, Van Auderkercke R *et al* 1996 Effect of salmon calcitonin on femoral bone quality in adult ovariectomized ewes *Calcif. Tissue Intl.* **59**(4) 315–20

[7] FDA 1994 Guidelines for preclinical and clinical evaluation of agents used in the prevention or treatment of postmenopausal osteoporosis (Bethesda, MD: FDA, Division of Metabolism and Endocrine Drug Products)

[8] Davidson M K, Lindsey J R and Davis J K 1987 Requirements and selection of an animal model *Isr. J. Med. Sci.* **23**(6) 551–5

[9] Miller S C, Bowman B M and Jee W S 1995 Available animal models of osteopenia—small and large *Bone* **17**(4 Suppl) S 117–23

[10] Danielsen C C, Mosekilde L and Andreassen T T 1992 Long-term effect of orchidectomy on cortical bone from rat femur: bone mass and mechanical properties *Calcif. Tissue Intl.* **50**(2) 169–74

[11] Wronski T J, Dann L M and Horner S L 1989 Time course of vertebral osteopenia in ovariectomized rats *Bone* **10**(4) 295–301

[12] Wronski T J, Schenck P A, Cintron M and Walsh C C 1987 Effect of body weight on osteopenia in ovariectomized rats *Calcif. Tissue Intl.* **40**(3) 155–9

[13] Wronski T J, Walsh C C and Ignaszewski L A 1986 Histologic evidence for osteopenia and increased bone turnover in ovariectomized rats *Bone* **7**(2) 119–23

[14] Kiebzak G M, Smith R, Howe J C, Gundberg C M and Sacktor B 1988 Bone status of senescent female rats: chemical, morphometric, and biomechanical analyses *J. Bone Min. Res.* **3**(4) 439–46

[15] Ismail F, Epstein S, Fallon M D, Thomas S B and Reinhardt T A 1988 Serum bone gla protein and the vitamin D endocrine system in the oophorectomized rat *Endocrinology* **122**(2) 624–30

[16] Heaney R P, Recker R R and Saville P D 1978 Menopausal changes in bone remodeling *J. Lab. Clin. Med.* **92**(6) 964–70

[17] Jiang Y, Zhao J, Van Audekercke R, Dequeker J and Geusens P 1996 Effects of low-dose long-term sodium fluoride preventive treatment on rat bone mass and biomechanical properties *Calcif. Tissue Intl.* **58**(1) 30–9

[18] Jee W S, Mori S, Li X J and Chan S 1990 Prostaglandin E2 enhances cortical bone mass and activates intracortical bone remodeling in intact and ovariectomized female rats *Bone* **11**(4) 253–66

[19] Jiang Y, Zhao J, Genant H K, Dequeker J and Geusens P 1997 Long-term changes in bone mineral and biomechanical properties of vertebrae and femur in aging, dietary calcium restricted, and/or estrogen-deprived/-replaced rats *J. Bone Min. Res.* **12**(5) 820–31

[20] Wink C S and Felts W J 1980 Effects of castration on the bone structure of male rats: a model of osteoporosis *Calcif. Tissue Intl.* **32**(1) 77–82

[21] Wronski T J, Yen C F, Burton K W, Mehta R C, Newman P S, Soltis E E *et al* 1991 Skeletal effects of calcitonin in ovariectomized rats *Endocrinology* **129**(4) 2246–50

[22] Mosekilde L 1995 Assessing bone quality—animal models in preclinical osteoporosis research *Bone* **17**(4 Suppl) S 343–52

[23] Geusens P, Schot L P, Nijs J and Dequeker J 1991 Calcium-deficient diet in ovariectomized dogs limits the effects of 17 beta-estradiol and nandrolone decanoate on bone *J. Bone Min. Res.* **6**(8) 791–7

[24] Jerome C P, Carlson C S, Register T C, Bain F T, Jayo M J, Weaver D S *et al* 1994 Bone functional changes in intact, ovariectomized, and ovariectomized, hormone-supplemented adult cynomolgus monkeys (*Macaca fascicularis*) evaluated by serum markers and dynamic histomorphometry *J. Bone Min. Res.* **9**(4) 527–40

[25] Newman E, Turner A S and Wark J D 1995 The potential of sheep for the study of osteopenia: current status and comparison with other animal models *Bone* **16**(4 Suppl) S277–84

[26] Jiang Y, Zhao J and Genant H K 2002 Macro and micro imaging of bone architecture. In *Principles of Bone Biology* 2nd edn, eds J P Bilezikian, L G Raisz and G A Rodan (San Diego: Academic Press) pp 1599–623

[27] Hildebrand T and Ruegsegger P 1997 Quantification of bone microarchitecture with the Structure Model Index. *Comput. Meth. Biomech. Biomed. Eng.* **1**(1) 15–23

[28] Muller R and Ruegsegger P 1997 Micro-tomographic imaging for the nondestructive evaluation of trabecular bone architecture. *Stud. Health Technol. Inform.* **40** 61–79

[29] Muller R and Ruegsegger P 1996 Analysis of mechanical properties of cancellous bone under conditions of simulated bone atrophy *J. Biomech.* **29**(8) 1053–60

[30] Geusens P, Dequeker J, Nijs J and Bramm E 1990 Effect of ovariectomy and prednisolone on bone mineral content in rats: evaluation by single photon absorptiometry and radiogrammetry *Calcif. Tissue Intl.* **47**(4) 243–50

[31] Grier S J, Turner A S and Alvis M R 1996 The use of dual-energy X-ray absorptiometry in animals *Invest. Radiol.* **31**(1) 50–62

[32] Gala Paniagua J, Diaz-Curiel M, de la Piedra Gordo C, Castilla Reparaz C and Torralbo Garcia M 1998 Bone mass assessment in rats by dual energy X-ray absorptiometry *Br. J. Radiol.* **71**(847) 754–8

[33] Ammann P, Rizzoli R, Slosman D and Bonjour J P 1992 Sequential and precise in vivo measurement of bone mineral density in rats using dual-energy X-ray absorptiometry *J. Bone Min. Res.* **7**(3) 311–6

[34] Vanderschueren D, Van Herck E, Schot P, Rush E, Einhorn T, Geusens P *et al* 1993 The aged male rat as a model for human osteoporosis: evaluation by nondestructive measurements and biomechanical testing *Calcif. Tissue Intl.* **53**(5) 342–7

[35] Fink C, Cooper H J, Huebner J L, Guilak F and Kraus V B 2002 Precision and accuracy of a transportable dual-energy X-ray absorptiometry unit for bone mineral measurements in guinea pigs *Calcif. Tissue Intl.* **70**(3) 164–9

[36] Sato M, Kim J, Short L L, Slemenda C W and Bryant H U 1995 Longitudinal and cross-sectional analysis of raloxifene effects on tibiae from ovariectomized aged rats *J. Pharmacol. Exp. Ther.* **272**(3) 1252–9

[37] Cointry G R, Mondelo N, Zanchetta J R, Montuori E and Ferretti J L 1995 Intravenous olpadronate restores ovariectomy-affected bone strength. A mechanical, densitometric and tomographic (pQCT) study *Bone* **17**(4 Suppl) S 373–8

[38] Rosen H N, Tollin S, Balena R, Middlebrooks V L, Beamer W G, Donohue L R *et al* 1995 Differentiating between orchidectomized rats and controls using measurements of trabecular bone density: a comparison among DXA, histomorphometry, and peripheral quantitative computerized tomography *Calcif. Tissue Intl.* **57**(1) 35–9

[39] Ma Y F, Ferretti J L, Capozza R F, Cointry G, Alippi R, Zanchetta J *et al* 1995 Effects of on/off anabolic hPTH and remodeling inhibitors on metaphyseal bone of immobilized rat femurs. Tomographical (pQCT) description and correlation with histomorphometric changes in tibial cancellous bone *Bone* **17**(4 Suppl) S321–7

[40] Ferretti J L, Gaffuri O, Capozza R, Cointry G, Bozzini C, Olivera M *et al* 1995 Dexamethasone effects on mechanical, geometric and densitometric properties of rat femur diaphyses as described by peripheral quantitative computerized tomography and bending tests *Bone* **16**(1) 119–24

[41] Qin L, Lu H B, Shi Y Y, Zhang G, Hung W Y and Leung P C 2001 Bone mineral density and structural evaluation on preventive effects of Chinese herbal preparation in ovariectomy induced osteoporosis in rats. In *International Symposium on Bone Biotechnology and Histotechnology* 7–10 March (Phoenix, Arizona) p 30

[42] Breen S A, Millest A J, Loveday B E, Johnstone D and Waterton J C 1996 Regional analysis of bone mineral density in the distal femur and proximal tibia using peripheral quantitative computed tomography in the rat in vivo *Calcif. Tissue Intl.* **58**(6) 449–53

[43] Guilak F 1994 Volume and surface area measurement of viable chondrocytes in situ using geometric modelling of serial confocal sections *J. Microsc.* **173** (Pt 3) 245–56

[44] Hildebrand T and Ruegsegger P 1997 A new method for the model independent assessment of thickness in three-dimensional images *J. Microsc.* **185** 67–75

[45] Odgaard A and Gundersen H J 1993 Quantification of connectivity in cancellous bone, with special emphasis on 3-D reconstructions *Bone* **14**(2) 173–82

[46] Parfitt A M, Mathews C H, Villanueva A R, Kleerekoper M, Frame B and Rao D S 1983 Relationships between surface, volume, and thickness of iliac trabecular bone in aging and in osteoporosis. Implications for the microanatomic and cellular mechanisms of bone loss *J. Clin. Invest.* **72**(4) 1396–409

[47] Hildebrand T, Laib A, Muller R, Dequeker J and Ruegsegger P 1999 Direct three-dimensional morphometric analysis of human cancellous bone: microstructural data from spine, femur, iliac crest, and calcaneus *J. Bone Min. Res.* **14**(7) 1167–74

[48] Wevers M, de Meester P, Lodewijckx M, Ni Y, Marchal G, Jiang Y *et al* 1993 Application of microfocus X-ray radiography in materials and medical research. *NDT* **26** 135–40

[49] Miller S C and Jee W S 1979 The effect of dichloromethylene diphosphonate, a pyrophosphate analog, on bone and bone cell structure in the growing rat *Anat. Rec.* **193**(3) 439–62

[50] Liens D, Delmas P D and Meunier P J 1994 Long-term effects of intravenous pamidronate in fibrous dysplasia of bone *Lancet* **343**(8903) 953–4

[51] Jiang Y 1995 *Radiology and Histology in the Assessment of Bone Quality* Leuven, PhD Thesis

[52] Fratzl P, Schreiber S, Roschger P, Lafage M H, Rodan G and Klaushofer K 1996 Effects of sodium fluoride and alendronate on the bone mineral in minipigs: a small-angle X-ray scattering and backscattered electron imaging study *J. Bone Min. Res.* **11**(2) 248–53

[53] Roschger P, Fratzl P, Klaushofer K and Rodan G 1997 Mineralization of cancellous bone after alendronate and sodium fluoride treatment: a quantitative backscattered electron imaging study on minipig ribs *Bone* **20**(5) 393–7

[54] Turner C H and Burr D B 1993 Basic biomechanical measurements of bone: a tutorial *Bone* **14**(4) 595–608

[55] Martens M, van Audekercke R, de Meester P and Mulier J C 1980 The mechanical characteristics of the long bones of the lower extremity in torsional loading *J. Biomech.* **13**(8) 667–76

[56] Torzilli P A, Takebe K, Burstein A H, Zika J M and Heiple K G 1982 The material properties of immature bone *J. Biomech. Eng.* **104**(1) 12–20

[57] Keller T S, Spengler D M and Carter D R 1986 Geometric, elastic, and structural properties of maturing rat femora *J. Orthop. Res.* **4**(1) 57–67

[58] Einhorn T A 1992 Bone strength: the bottom line *Calcif. Tissue Intl.* **51** 333–9

[59] Jiang Y, Zhao J, Augat P, Ouyang X, Lu Y, Majumdar S *et al* 1998 Trabecular bone mineral and calculated structure of human bone specimens scanned by peripheral quantitative computed tomography: relation to biomechanical properties *J. Bone Min. Res.* **13**(11) 1783–90

[60] Gasser J A 1995 Assessing bone quantity by pQCT *Bone* **17**(4 Suppl) S145–54

[61] Feldkamp L A, Goldstein S A, Parfitt A M, Jesion G and Kleerekoper M 1989 The direct examination of three-dimensional bone architecture in vitro by computed tomography *J. Bone Min. Res.* **4**(1) 3–11

[62] Kuhn J L, Goldstein S A, Feldkamp L A, Goulet R W and Jesion G 1990 Evaluation of a microcomputed tomography system to study trabecular bone structure *J. Orthop. Res.* **8**(6) 833–42

[63] Engelke K and Kalender W 1998 Beyond bone densitometry: Assessment of bone architecture by X-ray computed tomography at various levels of resolution. In *Bone Densitometry and Osteoporosis* eds H K Genant, G Guglielmi and M Jergas (Berlin: Springer) pp 417–47

[64] Kinney J H, Lane N E and Haupt D L 1995 In vivo, three-dimensional microscopy of trabecular bone *J. Bone Min. Res.* **10**(2) 264–70

[65] Goldstein S A, Goulet R and McCubbrey D 1993 Measurement and significance of three-dimensional architecture to the mechanical integrity of trabecular bone *Calcif. Tissue Intl.* **53**(Suppl 1) S127–32 [discussion S132–3]

[66] Goulet R W, Goldstein S A, Ciarelli M J, Kuhn J L, Brown M B and Feldkamp L A 1994 The relationship between the structural and orthogonal compressive properties of trabecular bone *J. Biomech.* **27**(4) 375–89

[67] Engelke K, Song S M, Gluer C C and Genant H K 1996 A digital model of trabecular bone *J. Bone Min. Res.* **11**(4) 480–9

[68] Smith C B and Smith M D 1994 Comparison between single slice CT and volume CT. In *International Symposium on Computerized Tomography for Industrial Applications* eds H C H Czichos and D Schnitger, 8–10 June, Berlin, Germany

[69] Holdsworth D W, Thornton M M, Drost D, Watson P H, Fraher L J and Hodsman A B 2000 Rapid small-animal dual-energy X-ray absorptiometry using digital radiography *J. Bone Min. Res.* **15**(12) 2451–7

[70] Bonse U, Busch F, Gunnewig O, Beckmann F, Pahl R, Delling G *et al* 1994 3D computed X-ray tomography of human cancellous bone at 8 microns spatial and 10(-4) energy resolution *Bone Miner.* **25**(1) 25–38

[71] Grodzins L 1983 Optimum energy for X-ray transmission tomography of small samples *Nucl. Instrum. Meth.* **206** 541–3

[72] Lane N E, Thompson J M, Strewler G J and Kinney J H 1995 Intermittent treatment with human parathyroid hormone (hPTH[1–34]) increased trabecular bone volume but not connectivity in osteopenic rats *J. Bone Min. Res.* **10**(10) 1470–7

[73] Lane N E, Thompson J M, Haupt D, Kimmel D B, Modin G and Kinney J H 1998 Acute changes in trabecular bone connectivity and osteoclast activity in the ovariectomized rat in vivo *J. Bone Min. Res.* **13**(2) 229–36

[74] Lane N E, Haupt D, Kimmel D B, Modin G and Kinney J H 1999 Early estrogen replacement therapy reverses the rapid loss of trabecular bone volume and prevents further deterioration of connectivity in the rat *J. Bone Min. Res.* **14**(2) 206–14

[75] Peyrin F, Salome M, Cloetens P, Laval-Jeantet A M, Ritman E and Ruegsegger P 1998 Micro-CT examinations of trabecular bone samples at different resolutions: 14, 7 and 2 micron level *Technol. Health Care* **6**(5–6) 391–401

[76] Peyrin F, Muller C, Carillon Y, Nuzzo S, Bonnassie A and Briguet A 2001 Synchrotron radiation microCT: a reference tool for the characterization of bone samples *Adv. Exp. Med Biol.* **496** 129–42

[77] Peyrin F, Salome M, Nuzzo S, Cloetens P, Laval-Jeantet A M and Baruchel J 2000 Perspectives in three-dimensional analysis of bone samples using synchrotron radiation microtomography *Cell Mol. Biol.* **46**(6) 1089–102

[78] Flynn M J, Seifert H A, Irving T C and Lai B 2001 Measurement of bone mineralization in whole trabeculae using 3D X-ray microtomography. In *International Symposium on Bone Biotechnology and Histotechnology* 7–10 March (Phoenix, Arizona) p 28

[79] Ruegsegger P, Koller B and Muller R 1996 A microtomographic system for the nondestructive evaluation of bone architecture *Calcif. Tissue Intl.* **58**(1) 24–9

[80] Zhao J, Jiang Y, Prevrhal S and Genant H K 2000 Effects of low dose longterm sodium on three dimensional trabecular microstructure, bone mineral and biomechanical properties of rat vertebral body. In *ASBMR 2000* (Toronto) p 816

[81] Zhao J, Jiang Y and Genant H K 2000 Three dimensional trabecular microstructure and biomechanicaal properties and their relationship in different bone quality models. In *Radiology 2000* p 411

[82] Turner C H, Hsieh Y F, Muller R, Bouxsein M L, Rosen C J, McCrann M E *et al* 2001 Variation in bone biomechanical properties, microstructure, and density in BXH recombinant inbred mice *J. Bone Min. Res.* **16**(2) 206–13

[83] Zhao J, Jiang Y, Shen V, Bain S and Genant H K 2000 MicroCT and pQCT assessment of a murine model of postmenopausal osteoporosis and estrogen therapy *Osteoporos. Intl.* **11** S3–11

[84] Jara H and Wehrli F W 1994 Determination of background gradients with diffusion MR imaging. *J. Magn. Reson. Imaging* **4**(6) 787–97

[85] Majumdar S, Newitt D, Jergas M, Gies A, Chiu E, Osman D *et al* 1995 Evaluation of technical factors affecting the quantification of trabecular bone structure using magnetic resonance imaging *Bone* **17**(4) 417–30

[86] Ouyang X, Selby K, Lang P, Engelke K, Klifa C, Fan B *et al* 1997 High resolution magnetic resonance imaging of the calcaneus: age-related changes in trabecular structure and comparison with dual X-ray absorptiometry measurements *Calcif. Tissue Intl.* **60**(2) 139–47

[87] Hipp J A, Jansujwicz A, Simmons C A and Snyder B D 1996 Trabecular bone morphology from micro-magnetic resonance imaging *J. Bone Min. Res.* **11**(2) 286–97

[88] White D, Schmidlin O, Jiang Y, Zhao J, Majumdar S, Genant H K *et al* 1997 MRI of trabecular bone in an ovariectomized rat model of osteoporosis. In *International Society for Magnetic Resonance in Medicine*, Vancouver, BC, Canada

[89] Kapadia R D, High W B, Soulleveld H A, Bertolini D and Sarkar S K 1993 Magnetic resonance microscopy in rat skeletal research. *Magn. Reson. Med.* **30**(2) 247–50

[90] Takahashi M, Wehrli F W, Wehrli S L, Hwang S N, Lundy M W, Hartke J *et al* 1999 Effect of prostaglandin and bisphosphonate on cancellous bone volume and structure in the ovariectomized rat studied by quantitative three-dimensional nuclear magnetic resonance microscopy *J. Bone Min. Res.* **14**(5) 680–9

[91] Jiang Y, White D, Zhao J, Peterfy C and Genant H K 1997 Menisectomy-induced osteoarthritis model in rabbits: MRI and radiographic assessments. In *Arthritis and Rheumatism 1997* p S89

[92] Zhao J, Jiang Y and Genant H K 1999 Preclinical studies of alendronate *J. Musculoskeletal Res.* **3** 209–16

[93] Mosekilde L 1993 Vertebral structure and strength in vivo and in vitro *Calcif. Tissue Intl.* 53 (Suppl 1) S121–5 [discussion S125–6]

[94] Mazess R B 1990 Fracture risk: a role for compact bone *Calcif. Tissue Intl.* **47**(4) 191–3

[95] Turner R T, Vandersteenhoven J J and Bell N H 1987 The effects of ovariectomy and 17 beta-estradiol on cortical bone histomorphometry in growing rats *J. Bone Min. Res.* **2**(2) 115–22

[96] Miller S C, Bowman B M, Miller M A and Bagi C M 1991 Calcium absorption and osseous organ-, tissue-, and envelope-specific changes following ovariectomy in rats *Bone* **12**(6) 439–46

[97] Bagi C M, Mecham M, Weiss J and Miller S C 1993 Comparative morphometric changes in rat cortical bone following ovariectomy and/or immobilization *Bone* **14**(6) 877–83

[98] Sogaard C H, Danielsen C C, Thorling E B and Mosekilde L 1994 Long-term exercise of young and adult female rats: effect on femoral neck biomechanical competence and bone structure *J. Bone Min. Res.* **9**(3) 409–16

[99] Peng Z, Tuukkanen J and Vaananen H K 1994 Exercise can provide protection against bone loss and prevent the decrease in mechanical strength of femoral neck in ovariectomized rats *J. Bone Min. Res.* **9**(10) 1559–64

[100] Peng Z, Tuukkanen J, Zhang H, Jamsa T and Vaananen H K 1994 The mechanical strength of bone in different rat models of experimental osteoporosis *Bone* **15**(5) 523–32

[101] Meng X W, Liang X G, Birchman R, Wu D D, Dempster D W, Lindsay R *et al* 1996 Temporal expression of the anabolic action of PTH in cancellous bone of ovariecto-mized rats *J. Bone Min. Res.* **11**(4) 421–9

[102] Hodgskinson R and Currey J D 1990 The effect of variation in structure on the Young's modulus of cancellous bone: a comparison of human and non-human material *Proc. Inst. Mech. Eng.* [H] **204**(2) 115–21

[103] Mundinger A, Wiesmeier B, Dinkel E, Helwig A, Beck A and Schulte Moenting J 1993 Quantitative image analysis of vertebral body architecture–improved diagnosis in osteoporosis based on high-resolution computed tomography *Br. J. Radiol.* **66**(783) 209–13

[104] Hansson T, Roos B and Nachemson A 1980 The bone mineral content and ultimate compressive strength of lumbar vertebrae *Spine* **5**(1) 46–55

[105] Hayes W C and Gerhart W C 1985 Biomechanics of bone: applications for assess-ment of bone strength. In *Bone and Mineral Research* vol 3 ed W A Peck (Amster-dam: Elsevier) pp 259–94

[106] Tran Van P T, Vignery A and Baron R 1982 Cellular kinetics of the bone remodeling sequence in the rat *Anat. Rec.* **202**(4) 445–51

[107] Baron R, Tross R and Vignery A 1984 Evidence of sequential remodeling in rat trabecular bone: morphology, dynamic histomorphometry, and changes during skeletal maturation *Anat. Rec.* **208**(1) 137–45

[108] Mosekilde L, Sogaard C H, Danielsen C C and Torring O 1991 The anabolic effects of human parathyroid hormone (hPTH) on rat vertebral body mass are also reflected in the quality of bone, assessed by biomechanical testing: a comparison study between hPTH-(1–34) and hPTH-(1–84) *Endocrinology* **66**(4) 1432–9

[109] Jiang Y, Wang Y, Zhao J, Marchal G, Wang Y, Shen Y, Xing S, Li R and Baert A L 1991 Bone remodeling in hypervitaminosis D$_3$: Radiologic-microangiographic-pathologic correlations *Invest. Radiol.* **26** 213–9

[110] Jiang Y, Wang Y, Xue D and Yin Y 2002 Microangiography, macrosectioning, and preparation for contact radiography. In *Handbook of Histology Methods for Bone and Cartilage* eds Y An and K Martin (Humana Press)

Index